Molecular Biotechnology

Molecular Biotechnology

Editor: Oscar Watson

CALLISTO REFERENCE
www.callistoreference.com

Callisto Reference,
118-35 Queens Blvd., Suite 400,
Forest Hills, NY 11375, USA

Visit us on the World Wide Web at:
www.callistoreference.com

ISBN: 978-1-63239-812-3 (Hardback)

The publisher's policy is to use permanent paper from mills that operate a sustainable forestry policy. Furthermore, the publisher ensures that the text paper and cover boards used have met acceptable environmental accreditation standards.

Trademark Notice: Registered trademark of products or corporate names are used only for explanation and identification without intent to infringe.

Printed in the United States of America.

Cataloging-in-publication Data

Molecular biotechnology / edited by Oscar Watson.
    p. cm.
Includes bibliographical references and index.
ISBN 978-1-63239-812-3
1. Molecular biotechnology. 2. Molecular genetics. 3. Genetic engineering. 4. Molecular biology. I. Watson, Oscar.
QH442 .M65 2017
572.8--dc23

Table of Contents

Permissions

List of Contributors

Index

Preface

Molecular biotechnology is the science and practice of using laboratory methods, in order to examine nucleic acids and proteins. These practices are used in many fields like medicinal science, veterinary medicine, agricultural science and environment. Molecular biotechnology is the amalgamation of many fields like, genetics, cell biology, molecular biology, microbiology, biochemistry, etc. This book present researches and studies performed by experts across the globe on the various fields related to molecular biotechnology. It picks up individual branches and explains their need and contribution in the context of the growth of this subject. Most of the topics introduced in the text cover new techniques and the applications of molecular biotechnology. It will help new researchers by foregrounding their knowledge in this branch. Students and scientists engaged in this field will find this book helpful as it compiles contributions made by experts from across the globe.

After months of intensive research and writing, this book is the end result of all who devoted their time and efforts in the initiation and progress of this book. It will surely be a source of reference in enhancing the required knowledge of the new developments in the area. During the course of developing this book, certain measures such as accuracy, authenticity and research focused analytical studies were given preference in order to produce a comprehensive book in the area of study.

This book would not have been possible without the efforts of the authors and the publisher. I extend my sincere thanks to them. Secondly, I express my gratitude to my family and well-wishers. And most importantly, I thank my students for constantly expressing their willingness and curiosity in enhancing their knowledge in the field, which encourages me to take up further research projects for the advancement of the area.

Editor

Lipid, Fatty Acid and Energy Density Profiles of White Sharks: Insights into the Feeding Ecology and Ecophysiology of a Complex Top Predator

Heidi R. Pethybridge[1]*, Christopher C. Parrish[2], Barry D. Bruce[1], Jock W. Young[1], Peter D. Nichols[1]

1 CSIRO Wealth from Ocean Flagship, Division of Marine and Atmospheric Research, Hobart, Australia, 2 Department of Ocean Sciences, Memorial University of Newfoundland, St. John's, Newfoundland, Canada

Abstract

Lipids are major sources of metabolic energy in sharks and are closely linked to environmental conditions and biological cycles, such as those related to diet, reproduction and migration. In this study, we report for the first time, the total lipid content, lipid class composition and fatty acid profiles of muscle and liver tissue of white sharks, *Carcharodon carcharias*, of various lengths (1.5–3.9 m), sampled at two geographically separate areas off southern and eastern Australia. Muscle tissue was low in total lipid content (<0.9% wet mass, wm) and was dominated by phospholipids (>90% of total lipid) and polyunsaturated fatty acids (34±12% of total fatty acids). In contrast, liver was high in total lipid which varied between 51–81% wm and was dominated by triacylglycerols (>93%) and monounsaturated fatty acids (36±12%). With knowledge of total lipid and dry tissue mass, we estimated the energy density of muscle (18.4±0.1 kJ g^{-1} dm) and liver (34.1±3.2 kJ g^{-1} dm), demonstrating that white sharks have very high energetic requirements. High among-individual variation in these biochemical parameters and related trophic markers were observed, but were not related to any one biological or environmental factor. Signature fatty acid profiles suggest that white sharks over the size range examined are generalist predators with fish, elasmobranchs and mammalian blubber all contributing to the diet. The ecological applications and physiological influences of lipids in white sharks are discussed along with recommendations for future research, including the use of non-lethal sampling to examine the nutritional condition, energetics and dietary relationships among and between individuals. Such knowledge is fundamental to better understand the implications of environmental perturbations on this iconic and threatened species.

Editor: David William Pond, Scottish Association for Marine Science, United Kingdom

Funding: Funding was provided by the Commonwealth Scientific and Industrial Research Organisation (CSIRO) Capability Development Fund Ernst Frohlich Fellowship. HP was supported by a CSIRO Office of the Executive Chief Postdoctoral Award. BDB's time on this paper was supported by the Australia Governments' National Environmental Research Program, Marine Biodiversity Hub. The funders had no role in study design, data collection and analysis, decision to publish, or preparation of the manuscript.

Competing Interests: The authors have declared that no competing interests exist.

* E-mail: Heidi.Pethybridge@csiro.au

Introduction

The white shark, *Carcharodon carcharias*, is considered the most voracious apex predator in temperate marine ecosystems world-wide [1], playing a key role in controlling ecosystem dynamics [2]. Despite this, limited quantitative dietary or energetic data for the species exist. Due to long standing declines in their population, white sharks are protected internationally; listed on Appendix II of the CITES and the Convention of Migratory Species [3]. This makes assessing contemporary biological parameters particularly difficult as investigative methods must be based on a limited number of samples, and be as non-invasive as possible. Consequently, much of our existing dietary knowledge on white sharks has been based on examining deceased sharks taken as incidental bycatch [4] or sharks taken as part of control programs [5]. These data have been supplemented by *in situ* observations of predation at aggregation sites (typically around seal colonies) or scavenging events on whale carcasses [4–8] which, although non-invasive, are likely to bias the recording of marine mammals in their diet. This

has fueled a general (public) perception that white sharks almost exclusively consume prey comprising these taxa.

Rapid advances in technology, data treatment and other numerical techniques, and our understanding of their tissue biochemistry, are allowing ecologists to gain a more complete understanding of white shark biology. Tracking studies have shown that white sharks engage in broad-scale inshore and offshore migrations [9,10]. Population genetics have revealed that white sharks exhibit fine-scale spatial structure and low effective population sizes [11]. Stable isotope data of carbon and nitrogen of white shark vertebrae and muscle tissue suggest large ontogenetic and among-individual variation in their foraging strategies [1,12]. Although not yet applied in studies of white sharks, signature lipid and fatty acid analysis is also gaining momentum to refine our ecological and physiological understanding of sharks [13,14]. This is mainly because lipids are major sources of metabolic energy and are closely linked to environmental conditions such as prey quality and quantity, and to biological cycles, such as those related to reproduction and migration. More recently, signature fatty acid analysis was validated as a technique to determine dietary changes in a controlled feeding study of Port Jackson sharks [15], and the extraction of fatty acids from muscle tissue biopsies of less than 0.1 g was used to analyze dietary variation in live, threatened manta rays and whale sharks [16]. The successful utilization of signature fatty acids is related to the well known expression 'you are what you eat', in that many fatty acids are transferrable from prey to predator with limited modification [17].

The aim of this study was to report the lipid content, lipid class composition and fatty acid profiles of the muscle and liver of white sharks sampled opportunistically from South Australia and New South Wales (central-east Australia). As tissue differences in lipids are well known in other species of shark [18], we compared muscle and liver tissues, and also muscle tissue sampled at different anatomical sites. Lipid content was used to calculate energy density profiles, and is an indicator of an individual's nutritional condition; lipid class composition helped shed light on an individual's physiology; fatty acid profiles were used to infer integrated dietary signatures. A particular focus was placed on examining intra-specific variation and determining the most appropriate way to apply lipid analysis to future ecological studies of white sharks and other protected, large oceanic sharks.

Materials and Methods

Ethics statement

All procedures were ratified by independent Animal Ethics Committees, including the New South Wales Department of Primary Industries (NSW DPI) Animal Research Authority (ACEC 12/07) and the Tasmanian Department of Primary Industries, Parks, Water and Environment (ACE 22/2011–2012). In South Australia, field work at the Neptune Islands was carried out under a South Australian S115 ministerial exemption 9902123. In NSW, field work was carried out under a NSW DPI Scientific Collection Permit (P07/0099-3.0 & P07/0099-4).

Sampling

Twenty one deceased white sharks of various lengths (1.5 to 3.9 m) were sampled opportunistically between 2001 to 2012 in waters off South Australia (SA) and along the New South Wales (NSW) coast, Australia (Table 1). Most specimens were taken either as fisheries bycatch or during the NSW Shark Control Program. The one basking shark included in this study was collected on the beach of Marrawah, north-western Tasmania, in 2012. Sub-samples of muscle (0.05–0.08 g) and liver (0.03–0.05 g) tissues were taken, stored in sterile cryotubes, and kept frozen at $-20°C$ until lipid analysis.

Lipid content and class composition

Total lipid content was extracted by the modified Bligh and Dyer [19] method using a one-phase $CHCl_3$:MeOH:milliQ H_2O solvent mixture (10:20:7.5 mL) which was left overnight. The following day, the solution was broken into two phases by adding 10 mL $CHCl_3$ and 10 mL saline milliQ H_2O (9 g NaCl L^{-1}) to give a final solvent ratio of 1:1:0.9. The lower layer was drained into a 50 mL round bottom flask and concentrated using a rotary evaporator. The extract was transferred in $CHCl_3$ into a pre-weighed glass 2 mL vial. The solvent was blown down under a constant stream of nitrogen gas. The round bottom flask was rinsed twice with $CHCl_3$ into the vial. The total lipid extract (TLE) was dried in the vial to constant weight and 1.5 mL of $CHCl_3$ added.

Lipid class composition of tissues was determined using an Iatroscan Mark V TH10 thin layer chromatograph coupled with a flame ionization detector (TLC-FID). For each sample, the TLE was spotted and developed in a polar solvent system (70:10:0.1 v/v/v hexane:diethyl ether: glacial acetic acid). Samples were also run in a non-polar solvent system (96:4 v/v hexane:ether) to resolve hydrocarbons (including squalene) from wax esters and diacylglyceryl ethers from triacylglycerols. All samples were run in duplicate along with standard solutions, which contained known quantities of common lipid classes. Chromarods were oven-dried for 10 min at 100°C and analyzed immediately. Peaks were quantified using SIC-480 Scientific Software.

Fatty acid proportions were determined using a direct methylation procedure [20]. Briefly, a small sub-sample (20–50 mg) of wet tissue was transferred to a glass tube, then directly transmethylated in methanol:dichloromethane:concentrated hydrochloric acid (10:1:1 v/v) for 2 hours at 80°C. Following the addition of 1.5 mL of Milli-Q water, a known concentration of internal injection standard (23:0 FAME) was added. After thorough mixing, the tube was centrifuged at 2000 rpm for 5 mins. The upper, organic layer was removed and reduced under a nitrogen gas stream and 0.2 μl from this solution was injected into an Agilent Technologies 7890B gas chromatograph (GC) (Palo Alto, California USA) equipped with an Equity-1 fused silica capillary column (15 m×0.1 mm internal diameter and 0.1 μm film thickness), a flame ionization detector, a splitless injector and an Agilent Technologies 7683B Series auto-sampler. At an oven temperature of 120°C, samples were injected in splitless mode and carried by helium gas. Oven temperature was raised to 270°C at 10°C per min, and then to 310°C at 5°C per min. Peaks were quantified using Agilent Technologies ChemStation software (Palo Alto, California USA). Confirmation of peak identifications was by GC-mass spectrometry (GC-MS), using a column of similar polarity to that described above and a Finnigan Thermoquest DSQ GC-MS system.

Dry mass and energy density of tissues

Water fraction (WF) of tissue mass was determined by taking weights before and after freeze-drying at $-82°C$ for 48 h. From this, and acquired values of the lipid fraction (LF), we determined the wet/dry mass ratio of each tissue and the proportion that consisted of protein and carbohydrate ($PCF_{dm} = 1 - LF_{dm}$ and $PCF_{wm} = 1 - LF_{wm} - WF$). Energy density (ED, kJ g^{-1}) of wet mass and dry mass of muscle and liver was then calculated using proximate composition data and published calorific values of 39.3 kJ g^{-1} for lipid and 17.8 kJ g^{-1} for protein and carbohydrates [21] with the following equations:

$$ED(kJ\ g^{-1}\ wm) = (1 - WF)([LF_{wm} \bullet 39.3] + [PCF_{wm} \bullet 17.8])$$

$$ED(kJ\ g^{-1}\ dm) = (LF_{dm} \bullet 39.3) + (PCF_{dm} \bullet 17.8)$$

Multivariate statistics

We used PRIMER (Plymouth Routines in Multivariate Ecological Research), a software package for analyzing multivariate ecological data, to assess the groupings and investigate relationships within the fatty acid data set. Specifically, we used parametric, principal component analysis (PCA) and non-parametric cluster analysis.

Table 1. Collection and biological information of the 21 white sharks (WS) and a single basking shark (BS1) analyzed in this study.

ID	Date	Length (m)	Sex	State	Port	Tissue analyzed					
						M	Mv	Md	Mdf	Mds	L
WS1	17/01/2011	2.25	F	NSW	Maroubra		a				
WS2	23/10/2003	1.85	Unk	NSW	Dee Why			a			
WS3	1/12/2005	1.9	Unk	NSW	Avalon		a				
WS4	24/09/2001	1.74	M	NSW	Stockton			a		b	
WS5	29/01/2003	3.81	M	SA	Port Lincoln		a				
WS7	24/09/2001	1.84	Unk	NSW	Stockton			a			
WS9	30/10/2002	3.9	F	SA	Port Lincoln		a				
WS10	1/01/2005	2.08	M	NSW	Wattamola		a				
WS11	20/02/2013	1.84	F	NSW	Coledale	c	e	b		a	d
WS12	2/05/2012	1.47	F	NSW	Coledale			a			b
WS13	2/05/2013	2.7	M	NSW	Cronulla				a		
WS14	2/05/2012	2.21	M	NSW	Warriewood			b			c
WS15	2/05/2012	2.7	M	NSW	Cronulla		c	a			b
WS16	2/05/2012	2.1	M	NSW	Wollongong			a			b
WS17	2/05/2012	1.78	F	NSW	Coledale			a			b
WS18	16/11/2012	1.9	F	NSW	Bondi			a			
WS19	27/11/2000	3.3	F	SA	Ceduna			a			
WS21	14/03/2005	2.5	M	NSW	Nth Cronulla		b			a	b
WS22	3/10/2002	2.38	M	NSW	Cath Hill Bay		a				
WS23	1/10/2002	2.54	Unk	NSW	Newcastle		a				
WS24	2/12/2002	2.0	M	SA	Port Lincoln		a				
BS1	11/06/2012	3.8	F	TAS	Marrawah			a			b

Sex: F- female, M- male, Unk - sex not identified. State: New South Wales (NSW), South Australia (SA), Tasmania (TAS). Tissues analyzed include the muscle (M), and muscle sampled from the: vertebrae (Mv), dorsal area (Md), dorsal fin not including skin (Mdf), and dorsal fin including the skin (Mds) and the liver (L).

Prey-predator and interspecies fatty acid comparisons

To assist in assessing broad scale food web groupings, a number of established fatty acid markers, or trophic indices, were calculated. These included 16:1ω7/16:0 and 22:6ω3/20:5ω3 to separate diatom versus dinoflagellate based energy channels [22]; 18:1ω9/18:1ω7 which have been linked to trophic position [23]; and 20:4ω6/20:5ω3 and ω6/ω3 which are nutritional condition indices, and may also provide information on benthic-derived inputs.

Of the total of fifty fatty acids detected in white sharks tissues, twenty two (with averages >0.05% of total fatty acids) were used in multivariate analyses to compare white shark muscle and liver profiles to other oceanic and coastal shark taxa with known dietary differences. Specifically, the mean fatty acid profiles of each of white shark muscle and liver, grouped according to cluster analyses, were compared to profiles of *in situ* caught school, gummy and broadnose sharks (*Galeorhinus galeus*, *Mustelus antarcticus*, *Notorynchus cepedianus*; [24], Australian catshark and shortnose spurdog (*Figaro boardmani* and *Squalus megalops*, [18], Port Jackson shark (*Heterodontus portusjacksoni* [15]), and liver profiles of tiger shark (*Galeocerdo cuvieri*) sampled off New South Wales in 2000 [25] and white shark sampled off South Africa [26].

Results and Discussion

Water, lipid content and class composition

White shark muscle was high in water (75.4±2.4%) and had a wet to dry mass ratio of 4.1±0.4 (Table 2). Muscle was low in total lipid content (consistently <0.9% wm) and varied marginally between different body sections (average coefficient of variation was only 6%). The proportion of wet mass that was protein and carbohydrate was 23.9%. The energetic density of white shark muscle was estimated at 4.5±0.5 kJ g^{-1} wm (Table 2). Phospholipids dominated the total lipid fraction (92.4±4.8%) with only small relative levels of sterols (6.1±4.1%) and free fatty acids (1.4±0.7%) present. In the only other study that has reported lipid components in white shark muscle [27], similar lipid content was observed in specimens (of unknown sizes) collected off South Africa (6.1±3.3 mg g^{-1} wm). In other shark species, similar total lipid content has been detected; typically accounting for less than 1% wm [28,18]. In another study more variable total lipid content was reported in muscle for 11 species of shark in the Caribbean (0.4–11% dm) with higher total lipid content occurring in hammerhead, seven-gills and six-gills [29]. The present study is the first to report the muscle lipid class composition of white sharks or any other large oceanic shark, but the observed profiles are comparable to those of demersal sharks [18] which have high relative levels of structural components (phospholipids) and nil to low levels of energy storage (neutral) lipids.

Liver had a wet to dry ratio of 1.2±0.1 (Table 2) and was lower in water content (16–24%) than that reported for the Atlantic sharpnose shark (25–70% [30]), the only other shark species for which such data is published (to the authors' knowledge). Total lipid content varied between 51.6–70.4% wm (67.8–84.1% dm) and was dominated by triacylglycerols (96.3±0.9%), followed by phospholipids (1.9±1.0%), wax and steryl esters (1.5±0.1%), sterols (0.4±0.4%), and free fatty acids (0.1±0.2%). The proportion of wet liver that was protein and carbohydrates was 4.1% and the liver-specific energy density was estimated at 27.3±2.8 kJ g^{-1} wm (Table 2). Total lipid content and lipid class composition of the liver of other large oceanic sharks has seldom been measured, and has yet to be reported for white sharks. In other oceanic sharks, total lipid content has been reported to be highly variable both inter-specifically (10 to 90% wm; [29,31,25])

and intra-specifically [32]. Nichols et al. [25] found that more than half the 17 species of pelagic sharks collected in northern Australian waters had liver lipid contents greater than 40% including the tiger shark, milk shark, school, gummy, whaler and blue sharks. Another study [27] reported very few inter-specific differences in total lipid content between blacktip and spinner sharks, but large differences between samples collected in the Atlantic and Indian Oceans. Furthermore, Davidson & Cliff [32] found large seasonal differences in grey nurse shark liver mass (10–16 kg) and lipid concentrations (326–572 mg g^{-1} wm) and related this to their migration and reproductive cycles. For most oceanic shark species analyzed, triacylglycerol seems to be the dominant class of lipid accounting for >90% total lipid [25,33]. In contrast, liver lipids of sharks living at more than 200 m depth are dominated by diacylglyceryl ethers and the isoprenoid hydrocarbon squalene [18].

Storage of large amounts of lipids which are high in energy and low in density are likely to assist white shark movement patterns [34]. Triacylglycerols (at 1 atmospheric pressure and 20°C) have an energy density of 0.93 g L^{-1} and the buoyancy in sea water (0.0103 g^{-1}) is +0.095 mL^{-1} lipid [35]. Because triacylglycerols are used for metabolism at higher rates than wax esters, diacylglyceryl ethers and squalene [36], they play a less important role in buoyancy and instead their major function is energy storage which can be used for energetically exhausting activities such as migration, reproduction or during extended periods of low prey availability. As both squalene and diacylglyceryl ethers, which have lower energy densities of 0.86 and 0.89 g L^{-1}, respectively [35], are more abundant in deep-sea sharks [18], the amount of lift generated in the white shark per unit weight of liver oil is lower than in deep-sea sharks, but high enough to achieve neutral buoyancy in seawater. Our results show that liver lipid reserves are unlikely to remain constant over time as was assumed by Raye et al. [34] and therefore, buoyant weight of the shark does not necessarily indicate a change in lipid volume. However, changes in lipid class composition could account for differences in buoyant mass.

White shark energetics

The estimated energy density of white shark muscle was noticeably lower than that of the liver (mean values: 4.5 kJ g^{-1} wm and 27.3 kJ g^{-1} wm, respectively, Table 2). This represents the total available energy from these tissues and supports the function of liver as an energy reserve. From the available liver weights of sharks analyzed in this study, liver was found to account for between 8.2–16.2% of total body mass. Using an estimate for muscle of 45% of total body mass, and the mean liver value of 12.2%, we calculated the muscle and liver energy content for a 428 kg individual to be 867 MJ wm and 1425 MJ wm, respectively. Few studies have investigated the energy density or content of elasmobranches. Based on bomb calorimeter, seasonal and inter-annual differences in the liver-specific energy content (which varied between 26.8 to 34.8 kJ g^{-1} dm) and energy content (590 to 4550 kJ dm) were observed in the Atlantic sharpnose shark in the Gulf of Mexico [36]. The only other reports (to the authors' knowledge) of energy density in shark tissue used a value of 5.41 kJ g^{-1} wm [37,38], based on consumption estimates given for lemon shark by Cortés & Gruber [39]. This value seems to be grossly underestimated for larger oceanic sharks as shown by the present study.

In comparison to other fish (demersal and pelagic) and mammals, white shark liver has a higher energy density, and therefore capacity to store energy. Drazen [40] reported high variability in the muscle and liver energy density between 22

Table 2. Biochemical parameters of white shark tissues collected from temperate waters off south east Australia.

Parameter			muscle	liver
Water content	%	wm	75.4±2.4	20.0±5.4
Wet/dry ratio	-		4.1±0.4	1.2±0.04
Total lipid content	%	dm	2.9±0.6	75.9±11.5
	%	wm	0.7±0.1	61.0±13.3
Total fatty acids	%	wm	0.5±0.1	13.0±0.1
Protein & carbohydrates^	%	dm	97.2±1.0	24.1±6.2
Energy density	kJ g^{-1}	dm*	18.4±0.1	34.1±3.2
	kJ g^{-1}	wm*	4.5±0.5	27.3±2.8
Lipid class composition				
Triacylglycerols (TAG)	%	wm	0.0±0.1	96.3±0.9
Phospholipids (PL)	%	wm	92.4±4.8	1.9±1.0
Sterols (ST)	%	wm	6.1±4.1	0.4±0.4
Wax esters (WE)	%	wm	0.0±0.0	1.5±0.1
Free fatty acids (FFA)	%	wm	1.4±0.7	0.1±0.2

^Estimated by subtracting the lipid fraction from 100%.
*For comparative studies, 1 kilocalorie = 4.184 kilojoules (kJ).
Values are in dry and wet mass (dm and wm).

species of demersal fish in the Pacific Ocean. They found that energy densities were consistently lower (between 1.6 and 13.6 times, and on average 4.3 times) in the muscle (1.2–5.5 kJ g^{-1} wm) than liver (4.2–17.9 kJ g^{-1} wm) with benthopelagic species, primarily gadiforms, having significantly larger lipid-rich livers than benthic species. In a comparative study of the energy density of 39 pelagic forage fishes [41], a five-fold difference (2.0–10.8 kJ g^{-1} wm fish) was reported with small, fast maturing species having higher relative values than those species maturing at a larger size. Energy density of muscle and liver of Antarctic fishes were respectively lower (21–25 kJ g^{-1} dm) and higher (25–30 kJ g^{-1} dm, [42]) than in the white sharks analyzed in the present study. Whereas sharks store energy in their livers, mammals store energy in adipose tissue with whale blubber having an estimated energy density of 27.5 to 30.6 kJ g^{-1} [43]. Such comparisons suggest that white sharks have much higher energy storage needs than whales or demersal and pelagic fishes. This is likely related to their high energetic requirements related to undertaking long migrations, reproduction (including courtship) and foraging.

A recent study [44] estimated daily energetic expenditure of adult white sharks at a seal colony off South Australia to be 28,200 kJ d^{-1}. Our study adds information on the static energy available in the muscle and liver of white sharks and gives additional information on their energy budget and allocation strategies. Such information is vital for the calibration and validation of bioenergetics models at individual, population and ecosystem scales. These models, which describe the mass balance relationship between energy acquired, transformed and allocated to organism productivity (growth, reproduction and survival), are increasingly being used to assist fisheries and conservation management [45]. As lipid content is the primary determinant of energy density, estimating tissue-specific energy density and content based on partial proximate composition is faster and cheaper than traditional bomb calorimetry. As lipid content variations are food-dependent, they can be directly related to body and energetic condition and are likely to be a robust way to predict features of their population dynamics, such as reproductive

potential and ability to sustain long-distance migrations. Indeed, positive associations between recruitment and total lipid energy have been proposed as a robust proxy for total egg production by fish stocks where 2.12 kJ of liver energy was found to be proportional to 1000 eggs [46]. Future research should seek to examine if a similar relationship and proxy approach can be reliably used for elasmobranches.

Fatty acid profiles

A total of fifty fatty acids were identified in the white shark tissues analyzed (Table S1 and S2), with thirty one fatty acids identified in relative levels greater than 0.1% (Table 3). White shark muscle was proportionally dominated by polyunsaturated fatty acids (PUFA, 35±11%) and saturated fatty acids (SFA, 35±6%), with slightly lower levels of monounsaturated fatty acids (MUFA, 27±10%). Dominant fatty acids (>5%) in decreasing order of relative importance included: 16:0, 22:6ω3, 18:0, 18:1ω9, 20:4ω6 and 18:1ω7. Limited variation in fatty acid proportions was found among muscle sampled at different anatomical sites (Table 1) with the mean coefficient of variation typically less than 16%. Greatest differences in fatty acid profiles were observed between muscle tissue taken from the vertebrae and from the dorsal fin and skin. White shark liver was dominated by MUFA (36±12%) followed by SFA (31±2%) and PUFA (30±12%). Dominant fatty acids in decreasing order of relative importance were similar, with the exception of the lower (twofold) levels of 20:4ω6, to those reported in the muscle and included: 16:0, 18:1ω9, 22:6ω3, 18:0, 16:1ω7 and 18:1ω7.

Muscle fatty acid profiles of white shark in South Africa [25]were similar to those reported in this study with comparable SFA (34.8±7.0%), MUFA (25.3±6.5%), and PUFA (34.5±12.2%PUFA) dominated by 22:6ω3, 20:4ω6, 22:4ω6. Similar to our results, no significant differences in fatty acid profiles of muscle taken from different anatomical sites were found in other large oceanic sharks, including tiger and white sharks sampled off South Africa [25]. Davidson & Cliff [47] reported similar liver fatty acid profiles for white sharks collected off South

Table 3. Fatty acid distribution of white shark muscle and liver (mean area % of total fatty acids ± standard deviation, and the coefficient of variation %) sampled off south and eastern Australia.

	Muscle			Liver		
n	21			7		
Length	1.74–3.9			1.84–3.3		
14:0	0.81	±0.66	82%	2.59	±1.90	73%
15:0	0.22	±0.15	69%	0.46	±0.30	66%
16:0	18.55	±3.48	19%	18.42	±2.29	12%
17:0	0.64	±0.22	34%	0.70	±0.20	29%
18:0	13.79	±3.64	26%	8.70	±3.93	45%
16:1ω9	0.44	±0.45	102%	0.51	±0.18	36%
16:1ω7	2.13	±1.81	85%	6.72	±5.72	85%
17:1ω8+a17:0	0.63	±0.37	58%	0.78	±0.32	41%
17:1	0.34	±0.31	93%	0.10	±0.09	91%
18:1ω9	11.90	±4.38	37%	16.19	±5.49	34%
18:1ω7	6.25	±1.50	24%	5.96	±1.16	19%
19:1	0.39	±0.11	28%	0.27	±0.06	21%
20:1ω11	0.14	±0.17	127%	0.37	±0.29	79%
20:1ω9	1.94	±0.91	47%	2.27	±1.33	59%
20:1ω7	0.23	±0.12	54%	0.33	±0.16	47%
22:1ω11	0.11	±0.16	138%	0.29	±0.21	74%
22:1ω9	0.45	±0.32	70%	0.62	±0.30	48%
24:1ω9	1.39	±2.26	163%	0.56	±0.47	83%
18:2ω6	0.70	±0.23	34%	0.98	±0.29	30%
20:4ω6 (AA)	9.22	±3.93	43%	4.86	±5.42	111%
20:5ω3 (EPA)	1.63	±1.05	64%	2.31	±0.81	35%
20:4ω3	0.17	±0.12	66%	0.32	±0.14	45%
22:5ω6	0.73	±0.42	57%	0.77	±0.22	29%
22:6ω3 (DHA)	15.52	±6.68	43%	13.62	±6.34	47%
22:4ω6	2.44	±1.17	48%	1.65	±1.24	75%
22:5ω3	2.75	±1.40	51%	3.40	±1.02	30%
Σ SFA	34.70	±5.71	16%	31.36	±2.34	7%
Σ MUFA	27.08	±9.60	35%	35.85	±11.91	33%
Σ PUFA	34.48	±12.19	35%	29.70	±11.83	40%
Σ ω3 PUFA	20.36	±8.60	42%	20.43	±6.99	34%
Σ ω6 PUFA	13.59	±4.90	36%	8.75	±6.46	74%
i15:0	0.33	±0.18	54%	0.42	±0.36	85%
i16:0	0.21	±0.15	71%	0.32	±0.29	92%
i17:0	0.95	±0.39	41%	0.62	±0.09	14%
16:0 FALD (75)	1.37	±0.78	57%	0.93	±0.77	83%
18:0 FALD (75)	0.43	±0.59	136%	0.21	±0.13	61%
Σ iso-SFA	1.70	±0.76	44%	1.72	±0.91	53%
Σ branched FA	2.05	±1.21	59%	1.36	±0.92	67%
Σ other (<0.2%)	2.94	±1.40	48%	3.77	±1.08	29%
Trophic markers						
16:1ω7/16:0	0.11	±0.10	85%	0.35	±0.31	87%

Lipid, Fatty Acid and Energy Density Profiles of White Sharks: Insights into the Feeding Ecology...

7

Table 3. Cont.

	Muscle			Liver		
n	21			7		
Length	1.74–3.9			1.84–3.3		
20:5ω3/22:6ω3	0.12	±0.10	86%	0.21	±0.13	62%
16:0/18:0	1.45	±0.51	35%	2.49	±1.01	40%
18:1ω9/18:1ω7	1.90	±0.59	31%	2.71	±0.85	31%
14:0+16:1ω7+20:5ω3	4.57	±2.84	62%	11.62	±7.62	66%
PUFA/SAT	1.04	±0.43	41%	0.96	±0.40	42%
20:4ω6/20:5ω3	11.95	±14.97	125%	2.59	±3.47	134%

SFA – saturated fatty acids, MUFA – monounsaturated fatty acids, PUFA – polyunsaturated fatty acids. The suffix *i* denotes branched fatty acids from the *iso*-series. FALD - fatty aldehyde analysed as dimethyl acetal.
Other fatty acids (that accounted for <0.2% of total fatty acids) are included in Table S1.
Data presented are for 31 components, with a cut off of 0.2%. For full fatty acid profiles of individual samples, see Table S1.

Africa with PUFA accounting for 26.6%. Long-chain ($\geq C_{20}$) polyunsaturated fatty acids have also been reported in large concentrations in the liver of other large, oceanic shark species [26,23,24]. In the present study, very high levels of 20:4ω6 were found in white shark muscle (9.2%) with moderate levels in the liver (4.9%) which are within the very upper range of that reported in other oceanic and predatory shark species [47,25,22,16]. Such variability could be related to differences in diet, with higher relative levels representative of benthic/coastal inputs. Another study [48] found that 20:4ω6 levels increased with size in juvenile bull sharks and low relative levels were likely to infer essential fatty acid deficiency.

In addition to the large differences occurring between the fatty acid profile of the muscle and liver, we detected large variations among individual sharks (Fig. 1 and Table S1). In the muscle, three clear groups of sharks were apparent based on 80% similarity of a non-parametric cluster analysis: group A consisted of four sub-adult sharks collected in New South Wales in May 2012 and September 2001 which had moderate proportions of 22:6ω3 and low 18:0; group B separated nine sharks of various lengths collected from various sampling months, years (2002, 2005, 2011–2013) and sites due their higher levels of 22:6ω3 and 18:0; while group C consisted of seven sharks high in 18:1ω9 and 18:0 (Fig. 1A). No single biological (size or maturation status) or environmental factor (collection site and month) stood out in the separation of the groups with juvenile and sub-adult sharks in all three groups, and larger (>3 m) sharks from South Australia identified in both groups B and C, although with noticeably higher levels of 18:0 (Fig. 1A). For the seven white shark liver samples, two groups separated based on 80% similarity with one group (D) having higher levels of 22:6ω3, 20:1ω9, and 22:5ω3, while the other group (E) had higher levels of 16:1ω7 and 14:0 (Fig. 1B). If we separated the two PCA axes at the central point, then only the liver sample of one large (adult >3 m female) shark collected from South Australian waters separates from the smaller (sub-adult) sharks collected from east Australia due to high levels of 16:0 and 16:1ω7.

Dietary inferences

Large intra-specific differences in fatty acid profiles of tissues have been experimentally shown to reflect an organism's diet at different temporal scales [13] and these differences can be directly related to its habitat utilization, nutritional condition and reproductive cycles [11,12,16,49]. Although caution should be taken when interpreting tissue-specific fatty acid profiles due to physiological factors that are not yet well understood in

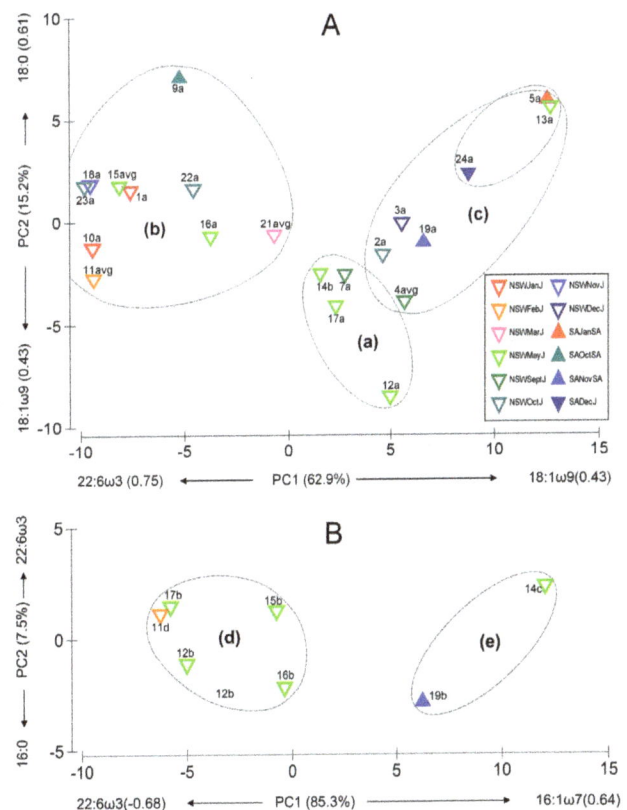

Figure 1. Principal component analysis (PCA) of the fatty acid profiles of juvenile (J) and sub-adult (SA) white shark (A) muscle, and (B) liver collected from New South Wales (NSW) and South Australia (SA) during various months and years. Eigenvalues in brackets represent the percent variance explained by each axis (PC1 and PC2). Fatty acids labeled on each of the axes represent the main coefficients (or eigenvectors) contributing to each PC. Black lines represent groups that have more than 80% similarity based on non-parametric cluster analysis complete linkages. Sample codes are listed in Table 1.

8

elasmobranchs [16], in this study the fatty acids responsible for separating muscle and liver groups on the PC1 axes were consistently 22:6ω3, 18:1ω9, and 16:1ω7 (Fig. 1). These differences are likely to reflect varying contributions of mammalian blubber and fish/cephalopods to the diets of individual white sharks, as blubber is typically high in 18:1ω9, 16:1ω7, and 20:5ω3 [50], whilst fish and cephalopods are higher in 22:6ω3, 16:0, 18:0 [22,16]. Previous dietary studies of white sharks show clear ontogenetic dietary differences between juveniles (<3 m, which commonly feed on large fish such as mulloway, salmon, snapper, redfish), and sub-adult (>3.0–3.6 m for males and >3.0–4.8 m for females) and adult (>3.6 m for males and >4.8 m for females) sharks which also feed on fur seals and occasionally on whales [8].

No clear separation among juvenile and adult sharks was found in the present study. This is likely related to the fact that individuals were caught over a long and variable sampling period (throughout the year between 2000 and 2006 and 2011–2013, with a maximum of 6 caught in any one year) and the animals were from a large size range (1.5–3.9 m). This variation, in association with the fact that white sharks engage in broad-scale inshore and offshore migrations [7,8], means that intra-specific

variation should be high. Future studies should seek to detail the fatty acid profiles of dominant prey species of white sharks and target research efforts on individuals with known diets of mammals such as those that occur for extended periods around seal colonies. Furthermore, in association with an increase in the use of telemetry and satellite tracking on white sharks [51], future work should combine telemetry with signature lipid analyses, through the recently developed and applied ability for researchers to obtain tissue biopsies of as little as 0.1 g [14] while placing tracking devices. For energetics studies, we recommend taking a very small (<0.01 g) biopsy of liver tissues over muscle, as unlike teleosts and mammals, sharks do not have adipose tissue and the liver is the main site for lipid storage and synthesis [52]. Our results show that muscle or dermal tissue can be taken from various anatomical sites as limited differences in their fatty acid profiles were observed. Future work should also seek to analyze fatty acid profiles of plasma, which is being increasingly withdrawn and used for genetic work. As shown by dietary studies of mammals [53,54], knowledge of the fatty acid profiles of white shark plasma will assist the acquisition of dietary data (on a shorter time scale to the

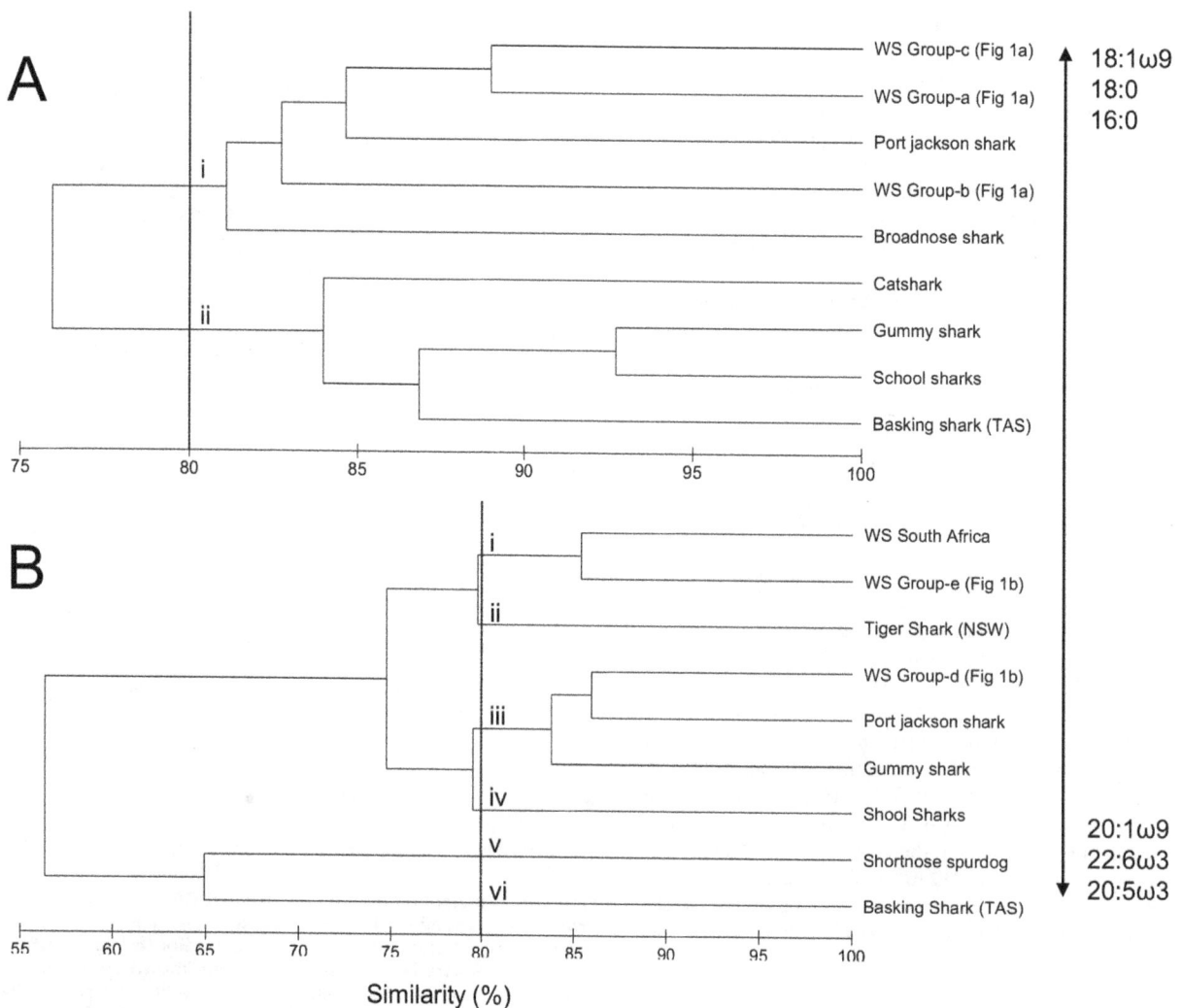

Figure 2. Dendrogram of cluster analysis (group averages) based on a Bray-Curtis similarity matrix for comparison of the fatty acid composition of the (A) muscle and (B) liver of white shark (WS) groups (identified in this study, Fig. 1) and from published data of white sharks collected off South Africa and to other shark species collected in Australian waters.

muscle and liver) and our current understanding of the cellular uptake and transport of lipids.

Inter-specific comparisons can assist in the interpretation of fatty acid data and making dietary inferences, particularly where sample size is low [11,16], such as in the present study. We therefore used non-parametric cluster analyses of the fatty acid profiles of muscle and liver of white shark groups identified in Fig. 1 and groupings with other shark species in Fig. 2. For the muscle tissue, two clear groups (containing individuals with >80% similarity) were identified with all white sharks identified in Fig. 1 grouping together with Port Jackson and broadnose sharks.

Cluster analyses of the liver fatty acid profiles showed a higher magnitude of dissimilarity between sharks than that in the muscle, with an 80% resemblance cut-off identifying six separate groups (Fig. 2B). White sharks identified in group E (Fig. 1) showed high similarity to the profiles of white sharks collected off South Africa and tiger shark from north Australia, with the latter being known top predators of mid-trophic fish, elasmobranchs, cephalopods, mammals and seabirds [55]. Group D white sharks grouped with Port Jackson sharks. For muscle and liver, the fatty acids of basking shark, known planktivores [56], were the most dissimilar to white sharks. The liver fatty acid profiles of gummy, school and spurdog sharks, all prominent pelagic fish feeders [57,11] were also dissimilar to the acid profiles of white sharks.

The close similarity between muscle profiles of white shark in the present study and broadnose shark is fitting; the latter is a dominant predator in coastal systems, commonly consuming mammalian blubber and small elasmobranchs including spurdogs, school and gummy sharks [58]. The grouping with the muscle and liver of Port Jackson shark, a tertiary consumer of largely molluscs, teleosts and cephalopods [59], was surprising and may suggest a high intake of these benthic and coastal prey, or alternatively (and most likely) a high consumption of these or similar shark species. The fatty acids mostly responsible for grouping Port Jackson and white sharks were those that are known to reflect trophic position, 16:0 and 18:1ω9 [21], and that are abundant in dinoflagellates (22:6ω3 [20]). In comparison with large predatory fish such as albacore and skipjack tuna [60], white sharks have higher levels of 20:4ω6, 22:4ω6 and 18:1ω7 fatty acids. Collectively, these results suggest that white sharks are more of a generalist species, including benthic/coastal food-chain linkages, than a mammal specialist.

Of particular interest in understanding the diets of white sharks is the contribution of oceanic relative to coastal prey. In a study of pelagic fishes in oceanic waters off eastern Australia, mako shark had the highest nitrogen isotopic value indicating their position as top predator in the system [61]. White sharks are known to move between coastal and oceanic waters, in some cases travelling large distances within and between oceanic basins [7,8]. Whether these migrations are driven by changes in food, reproductive cycles or other reasons is yet to be determined, although the high energetics related to them means that maintaining lipid reserves in the liver is

a crucial, and perhaps, regulatory component. We are hopeful that signature lipid and fatty acid analyses can assist in obtaining a better understanding of the trophic ecology of white sharks.

Conclusions

In this study we used lipid analyses to directly extract energetic and feeding related information on white sharks. We found that while the muscle is a site of structural energy, the liver has an even higher energy density than whale blubber and serves as a storage unit to fuel white shark migrations, growth and reproduction. Among-individual variation in fatty acid profiles suggest individuality in foraging strategies with some sort of a gradient between dominant or partial consumption of fish and mammals. Such information, although currently very limited in elasmobrachs, is required for the development of bioenergetics-based models including dynamic energy budget models and ecosystem models increasingly used in fisheries and conservation management. Indeed, measuring total lipid energy is appealing because lipids represent a common currency, linking food quality and quantity to many aspects of fish nutrition, reproduction and population dynamics. Future studies should include targeted sampling and utilization of this novel approach on live specimens, along with ongoing telemetry studies, to explore the relative contributions of biological and environmental factors influencing intra-specific variation. Such data will allow us to better understand the ecology of white shark and predict how they might respond to environmental perturbations.

Supporting Information

Table S1 Muscle full fatty acid profiles of individual white sharks and a single basking shark (BS1) analyzed in this study.

Table S2 Liver full fatty acid profiles of individual white sharks and a single basking shark (BS1) analyzed in this study.

Acknowledgments

We thank the New South Wales Department of Primary Industries Shark Meshing Program for collecting and storing tissue material. We thank Russell Bradford and Ross Daley for their helpful comments on the manuscript. Danny Holdsworth managed the CSIRO GC and GC-MS facility.

Author Contributions

Conceived and designed the experiments: CCP HRP PDN BB JWY. Performed the experiments: CCP HRP. Analyzed the data: HRP CCP. Contributed reagents/materials/analysis tools: PDN BB JWY. Wrote the paper: HRP CCP PDN BB JWY.

References

1. Hussey NE, McCann HM, Cliff G, Dudley SFJ, Wintner SP, et al. (2012) Size based analysis of diet and trophic position of the white shark, *Carcharodon carcharias*, in South African waters. In: Domeier ML, editor. Global perspectives on the biology and life history of the white shark. CRC Press. pp. 27–49.

2. Myers RA, Baum JK, Shepherd TD, Powers SP, Peterson CH (2007) Cascading effects of the loss of apex predatory sharks from a coastal ocean. Science, 315(5820): 1846–1850.

3. Dulvy NK, Baum JK, Clarke S, Compagno LJ, Cortés E, et al. (2008) You can swim but you can't hide: the global status and conservation of oceanic pelagic sharks and rays. Aquat Conservat Mar Freshw Ecosyst 18(5):459–482.

4. Casey JG, Pratt HL (1985)- Distribution of the white shark, *Carcharodon carcharias*, in the western North Atlantic. Memoirs of the Southern Californian Academy of Science 9: 2–14.

5. Cliff G., Dudley, S. F. J., & Davis B. (1989) Sharks caught in the protective gill nets off natal, South Africa. 2. The great white shark *Carcharodon carcharias* (linnaeus) South African Journal of Marine Science 8: 131–144.

6. Bruce BD (1992). Preliminary observations on the biology of the white shark, *Carcharodon carcharias*, in south Australian waters. Mar Freshw Res 43(1): 1–11.

7. Taylor JKJ, Mandelman JW, McLellan WA, Moore MJ, Skomal GB, et al. (2012) Shark predation on North Atlantic right whales (*Eubalaena glacialis*) in the southeastern United States calving ground. Mar Mamm Sci 29(1):204–212.

8. Fallows C, Gallagher AJ, Hammerschlag N (2013) White Sharks (*Carcharodon carcharias*) scavenging on whales and its potential role in further shaping the ecology of an apex predator. PLoS ONE 8(4):e60797.

9. Bonfil R, Meyer MA, Scholl MC, Johnson RL, O'Brian S, et al. (2005) Transoceanic migration, long-distance return migration and local movement patterns in the great white shark. Science 301: 100–103.

10. Bruce BD, Bradford RW (2012) Habitat use and spatial dynamics of juvenile white sharks, *Carcharodon carcharias*, in Eastern Australia. In: Domeier M, editor. Global perspectives on the biology and life history of the Great White Shark. Boca Raton: CRC Press. pp. 225–254.

11. Blower DC, Pandolfi JM, Bruce BD, Gomez-Cabrera MDC, Ovenden JR (2012) Population genetics of Australian white sharks reveals fine-scale spatial structure, transoceanic dispersal events and low effective population sizes. Mar Ecol Prog Ser 455: 229–244.

12. Kim SL, Tinker MT, Estes JA, Koch PL (2012) Ontogenetic and among-Individual variation in foraging strategies of Northeast Pacific white sharks based on stable isotope analysis. PLoS ONE 7(9): e45068.

13. Pethybridge H, Daley R, Virtue P, Nichols P (2011) Diet composition of demersal sharks and chimaeras inferred by fatty acid profiles and stomach content analysis. J Exp Mar Biol Ecol 409(1–2): 290–299.

14. McMeans BC, Arts MT, Fisk AT (2012) Similarity between predator and prey fatty acid profiles is tissue dependent in Greenland sharks (*Somniosus microcephalus*): Implications for diet reconstruction. J Exp Mar Biol Ecol 429:55–63.

15. Beckmann CL, Mitchell JG, Stone DAJ, Huveneers C (2013) A controlled feeding experiment investigating the effects of a dietary switch on muscle and liver fatty acid profiles in Port Jackson sharks *Heterodontus portusjacksoni*. J Exp Mar Biol Ecol 448:10–18.

16. Couturier LIE, Rohner CA, Richardson AJ, Pierce SJ, Marshall AD, et al. (2013) Unusually high levels of n-6 polyunsaturated fatty acids in whale sharks and reef manta rays. Lipids 48(10):1029–1034.

17. Iverson SJ, Field C, Don Bowen W, Blanchard W (2004) Quantitative fatty acid signature analysis: a new method of estimating predator diets. Ecol Monogr 74(2): 211–235.

18. Pethybridge H, Daley R, Virtue P, Nichols PD (2010) Lipid composition and partitioning of deepwater chondrichthyans: inferences of feeding ecology and distribution. Mar Biol 157(6): 1367–1387.

19. Bligh EG, Dyer WJ (1959) A rapid method of total lipid extraction and purification. Can J Biochem Physiol 37: 911–917.

20. Parrish CC, Nichols PD, Young JW (In prep) Direct determination of fatty acids in fish tissue in order to define and then quantify top predator trophic connections in the marine environment. Oecologia.

21. Schmidt-Nielsen K (1997) Animal physiology: Adaptation and environment (5th ed.), Cambridge University Press, New York.

22. Graeve M, Kattner G, Hagen W (1994) Diet-induced changes in the fatty acid composition of Arctic herbivorous copepods: experimental evidence of trophic markers. J Exper Mar Biol Ecol 182(1): 97–110.

23. El-Sabaawi R, Dower JF, Kainz M, Mazumder A (2009) Characterizing dietary variability and trophic positions of coastal calanoid copepods: insight from stable isotopes and fatty acids. Mar Biol 156(3): 225–237.

24. Nichols PD, Bakes MJ, Elliott NG (1998) Oils rich in docosahexaenoic acid in livers of sharks from temperate Australian waters. Mar Freshw Res 49: 763–767

25. Nichols PD, Rayner M, Stevens J (2001) A pilot investigation on Northern Australian shark liver oils: characterization and value-adding. FRDC Project Report 99/369. CSIRO Marine Research and Fisheries Research and Development Corporation, Australia.

26. Davidson BC, Rottanburg D, Prinz W, Cliff G (2007) The influence of shark liver oils on normal and transformed mammalian cells in culture. In Vivo 21(2): 333–337.

27. Davidson B, Sidell J, Rhodes J, Cliff G (2011) A comparison of the heart and muscle total lipid and fatty acid profiles of nine large shark species from the east coast of South Africa. Fish Physiol Biochem 37(1): 105–112.

28. Sargent JR, Gatten RR, McIntosh R (1973) The distribution of neutral lipids in shark tissues. J Mar Biol Ass UK 53: 649–656.

29. Van Vleet ES, Candileri S, McNeillie J, Reinhardt SB, Conkright ME, et al. (1984) Neutral lipid components of eleven species of Caribbean sharks. Comp Biochem Physiol B 79(4): 549–554.

30. Hoffmayer ER, Parsons GR, Horton J (2006) Seasonal and interannual variation in the energetic condition of adult male Atlantic sharpnose shark *Rhizoprionodon terraenovae* in the northern Gulf of Mexico. J Fish Biol 68(2): 645–653.

31. Jayasinghe C, Gotoh N, Tokairin S, Ehara H, Wada S (2003) Inter species changes of lipid compositions in liver of shallow-water sharks from the Indian Ocean. Fisheries Sci 69(3): 644–653.

32. Davidson B, Cliff G (2011) Liver lipids of female *Carcharias taurus* (spotted raggedtooth) sharks: a comparison between seasons. Fish Physiol Biochem 37(3): 613–618.

33. Navarro-Garcia G, Pacheco-Aguilar R, Vallejo-Cordova B, Ramirez-Suarez JC, Bolanos A (2000) Lipid composition of the liver oil of shark species from the Caribbean and Gulf of California waters. J Food Comp Anal 13: 791–798.

34. Del Raye G, Jorgensen SJ, Krumhansl K, Ezcurra JM, Block BA (2013) Travelling light: white sharks (*Carcharodon carcharias*) rely on body lipid stores to power ocean-basin scale migration. Proceed Royal Soc B, 280(1766).

35. Phleger CF (1998) Buoyancy in marine fishes: direct and indirect role of lipids. American Zoo 38(2): 321–330.

36. Hoffmayer ER, Parsons GR, Horton J (2006) Seasonal and interannual variation in the energetic condition of adult male Atlantic sharpnose shark *Rhizoprionodon terraenovae* in the northern Gulf of Mexico. J Fish Biol 68(2): 645–653.

37. Schindler DE, Essington TE, Kitchell JF, Boggs C, Hilborn R (2002) Sharks and tunas: fisheries impacts on predators with contrasting life histories. Ecolog App 12(3): 735–748.

38. Hammerschlag N, Gallagher AJ, Carlson JK (2013) A revised estimate of daily ration in the tiger shark with implication for assessing ecosystem impacts of apex predators. Funct Ecol 27(5): 1273–1274.

39. Cortés E, Gruber SH (1990) Diet, feeding habits and estimates of daily ration of young lemon sharks, *Negaprion brevirostris* (Poey). Copeia 1: 204–218.

40. Drazen JC (2007) Depth related trends in proximate composition of demersal fishes in the eastern North Pacific. Deep Sea Res I 54(2): 203–219.

41. Anthony JA, Roby DD, Turco KR (2000) Lipid content and energy density of forage fishes from the northern Gulf of Alaska. J Exp Mar Biol Ecol 248(1): 53–78.

42. Vanella FA, Calvo J, Morriconi ER, Aureliano DR (2005) Somatic energy content and histological analysis of the gonads in Antarctic fish from the Scotia Arc. Scientia Marina 69(S2): 305–316.

43. Nordoy ES, Folkow LP, Martensson P-E, Blix AS (1995) Food requirements of northeast Atlantic minke whales In: Baker CS, Herman LM, Perry A, et al, editors. Whales, Seals, Fish and Man. Developments in Marine Biology 4. Amsterdam: Elsevier Science. pp. 307–317.

44. Semmens JM, Payne NL, Huveneers C, Sims DW, Bruce BD (2013) Feeding requirements of white sharks may be higher than originally thought. Sci Rep 3:1471.

45. Chipps SR, Wahl DH (2008) Bioenergetics modeling in the 21st century: reviewing new insights and revisiting old constraints. Trans Americ Fish Soc 137(1): 298–313.

46. Marshall CT, Yaragina NA, Lambert Y, Kjesbu OS (1999) Total lipid energy as a proxy for total egg production by fish stocks. Nature 402(6759): 288–290.

47. Davidson BC, Cliff G (2002) The liver fatty acid profiles of seven Indian Ocean shark species. Fish Physiol Biochem 26(2): 171–175.

48. Belicka LL, Matich P, Jaffé R, Heithaus MR (2012) Fatty acids and stable isotopes as indicators of early-life feeding and potential maternal resource dependency in the bull shark *Carcharhinus leucas*. Mar Ecol Prog Ser 455: 245–256.

49. Pethybridge H, Daley R, Virtue P, Nichols PD (2011) Lipid (energy) reserves, utilisation and provisioning during oocyte maturation and early embryonic development of deepwater chondrichthyans. Mar Biol 158(12): 2741–2754.

50. Waugh CA, Nichols PD, Noad MC, Bengtson Nash S (2012) Lipid and fatty acid profiles of migrating Southern Hemisphere humpback whales *Megaptera novaeangliae*. Mar Ecol Prog Ser 471: 271–281.

51. Bruce BD, Stevens JD, Malcolm H (2006) Movements and swimming behaviour of white sharks (*Carcharodon carcharias*) in Australian waters. Mar Biol 150(2): 161–172.

52. Ballantyne JS (1997) Jaws: the inside story. The metabolism of elasmobranch fishes. Comp Biochem Physiol B 118(4): 703–742.

53. Tierney M, Nichols PD, Wheatley KE, Hindell MA (2008) Blood fatty acids indicate inter-and intra-annual variation in the diet of Adélie penguins: comparison with stomach content and stable isotope analysis. J Exp Mar Biol Ecol 367(2): 65–74.

54. Cooper MH, Iverson SJ, Heras H (2005) Dynamics of blood chylomicron fatty acids in a marine carnivore: implications for lipid metabolism and quantitative estimation of predator diets. J Comp Phys B 175(2): 133–145.

55. Simpfendorfer CA, Goodreid AB, McAuley RB (2001) Size, sex and geographic variation in the diet of the tiger shark, *Galeocerdo cuvier*, from Western Australian waters. Env Biol Fish 61: 37–46.

56. Sims DW, Quayle VA (1998) Selective foraging behaviour of basking sharks on zooplankton in a small-scale front. Nature 393(6684):460–464.

57. Walker T (1989) Stomach contents of gummy shark, *Mustelus antarcticus* Gunther, and school shark, *Galeorhinus galeus* (Linnaeus), from south-eastern Australian waters. In: 'Southern Shark Assessment Project: Final FIRTA Report - March, Marine Science Laboratories.

58. Barnett A, Abrantes K, Stevens JD, Yick J, Frusher SD, et al. (2010) Predator-prey relationships and foraging ecology of a marine apex predator with a wide temperate distribution. Mar Ecol Prog Ser 416: 189–200.

59. Powter DM, Gladstone W, Platell M (2010) The influence of sex and maturity on the diet, mouth morphology and dentition of the Port Jackson shark, *Heterodontus portusjacksoni*. Mar Freshw Res 61(1): 74–85.

60. Parrish CC, Pethybridge H, Young JW, Nichols PD (2014) Spatial variation in fatty acid trophic markers in albacore tuna from the southwestern Pacific Ocean-a potential 'tropicalization' signal. Deep Sea Res II: In Press: http://dx.doi.org/10.1016/j.dsr2.2013.12.003

61. Revill AT, Young JW, Lansdell M (2009) Stable isotopic evidence for trophic groupings and bio-regionalization of predators and their prey in oceanic waters off eastern Australia. Mar Biol 156(6): 1241–1253.

2

Non-Replicating *Mycobacterium tuberculosis* Elicits a Reduced Infectivity Profile with Corresponding Modifications to the Cell Wall and Extracellular Matrix

Joanna Bacon[1]*, Luke J. Alderwick[2], Jon A. Allnutt[1], Evelina Gabasova[3], Robert Watson[1], Kim A. Hatch[1], Simon O. Clark[1], Rose E. Jeeves[1], Alice Marriott[1], Emma Rayner[1], Howard Tolley[1], Geoff Pearson[1], Graham Hall[1], Gurdyal S. Besra[2], Lorenz Wernisch[3], Ann Williams[1], Philip D. Marsh[1]

1 Public Health England, Microbiology Services, Porton Down, Salisbury, Wiltshire, United Kingdom, 2 Institute of Microbiology and Infection, School of Biosciences, University of Birmingham, Birmingham, United Kingdom, 3 MRC Biostatistics Unit, Institute of Public Health, Cambridge, United Kingdom

Abstract

A key feature of *Mycobacterium tuberculosis* is its ability to become dormant in the host. Little is known of the mechanisms by which these bacilli are able to persist in this state. Therefore, the focus of this study was to emulate environmental conditions encountered by *M. tuberculosis* in the granuloma, and determine the effect of such conditions on the physiology and infectivity of the organism. Non-replicating persistent (NRP) *M. tuberculosis* was established by the gradual depletion of nutrients in an oxygen-replete and controlled environment. In contrast to rapidly dividing bacilli, NRP bacteria exhibited a distinct phenotype by accumulating an extracellular matrix rich in free mycolate and lipoglycans, with increased arabinosylation. Microarray studies demonstrated a substantial down-regulation of genes involved in energy metabolism in NRP bacteria. Despite this reduction in metabolic activity, cells were still able to infect guinea pigs, but with a delay in the development of disease when compared to exponential phase bacilli. Using these approaches to investigate the interplay between the changing environment of the host and altered physiology of NRP bacteria, this study sheds new light on the conditions that are pertinent to *M. tuberculosis* dormancy and how this organism could be establishing latent disease.

Editor: Anil Kumar Tyagi, University of Delhi, India

Funding: This report is work commissioned by the National Institute of Health Research. The views expressed in this publication are those of the authors and not necessarily those of the NHS, the National Institute for Health Research, or the Department of Health. The funders had no role in study design, data collection and analysis, decision to publish, or preparation of the manuscript.

* E-mail: joanna.bacon@phe.gov.uk

Introduction

Tuberculosis (TB) is characterised by long term persistence in a latent state, which after decades, can be reactivated and lead to further spread of the disease. *Mycobacterium tuberculosis* is thought to adapt and thrive in diverse environmental niches *in vivo* during latency [1]. However, the location and physiology of the bacterium during this phase of the disease remains unclear [2] [3]. It is generally believed that during latency *M. tuberculosis* resides within the solid granulomas which are characteristic of latent TB infection. It is thought that the tubercle bacilli located in these regions reside in a slow growing or non-replicating dormant-like state, which could be achieved by exposure to perturbations in the availability and supply of oxygen and the sources of available nutrients [1] [4]. The dormant-like state has been extensively investigated using *in vitro* models in an attempt to simulate the granuloma environment with a particular focus on hypoxia-induced non-replicating persistent (NRP) states, which have demonstrated that *M. tuberculosis* is able to survive for extended periods of time [5] [6]. The DosR regulon is implicated in the hypoxic adaptation of *M. tuberculosis* and subsequent virulence profiles. During infection studies, *M.*

tuberculosis dosR mutants exhibit a variably attenuated phenotype [7] [8] [9]. Differences in the methodologies employed to study *M. tuberculosis* dormancy such as the choice of animal species, the disease-stage, and the parameters used to define attenuation, make interpretation of these variable findings difficult to reconcile. However, these studies highlighted the fact that there are other environmental factors, such as the availability of nutrients, which could be triggers for establishing latent TB infection [1].

The focus of this study was to model environmental conditions other than hypoxia, such as nutrient-depletion, that will be encountered by *M. tuberculosis* during chronic infection. We exploited the advantages of controlled batch fermenter cultures of *M. tuberculosis* utilising fatty acids as the primary carbon source, which were gradually depleted over an extended period of time. The physiological and pathogenic responses of *M. tuberculosis* to nutrient depletion have not been investigated fully in previous studies. Therefore, the role of cell wall re-modeling in the establishment of NRP and the impact of these different physiological states on infectivity in the guinea pig were explored.

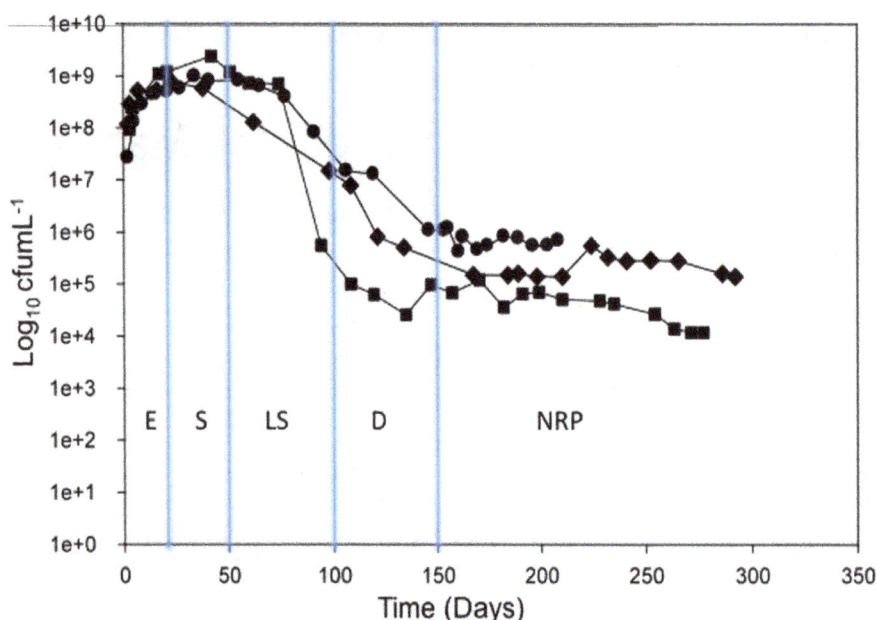

Figure 1. Growth curves for *M. tuberculosis* grown in medium containing Tween 80 as the primary carbon source. Total viable counts for Culture 1 (circles), Culture 2 (diamonds), and Culture 3 (squares), were measured over an extended period of at least 200 days. An estimation of each growth phase is indicated on the graph: exponential phase (E), stationary phase (S), late stationary phase (LS), death phase (D), and non-replicating persistent phase (NRP).

Results

NRP derived by nutrient depletion

Three independent replicate cultures of *M. tuberculosis*, (Cultures 1, 2, and 3) were established using a fermenter-controlled batch growth model based on a modified continuous culture vessel that we have previously reported [10]. This approach has enabled us to establish non-replicating persistent (NRP) populations of *M. tuberculosis* in culture by the gradual depletion of nutrients over time in a controlled oxygen-replete environment. The medium was CMM Mod6 (medium recipe, Table S1, supporting information), which contains the primary carbon source Tween 80; hydrolysis to oleate provided an indirect source of fatty acids [11] [12]. The viability of organisms was monitored over an extended period of at least 200 days. In each of the three independent experiments, exponential phase (E) occurred during the first 10 days of culture (total viable counts reached approximately 10^9 cfu mL^{-1}) and proceeded into stationary phase (S) which lasted for approximately 40 days (Figure 1). Over the subsequent period, cells remained in a late stationary phase (LS) followed by a significant drop of 10^3 cfu mL^{-1} in cell count viability (D) (Figure 1). A final phase, that we termed NRP, lasted for a period of at least 60 days with a concomitant viable count of 10^5–10^6 cfu mL^{-1} (Figure 1). Multiple culture samples were removed at growth stages E, S, LS, and NRP and used for subsequent physiological and biochemical analyses. The death phase (D) was not included in the analyses. There was a finite amount of biomass from each time-course and so each type of analysis could not be performed on all three cultures. For clarification, Culture 1 was sampled for transcriptomics, Culture 2 was sampled for transcriptomics and lipid/carbohydrate analyses, and Culture 3 was used for lipid/carbohydrate analyses and infectivity studies. Tween 80 was confirmed as depleted and restricting growth by the analysis of free fatty acids in the spent culture supernatant; eighty percent of the Tween 80 was consumed by the NRP phase for both Culture 2 and Culture 3 (Figure S1, supporting information). Tween 80-depletion was limiting growth as shown by the addition of Tween 80 (0.2% v/v in water) to a further NRP culture, which initiated re-growth over a period of 12 days and viability levels rose to 10^8 cfu ml^{-1} (unpublished results).

The morphology of M. tuberculosis isolated from each growth phase

Electron microscopy revealed the presence of extracellular material surrounding *M. tuberculosis* cells sampled from each of the growth phases (Figure 2), which was particularly pronounced in late stationary phase and NRP phase (Figure 2C and D). Samples from late stationary phase were stained with either Alcian Blue or Sudan Black in order to selectively identify the presence of carbohydrates or lipids, respectively [13]. In each case, the Ziehl-Neelsen acid fast stain was used in parallel to detect *M. tuberculosis* bacilli. Staining with Sudan Black gave a negative result for each of the samples tested, which suggested that the extracellular material was unlikely to consist predominantly of lipid. However, upon staining with Alcian Blue followed by visualisation using light microscopy, at x100 magnification, we observed a clear selective blue-coloration of the extracellular material surrounding the *M. tuberculosis* bacilli (Figure 3). These initial findings indicated that, during the latter phases of growth, these cultures over-produced an extracellular material which is, at least in part, composed of carbohydrate, which is a characteristic of biofilm formation in bacteria survival. The content of the following sections were focused on the analyses of the extracellular material and the re-modeling of the cell wall as these have implications for the survival of *M. tuberculosis* in NRP.

Re-modeling of the cell wall

Mycobacteria have an unusual lipid-rich cell wall, which is complex in structure and vital for cell survival and host-pathogen interactions. Polar and apolar lipids were selectively extracted

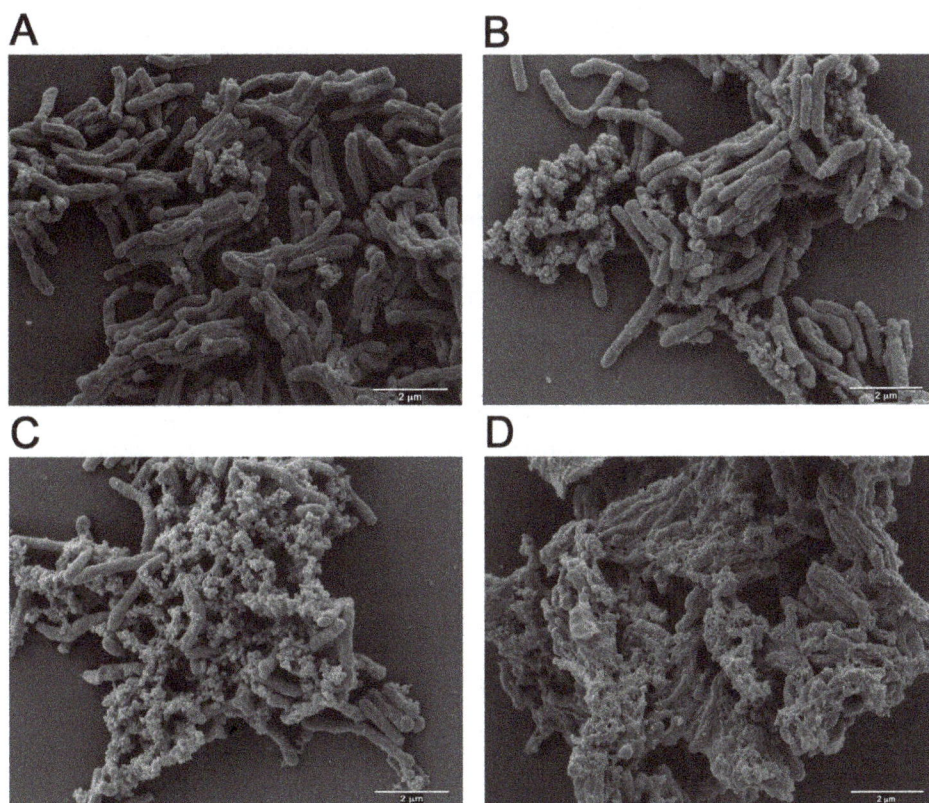

Figure 2. Scanning electron micrograph images of formaldehyde-fixed cells sampled from Culture 2. Panel A) exponential phase at day 8; Panel B) stationary phase at day 44; Panel C) late stationary phase at day 107; and Panel D) NRP at day 292.

using organic solvents from cells that had been harvested from each of the four phases of growth, E, S, LS, and NRP (Figure 1). Lipid fractions were analysed using thin layer chromatography (TLC) to detect individual species of cell wall lipids as compared to known standards [14]. 1D TLC analyses of freely extractable lipids from Culture 2 are illustrated (Figure 4) and show a

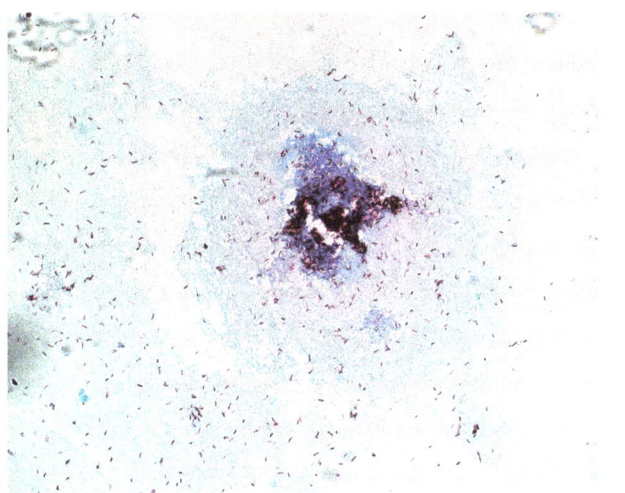

Figure 3. Cells sampled during late stationary phase. Bacilli stained with Ziehl Neelson and the surrounding extracellular material was stained with Alcian Blue, pH 1.0, and visualised at a magnification of ×100.

gradual increase in the cellular content of free mycolate and a corresponding depletion in the content of trehalose dimycolate (TDM) and trehalose monomycolate (TMM). The data are representative of Cultures 2 and 3. Apolar lipids extracted from Culture 2 and Culture 3 were subjected to 2D TLC analysis to enable further comparisons of how the lipid profile alters at each growth stage over the duration of the culture (Figure 5). There was a significant accumulation of a spot on each TLC plate throughout the time-course (using solvent system C) which migrated to a position corresponding to free mycolic acid (MA) [14], for Culture 2 and Culture 3 (Figure 5). In order to quantify this change in lipid content over time, we conducted a densitometry analysis of the fast-migrating spots on the TLC (corresponding to MA) normalised against the non-migrating lipids at the origin of each TLC plate. For each TLC (representing the day at which samples were collected from Culture 2), the ratio of lipids (origin: mycolate) increased gradually as follows, day 8 (1:0.95), day 44 (1:0.96), day 107 (1:0.99) and day 292 (1:1.27). We also observed an almost identical pattern of MA accumulation in Culture 3, as demonstrated by the following increase in the ratio of lipids (origin: mycolate), day 4 (1:0.75), day 29 (1:0.90), day 72 (1:0.92) and day 277 (1:1.05). No changes in any of the other lipids were found [14].

Adaptation through alterations in cell wall carbohydrates

Our initial Alcian blue staining of the extracellular material accumulating during growth (Figure 3) directed us to investigate the hypothesis that polysaccharide material derived from *M. tuberculosis* was accumulating in the spent culture medium. Highly

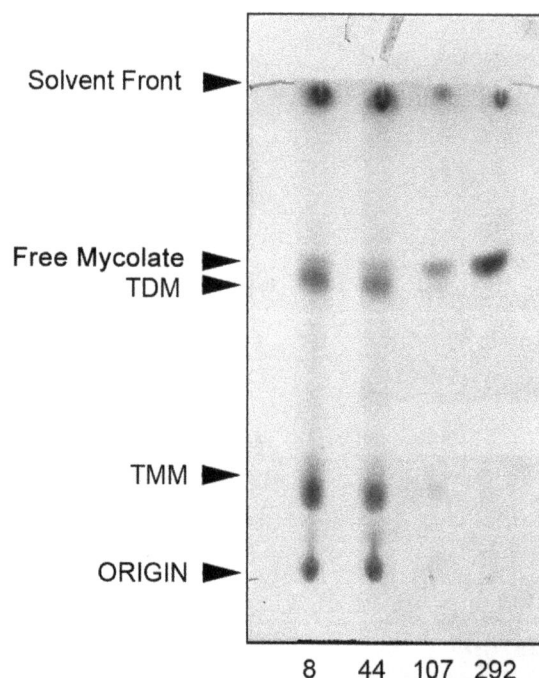

Figure 4. 1D TLC of freely extractable lipids from Culture 2. Cell biomass was sampled during exponential phase (day 8), stationary phase (day 44), late stationary phase (day 107) and non-replicating persistent phase (day 292). Lipids migrating on the TLC were visualised by staining with MPA and compared to known standards [14].

lipoglycans and polysaccharides separated by molecular mass (Figure 6) [15]. There was a gradual increase over time in the apparent molecular mass of the band corresponding to lipoarabinomannan (LAM) (Figure 6A and 6B). Interestingly, the position of the lower band, representing lipomannan (LM), remained unchanged over time. This lipoglycan profile correlated with an increase in the molecular mass of LAM (with respect to its initial mass at the start of culture). In order for us to determine the total sugar composition in each of the fractions presented in Figure 4, highly purified lipoglycan material was chemically modified in order to produce alditiol acetate derivatives which were subsequently analysed via gas chromatography (GC) as described previously [16] [15]. Both cultures (2 and 3) displayed a similar phenotype in terms of the overall total sugar content, with a relative increase in the amount of Ara, with respect to Man, over the duration of the two independent culture experiments. An increase in the ratio of Ara: Man was observed as follows: Culture 2, day 8 (1.03), day 44 (3.62), day 107 (4.03) day 292 (5.03) and Culture 3, day 4 (1.21), 29 (3.00), 72 (5.76) and 277 (5.27). Cell wall extracts from Culture 3 were also chemically modified in order to produce alditiol acetate derivatives, which were analysed by GC. An increase in the ratio of Ara: Man was observed as follows: day 4: 0.96, day 29: 0.87, day 72: 1.13 and day 277: 1.63, reflecting the increased Ara:Man ratios also observed in the cell biomass. The ratio of Ara:Man has been shown to have an impact on the immunological properties of LAM [15].

Global gene expression analysis

Gene expression analyses were applied to understand more about the molecular genetics underlying the transition between each growth phase in culture and particularly to the biochemical changes observed in the cell wall. Whole genome gene expression analyses were performed throughout the time-course for Cultures 1 and 2. Probabilistic analyses using Gaussian process regression and Bayesian model selection were applied to identify genes which showed similar gene expression trends in both cultures [17]. Using the probabilistic model, expression profiles of approximately fifty

purified carbohydrate extracts (representing glycolipid and lipoglycan macromolecules shed from *M. tuberculosis* in liquid medium) were extracted from spent culture medium that had been collected from Culture 2 and Culture 3 in uniform volumes during each of the four phases of growth. Each fraction was subjected to SDS-PAGE analysis and specifically stained to visualise species of

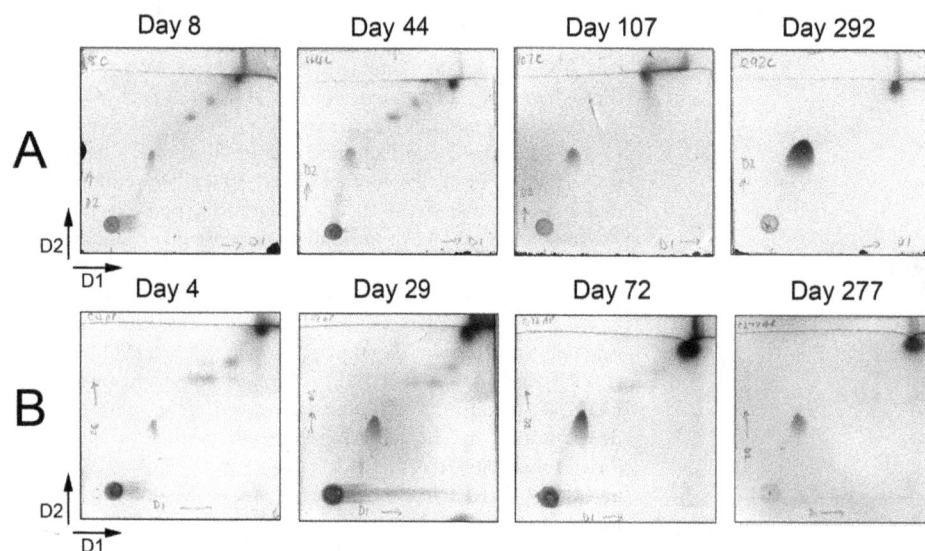

Figure 5. 2D TLC analyses of apolar lipids extracted from Culture 2 and Culture 3. Panel A, cell biomass was sampled from culture 2 during exponential phase (day 8), stationary phase (day 44), late stationary phase (day 107), non-replicating persistent phase (day 292). Panel B, cell biomass was sampled from culture 3 during exponential phase (day 4), stationary phase (day 29), late stationary phase (day 72), NRP phase (day 277). Lipids migrating on the TLC (using solvent system C) were visualised by staining with MPA and compared to known standards [14].

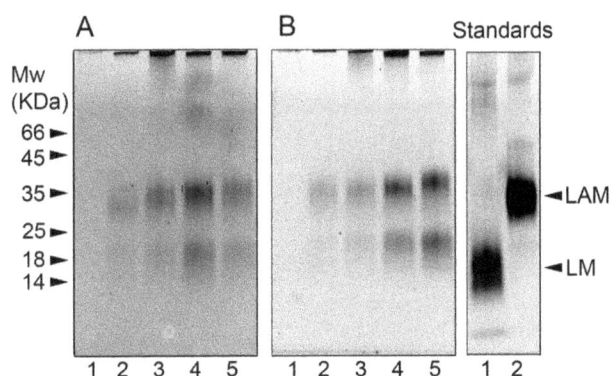

Figure 6. SDS-PAGE analysis of polysaccharide material isolated from the culture medium during the four phases of growth in culture 2 and culture 3. Culture 2 in Panel A; polysaccharide material was purified from liquid medium prior to inoculation in lane 1, exponential phase growth (day 8) in lane 2, stationary phase growth (day 44) in lane 3, late stationary phase growth (day 107) in lane 4 and NRP phase growth (day 292) in lane 5. Culture 3 in Panel B, polysaccharide material was purified from liquid medium prior to inoculation in lane 1, exponential phase growth (day 4) in lane 2, stationary phase growth (day 29) in lane 3, late stationary phase growth (day 72) in lane 4 and NRP phase growth (day 278) lane 5. For comparison, the panel on the right shows LM (lane 1) and LAM (lane 2) previously isolated from *M. tuberculosis* H37Rv cultured exponentially in liquid medium.

percent of the genes showed similar dynamics profiles across the time-courses in the two cultures and therefore their profiles could be merged. The remaining genes showed either different gene expression trends in both cultures or the signal present in the data was obscured (in one of the cultures or in both cultures) by the noise level. Details of the probabilistic analysis applied to identify consistent genes between the two cultures are given in (Methods S1) in the supporting information. The merged profiles were arranged in a total of 55 clusters based on the trend in their expression levels, using Bayesian hierarchical clustering of curves [18]. Only those genes that could be merged in terms of their expression profiles in the two cultures were included in the analysis described below. (All 55 cluster profiles can be found at http://xenakis.mrc-bsu.cam.ac.uk/wernisch/enrichment/html/ "An overview of all clusters"). The gene expression data were not normalised to account for the potential reduction of RNA in stationary phase and NRP because there was likely to be a heterogeneous mixture of cells that were in different phenotypic states and therefore an assumption could not be made that all the cells in the stationary and NRP phases were equivalent in their total RNA levels; this could have added further bias and inaccuracies to the data analysis.

Most of the 55 clusters were down-regulated in the NRP after day 150 and a huge downward shift in the metabolic response of the bacteria as they enter an NRP state. Eleven clusters containing 561 genes were up-regulated in stationary phase and late stationary phase (compared with exponential phase) followed by down-regulation at approximately day 150 at the start of the NRP phase. Twenty-one clusters containing 315 genes followed a trend of down-regulation from exponential phase right through to the end of the NRP phase. Eight clusters containing 157 genes revealed a profile of flat expression in exponential phase and through stationary phase followed by down-regulation from day 150 in NRP. For eight of the clusters the expression profiles remained flat throughout the time-course. Enrichment for

function was applied, using all the gene annotations provide by the Sanger Institute and the MTB-GOA server, to determine which functional groups of genes were important for establishment of an NRP state. The complete results of the enrichment analysis are provided as hyperlinked HTML files as supporting information http://xenakis.mrc-bsu.cam.ac.uk/wernisch/enrichment/html/. "Sanger classification enrichment" and "MTB-GOA classification enrichment". Some of the gene clusters showed a marked enrichment in fatty acid metabolism, lipid degradation and cell wall re-modeling (Figure 7). Details of the clusters that show enrichment for functional category are presented in Table 1. A list of all the genes that have been enriched for a functional class can also be found at http://xenakis.mrc-bsu.cam.ac.uk/wernisch/enrichment/html/ "Genes of cluster enrichment analysis".

We have observed changes in the LM, LAM, and free mycolate in relation to an NRP state and so we focused specifically on the genes associated with the biosynthesis of these molecules. We constructed a comprehensive list of genes (Table S2, supporting information) that have been implicated in the PIM→LM→LAM pathway [19] [20]. Using the enrichment analysis method, genes Rv3257c (*pmmA*), Rv3806c (*ubiA*), and Rv3793 (*embC*), were the only genes from the list of LM/LAM biosynthetic genes that were enriched. All three of these genes were found in cluster 55, showing that the dynamics in the expression profiles for these genes were very similar with sustained up-regulation from early stationary phase until day 150 (Figure 8). *PknH* (Rv1266), was revealed to be enriched for in cluster 9 (Figure 7) and evidence suggests that it is directly implicated in the regulation of LAM biosynthesis [21] and was also induced in chronically infected mice (Table 2) [22]. The GO:0071767 annotation for "mycolate acid metabolic process" was used to look for the enrichment of genes involved in the accumulation of free mycolate. Cluster 5 was enriched for genes of the mycolic acid metabolic process and the profile for this cluster consisted of induction early in stationary phase at day 30 followed by further induction at day 116 and day 146. More specifically, genes involved in the synthesis and processing of mycolates were up-regulated early in stationary phase and these were Rv0643 (*mmaA3*, methoxy mycolic acid synthase 3), Rv1273c (transmembrane ABC transporter), Rv1349 (*irtB*, iron-regulated transmembrane ABC transporter), Rv2006 (*otsB1*, trehalose-6-phosphate phosphatase), and Rv3801c (*fadD32*, fatty-acid-AMP synthetase).

Adaptation of central metabolism

Gene expression profiling was also performed to investigate energy metabolism in *M. tuberculosis* during its transition into an NRP state and how the cells maintained their energy levels in NRP. Previous studies have observed that nutrient-starved, non-replicating bacilli undergo a global down-regulation of metabolic genes involved in respiration [23] [10] [24]. Genes involved in the degradation of fatty acids via the β-oxidation pathway (Rv0914c, keto acyl-CoA thiolase) were up-regulated early around day 41 in Cluster 9 (Figure 7) and showed sustained up-regulation until about day 150. Cluster 16 (Figure 7) is one of the largest clusters containing 125 genes and had several time-points at which the induction of gene expression occurred in exponential phase, stationary phase and late stationary phase, followed by down-regulation throughout NRP. Cluster 16 was enriched for in several Sanger and GO functional groups and contained genes involved in energy metabolism such as ATP synthases (Rv1305, Rv1306, Rv1308, Rv1309, and Rv1310). Cluster 16 is generally enriched in genes for biosynthesis of lipids such as Rv0242c (*fabG4*), Rv0673 (*echA4*), Rv0271c (*FadE6*), or Rv0672 (*FadE8*), or lipopolysaccharides and phospholipids (Rv3032, Rv0062 (*celA1*), and Rv0315).

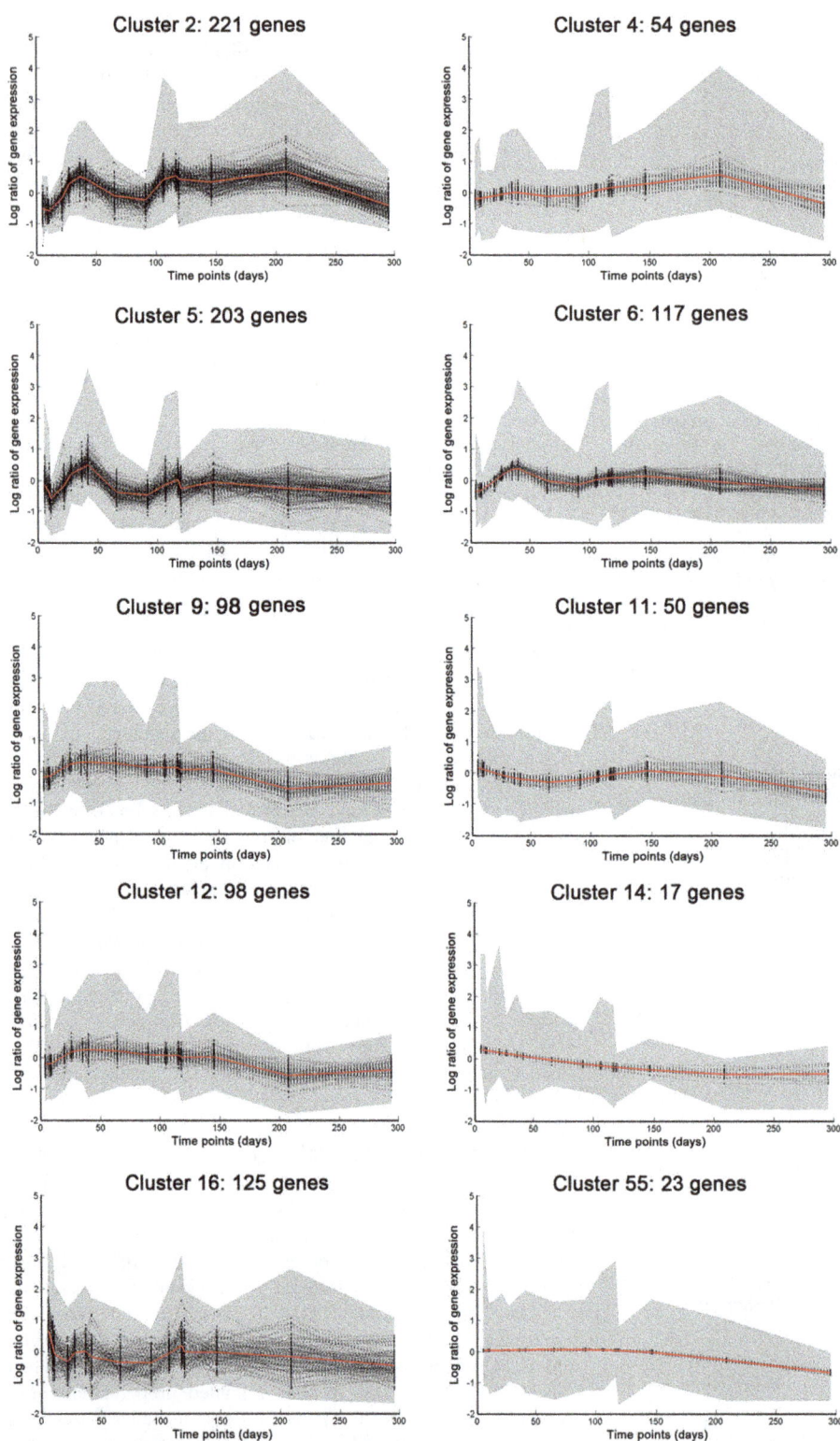

Figure 7. Gene expression clusters with a marked enrichment in fatty acid metabolism, lipid degradation, and cell wall remodeling. Graphs show normalised log expression values (centered around a mean value of zero) over the time-courses in Cultures 1 and 2. The shaded area represents the 99% confidence interval for the curves.

In Cluster 2 (221 genes) (Figure 7), the induction of gene expression occurred at around day 41; following this was a reduction in gene expression with a further induction at day 208.

Genes included in this cluster were Rv1169c (*lipX*) and Rv0467 (*icl1*), which confirmed that the induction of the β-oxidation of fatty acids and lipid degradation are important for the survival of

Table 1. Annotation of clusters that were enriched for functional category.

Cluster	Function	Description	P	Ratio	f0-c0	f1-c0	f0-c1	f1-c1
14	I.B.6.a	Aerobic	0	33.7	1883	17	13	4
9	I.H.3	Acyltransferases and Mycolyltransferases	0.91	5.9	1806	13	94	4
55	-	LAM/LM	0.23	13.98	1874	20	20	3
51	II.A.1	Ribosomal protein synthesis, modification	0.48	10.56	1865	33	16	3
16	I.B.8	ATP-proton motive force	0	74.16	1791	1	120	5
16	GO:0050896	Response to stimulus	0.47	5.69	1779	13	120	5
16	GO:0044464	Cell part	0.33	1.99	1547	245	95	30
16	III.B	Chaperones/Heat shock	0.1	14.72	1788	4	121	4
16	II.B.4	Polysaccharides, lipopolysaccharides	0.25	21.91	1790	2	122	3
16	III.D	Protein and peptide secretion	0.14	4.46	1765	27	117	8
16	GO:0032940	Secretion by the cell	0.25	21.91	1790	2	122	3
11	GO:0009072	Aromatic amino acid process	0.39	38.45	1865	2	48	2
11	GO:0042401	Metabolic process	0.39	38.45	1865	2	48	2
11	GO:0003677	DNA binding	0.31	13.12	1858	9	47	3
6	GO:0050662	Coenzyme binding,	0.45	7.91	1792	8	113	4
6	GO:0000166	Nucleotide binding	0.45	7.91	1792	8	113	4
13	GO:0005886	Plasma membrane	0.12	2.57	1266	600	23	28
5	GO:0071767	Mycolic acid metabolic process	3.44	3.3	1701	13	198	5
44	II.A.1	Ribosomal protein synthesis, modification	0.06	12.87	1863	32	18	4
4	I.G.2	Folic acid biosynthesis	0.22	15.51	1856	7	51	3
4	GO:0051234	Establishment of localisation	0.23	70.7	1862	1	52	2
4	GO:0045927	Positive regulation of growth	0.08	12.29	1851	12	50	4
2	IV.C.1.a	PE subfamily	0.16	4.19	1679	17	212	9
2	VI	Unknowns	0.1	1.95	1533	163	183	38
2	GO:0075136	Response to the host	0.38	3.55	1676	20	212	9
46	VI	Unknowns	0.4	5.24	1706	195	10	6
46	GO:0010608 GO:0032268	Post-transcriptional regulation	0.1	66.26	1897	4	14	2
46	GO:0009891	Positive regulation	0.34	29.7	1892	9	14	46

Annotation by the Sanger Institute is indicated by roman literals, GO annotation by the GO label, p-values are in percent (not corrected for multiple testing), ratio is the Fisher estimate of the odds ratio and membership of a gene for a cluster (c) or a function (f) is indicated by 1 for membership and 0 for non-members. Lists of genes from the f1-c1 category (joint members of functional class) and are provided in the supporting information at http://xenakis.mrc-bsu.cam.ac.uk/wernisch/enrichment/html/.

M. tuberculosis on diminishing levels of fatty acid. Cluster 6 (Figure 7), showed early induction at day 41 with a second peak of expression at day 146, and comprised genes coding for nucleotide binding proteins involved in arginine biosynthesis Rv1652 (*argC*) and NADPH requiring oxidoreductases (Rv3106 (*fprA*), Rv3303 (*lpdA*)) involved in energy metabolism. Genes involved in the regulation of lipid metabolism, (Rv3574 (*kstR*) [25] were enriched for in Cluster 11 (Figure 7), which showed late gene induction at day 146 followed by down-regulation thereafter. Furthermore, genes involved in the biosynthesis of aromatic amino acids also fell into this cluster (Figure 7). We were surprised to see a few clusters with genes induced in NRP at around day 208. Cluster 4 (Figure 7) contained folic acid biosynthetic genes, Rv3607c (*folB*), Rv2447c (*folC*) and Rv3608c (*folP1*), which were up-regulated at this late stage. It also contained genes involved in transport and ion channels, Rv3065 (*mmr*) and Rv0985c (*mscL*), as well as the toxin-antitoxin *vapBC* family under the GO term for positive regulation of growth. Cluster 14 (Figure 7) contains genes that were down-regulated throughout the time-course. The profiles of genes which are involved in energy metabolism, in particular, NADH

dehydrogenases (Rv3156, Rv3158, Rv3154, and Rv2194), indicate a reduction in aerobic respiration from exponential phase. Two acyltransferases Rv2482c (*plsB2*) and Rv2881 (*cdsA*) were also enriched for in this cluster (Figure 7).

Comparisons with other gene expression studies

Comparisons were made between genes that were more highly expressed in stationary phase/late stationary phase (than in the exponential phase, 561 genes, (Table S3, supporting information). in our nutrient-starved cultures and the genes induced in previously described *in vivo* models of *M. tuberculosis* infection and *in vitro* models of nutrient-starvation [22] [26] [27] [23] [10] Table 2 displays genes that were induced in stationary phase in our cultures and at 60 days post-infection in chronically infected mice [22]. One fifth of the stationary phase-induced genes from our cultures were induced at 48 hours post-infection in the macrophage (Table S3, supporting information) [26]. Of these genes, nine genes were found to be induced in at least one of the other *in vivo* or *in vitro* data sets used for this comparison (Table 3). Genes Rv3555c, Rv2642, Rv0083 were induced in the human granulo-

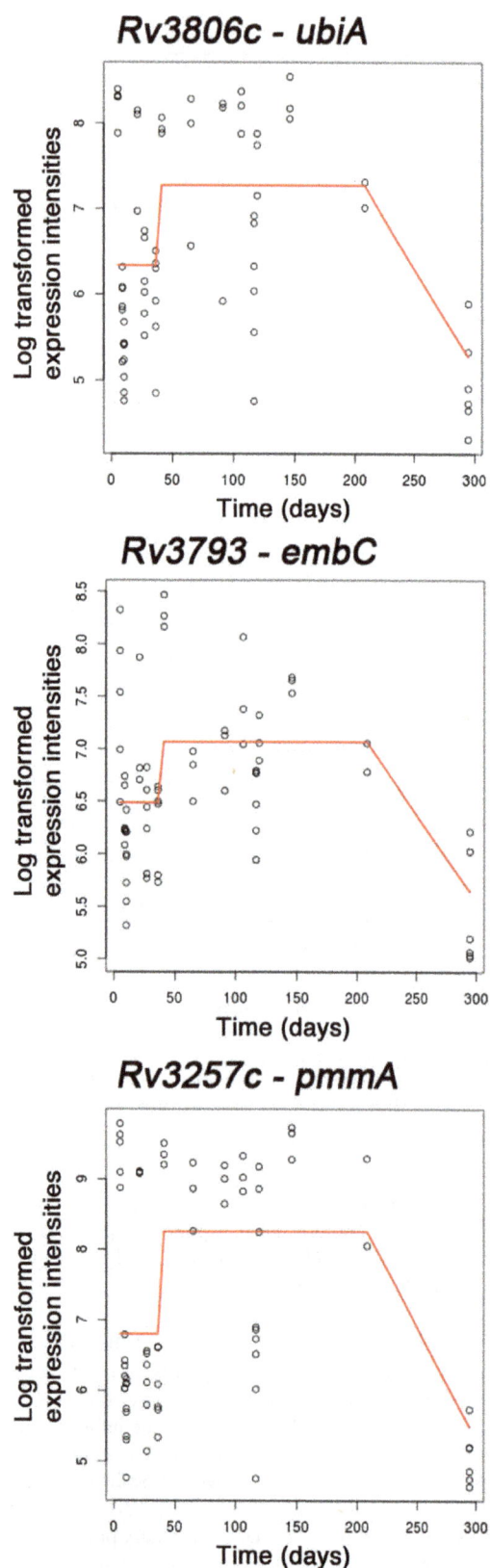

Figure 8. Gene expression profiles for LAM biosynthetic genes, Rv3257c (*pmmA*), Rv3793 (*embC*), Rv3806c (*ubiA*). The lines on the graphs show the average normalised log ratios of gene expression over the time-courses in Cultures 1 and 2.

ma [27] and gene Rv0251c, encoding heat shock protein *hsp*, was the only gene induced under all the conditions compared.

Infectivity of M. tuberculosis in the guinea pig

A guinea pig study was performed to determine whether the organism's ability to infect was improved or impaired by the different physiological states we have observed. Bacteria were removed at day 4 (exponential phase), and day 277 (NRP phase) from Culture 3 and used to infect guinea pigs by the aerosol route. In order to achieve a low-dose infection (less than 20 bacilli implanted in the lung), a concentration of approximately 1×10^5 cfu mL^{-1} was needed in the aerosol generator. The culture samples were adjusted to this concentration using an estimation based on OD which was retrospectively verified by plating on solid agar. The concentration of *M. tuberculosis* in each of the nebuliser solutions was 9.33×10^4 cfu mL^{-1} (exponential phase) and 2.09×10^4 cfu mL^{-1} (NRP). Verification of the very low number of organisms implanted in the lungs was not feasible due to the inability to enumerate small numbers of cells in a large sample of lung tissue. However, historical data enabled a relationship between the nebuliser concentration and the delivered dose to be established [28] [29]. On this basis the animals infected with the exponential phase and NRP bacteria received similar doses. The bacterial load (Figure 9) and histopathological changes (Figure 10 and Figure 11) resulting from these infections were determined in lungs and spleens at days 16 and 42 post-challenge.

At day 16 post-challenge, the bacterial load in the lungs and spleens of animals infected with the exponential phase bacteria was significantly higher than in the organs of animals infected with NRP ($P = <0.001$) (Figure 9). Bacteria were not detected (limit of detection $= 0.7$ log$_{10}$ cfu mL^{-1}, Figure 9) in the spleens of NRP-challenged animals. By day 42 post-challenge, the lung bacterial load had increased (relative to day 16) in animals infected with NRP bacteria, in contrast to the animals infected with exponential phase bacteria where the mean value was lower at day 42 compared to day 16. Statistical analysis showed no significant differences between the mean values of cfu in lungs at day 42 between both groups of guinea pigs ($P = 0.471$). In the spleens, the bacterial load had increased in both groups of animals relative to day 16 but the increase was more pronounced in the NRP group. The mean values for cfu in the spleen at day 42 were not significantly different from each other by statistical analysis ($P = 0.163$).

Lesions in the lung and spleen consisted of variable sized granulomatous foci with mainly macrophages and lymphoid cells. Occasional foci of necrosis and calcification were observed (Figure 10). The subjective histopathology scores are summarised in Figure 11; a mean score of all animals in the group is given in the text below and the range is shown in curly brackets. In all animals challenged with exponential phase cells, at day 16, the lungs exhibited histopathological changes with a score of 6.7 {2–13}. In comparison, there was much reduced pathology in the lungs of the animals challenged with NRP phase bacteria (score $= 0.6$ {0–2}) and 4 out of 8 animals had no lesions. At 42 days post-challenge, there was more extensive lung pathology (than at day 16) in both of the groups but the lowest pathology was observed in animals challenged with the NRP phase bacteria (4.75; {3–9}). This level of pathology at day 42 was similar to that observed in the animals infected with exponential phase at day 16 post challenge. Histopathological changes were not observed in the spleen in any animal at 16 days post-challenge. At day 42, the pathology in animals infected with NRP bacteria (3.6 {0–7}) was lower than exponential phase (6.4 {5–9}) infected animals.

Non-Replicating Mycobacterium tuberculosis Elicits a Reduced Infectivity Profile with Corresponding...

19

Table 2. Genes up-regulated in late stationary phase and during chronic infection in mice.

Gene ID (Rv no.)	Gene Name	Cluster no.	Function	Fold up-regulation [22]
Rv2667	clpX'	2	ATP-dependent protease ATP-binding subunit	1.92
Rv2853	PE_PGRS	4	PGRS subfamily of gly-rich proteins	2.37
Rv2612c	pgsA	16	PIsynthase/CDP-diacylglyceride – inositol phosphatidyltransferase	3.47
Rv1266c	pknH	2	Serine/threonine-protein kinase	5.4
Rv2439c	proB	4	Glutamate 5-kinase protein	3.55
Rv0612	Rv0612	2	Conserved hypothetical protein	2.2
Rv0666	Rv0666	2	Membrane protein	6.26
Rv2850c	Rv2850c	4	Magnesium-chelatase	3.53
Rv2891	Rv2891	2	Conserved hypothetical protein	3.63

Genes up-regulated in late stationary phase in Cultures 1 and 2 of M. tuberculosis and at 60 days post-infection in the BALB/C mice [22]. The cluster profiles for the genes listed can be found in the supporting data at http://xenakis.mrc-bsu.cam.ac.uk/wernisch/enrichment/html/.

Discussion

NRP bacilli can be generated by extended nutrient-depletion

The effects of nutrient-depletion on the persistence and survival of M. tuberculosis were assessed independently of the effects of low oxygen in employing controlled batch cultures [30]. An extended period of nutrient-depletion resulted in a population of bacteria that were culturable on agar and maintained a consistent viable count; a population that we have termed NRP. We recognise that during this phase, the cultures were likely to contain a heterogeneous mixture of organisms, some of which were non-replicating, whilst other bacilli could be dividing or viable but not culturable, thus reflecting the heterogeneity of cell states within the granuloma [31]. This study involved growing M. tuberculosis in a medium that contained oleic acid (derived from Tween 80), as the primary carbon source, which reflects the nutrient sources available in vivo [4]. We have previously shown that M. tuberculosis and Mycobacterium bovis will utilise Tween 80 in the absence of glycerol and glucose in continuous culture [11]. Tween 80 is cleaved to liberate oleic acid and a polyethylene derivative of sorbitol; each component is then subsequently absorbed and metabolised through the glyoxylate shunt and fatty acid degradation biochemical pathways [32]. Eighty percent of the available Tween 80 was being metabolised by our cultures in the current study (assuming that one molecule of Tween 80 was hydrolysed to one molecule of oleic acid) (Figure S1, supporting information). It seems unlikely that an NRP phase would be established if free fatty

Table 3. Genes up-regulated in stationary phase in Cultures 1 and 2 and in further conditions in vitro or in vivo.

Gene ID (Rv no.)	Gene Name	Clusterno.	Function	Also up-regulated in the following studies:
Rv0083		2	Oxidoreductase	Human granuloma v in vitro culture in 7H9; [27]
Rv0122		2	Hypothetical protein	96h starvation in PBS [23]
Rv0251c	hsp	16	Heat shock protein	Human granuloma v in vitro [27] 96h starvation in PBS [23] 75 days starvation [10]
Rv0284		50	Conserved membrane protein	75 days starvation [10]
Rv1072		11	Conserved membrane protein	75 days starvation [10]
Rv1285	cysD	2	Sulfate adenylyltransferase	75 days starvation [10]
Rv1461		15	Conserved hypothetical protein	75 days starvation [10]
Rv1462		3	Conserved hypothetical protein	75 days starvation [10]
Rv2497c	pdhA	2	Pyruvate dehydrogenase	75 days starvation [10]
Rv2642		14	Transcriptional regulator, arsR	Human granuloma v in vitro culture in 7H9 [27]
Rv2710	sigB	16	RNA polymerase sigma factor	75 days starvation [10]
Rv3139	fadE24	3	Acyl-CoA dehydrogenase	75 days starvation [10]
Rv3173c		4	Transcriptional regulator, tetR acrR-family	75 days starvation [10]
Rv3555c		3	Conserved hypothetical protein	Human granuloma v in vitro culture in 7H9, Human granuloma v pericavity [27]

Table 3 Genes that were up-regulated at 48 h in macrophages [26] and in Cultures 1 and 2 of M. tuberculosis in late stationary phase (Table S3, supporting information), and in at least one other condition either in vitro or in vivo model as indicated in Table 3. The details of cluster profiles can be found in the supporting information at http://xenakis.mrc-bsu.cam.ac.uk/wernisch/enrichment/html/.

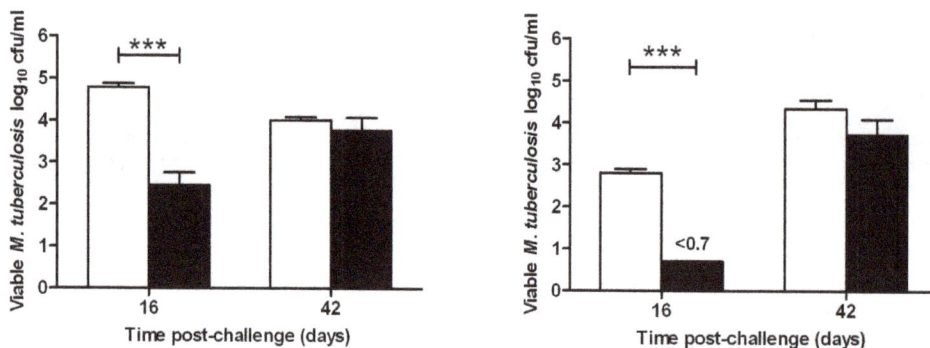

Figure 9. Bacterial load at day 16 and day 42 in guinea pigs post aerosol challenge with either log phase culture or NRP culture. Lungs in Panel A) and spleen in Panel B), challenged with either exponentially growing culture (white) or NRP culture of *M. tuberculosis* (black). Bars indicate group mean $\log_{10}$ cfu mL^{-1} +/− S.E.M of 8 guinea pigs. Statistical analysis was performed using paired T-tests. *** indicates $P = <0.001$.

acids were still available. However, the assay used (Free fatty acid half micro kit, Roche) is a non-discriminate measure of different free fatty acids levels; oleic acids in the supernatant could have been further hydrolysed to alternative fatty acids that were not available to *M. tuberculosis* as a carbon source.

The current paradigm concerning non-replicating, dormant and persistent sub-populations of *M. tuberculosis* relies heavily upon the notion that adaptation of the bacilli to anaerobiosis is key to maintaining cell viability for a prolonged period of time, which is mediated by the DosR-regulated dormancy regulon [33]. Our previous findings using steady-state chemostat cultures show that the DosR regulon is also up-regulated in actively dividing bacilli (at a doubling time of 23 h) growing in very low levels of oxygen (0.2% dissolved oxygen tension) [30]. These data highlight that there are likely to be environmental cues *in vivo* in addition to hypoxia that are encountered by *M. tuberculosis* in the granuloma, such as nutrient-limitation and alternative carbon sources [4]. However, there are a limited number of *in vitro* studies that have investigated nutrient-depletion as a potential stimulus for triggering the transition of *M. tuberculosis* into an NRP state. Loebel *et al.* [34] investigated the effect of nutrients

predicted to be available in a granuloma on the metabolism of *M. tuberculosis* by transferring cells from nutrient-rich medium into phosphate-buffered saline (PBS) and measuring the respiration rate. Nutrient-starvation resulted in a gradual shutdown of respiration to minimal levels, but bacilli remained viable and were later able to recover on rich medium. Betts et al. (2002) used a similar approach of starvation in PBS in static, sealed bottles to observe the effects of nutrient depletion and hypoxia. This study, and other *in vitro* models of dormancy, combined with post-genomic approaches, have provided further evidence of Loebel's early work by showing that nutrient-starved, non-replicating bacilli undergo a global down-regulation of metabolic genes involved in respiration [23] [10] [24].

Lower expression levels of genes involved in ATP synthesis were observed in our NRP cultures whereas there were increases in the expression of genes such as *icl1* (isocitrate lyase) and genes associated with the β-oxidation of fatty acids. This response was also reflected in the nutrient-starved cultures described in this study, since *icl1* and other genes involved in the degradation of fatty acids *via* the β-oxidation pathway were induced (Rv0914c, keto acyl-CoA thiolase). The induction of *icl1* is an important observation as it is known to be essential for *M. tuberculosis* to persist in mouse macrophages [35] and has more recently been shown to be important for utilisation of cytotoxic propionyl-CoA and its conversion to succinate [36]. These observations and the finding that a large proportion of genes induced in stationary phase were also induced in the macrophage (Schnappinger *et al.* 2003) (Table S3, supporting information) provide supporting evidence that adaptation to nutrient-limitation and the metabolism of fatty acids as a carbon source are important *in vivo*.

Folic acid biosynthetic genes, Rv3607c (*folB*), Rv2447c (*folC*) and Rv3608c (*folP1*), were up-regulated throughout late stationary phase and into the NRP phase, only to drop in their expression level at the end of the NRP phase. Intermediates from the folic acid biosynthetic pathway are known to be incorporated into the molybdopterin biosynthetic pathway [37], and molybdopterin synthesising enzymes have been shown to catalyse important redox reactions during dormancy regulation, the metabolism of energy sources, and nitrogen sources. The importance of molybdopterin biosynthesis as a cofactor in *M. tuberculosis* is a relatively unexplored area, but warrants further investigation [38].

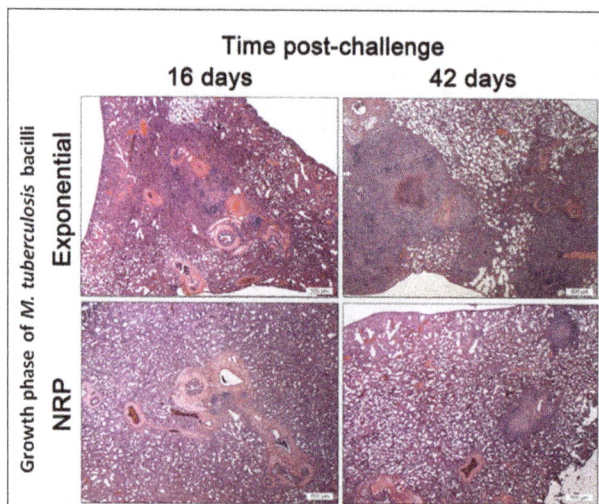

Figure 10. Histopathological changes in guinea pig lung following aerosol challenge with *M. tuberculosis* cultured from different growth phases of nutrient starvation at days 16 and 42 post challenge. H&E. Magnification bar, 500 μm.

The composition of extracellular material

Previously, it has been demonstrated that an increased production of an amorphous material composed of protein and polysaccharides, arabinomannan and glucan followed the growth

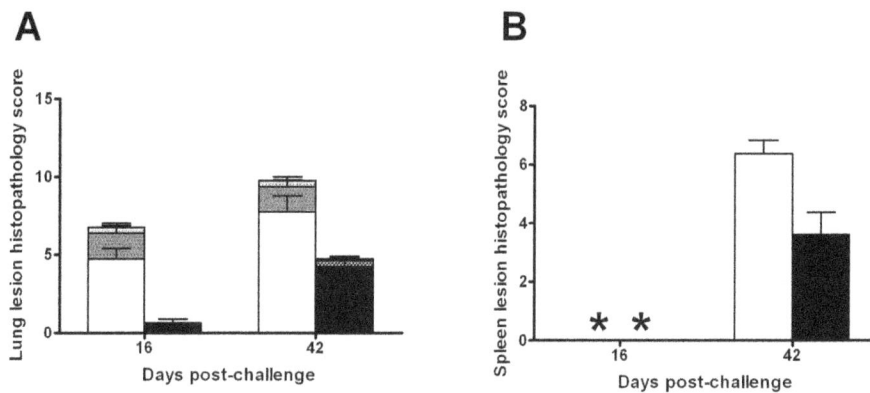

Figure 11. Subjective histopathology scores of *post mortem* at 16 and 42 days post aerosol challenge. Lung in Panel A) and spleen in Panel B) challenged with either exponentially growing culture (white) or NRP cultures of *M. tuberculosis* (black). Bars indicate group mean of 8 animals. Histopathological changes in the lung were recorded either as consolidated (no-shaded bars), necrotic (heavy-shaded bars) or calcified (light shaded bars). Spleen scores for each animal comprise number, size, and foci of necrosis and calcification.

curve of *M. tuberculosis*, [39] [40] [41]. We also observed a similar appearance of an extracellular material which accumulated over the duration of the experiment (Figure 2). A gradual accumulation of free mycolates in the cell wall lipids was clearly observed, which was accompanied by a concomitant reduction in the levels of TDM and TMM (Figure 4). The induction of a number of mycolate biosynthetic and processing genes in early stationary phase supports the finding that free mycolates gradually accumulated in culture, particularly Rv2006 (*otsB1*), which is involved in the final steps of mycolic acid biosynthesis and serves to activate meromycolic acid into meromycolyl-AMP thereby transferring the meromycolic acyl chains onto *pks13* [42]. Ojha *et al.* [43] previously showed that free mycolates played a key role in the formation of mycobacterial pellicle biofilms and these authors also demonstrated that TDM is directly cleaved, thus liberating free mycolate [44]. However, until now, the environmental factors that stimulate *M. tuberculosis* to remodel its cell wall and induce production of free mycolate have not been investigated fully. Conventionally, biofilm formation starts with microbial attachment to a surface [45]. However, freely dispersed aggregates or "flocs" of bacteria have been described for other bacterial pathogens, similar to the aggregates of bacilli observed in our cultures [46]. *M. tuberculosis* biofilms, *in vivo*, have been described by Canetti who observed dense sheets of bacilli, that were not adhered to a surface, within the caseum of the granuloma [47].

We observed an increase over time not only in the quantity of LM and LAM, but also in the size and arabinose/mannose ratio of the LAM (Figure 6). It appeared that *M. tuberculosis* was altering the structure of LAM by increasing the level of arabinosylation of the mannan domain in response to nutrient-depletion. This is in contrast to the findings for *Mycobacterium smegmatis*, in which Dhiman *et al.* [48] showed that the Ara to Man ratio decreased in the cell biomass over time. This is the first time that LAM has been shown to be associated with adaptation of *M. tuberculosis* to nutrient-depletion *in vitro*. This finding is also supported by the transcriptomic analyses showing that several genes involved in the key stages of LAM biosynthesis such as Rv3257c (*pmmA*), Rv3806c (*ubiA*), and Rv3793 (*embC*) are up-regulated in early stationary phase and through late stationary phase (Figure 8). Gene *ubiA* encodes for decaprenyl-phosphate 5 phospho-ribosyltransferase, which is required to produce a key intermediate leading to the biosynthesis of decaprenylmonophosphoarabinose (DPA) [49], which is the sole substrate utilised by the membrane bound

arabinosyltransferases (AraTs) in the formation of D-arabinan in mycobacteria [50]. In this regard, EmbC is an α(1→5)-arabinofuranosyltransferase which serves to elongate the 5-Ara*f* linkages of the arabinan domain of LAM [15]. The function of EmbC is specific to LAM biosynthesis and is regulated by the action of PknH [51]. Our microarray data indicate that *pknH* is up-regulated in stationary phase (Cluster 9, Figure 7) and is the cognate kinase that phosphorylates EmbR, which in turn regulates *embC* expression [52]. It is therefore plausible to speculate that PknH senses nutrient-depletion as a stimulus by an as yet unknown mechanism. The resulting signal is then transduced via the Ser/Thr kinase response network into a response which increases the expression of *embC* thereby increasing the arabinose content of LAM. Apart from the increase in free mycolates and LAM/LM ratio observed in culture, no other changes in any of the other lipids (including triacylglycerols (TAG)) or carbohydrates could be observed. Previously reported dormancy models have shown an apparent increase in TAG, but these have been under alternative growth conditions, which serve to highlight the contributions made by different microenvironments with respect to non-replicating persistence [53] [54] [55] [56].

We have previously shown that a deletion in *aftC* (an α(1→3) arabinofuranosyltransferase) resulted in the truncation of the arabinan domain of LAM, which drastically altered the immunological properties of this truncated molecule (termed AftC-LAM) making it more pro-inflammatory in comparison to "wild type" LAM [15]. This key difference has been attributed to the way in which the arabinan domains of AftC-LAM have a reduced effect of "masking" the pro-inflammatory nature of the mannose core of the LAM molecule, thereby altering its immunogenic properties. More recent evidence showed that AftE is involved in the biosynthesis of single arabinans of LAM. Deletion of the *aftE* gene resulted in hyper-mannosylated LAM, which is a stronger inducer of cytokine production *in vitro* than LAM [57]. We observed a clear increase in the expression of *embC*, which is likely to be responsible for the observed increased arabinosylation of LAM over the time-course (Figure 8). Therefore, it is plausible that the LAM molecules matured towards the end of culture period, particularly during NRP and have increased the "masking" of the mannan domain, which in turn is likely to exert a transient effect on the pathogenicity of *M. tuberculosis in vivo*, due to the altered immunogenic properties of the lipoglycans being produced. The mycolic acid layer provides a hydrophobic mesh-like structure for the intercalation of additional complex lipids and lipoglycans; therefore, we hypothesise that stationary phase and NRP phase bacilli are accumulating additional quantities of free mycolate through increased cleavage of TDM to serve as a scaffold for the production and export of LM and LAM into the extracellular matrix. TDM is required for a pro-inflammatory response and the formation of granulomas via the mincle pathway [58] [4]; the proportions of cell-associated TDM/TMM/free mycolate, combined with the altered inflammatory properties of hyper-arabinosylated LAM, could be having an important impact on the outcome of the early stages of infection. These findings might also have wider reaching implications as EmbC is the target of the front-line drug ethambutol [59].

Effect of nutrient depletion on pathogenicity

NRP bacilli were able to successfully infect guinea pigs. We compared the pathogenicity of bacilli from exponential and NRP phases in guinea pigs when delivered via the aerosol route, using bacterial replication and organ pathology as measures of infectivity. Bacilli from the NRP phase were different in each of these parameters compared to the exponential phase bacilli. In the early stages of infection, NRP infected animals showed a significantly lower bacterial load than the exponential phase infected animals and minimal pathological changes were observed. However, by day 42 post-challenge, the bacterial load in the NRP group was similar to the animals infected with exponential phase bacilli. In contrast, the pathology observed in the NRP infected animals at the later time point, although increased compared to day 16, did not reach equivalent levels to that caused by the exponential phase bacteria. Indeed, the extent and characteristics of the pathology in the NRP-infected animals was very similar to that observed in the exponential phase group but at the earlier time-point. The usual course of infection following low-dose aerosol challenge of guinea pigs is an initial replication in the lungs with a peak in the bacterial load at around 3–4 weeks, followed by a slight decrease to a level which is then sustained for a prolonged period of 15–20 weeks [60]. This control of bacterial replication is mediated by an on-going immune response which results in a steady increase in pathological features such as cellular infiltration and granuloma formation. The animals infected with exponential phase bacteria demonstrated this pattern whereby the bacterial load had stabilised by day 42 but the pathology continued to increase. In contrast, the observations in the animals infected with the NRP bacilli were consistent with a delay in the development of disease; at day 16 post-infection the bacteria were in the early stages of replication and there was little or no immunopathology observed. By day 42, the bacterial load had reached a plateau that was at a similar level to those animals that were infected with exponential phase organisms. However, at this time-point the pathological features were consistent with a less advanced stage of immune interaction. We propose that the NRP cells remained in a non-replicating state for a period of possibly up to 2 weeks after aerosol challenge. Adaptation to the *in vivo* environment allowed reversion to a phenotype which resulted in progressive infection and disease. The host and bacterial changes which trigger this replication may be similar to those which occur during reactivation of latent infection in humans. The only other published study reporting on the virulence properties of NRP cells utilised a 'Wayne-type' hypoxia to generate NRP bacilli which were used to infect mice via the intranasal route [61]. Despite many differences between this study and ours, there was a similarity in the finding that the bacterial load was lower in the early stages post-inoculation when compared to "regular" cultures. Therefore, whether induced by hypoxia or nutrient starvation, it appears that NRP bacteria retain their capacity to establish an infection in the susceptible host but with a reduced infectivity.

Concluding remarks

In comparison to exponentially growing cells, NRP bacteria exhibit reduced infectivity for guinea pigs, which coincides with significant alterations to the cell wall components known to be associated with host-pathogen interactions. Gene expression analysis of the biochemical pathways leading to the assembly of these important molecules lends additional support to our hypothesis that, upon prolonged nutrient limitation, NRP bacilli exhibit a drastically altered cell wall phenotype which corresponds with a reduced infectivity profile. It could be that NRP phase organisms were unable to initiate replication and interaction with the immune system due in part to alterations in the composition of free mycolate and LAM. This now warrants further investigation as physiological changes such as these could have a key role in the establishment of latent disease. Similarly, the changes which occur in both the bacteria and the host as replication re-establishes will provide insights to the mechanisms associated with reactivation of tuberculosis disease.

Materials and Methods

Ethics Statement

The studies were conducted according to UK Home Office Legislation for animal experimentation and were approved by a local ethical committee at the Health Protection Agency, Porton Down, UK. The project licence number under which the work was completed was PPL 30/2704.

Strains and medium

Studies were performed with *M. tuberculosis* strain H37Rv (NCTC cat. no. 7416). Stock cultures were grown on Middlebrook 7H10+ OADC for 3 weeks at $37 \pm 2°C$.

In vitro model of mycobacterial non-replicating persistence under nutrient-starved conditions

Cultures were established in CAMR Mycobacterium Medium Mod 6 (CMM Mod 6) [12]. The first two components of the medium were added to the first volume of water. The remaining components were added in the order listed (Table S1, supporting information). The pH was adjusted to 6.5 using 20% potassium hydroxide solution (w/v in distilled water). The medium was filter sterilised by passage through a 0.1 μm pore size cellulose acetate membrane filter capsule (Sartorius Ltd). Middlebrook 7H10+ OADC agar was used to prepare colonies for inoculation of the cultures and for enumeration of viable bacteria in the cultures.

Inoculation and culture of M. tuberculosis

The cultures were established following a modification of the method described previously [10]. Culture experiments were performed in a two litre glass vessel operated at a working volume of 1800 mL. The culture was agitated by a magnetic bar placed in the culture vessel coupled to a magnetic stirrer positioned beneath the vessel. Culture conditions were continuously monitored by an Anglicon Microlab Fermentation System (Brighton Systems, Newhaven), linked to sensor probes inserted into the culture through sealed ports in the top plate. The vessel was filled with 1800 ml of sterile culture medium (CMM Mod6) and parameters were allowed to stabilise at $37°C \pm 2°C$, pH 6.9 ± 0.3 and a dissolved oxygen tension of approximately 50% air saturation (10% DOT). A dense inoculum was prepared by re-suspending colonies from 5 Middlebrook agar cultures (grown at $37^{\circ}C \pm 2^{\circ}C$ for 3 weeks) in sterile deionised water. The inoculum was aseptically transferred to the culture vessel, to provide an initial culture turbidity of approximately 0.25 at 540_{nm}. The culture temperature was monitored by an Anglicon temperature probe, and maintained at $37^{\circ}C$ by a heating pad positioned beneath the culture vessel. The culture was stirred at an agitation rate of 500 to 750 rpm. The oxygen concentration was monitored with a galvanic oxygen electrode (Uniprobe, Cardiff) and the air saturation was maintained at 50% (10% dissolved oxygen tension). The initial culture pH was set at 6.7 and was monitored through-out the experiment using an Ingold pH electrode (Mettler-Toledo, Leicester). Each culture was maintained for at least 200 days and samples were removed regularly to monitor growth and survival and for lipid/carbohydrate analysis, infectivity studies and gene expression.

Growth and survival

Bacterial growth and survival was assessed by determining the number of viable cells in the culture system at specific time-points selected in each phase of growth (exponential, stationary phase, late stationary phase, and NRP phase; Figure 1). This was achieved by preparing a decimal dilution series of the sample in sterile water and plating 100 μL aliquots onto Middlebrook 7H10+ OADC plates in triplicate. The plates were incubated at $37^{\circ}C$ for up to 4 weeks before enumerating the number of colonies formed.

Analysis of Tween 80 levels in spent culture supernatant

Samples of spent supernatant and a sample of the starting medium were hydrolysed to free fatty acids by heating 0.5 mL of sample with 0.5 mL of methanol and 0.2 mL of 25% potassium hydroxide (w/v in water), in an eppendorf tube, to $100^{\circ}C$ for an hour. The pH of each sample was adjusted to pH 7.0 (using concentrated hydrochloric acid and 25% potassium hydroxide) prior to fatty acid determination, which was then performed using the "Free fatty acids, Half-micro test" by following the manufacturer's instructions (supplied by Roche, Welwyn Garden City, UK) [62].

EM analysis

Culture samples (5 mL volumes) were fixed in 4% formaldehyde for at least 16 hours. Fixed cells were immobilised by allowing them to settle onto a poly-l-lysine coated 10 mm glass cover slip overnight in a moist chamber. Immobilised cells were further fixed in 1% v/v osmium tetroxide for 1–2 hours at room temperature. Samples were dehydrated at room temperature through a graded ethanol series from 25% ethanol (v/v) to 100% ethanol (v/v) in steps of 25%. Each dehydration step was performed for 15 minutes. Cover slips were then washed twice in hexamethyldilsilazane (HMDS) for 15 minutes and air dried. The cover slip was mounted onto an SEM stub and a conductive gold coating was applied using an Atom Tech ion beam Z705 ultra fine grain coating unit (approx 10 nm thickness). Specimens were examined using a FEI XL30FEG scanning electron microscope at an accelerating voltage of 4 kV and a working distance of 10 mm.

Staining for exopolymer production

A few drops of the culture sample were added to silanised slides, spread over the middle of the slide and left to dry overnight in a damp chamber. The slides were pre-treated with 100% (v/v) industrial methylated spirits (IMS) by rinsing the slide for a few minutes in the alcohol and then rinsing with distilled water before the staining method was undertaken. The slides were stained with carbol fuchsin for 10 minutes, rinsed in tap water and de-colorised with 1% hydrochloric acid in 70% IMS (v/v). The slides were then rinsed in tap water and then rinsed in distilled water. The alcian blue solution was applied for 5 minutes at a pH of 1.0, to stain for polysaccharides. Slides were rinsed with distilled water followed by a rinse in 100% IMS (v/v), cleared in xylene, and mounted for light microscopy. Slides were visualised and photographed at a magnification of $\times 100$.

Extraction of polar and apolar lipids

Polar and apolar lipids were extracted from cell pastes (adjusted to approximately 2.5×10^{7} cells, in order to normalise our analysis) from each culture time-point as described previously [14]. Apolar lipids were re-suspendend in 50 μL of 2:1 $CHCl_3/MeOH$ and equivalent volumes were subjected to 1D-TLC with silica gel plates (5734 silica gel $60F_{254}$; Merck, Darmstadt, Germany), developed in $CHCl_3/MeOH/H_2O$ (60:16:2, v/v/v) [14]. Equivalent volumes of apolar lipids were also subject to 2D-TLC analysis developed in $CHCl_3/MeOH$ (96:4, v/v) for direction 1 and toluene/acetone (82:2, v/v) for direction 2 [14]. TLC plates were sprayed with 5% ethanolic molybdophoshoric acid (MPA) and charred to visualise the lipids.

Extraction of lipoglycans from culture supernatant

In order to extract polysaccharide material from a normalised culture supernatant, 50 mL of supernatant fractions collected at each time-point from Culture 2 (d0, d4, d29, d72, and d277) and Culture 3 (d0, d8, d44, d107, d292) were spun down by centrifugation at $3000 \times g$ to remove the cells. This step was repeated twice. The supernatants were then stored at $-80^{\circ}C$ until the end of the culture time-course. The frozen supernatant samples were irradiated for two hours, in a Gammacell 220 instrument, using gamma rays at an energy level of 1.33 MeV. The samples remained frozen throughout the procedure. The supernatants were subsequently defrosted and added to 200 mL of ice cold acetone and stored overnight at $-80^{\circ}C$. The resulting precipitate was collected by centrifugation at $5,000 \times g$ for 1 hr at $4^{\circ}C$. The pelleted material was subjected to treatment with 6 mL of 90% phenol heated to $80^{\circ}C$ for 2 hr. After cooling, each sample was centrifuged at $3,500 \times g$ for 30 min to induce a bi-phase. The upper layer (~3 mL) was recovered and a further 3 mL of PBS was added to the lower phenol layer, mixed and recovered by centrifugation at $3,500 \times g$ for 30 min. The upper layer (~3 mL) was pooled with the previously collected upper layer giving a total sample of 6 mL. After extensive dialysis against water using a MWCO 3,500 dialysis membrane (Spectrum Laboratories) each of the samples was dried and re-suspended in 400 μL of water and treated at $37^{\circ}C$ for 5 hours with proteinase K (Sigma). Following this, samples were re-extracted with phenol and dialysed overnight against water as described above. Samples were analysed by 15% SDS-PAGE and stained using Pro-Q emerald glycoprotein stain (Invitrogen).

Cell walls extractions

Cell walls were extracted from cell pastes (adjusted to approximately 2.5×10^{7} cells in order to normalise our analysis) from each culture time-point. Bacterial cell pastes were re-suspended in 0.3% NaCl and re-fluxed in 50% ethanol at $80^{\circ}C$ overnight and spun down at 4000 rpm for 20 minutes. The supernatant was transferred to a glass tube, dialysed overnight, and dried down. To extract the cell walls, the pellets were re-suspended in 5 mL of PBS and sonicated. 5 mL of 4% SDS was then added and this was heated at 85°C for 2 hours. The pellet was washed 3 times with water followed by two washes with 80% acetone and two washes with 100% acetone. This was left to air dry.

Carbohydrate analysis

Lipoglycans and cell wall material were chemically modified to alditol acetate derivatives as described previously [16] [15]. Gas chromatography (GC) analysis was performed using a Thermo-quest Trace GC 2000. Samples were injected in the splitless mode. The column used was a DB225 (Supelco). The oven was programmed to hold at an isothermal temperature of 275°C for a run time of 15 min. All the data were collected and analysed using Xcaliber (v.1.2) software.

RNA extraction and amplification

Bacterial cell samples were collected from Culture 1 and Culture 2 for RNA extraction throughout the time-course. The sample size depended on the growth phase and ranged between 40–500 mL; larger samples were required later in the culture because of the reduced cell density of the culture (details of samples can be found at http://bugs.sgul.ac.uk/bugsbase, experiment accession no. E-BUGS-142). Each sample was collected directly into 4 volumes of lysis solution (5 M guanidium thiocyanate, 0.5% (w/v) sodium N-lauroyl sarcosine, 25 mM tri-sodium citrate, 0.1 M 2-mercaptoethanol and 0.5% (w/v) Tween 80 in RNAse-free water (Sigma-Aldrich, Poole, UK)) and incubated at room temperature for 1 hour. After centrifugation, the cells were re-suspended in Trizol (Life Technologies Inc., Paisley, UK), and were disrupted for 45 seconds (setting of 6.5) using a ribolyser (Hybaid). RNA was extracted with chloroform and precipitated with isopropanol and 0.3M sodium acetate. The RNA was treated with deoxyribonuclease 1 (DNase I amplification grade, Life Technologies Inc., Paisley, UK) and purified using a Qiagen RNeasy clean up kit (Qiagen, Crawley, UK) following the manufacturer's guidelines. Amplification was performed on 100 ng of total RNA from each time-point, following the method described by the MessageAmp™ II-Bacteria kit (Life Technologies Inc., Paisley, UK) to achieve aRNA concentrations that were approximately 200 fold higher than the starting material.

Microarray procedures

Three separate labelling reactions were performed on each aRNA sample. Each aliquot of Cy5-labelled cDNA generated from aRNA (test sample) was co-hybridised with Cy3-labelled DNA generated from genomic DNA (control sample). Total RNA (8 μg) was used as a template for reverse transcriptase (Superscript II RNAse H, 200 U μL^{-1}; Life Technologies Inc., Paisley, UK) in the presence of random primers and cyanine 5 (Cy5)-labelled dCTP. Genomic DNA was extracted from a cell pellet of *M. tuberculosis* H37Rv harvested from a steady-state culture using the procedure described previously [30]. DNA (1 μg) was used as a template for DNA polymerase (Klenow, 5 U μL^{-1}; Life Technologies Inc., Paisley, UK) in the presence of random primers and Cy3-labelled dCTP.

Array hybridisation

Arrays were performed on 8 time-points for Culture 1 (days 5, 21, 41, 65, 91, 119, 146, and 208) and 6 time-points for Culture 2

(days 2, 3, 16, 23, 85, 222) using whole genome arrays for *M. tuberculosis*; TBv1.1.0 arrays (A-BUGS-1) for Culture 1 and TBv2.1.1 arrays (A-BUGS-23) for Culture 2. The array designs are available in BμG@Sbase (Accession No. A-BUGS-1 http://bugs.sgul.ac.uk/A-BUGS-1 and A-BUGS-23; http://bugs.sgul.ac.uk/A-BUGS-23).

The Cy3 and Cy5 labelled products for each array were combined and purified using a MinElute PCR purification kit (Qiagen). The microarray slides were incubated in pre-hybridisation solution (3.5× SSC, 0.1% (w/v) SDS, bovine serum albumin (BSA 10 mg ml^{-1} Fraction V 96–99%, Sigma-Aldrich) at 65°C for 30 minutes. The slides were rinsed thoroughly in distilled water followed by isopropanol and dried by centrifugation at 1,500 rpm for 5 minutes. The purified Cy3/Cy5 labelled DNA was mixed with hybridisation solution (10.5 μL Cy3/Cy5 labelled DNA, 3.2 μL filtered 20× SSC, and 2.3 μl filtered 2% (v/v) SDS) and heated at 95°C for 2 minutes. The reaction was cooled slightly and centrifuged before being added to the slide and covered with a cover slip. The hybridisation cassette (Telechem International, Sunnyvale, USA) was sealed and submerged in a water-bath at 65°C in the dark for 16–20 hr. After hybridisation, the slides were washed gently in wash solution (1× SSC with 0.05% (w/v) SDS). The slides were rinsed in 0.06× SSC in distilled water and were dried by centrifugation at 1,500 rpm for 5 minutes. Scanning was performed using a dual-laser scanner (Affymetrix 428, MWG-Biotech) at a level (or gain) just below saturation of the most intensely fluorescent spots on each array. The images were quantified using Bluefuse software (https://www.msi.umn.edu/sw/bluefuse).

Statistical analyses

Transcriptomic analyses were performed on Culture 1 and Culture 2 across the two time-courses. Fully annotated microarray data have been deposited in BμG@Sbase (accession number E-BUGS-142; http://bugs.sgul.ac.uk/E-BUGS-142) and also ArrayExpress (accession number E-BUGS-142). A probabilistic model based on Gaussian process regression and Bayesian model selection was used to analyse and identify consistent genes from which expression profile observations could be merged. The growth curves of Cultures 1 and 2 showed the same dynamics in their growth curves, as determined by total viable count throughout culture (Figure 1). To allow direct comparison of gene expression measurements taken at different time points in each culture, the growth curves were synchronised by applying a linear transformation on the time-scale of one of the cultures. The same transformation was applied to transcriptomic time-points to take the observations onto the same time-scale. The linear transformation was computed by minimising the sum of squared errors between the growth curves of Cultures 1 and 2. The time-point of each RNA sample for Culture 2 was linearly transformed as $\hat{t} = at + b$ where t represents the time-point when the culture was sampled and $a = 1.299$, and $b = 6.152$.

The observed gene expressions were log-transformed and normalised to zero mean. All replicate measurements were used in the Bayesian model as the probabilistic method takes into account the uncertainty and noise in data, which would be otherwise be lost by averaging. Probabilistic models for gene expression of each gene were created using Gaussian process model [17], which computes a nonlinear regression approximation of the data. A Bayesian model selection procedure was applied to each gene to decide whether the expression profiles for a gene were similar enough in both cultures to be merged into a single time series. Clustering was performed using Bayesian Hierarchical clustering. For technical details see Methods S1 in the supporting information.

The clustering analysis was performed on genes that were identified as being consistent between Cultures 1 and 2. Expression profiles of genes that showed the same expression pattern profile in both cultures were merged together. Replicate expression values were summarised for each time point using a fitted Gaussian process model. Time series clustering of the whole genome data was performed using Bayesian clustering of time series using Gaussian processes with basis function representations, implemented in package SplineCluster [18]. The algorithm automatically determines the optimal number of clusters by maximising likelihoods of different cluster divisions.

Enrichment analysis for clusters was performed using Fisher's exact test for 2×2 contingency matrices as implemented in the R statistical software [63]. Significance of enrichment for an over or under representation of a cluster with respect to a functional class was measured by p-values. A cutoff value of $p = 0.01$ was used with the exception of the lipid analysis (GO:0071767 and GO:9999999) for which the cut off was set to 0.05. The p-values were not corrected for multiple testing; therefore the enrichment analysis discussed in this work is taken as a heuristic indication of the relevance of certain biological processes. Enrichment was assessed with respect to two different functional annotations. A classification for genes of *M. tuberculosis* has been obtained from the Sanger Institute ftp://ftp.sanger.ac.uk/pub/pathogens/Mycobacterium/tuberculosis/functional_classes/ while Gene Ontology http://www.geneontology.org/GO.downloads.ontology.shtml annotations for *M. tuberculosis* were obtained from the MTB GOA project http://www.ark.in-berlin.de/Site/MTB-GOA.html. The Sanger functional annotation provided three levels of increasing detail, through from Level 1 to Level 3. Similar level numbers were also assigned to the Gene Ontology terms.

Guinea pig aerosol infection

Animals were infected with a low aerosol dose of *M. tuberculosis* H37Rv using a fully contained Henderson apparatus as previously described [64]. Fine particle aerosols of *M. tuberculosis* H37Rv, with a mean diameter of 2 μm (diameter range, 0.5–7 μm) [65], were generated using a Collison nebuliser and delivered directly to the animal snout. The aerosol was generated from a suspension of cells that had been diluted in spent culture medium at each time-point and adjusted to approximately 1×10^5 cfu mL^{-1} in order to obtain an estimated retained, inhaled dose of approximately 10 cfu/lung. The Henderson apparatus allows controlled delivery of aerosols to the animals and the reproducibility of the system and relationship between inhaled cfu and the concentration of organisms in the nebuliser has been described previously [28] [29]. The studies were conducted according to UK Home Office Legislation for animal experimentation and were approved by a local ethical committee at the Health Protection Agency, Porton Down, UK. The project licence number under which the work was completed was PPL 30/2704.

Bacteriology and histopathology of infected organs

At 16 days and 42 days post-challenge, guinea pigs were killed humanely by intraperitoneal injection of pentabarbitone (Euthatal). Tissues were removed aseptically *post mortem* for bacteriological (cfu counts) and histopathological examination.

Tissues for bacterial counts in organs were homogenised in 10 ml (lungs) or 5 mL (spleens) of sterile distilled water using a rotating blade macerator system (Ystral, UK). Viable counts were performed on the macerate by preparing serial dilutions in sterile water and 100 μl aliquots were plated onto Middlebrook 7H11+ OADC (Oleate, Albumin, Dextrose & Catalase) agar (BioMerieux, UK). Plates were incubated at 37°C for 3 weeks before counting the number of *M. tuberculosis* colonies (cfu).

For histopathological examination, samples of individual lung lobes and spleen were collected, fixed in 10% (v/v) Neutral Buffered Formalin and processed to paraffin wax. Sections cut at 4 μm were stained with haematoxylin and eosin. The nature and severity of the microscopic lesions was evaluated subjectively and scored by a pathologist; evaluations were blinded. Lung lobes were assigned a score as follows: no abnormality = 0; very small, very few lesions, <10% consolidation = 1; few or small lesions, 10–20% consolidation = 2; medium sized lesions, 20–33% consolidation = 3; moderately sized lesions, 33–50% consolidation = 4; large lesions, moderately extensive pneumonia, 50–80% consolidation = 5; extensive pneumonia >80% consolidation = 6. A mean consolidation score per lobe was calculated for each group. The number of foci of necrosis/caseation and the number of calcified lesions was recorded and a mean score per lobe was calculated for each group. For the spleen, the number of lesions, lesion size and foci of necrosis and calcification were recorded subjectively. For lesion number, >10 = 1, 11–30 = 2, >30 = 4. For lesion size, small lesions = 1, medium lesions = 2 and large lesions = 3. For necrotic and calcified lesions, <5 = 1, 6–10 = 2, >10 = 3.

Supporting Information

Figure S1   Depletion of Tween 80 measured in samples of spent supernatant taken from Culture 2 and Culture 3 throughout the time-course.

Table S1   Recipe for CMM Mod6 medium.

Table S2   Lipomannan and Lipoarabinomannan biosynthetic genes.

Table S3   *M. tuberculosis* genes induced under different *in vivo* and *in vitro* conditions.

Method S1   Probabilistic analysis applied to identify consistent genes between the two cultures.

Acknowledgments

We thank Mark Broughton at PHE Colindale for irradiation of the culture supernatants for carbohydrate analyses and the Biological Investigations Group at PHE, Porton Down, for support on the infectivity studies. Thanks to the Bugs Microarray Users Group at St Georges, University of London for technical help with the microarray and to Susan Gray at PHE, Porton Down, for her contribution to the cell staining.

Author Contributions

Conceived and designed the experiments: JB LJA JA SOC GSB AW PDM. Performed the experiments: JB LJA JAA RW KAH SOC AM ER HT GP GH. Analyzed the data: JB LJA JAA EG RW KAH SOC REJ AM ER HT GP GH GSB LW AW PDM. Contributed reagents/materials/analysis tools: LJA EG GSB LW. Wrote the paper: JB LJA JAA EG RW KAH SOC REJ AM ER HT GP GH GSB LW AW PDM.

References

1. Russell DG, VanderVen BC, Lee W, Abramovitch RB, Kim M, et al. (2010) Mycobacterium tuberculosis wears what it eats. Cell Host and Microbe 8: 68–76.

2. Ulrichs T, Kaufmann SHE (2006) New insights into the function of granulomas in human tuberculosis. J Pathol 208: 261–269.

3. Lenaerts AJ, Hoff D, Aly S, Ehlers S, Andries K, et al. (2007) Location of persisting mycobacteria in a guinea pig model of tuberculosis revealed by r207910. Antimicrob Agents Chemother 51: 3338–3345.

4. Russell DG (2011) Mycobacterium tuberculosis and the intimate discourse of a chronic infection. Immunol Rev 240: 252–268.

5. Wayne LG, Lin KY (1982) Glyoxylate metabolism and adaptation of Mycobacterium tuberculosis to survival under anaerobic conditions. Infect Immun 37: 1042–1049.

6. Wayne LG, Hayes LG (1996) An in vitro model for sequential study of shiftdown of Mycobacterium tuberculosis through two stages of non-replicating persistence. Infect Immun 64: 2062–2069.

7. Parish T, Smith DA, Kendall S, Casali N, Bancroft GJ, et al. (2003) Deletion of two-component regulatory systems increases the virulence of Mycobacterium tuberculosis. Infect Immun 71: 1134–1140.

8. Converse PJ, Karakousis PC, Klinkenberg LG, Kesavan AK, Ly LH, et al. (2009) Role of the dosR-dosS two-component regulatory system in Mycobacterium tuberculosis virulence in three animal models. Infect Immun 77: 1230–1237.

9. Majumdar S De, Vashist A, Dhingra S, Gupta R, Singh A, et al. (2012) Appropriate DevR (DosR)-mediated signaling determines transcriptional response, hypoxic viability and virulence of Mycobacterium tuberculosis. PloS One 7: e35847.

10. Hampshire T, Soneji S, Bacon J, James BW, Hinds J, et al. (2004) Stationary phase gene expression of Mycobacterium tuberculosis following a progressive nutrient depletion: a model for persistent organisms? Tuberculosis (Edinb) 84: 228–238.

11. Golby P, Hatch KA, Bacon J, Cooney R, Riley P, et al. (2007) Comparative transcriptomics reveals key gene expression differences between the human and bovine pathogens of the Mycobacterium tuberculosis complex. Microbiology 153: 3323–3336.

12. Bacon J, Hatch KA (2009) Continuous culture of mycobacteria. Methods Mol Biol (Clifton, NJ) 465: 153–171.

13. Karlyshev AV, Wren BW (2001) Detection and initial characterization of novel capsular polysaccharide among diverse Campylobacter jejuni strains using alcian blue dye. J Clin Microbiol 39: 279–284.

14. Dobson G, Minnikin DE, Minnikin SM, Partlett JH, Goodfellow M, et al. (1985) Systematic Analysis of Complex Mycobacterial Lipids. London: Academic Press.

15. Birch HL, Alderwick LJ, Appelmelk BJ, Maaskant J, Bhatt A, et al. (2010) A truncated lipoglycan from mycobacteria with altered immunological properties. Proc Natl Acad Sci USA 107: 2634–2639.

16. Alderwick LJ, Radmacher E, Seidel M, Gande R, Hitchen PG, et al. (2005) Deletion of Cg-emb in corynebacterineae leads to a novel truncated cell wall arabinogalactan, whereas inactivation of Cg-ubiA results in an arabinan-deficient mutant with a cell wall galactan core. J Biol Chem 280: 32362–32371.

17. Carl Edward Rasmussen, Christopher KI Williams (2006) Guassian Processes for Machine Learning. Cambridge MA: MIT Press.

18. Heard NA, Holmes CC, Stephens DA (2006) A quantitative study of gene regulation involved in the immune response of Anopheline Mosquitoes: An application of Bayesian Hierarchical Clustering Curves. J Am Stat Assoc 101: 18–29.

19. Jankute M, Grover S, Rana AK, Besra GS (2012) Arabinogalactan and lipoarabinomannan biosynthesis: structure, biogenesis and their potential as drug targets. Future Microbiol 7: 129–147.

20. Mishra AK, Driessen NN, Appelmelk BJ, Besra GS (2011) Lipoarabinomannan and related glycoconjugates: structure, biogenesis and role in Mycobacterium tuberculosis physiology and host-pathogen interaction. FEMS Microbiol Rev 35: 1126–1157.

21. Papavinasasundaram KG, Chan B, Chung JH, Colston MJ, Davis EO, et al. (2005) Deletion of the Mycobacterium tuberculosis pknH gene confers a higher bacillary load during the chronic phase of infection in BALB/c mice. J Bacteriol 187: 5751–5760.

22. Talaat AM, Ward SK, Wu C-W, Rondon E, Tavano C, et al. (2007) Mycobacterial bacilli are metabolically active during chronic tuberculosis in murine lungs: insights from genome-wide transcriptional profiling. J Bacteriol 189: 4265–4274.

23. Betts JC, Lukey PT, Robb LC, McAdam RA, Duncan K (2002) Evaluation of a nutrient starvation model of Mycobacterium tuberculosis persistence by gene and protein expression profiling. Mol Microbiol 43: 717–731.

24. Gengenbacher M, Rao SPS, Pethe K, Dick T (2010) Nutrient-starved, non-replicating Mycobacterium tuberculosis requires respiration, ATP synthase and isocitrate lyase for maintenance of ATP homeostasis and viability. Microbiol 156: 81–87.

25. Kendall SL, Withers M, Soffair CN, Moreland NJ, Gurcha S, et al. (2007) A highly conserved transcriptional repressor controls a large regulon involved in lipid degradation in Mycobacterium smegmatis and Mycobacterium tuberculosis. Mol Microbiol 65: 684–699.

26. Schnappinger D, Ehrt S, Voskuil MI, Liu Y, Mangan JA, et al. (2003) Transcriptional Adaptation of Mycobacterium tuberculosis within Macrophages: Insights into the Phagosomal Environment. J Exp Med 198: 693–704.

27. Rachman H, Strong M, Ulrichs T, Grode L, Schuchhardt J, et al. (2006) Unique transcriptome signature of Mycobacterium tuberculosis in pulmonary tuberculosis. Infect Immun 74: 1233–1242.

28. Clark SO, Hall Y, Kelly DLF, Hatch GJ, Williams A (2011) Survival of Mycobacterium tuberculosis during experimental aerosolization and implications for aerosol challenge models. J Appl Microbiol 111: 350–359.

29. Chambers MA, Williams A, Gavier-Widén D, Whelan A, Hall G, et al. (2000) Identification of a Mycobacterium bovis BCG auxotrophic mutant that protects guinea pigs against M. bovis and hematogenous spread of Mycobacterium tuberculosis without sensitization to tuberculin. Infect Immun 68: 7094–7099.

30. Bacon J, James BW, Wernisch L, Williams A, Morley KA, et al. (2004) The influence of reduced oxygen availability on pathogenicity and gene expression in Mycobacterium tuberculosis. Tuberculosis (Edinb) 84: 205–217.

31. Gengenbacher M, Kaufmann SHE (2012) Mycobacterium tuberculosis: success through dormancy. FEMS Microbiol Rev 36: 514–532.

32. Miller RM, Tomaras AP, Barker AP, Voelker DR, Chan ED, et al. (2008) Pseudomonas aeruginosa twitching motility-mediated chemotaxis towards phospholipids and fatty acids: specificity and metabolic requirements. J Bacteriol 190: 4038–4049.

33. Park H-D, Guinn KM, Harrell MI, Liao R, Voskuil MI, et al. (2003) Rv3133c/dosR is a transcription factor that mediates the hypoxic response of Mycobacterium tuberculosis. Mol Microbiol 48: 833–843.

34. Loebel RO, Shorr E, Richardson HB (1933) The Influence of Adverse Conditions upon the Respiratory Metabolism and Growth of Human Tubercle Bacilli. J Bacteriol 26: 167–200.

35. McKinney JD, Höner zu Bentrup K, Muñoz-Elías EJ, Miczak a, Chen B, et al. (2000) Persistence of Mycobacterium tuberculosis in macrophages and mice requires the glyoxylate shunt enzyme isocitrate lyase. Nature 406: 735–738.

36. Savvi S, Warner DF, Kana BD, McKinney JD, Mizrahi V, et al. (2008) Functional characterization of a vitamin B12-dependent methylmalonyl pathway in Mycobacterium tuberculosis: implications for propionate metabolism during growth on fatty acids. J Bacteriol 190: 3886–3895.

37. Irby RB, Adair WL (1994) Intermediates in the folic acid biosynthetic pathway are incorporated into molybdopterin the yeast, Pichia canadensis. J Biol Chem 269: 23981–23987.

38. Shi T, Xie J (2011) Molybdenum enzymes and molybdenum cofactor in mycobacteria. J Cell Biochem 112: 2721–2728.

39. Lemassu A, Daffé M (1994) Structural features of the exocellular polysaccharides of Mycobacterium tuberculosis. Biochem J 297 (Pt 2): 351–357.

40. Ortalo-Magné A, Lemassu A, Lanéelle MA, Bardou F, Silve G, et al. (1996) Identification of the surface-exposed lipids on the cell envelopes of Mycobacterium tuberculosis and other mycobacterial species. J Bacteriol 178: 456–461.

41. Ortalo-Magné A, Dupont MA, Lemassu A, Andersen AB, Gounon P, et al. (1995) Molecular composition of the outermost capsular material of the tubercle bacillus. Microbiol 141: 1609–1620.

42. Takayama K, Wang C, Besra GS (2005) Pathway to synthesis and processing of mycolic acids in Mycobacterium tuberculosis. Clin Microbiol Rev 18: 81–101.

43. Ojha AK, Baughn AD, Sambandan D, Hsu T, Trivelli X, et al. (2008) Growth of Mycobacterium tuberculosis biofilms containing free mycolic acids and harbouring drug-tolerant bacteria. Mol Microbiol 69: 164–174.

44. Ojha AK, Trivelli X, Guerardel Y, Kremer L, Hatfull GF (2010) Enzymatic hydrolysis of trehalose dimycolate releases free mycolic acids during mycobacterial growth in biofilms. J Biol Chem 285: 17380–17389.

45. Hall-Stoodley L, Costerton JW, Stoodley P (2004) Bacterial biofilms: from the natural environment to infectious diseases. Nature reviews Microbiology 2: 95–108.

46. Joshua GWP, Guthrie-Irons C, Karlyshev A V, Wren BW (2006) Biofilm formation in Campylobacter jejuni. Microbiol 152: 387–396.

47. Canetti G (1965) Present aspects of bacterial resistance in tuberculosis. Am Rev Resp Dis 92: 687–703.

48. Dhiman RK, Dinadayala P, Ryan GJ, Lenaerts AJ, Schenkel AR, et al. (2011) Lipoarabinomannan localization and abundance during growth of Mycobacterium smegmatis. J Bacteriol 193: 5802–5809.

49. Huang H, Scherman MS, D'Haeze W, Vereecke D, Holsters M, et al. (2005) Identification and active expression of the Mycobacterium tuberculosis gene encoding 5-phospho-{alpha}-d-ribose-1-diphosphate: decaprenyl-phosphate 5-phosphoribosyltransferase, the first enzyme committed to decaprenylphosphoryl-d-arabinose synthesis. J Biol Chem 280: 24539–24543.

50. Alderwick LJ, Lloyd GS, Ghadbane H, May JW, Bhatt A, et al. (2011) The C-terminal domain of the Arabinosyltransferase Mycobacterium tuberculosis EmbC is a lectin-like carbohydrate binding module. PLoS Path 7: e1001299.

51. Sharma K, Gupta M, Pathak M, Gupta N, Koul A, et al. (2006) Transcriptional control of the mycobacterial embCAB operon by PknH through a regulatory protein, EmbR, in vivo. J Bacteriol 188: 2936–2944.

52. Molle V, Kremer L, Girard-Blanc C, Besra GS, Cozzone AJ, et al. (2003) An FHA phosphoprotein recognition domain mediates protein EmbR phosphorylation by PknH, a Ser/Thr protein kinase from Mycobacterium tuberculosis. Biochem 42: 15300–15309.

53. Daniel J, Maamar H, Deb C, Sirakova TD, Kolattukudy PE (2011) Mycobacterium tuberculosis uses host triacylglycerol to accumulate lipid droplets and acquires a dormancy-like phenotype in lipid-loaded macrophages. PLoS path 7: e1002093.

54. Deb C, Lee CM, Dubey VS, Daniel J, Abomoelak B, et al. (2009) A novel in vitro multiple-stress dormancy model for Mycobacterium tuberculosis generates a lipid-loaded, drug-tolerant, dormant pathogen. PloS One 4: e6077.

55. Sirakova TD, Dubey VS, Deb C, Daniel J, Korotkova TA, et al. (2006) Identification of a diacylglycerol acyltransferase gene involved in accumulation of triacylglycerol in Mycobacterium tuberculosis under stress. Microbiol 152: 2717–2725.

56. Bacon J, Dover LG, Hatch KA, Zhang Y, Gomes JM, et al. (2007) Lipid composition and transcriptional response of Mycobacterium tuberculosis grown under iron-limitation in continuous culture: identification of a novel wax ester. Microbiol 153: 1435–44.

57. Mishra AK, Alves JE, Krumbach K, Nigou J, Castro AG, et al. (2012) Differential arabinan capping of lipoarabinomannan modulates innate immune responses and impacts T helper cell differentiation. J Biol Chem 287: 44173–83.

58. Lee W-B, Kang J-S, Yan J-J, Lee MS, Jeon BY, et al. (2012) Neutrophils Promote Mycobacterial Trehalose Dimycolate-Induced Lung Inflammation via the Mincle Pathway. PLoS Path 8: e1002614.

59. Telenti A, Philipp WJ, Sreevatsan S, Bernasconi C, Stockbauer KE, et al. (1997) The emb operon, a gene cluster of Mycobacterium tuberculosis involved in resistance to ethambutol. Nat Med 3: 567–570.

60. Turner OC, Basaraba RJ, Orme IM (2003) Immunopathogenesis of pulmonary granulomas in the guinea pig after infection with Mycobacterium tuberculosis. Infect Immun 71: 864–871.

61. Woolhiser L, Tamayo MH, Wang B, Gruppo V, Belisle JT, et al. (2007) In vivo adaptation of the Wayne model of latent tuberculosis. Infect Immun 75: 2621–2625.

62. Beste DJ V, Bonde B, Hawkins N, Ward JL, Beale MH, et al. (2011) ^{13}C metabolic flux analysis identifies an unusual route for pyruvate dissimilation in mycobacteria which requires isocitrate lyase and carbon dioxide fixation. PLoS Path 7: e1002091.

63. R Core Team (2012). R: A language and environment for statistical computing. R Foundation for Statistical Computing http://www.R-project.org/.

64. Williams A, Davies A, Marsh PD, Chambers MA, Hewinson RG (2000) Comparison of the protective efficacy of bacille calmette-Guérin vaccination against aerosol challenge with Mycobacterium tuberculosis and Mycobacterium bovis. Clin Infect Dis: 30 299–301.

65. Lever MS, Williams A, Bennett AM (2000) Survival of mycobacterial species in aerosols generated from artificial saliva. Lett Appl Microbiol 31: 238–241.

3

Plasma and Serum Lipidomics of Healthy White Adults Shows Characteristic Profiles by Subjects' Gender and Age

Masaki Ishikawa[1], Keiko Maekawa[1]*, Kosuke Saito[1], Yuya Senoo[1], Masayo Urata[1], Mayumi Murayama[1], Yoko Tajima[1], Yuji Kumagai[2], Yoshiro Saito[1]

1 Division of Medicinal Safety Science and Disease Metabolome Project, National Institute of Health Sciences, Setagaya, Tokyo, Japan, 2 Clinical Trial Center, Kitasato University East Hospital, Sagamihara, Kanagawa, Japan

Abstract

Blood is a commonly used biofluid for biomarker discovery. Although blood lipid metabolites are considered to be potential biomarker candidates, their fundamental properties are not well characterized. We aimed to (1) investigate the matrix type (serum vs. plasma) that may be preferable for lipid biomarker exploration, (2) elucidate age- and gender-associated differences in lipid metabolite levels, and (3) examine the stability of lipid metabolites in matrix samples subjected to repeated freeze-thaw cycles. Using liquid chromatography-mass spectrometry, we performed lipidomic analyses for fasting plasma and serum samples for four groups (15 subjects/group) of young and elderly (25–34 and 55–64 years old, respectively) males and females and for an additional aliquot of samples from young males, which were subjected to repeated freeze-thaw cycles. Lysophosphatidylcholine and diacylglycerol levels were higher in serum than in plasma samples, suggesting that the clotting process influences serum lipid metabolite levels. Gender-associated differences highlighted that the levels of many sphingomyelin species were significantly higher in females than in males, irrespective of age and matrix (plasma and serum). Age-associated differences were more prominent in females than in males, and in both matrices, levels of many triacylglycerols were significantly higher in elderly females than in young females. Plasma and serum levels of most lipid metabolites were reduced by freeze-thawing. Our results indicate that plasma is an optimal matrix for exploring lipid biomarkers because it represents the original properties of an individual's blood sample. In addition, the levels of some blood lipid species of healthy adults showed gender- and age-associated differences; thus, this should be considered during biomarker exploration and its application in diagnostics. Our fundamental findings on sample selection and handling procedures for measuring blood lipid metabolites is important for ensuring the quality of biomarkers identified and its qualification process.

Editor: Angelo Scuteri, INRCA, Italy

Funding: This work was supported by the Health Labour Sciences Research Grants (Grant number 028) from the Ministry of Health, Labour and Welfare, and by the Advanced Research for Products Mining Program (Grant number 10–45) from the National Institute of Biomedical Innovation of Japan. The funders had no role in study design, data collection and analysis, decision to publish, or preparation of the manuscript.

Competing Interests: The authors have declared that no competing interests exist.

* E-mail: maekawa@nihs.go.jp

Introduction

Metabolomics is one of the "omics" platforms for analyzing comprehensive profiles of small molecule metabolites in cells, tissues, or biofluids such as blood and urine. Metabolomics provides a useful tool to analyze metabolite levels in physiological and biological states, and is therefore applied to explore biomarkers for disease diagnosis [1–4], and drug responses and toxicity [5–8]. Various metabolomic approaches are used, and among these, lipidomics includes the comprehensive analysis of lipid metabolites [9,10]. Lipid metabolites are not only components of cell membranes but are also involved in signal transduction [11,12]. Lipid metabolites are therefore considered potential biomarker candidates for disease diagnosis and drug responses. Indeed, recent lipidomic studies have shown that lipid metabolites such as eicosanoids and sphingolipids are biomarker candidates for cardiovascular events [1], traumatic brain injury

[13], Alzheimer's disease [3,14], type 2 diabetes [15], and depression [16].

Blood is a commonly used biofluid for biomarker discovery because it is a "data-rich" source containing several thousands of hydrophilic and hydrophobic metabolites [17] that likely reflect many complex biological process in the body. In addition, collection of blood samples is a minimally invasive procedure as compared with collection of tissue samples by biopsy. Serum and plasma are two distinct matrices separated from blood after phlebotomy. Serum is prepared from whole blood following a clotting process. Plasma is obtained from whole blood in the presence of an anticoagulant, so that coagulation factors are not activated and thus no blood clot is formed. Because of the differences in the preparation of the two blood matrices, metabolite levels are expected to differ between plasma and serum. Indeed, previous studies, focusing primarily on hydrophilic metabolites, have found differences in the metabolite profiles of plasma and serum, and discussed the advantage of each matrix for metabolite analysis [18–21]. As for lipid (hydrophobic) metabolites, we recently investigated the different levels of these molecules in plasma and serum by using non-fasting blood samples from healthy human subjects [21]. In that study, we found that the levels of thromboxane B_2 (TXB_2), 12-hydroxy-eicosatetraenoic acid (12-HETE), and 12-hydroxy-eicosapentaenoic acid (12-HEPE) tended to be higher in serum than in plasma, suggesting their release from activated platelets by the clotting process. In the non-fasted condition, however, relatively large inter-individual differences were observed for lipid metabolites levels in blood. On the other hand, matrix-associated differences in lipid metabolite levels remain to be evaluated in fasted subjects. Such data would be valuable because fasting blood is commonly used for biomarker discovery studies and dietary factors are known to affect lipid metabolite levels in blood [22].

To date, potential confounding factors that may affect lipid metabolite levels in the blood for biomarker exploration studies have not been thoroughly evaluated. Although it has been reported that the levels of lipid metabolites such as sphingomyelins (SMs) and phosphatidylcholines (PCs) in blood differ between genders [23,24], comprehensive data are lacking, especially on the exact molecular species and the extent of differences in their levels among samples derived from subjects from various backgrounds. In order to prevent false-positive or false-negative results in biomarker discovery, gender- and age-associated differences in the basal levels of individual lipid species should be analyzed in advance. Furthermore, it is important to determine an optimal matrix for analyzing lipid biomarkers and to clarify the stability of metabolites subjected to various sample handling and processing procedures. In this study, to obtain fundamental information on lipid metabolites in blood, we aimed to (1) investigate which of the two matrices, plasma or serum, is more suitable for lipid analysis, (2) elucidate gender- and age-associated differences in basal lipid metabolite levels, and (3) examine the stability of lipid metabolites following repeated freeze-thaw cycles of sample matrices. Toward these aims, we performed a lipidomic analysis by using liquid chromatography-mass spectrometry (LC-MS) and liquid chromatography-tandem mass spectrometry (LC-MS/MS) in fasting human plasma and serum samples for four groups (15 subjects/group) consisting of young (25–34 years old) and elderly (55–64 years old) subjects of both genders, and in repeatedly frozen and thawed plasma and serum samples from young male subjects. Because phosphoglycerolipids (PLs) and sphingolipids (SLs) have different physiological functions depending on their classes and fatty acid composition, we identified the exact species of each metabolite to understand gender- and age-associated differences in their levels.

Materials and Methods

Collection of Human Blood and Preparation of Plasma and Serum

Blood samples from healthy adults were purchased from PromedDX (Norton, MA). The samples were collected after obtaining written informed consent from all subjects. The ethics committee of the National Institute of Health Sciences authorized PromedDX as a validated provider of blood samples and exempted us from the committee's approval for use of the purchased blood samples. Venous blood was collected from 60 white subjects on the morning after fasting for 14 h. Participants were divided into four groups of 15 subjects each: young males (25–33 years old), elderly males (55–64 years old), young females (25–34 years old), and elderly females (55–63 years old) (Table 1).

Fresh blood from each individual was collected and simultaneously drawn into 10 ml Vacutainer Serum Separator Tubes with a clot activator for serum and 10 ml Vacutainer Plasma Separator Tubes containing K_2-EDTA for plasma separation. Vacutainer tubes were purchased from Becton Dickinson (Franklin Lakes, NJ). Samples were centrifuged according to the manufacturer's instructions, and serum and plasma were separated within 2 h after collection of blood samples. The plasma and serum samples were immediately frozen and stored at $-80°C$. After shipment with dry ice from PromedDX, all frozen samples were thawed once on ice and divided into small aliquots before storing at $-80°C$ until lipid extraction.

Extraction and Measurements of Lipid Metabolites

Lipid extraction and measurement of lipid metabolites by LC-MS(/MS) was performed as reported previously [21]. In brief, small aliquots of frozen plasma and serum were thawed on ice for 2 h. Our normal samples were thus frozen and thawed twice in total, including the dispensing process described above. Lipid metabolites were extracted from 100 μl of plasma or serum by using the method described by Bligh and Dyer (BD) [25] with a few modifications [21]. Lower organic layers were measured by ultra-performance liquid chromatography-time of flight mass spectrometry (UPLC-TOFMS; LCT Premier XE; Waters Micro-mass, Waters, Milford, MA) for analysis of phosphoglycerolipids (PLs), sphingolipids (SLs), and neutral lipids (NLs). To distinguish alkenylacyl and alkyl PL species with the same exact mass, a small aliquot of each BD sample was acid-hydrolyzed [26] and analyzed by UPLC-TOFMS. Upper aqueous layers were subjected to solid extraction to obtain polyunsaturated fatty acids (PUFAs) and their oxidative fatty acids (oxFAs), and then measured by UPLC-MS/MS using a 5500QTRAP quadrupole-linear ion trap hybrid mass spectrometer (AB Sciex, Framingham, MA) interfaced with an ACQUITY UPLC System (Waters, Milford, MA). Structural analysis of PLs and SLs was performed by LC-Fourier Transform Mass Spectrometry (LC-FTMS; LTQ Orbitrap XL, Thermo Fisher Scientific, Waltham, MA) as previously described [26], with a few modifications. Data-dependent MS^3 analysis was performed in the positive-ion mode to identify the long chain base of ceramides and cerebrosides.

Effect of Freeze-thawing on the Stability of Lipid Metabolites

To investigate the stability of lipid metabolites in the plasma and serum, we performed 8 additional freeze-thaw cycles (10 cycles in total) in plasma and serum samples of young males. Frozen plasma and serum samples were thawed on ice for 2 h and then re-frozen at $-80°C$ for 30 min. After 10 cycles of freeze-thawing, lipid metabolites were extracted from the samples and analyzed by UPLC-TOFMS and UPLC-MS/MS as described above for normal samples.

Data Processing

UPLC-TOFMS data were processed using the 2DICAL software (Mitsui Knowledge Industry, Tokyo, Japan) [27]. The extracted ion peaks were normalized using internal standards (ISs). Metabolites eluted from 0.1 to 38.0 min (PLs, SLs, diacylglycerol [DG], and cholesterol [Ch]), and from 37.5 to 60 min (cholestryl ester [ChE], coenzyme Q10 [CoQ10], and triacylglycerol [TG]) by UPLC were separately normalized to 1,2-dipalmitoyl-$[^2H_6]$-sn-glycero-3-phosphocholine (16:0–16:0 PC-d6; Larodan Fine Chemicals, Malmo, Sweden) and 1,2-dioctanoyl-3-linoleoyl-sn-glycerol (8:0–8:0–18:2 TG, Larodan), respectively. Data for

Table 1. Subject information (fasted white subjects, n = 15 in each group).

Group	Young male (YM)	Young female (YF)	Elderly male (EM)	Elderly female (EF)
Age range (yr)	25–33 (median 29)	25–34 (median 28)	55–64 (median 59)	55–63 (median 59)
Height (cm)	154.9–185.4 (median 172.7)	149.9–182.9 (median 162.6)[†]	165.1–190.5 (median 177.8)[††]	152.4–175.3 (median 162.6)[†]
Weight (kg)	52.2–113.9 (median 78.0)	59.9–147.4 (median 93.4)	63.5–116.1 (median 75.8)	62.6–114.3 (median 90.7)
BMI [kg/m^2]	18.0–36.6 (median 26.2)	24.9–49.7 (median 35.4)[†]	19.5–34.9 (median 24.5)	26.1–43.3 (median 32.7)[†]

(BMI, body mass index).
[†]Heights and BMIs are significantly different between males and females of corresponding age groups ($p<0.05$ by Man-Whitney U-test).
[††]Heights of elderly males are significantly higher than those of young males ($p<0.05$ by Mann-Whitney U-test).

PUFAs and oxFAs from UPLC-MS/MS were processed using the MultiQuant™ Software (Version 2.1, AB Sciex, Framingham, MA). The integrated peak area of each metabolite was normalized to deuterated leukotriene B4 (LTB$_4$-d4, Cayman Chemical, Ann Arbor, MI).

Statistical Analysis

The data were analyzed statistically using the Wilcoxon matched-pairs signed-rank tests for comparison of serum and plasma levels for each metabolite for the same group of subjects, and the Mann-Whitney U-test test for comparison of metabolite levels among the four groups. Statistical analyses were carried out using R statistical environment (http://r-project.org/) software. Differences with p values of less than 0.05 were considered statistically significant.

Results

Lipid Metabolite Profiles in Human Blood Samples

Lipidomic analyses were performed for both plasma and serum samples collected from 60 white subjects, and 253 lipid metabolites were thus identified. The metabolites included nine lysophosphatidylcholines (lysoPCs), 34 phosphatidylcholines (PCs), 20 ether-type PCs, two lysophosphatidylethanolamines (lysoPEs), nine phosphatidylethanolamines (PEs), nine ether-type PEs, eight phosphatidylinositols (PIs), 22 sphingomyelins (SMs), seven Cers, eight cerebrosides (CBs), one Ch, 12 ChEs, one CoQ10, seven DGs, 79 TGs (Table S1), three free PUFAs, and 22 oxFAs (Table S2).

Differences in Lipid Metabolite Profiles between Plasma and Serum Samples

Matrix choice is a fundamental consideration for biomarker exploration in blood. We thus first examined the differences in lipid metabolite profiles of plasma and serum in all the groups. Of 253 metabolites, significant differences in plasma and serum levels of 34 metabolites were observed in young males, 82 in young females, 107 in elderly males, and 74 in elderly females ($p<0.05$; Table S3). Levels of PLs and NLs with marked fold changes (serum/plasma ratio >1.5) and lower p-values ($p<0.01$) in any of the four groups are shown in Figure 1. Representative species with matrix-associated differences (18:0 lysoPC and 36:3 DG) are depicted as scatter plots in Figures 2A and 2B, respectively.

Of the PUFAs and oxFAs, the levels of arachidonic acid (AA), 12-hydroxyheptadecatrienoic acid (12-HHT), 12-hydroxyeicosatetraenoic acid (12-HETE), 5,6-dihydroxyeicosatrienoic acid (5,6-diHETrE), eicosapentaenoic acid, and 4-hydroxyldocosahexaenoic acid (4-HDoHE) were significantly higher in serum than in plasma, whereas four metabolites (8-HETE, 15-HETE, 10-

HDoHE, and 20-HDoHE) were present at significantly lower levels in serum than in plasma ($p<0.05$; Table 2) in all the four groups. In particular, the levels of 5,6-diHETrE (Figure 2C) were more than 3.2-fold higher in serum than in plasma. In contrast, the levels of 8-HETE (Figure 2D) in serum were less than half of those in plasma.

Gender-associated Differences in Lipid Metabolite Levels in Blood Samples

To clarify whether gender is a confounding factor for biomarker exploration studies, we investigated gender-associated differences in lipid metabolite levels in both plasma and serum samples from subjects of both age groups. Of the 253 lipid metabolites analyzed, significant gender-associated differences were observed for 16 metabolites in plasma and 20 in serum samples of the young age group, as well as for 61 metabolites in plasma and 33 in serum samples of the elderly age group ($p<0.05$; Table S4). Figure 3 summarizes the molecular species found to have marked fold changes in their levels (female/male ratio >1.5 or <0.67) and lower p-values ($p<0.01$) in either plasma or serum. In the young age groups (Figure 3A), the levels of SM species were markedly higher in females than in males. Differences in the levels of a representative SM species (d18:1–18:0 SM) among all four groups are precisely depicted in Figure 4A. In the elderly age groups (Figure 3B), not only SMs but also species of various other classes (PCs, PEs, ether PEs, Cers, ChE, TGs, and oxFA) of lipid metabolites showed markedly different levels between genders, indicating that gender-associated differences were more prominent in elderly subjects than in young subjects. Notably, in the elderly age group, the levels of five lipid metabolites containing docosahexaenoic acid (DHA), including 18:0–22:6 PC (Figure 4B), 22:6 ChE, and 18:0:22:6 PE, were remarkably higher in females than in males.

No PUFA and oxFA metabolites were found to have significant gender-associated differences in both young and elderly age groups; however, the levels of 18-HETE (Figure 4C) and 17,18-diHETE were markedly lower in elderly females than in elderly males (Figure 3B). These changes were observed in both plasma and serum samples.

Age-associated Differences in Lipid Metabolite Levels in Human Blood Samples

The lipid metabolite levels were compared between the young and elderly subjects of both genders. Of the 253 metabolites, eight metabolites in plasma and 29 in serum samples of males, and 81 metabolites in plasma and 59 in serum samples of females showed significant age-associated differences ($p<0.05$; Table S5). Of these, the metabolites with marked fold changes (elderly/young ratio > 1.5 or <0.67) and lower p-values ($p<0.01$) in either plasma or

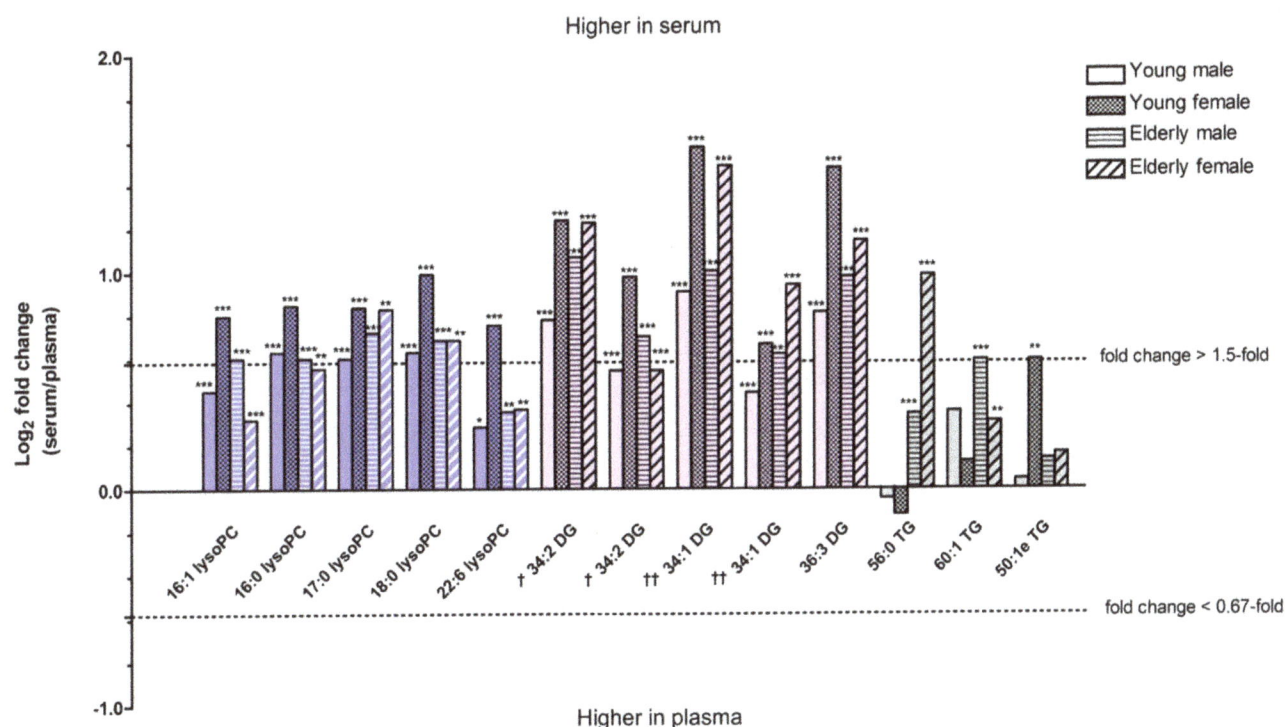

Figure 1. Differences in lipid metabolite levels in human blood samples between plasma and serum. Lipid metabolites with marked fold changes (serum/plasma >1.5 or <0.67) and $p<0.01$ values between plasma and serum in any of the four subject groups are plotted for each group. Statistical significance was determined by Wilcoxon matched-pairs signed-rank test (*$p<0.05$, **$p<0.01$, ***$p<0.001$). Some metabolites ([†, ††]) with same exact mass were eluted with different retention time (Table S1) and therefore seem to be different molecular species of DG. LysoPC, lysophosphatidylcholine; DG, diacylglycerol; TG, triacylglycerol.

serum are shown in Figure 5. In males (Figure 5A), the levels of metabolites in the young age group were generally comparable to those in the elderly age group, with a few exceptions: one metabolite in plasma (5-hydroxyeicosapentaenoic acid [5-HEPE], Figure 4D) and five in serum (16:0–20:5 PC, d18:1–22:0 Cer, 20:5 ChE, 60:1 TG, 5-HEPE) were present at remarkably higher levels (fold changes >1.5; $p<0.01$) in the elderly age groups than in the young age groups.

Age-associated differences were more prominent in females than in males (Figure 5B), and many TGs were measured at significantly higher levels in elderly females than in young females in both matrices. In females (Figure 5B), the levels of 28 metabolites (two lysoPCs, three PCs, two ChEs, 20 TGs, and one oxFA) in plasma and 23 (22 TGs and one oxFA) in serum differed markedly between young and elderly age groups (fold change >1.5 or <0.67; $p<0.01$). It should be noted that female-specific age-associated changes in the levels of lysoPCs were clearly observed in plasma but were markedly less in serum (Figure 5B), suggesting that increases in the levels of lysoPCs in serum caused by the clotting process during sample preparation may obscure the original variations between the different aged groups. As for oxFAs, the levels of the cytochrome P450 (CYP) metabolite 18-HETE (Figure 4C), were remarkably lower in elderly than in young females for both matrices.

Freeze-thawing Effect on Lipid Metabolite Stability

The stability of lipid metabolites was examined using 10 cycles of freezing and thawing in plasma and serum. The levels of most lipid metabolites in the samples subjected to repeated freeze-thaw cycles were greatly decreased compared with those in the normal

samples (Tables S6 and S7). Representative metabolites for which marked differences were observed are shown in Figure 6. The effects of freeze-thawing were comparable between plasma and serum for almost all metabolites. For example, the levels of one major phosphoglycerolipid (18:0–18:2 PC) were decreased by about 79% and 88% in plasma and serum, respectively, by repeated freeze-thawing cycles (Figure 6A). In particular, PUFA and oxFAs were markedly decreased by 10 cycles of freeze-thawing (Table S7; arachidonic acid is shown in Figure 6B). Remarkably, the levels of PUFAs containing ChEs, such as 20:4 ChE, were significantly increased by 10 cycles of freeze-thawing (Figure 6D), whereas those of free Ch as a substrate of ChEs was significantly decreased (Figure 6C).

Discussion

We performed a comprehensive lipidomic analysis of blood from healthy adults under various conditions on sample preparation (plasma collection versus serum collection), sample selection (subjects' age and gender) and storage (few versus more number of freeze-thaw cycles). First, our results demonstrated that, compared with serum, plasma is a more suitable matrix for exploring lipid biomarkers. We found matrix-associated differences among the levels of lysoPCs, DGs, free PUFAs, and several oxFAs, irrespective of the subjects' backgrounds. In the preparation of serum from whole blood, the clotting process results in thrombin-stimulated platelets releasing DG and inositol 1,4,5-phosphate (IP_3) by the degradation of phosphatidylinositol 4,5-bisphosphate (PIP_2) via the activation of phospholipase C (PLC) [28]. Furthermore, lysoPLs and PUFA are released from membrane PLs by activated phospholipase A_2 (PLA_2) via IP_3 signaling.

A 18:0 lysoPC

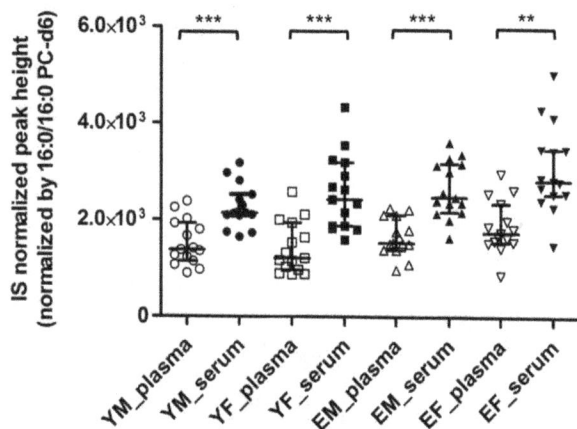

B 36:3 DG

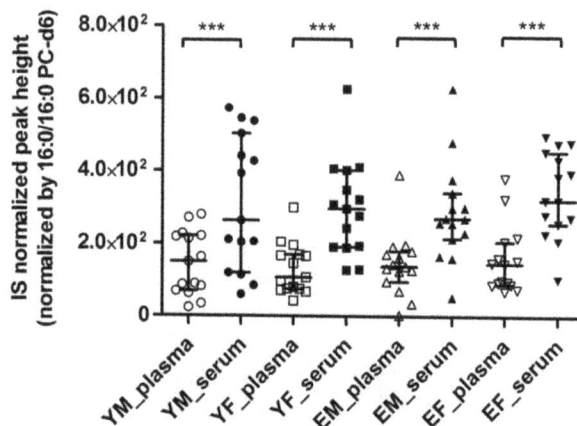

C 5,6-diHEtrE

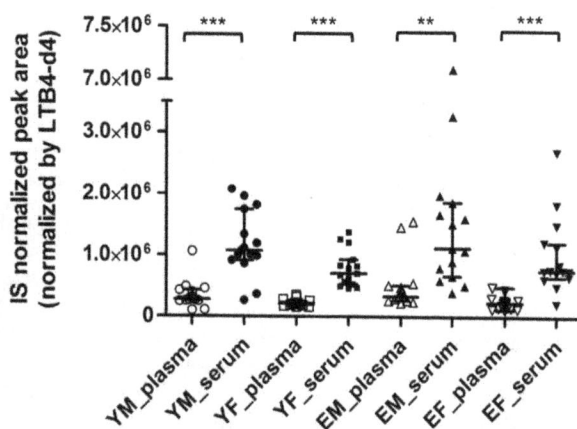

D 8-HETE

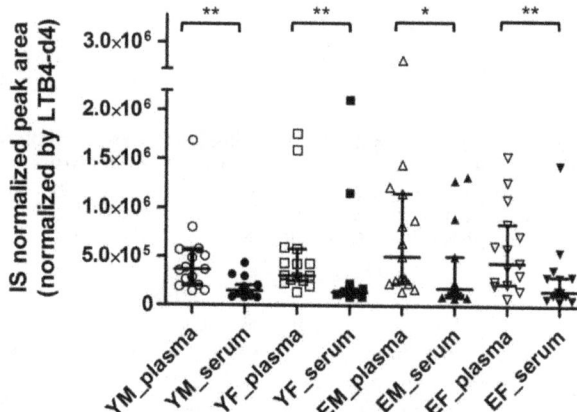

Figure 2. Levels of representative lipid metabolites with significant matrix-associated differences. (A) 18:0 lysoPC, (B) 36:3 DG, (C) 5,6-diHEtrE, (D) 8-HETE. The graph shows medians and interquartile ranges. Statistical significance was determined by Wilcoxon matched-pairs signed-rank test (*$p<0.05$, **$p<0.01$, ***$p<0.001$). YM, young male; YF, young female; EM, elderly male; EF, elderly female. LysoPC, lysophosphatidylcholine; DG, diacylglycerol; diHEtrE, dihydroxyeicosatrienoic acid; HETE, hydroxyeicosatetraenoic acid.

Several oxFAs such as TXB$_2$, 12-HHT, and 12-lipoxygenase products (12-HETE, 12-HEPE and 14-HDoHE) are also known to be released from activated platelets, thereby exhibiting a variety of biological effects [29]. The levels of these metabolites in serum are therefore unlikely to reflect normal biological processes in the body. Furthermore, because the clotting process is difficult to control strictly *in vitro*, it is possible that serum conceals the changes in lipid metabolism that have occurred in the body, thus obscuring variations between samples/groups. Indeed, we observed that age-associated differences in lysoPCs levels in females were less represented in serum than in plasma (Figure 4B), suggesting that the elevation of lysoPC levels in serum by the clotting process conceals the underlying age-associated differences in blood.

Mas et al. reported that the concentrations of 18-HEPE and 17-HDoHE in plasma were approximately two-fold greater than those in serum following 3 weeks of oral supplementation with n-3 fatty acids [30]. Although the metabolites detected in our study and the study by Mas and colleagues were not identical because we did not carry out a dietary intervention in our study, our results

reveal, for the first time, that plasma contains higher levels of some hydroxy-fatty acids (8-HETE, 15-HETE, 10-HDoHE, and 20-HDoHE) than does serum. Degradation of these lipid metabolites in serum may occur during the clotting process. When comparing matrix-associated differences, we found, for the first time, that 5,6-diHETrE was present at markedly (more than 3-fold) higher levels in serum than in plasma (Figure 2C, Table 2). In platelets, 5,6-epoxyeicosatrienoic acid (EET), a precursor of 5,6-diHETrE, is proposed to be a Ca^{2+} entry activator following depletion of intracellular Ca^{2+} stores caused by the activation of IP$_3$ cascades [31]. This finding therefore strongly supports the hypothesis that increased levels of 5,6-diHETrE are associated with the clotting process.

Compared with our previous studies using non-fasting plasma and serum of young males and females [21], a higher number of metabolites that show matrix-associated differences was detected in fasting blood than in non-fasting blood samples. For example, significantly higher levels of lysoPCs and free PUFAs such as arachidonic acid and eicosapentaenoic acid were observed in serum than in plasma from fasting blood samples, but not from

Table 2. PUFA and oxFA levels showing >1.5 or 0.67-fold changes and $p<0.01$ values between plasma and serum samples of either male or female subjects.

Molecular species	Plasma vs Serum (YM)		Plasma vs Serum (YF)		Plasma vs Serum (EM)		Plasma vs Serum (EF)	
	MFC (S/P)	p value	MFC (S/P)	p value	MFC (S/P)	p value	MFC (S/P)	p value
Arachidonic acids (20:4 FA) & its metabolites								
Arachidonic acid	1.22	6.10E-05***	1.40	1.83E-04***	1.44	6.71E-03**	1.63	3.05E-04***
12-HHT	1.75	4.27E-04***	2.43	3.36E-03**	1.38	4.27E-04***	2.25	3.36E-03**
8-HETE	0.41	1.53E-03**	0.45	4.27E-03**	0.36	1.25E-02*	0.35	4.27E-03**
12-HETE	1.61	8.36E-03**	2.16	1.83E-04***	1.31	3.02E-02*	1.66	6.10E-04***
15-HETE	0.72	2.62E-03**	0.55	6.71E-03**	0.62	2.56E-02*	0.46	6.71E-03**
5,15-diHETE	0.79	1.22E-04***	0.64	5.37E-03**	0.64	1.35E-01	0.48	6.71E-03**
5,6-diHETrE	3.88	6.10E-05***	3.29	6.10E-05***	3.33	1.53E-03**	3.23	6.10E-05***
Eicosapentaenoic acid (20:5 FA) & its metabolites								
Eicosapentaenoic acid	1.38	6.10E-05***	1.35	8.54E-04***	1.39	3.36E-03**	1.79	8.54E-04***
12-HEPE	1.38	1.81E-02*	1.60	2.62E-03**	1.48	2.52E-01	1.67	4.13E-02*
Docosahexaenoic acid (22:6 FA) & its metabolites								
4-HDoHE	3.25	6.10E-05***	2.56	6.10E-05***	1.90	4.27E-04***	4.09	6.10E-05***
10-HDoHE	0.52	3.36E-03**	0.41	6.71E-03**	0.44	1.25E-02*	0.34	6.71E-03**
14-HDoHE	1.51	4.79E-02*	1.66	2.62E-03**	1.20	2.77E-01	2.14	6.71E-03**
20-HDoHE	0.51	3.05E-04***	0.62	1.22E-04***	0.50	5.37E-03**	0.53	5.37E-03***

(MFC (S/P), median fold change in serum/plasma; YM, young male; YF, young female; EM, elderly male; EF, elderly female; HHT, hydroxyheptadecatrienoic acid; HETE, hydroxyeicosatetraenoic acid; diHETE, dihydroxyeicosatetraenoic acid; diHETrE, dihydroxyeicosatrienoic acid; HEPE, hydroxyeicosapentaenoic acid; HDoHE, hydroxydocosahexaenoic acid). Statistical significance was determined by Wilcoxon matched-pairs signed-rank test (*$p<0.05$, **$p<0.01$, ***$p<0.001$).

A Young

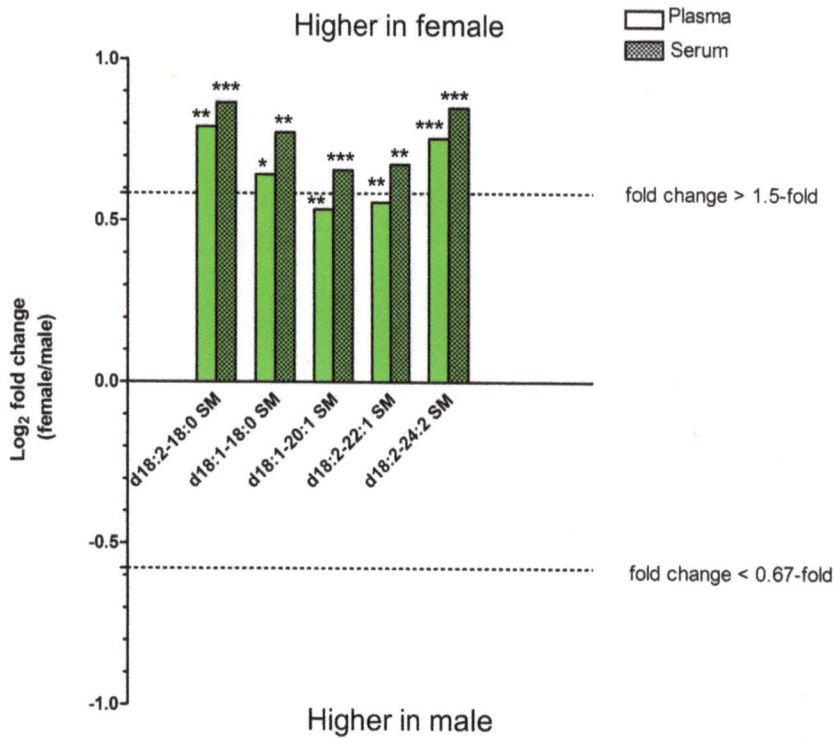

B Elderly

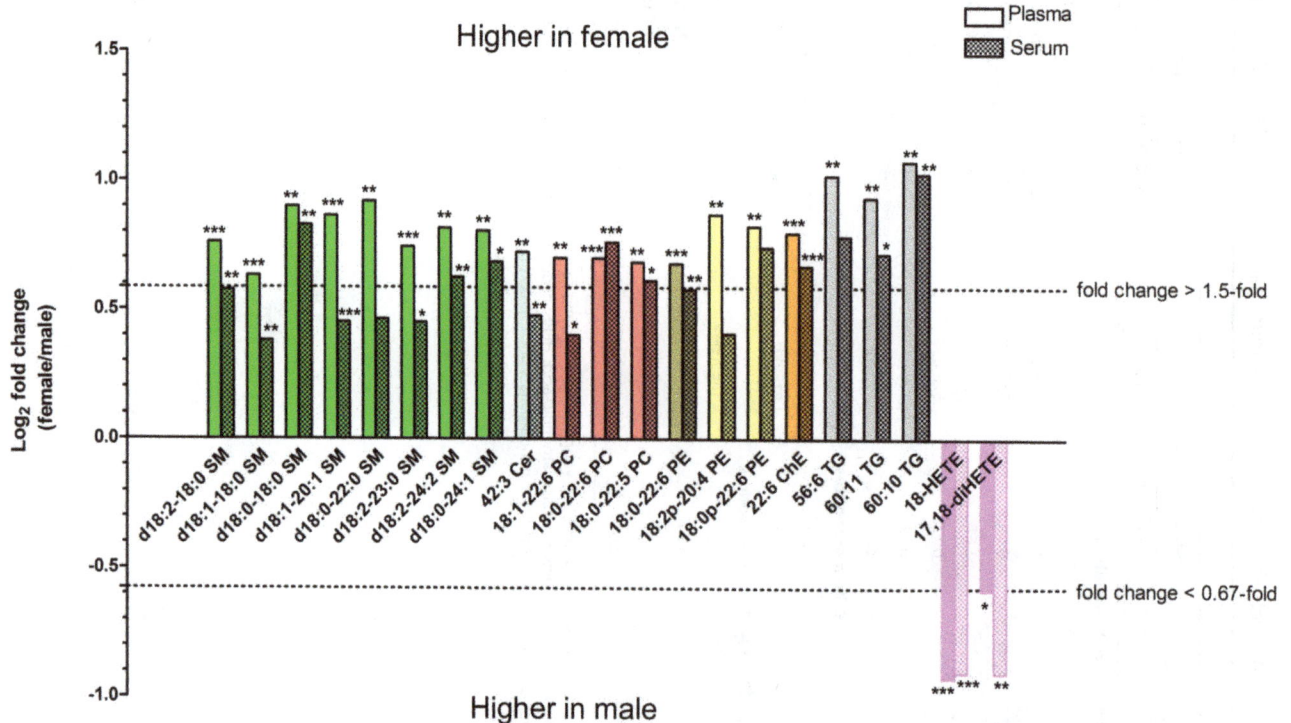

Figure 3. Differences in lipid metabolite levels in human blood samples between males and females. Lipid metabolites with marked fold changes (female/male >1.5 or <0.67) and $p<0.01$ values between males and females in either plasma or serum are plotted for young (A) and elderly (B) age groups. Statistical significance was determined by Mann-Whitney U-test (*$p<0.05$, **$p<0.01$, ***$p<0.001$). SM, sphingomyelin; Cer, ceramide; PC, phosphatidylcholine; PE, phosphatidylethanolamine; ChE, cholesteryl ester; TG, triacylglycerol; HETE; hydroxyeicosatetraenoic acid; diHETE, dihydroxyeicosatetraenoic acid.

A d18:1-18:0 SM

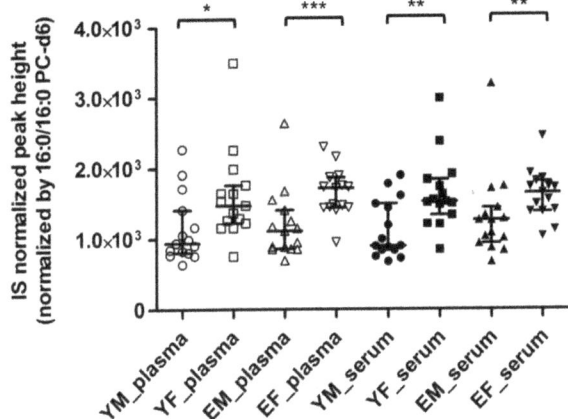

B 18:0-22:6 PC

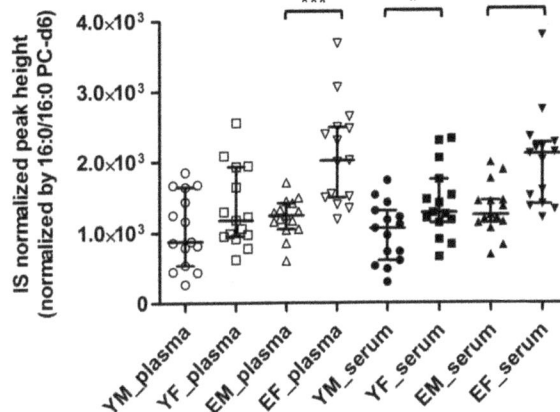

C 18-HETE

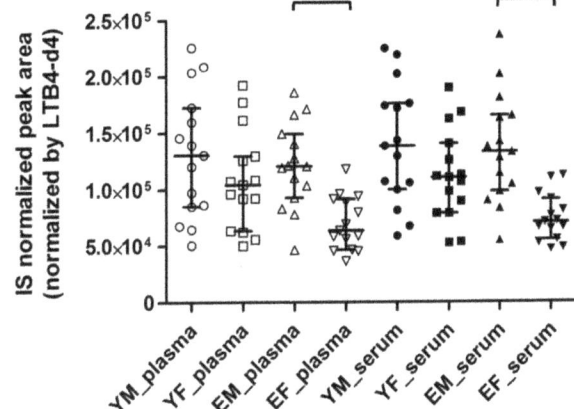

D 5-HEPE

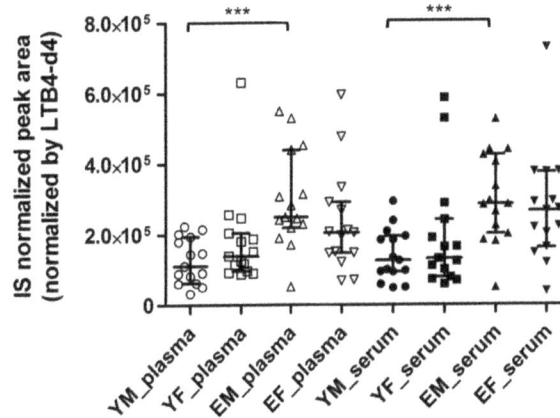

Figure 4. Levels of representative lipid metabolites with significant gender- and age-associated difference. (A) d18:1–18:0 SM, (B) 18:0–22:6 PC, (C) 18-HETE, (D) 5-HEPE. The graph represents medians and interquartile ranges. Statistical significance was determined by Mann-Whitney U-test (*$p<0.05$, **$p<0.01$, ***$p<0.001$). YM, young male; YF, young female; EM, elderly male; EF, elderly female. SM, sphingomyelin; PC, phosphatidylcholine; HETE, hydroxyeicosatetraenoic acid; HEPE, hydroxyeicosapentaenoic acid.

non-fasting blood samples, suggesting that fasting and non-fasting blood should be treated as different samples when exploring biomarkers in the two matrices.

Next, our results demonstrated that gender and age influence the levels of several metabolites in blood and therefore are confounding factors in exploring lipid biomarkers. Gender-associated differences highlighted that the levels of SMs are remarkably higher (fold change >1.5-fold and $p<0.01$) in females than in males irrespective of the subjects' age (Figure 3). Moreover, body mass index (BMI) was found to be significantly higher in females than in males (Table 1), raising the possibility that differences in the levels of SMs are associated with BMI rather than with gender. However, gender-associated differences in SMs were observed even when subgroups of males and females with comparable BMI levels were compared (data not shown), indicating that the influence of BMI on SM levels is marginal. It was reported in a prospective study that the levels of SM species in plasma were significantly lower ($p<0.05$) in AD subjects than in cognitively normal control subjects [14], and are thus of interest as AD biomarkers. Our results suggest that gender-associated

differences in SM levels should be paid attention for their application as biomarkers in a clinical setting in the future.

Previous research suggests that estrogen may be involved in the regulation of SM metabolism [32]. It has been also reported that young children with low levels of sex hormones exhibit gender-associated differences in plasma SM levels [23]. The potential role of estrogen in SM metabolism therefore remains controversial. In the present study, we show that SMs are present at higher levels in elderly females than in elderly males, and at comparable levels in young and elderly females. This finding was observed despite the levels of estrogen being markedly decreased in postmenopausal elderly females, usually lower even than in elderly males [33]. This result suggests that factors other than estrogen are probably responsible for gender-associated differences in SM levels. The exact mechanisms of the regulation of SM metabolism have not been elucidated thus far.

Upon further investigation of fatty acyl chains of PLs, our analysis revealed that DHA-containing lipid metabolites such as 18:1–22:6 PC and 18:0–22:6 PC (Figure 4B) showed both gender- and age-associated differences in matrices of elderly females

A   Male

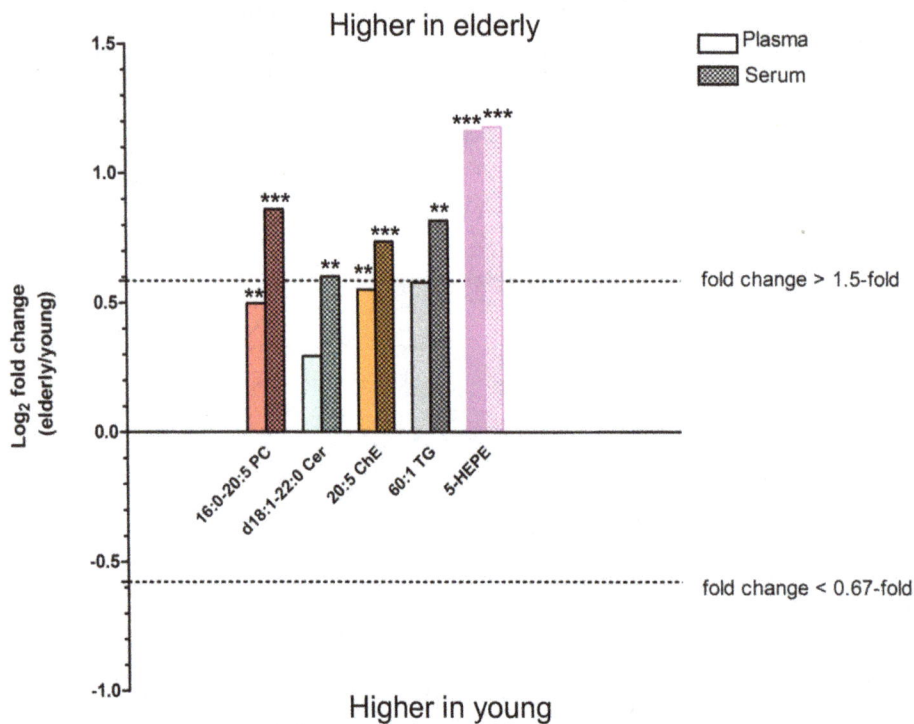

B   Female

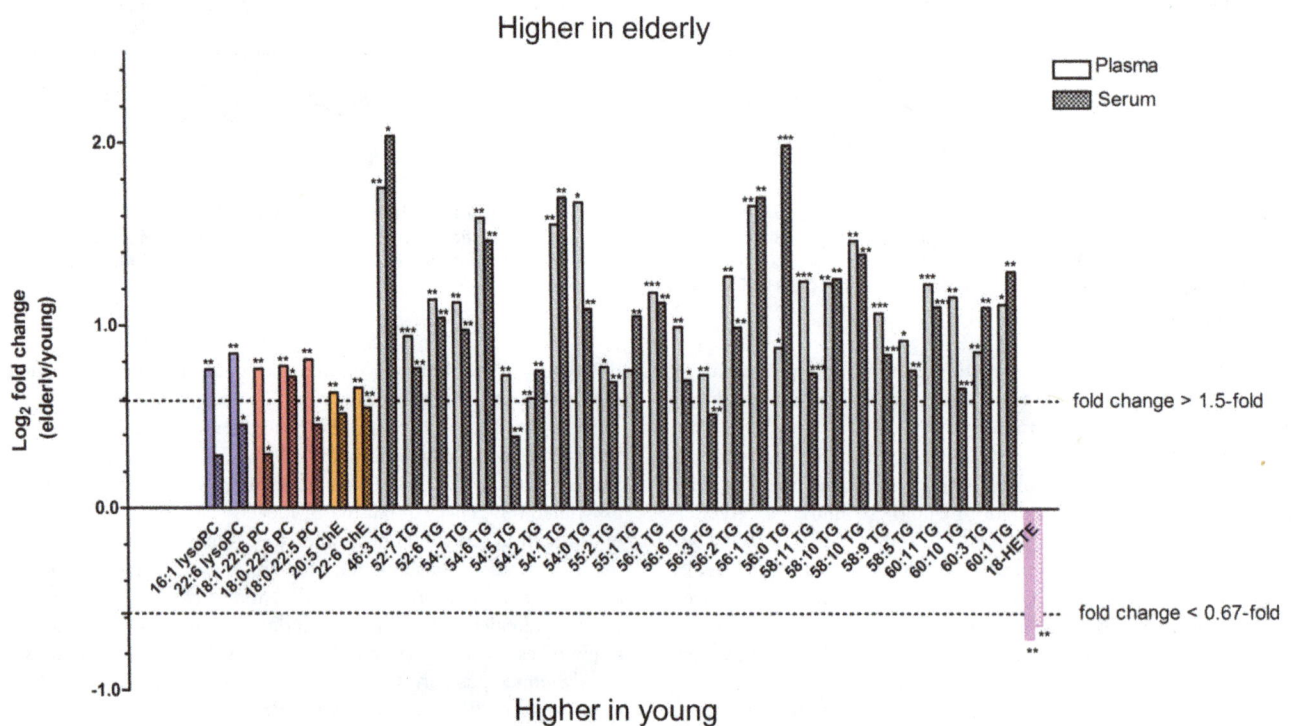

Figure 5. Differences in lipid metabolite levels in human blood samples between young and elderly age groups. Lipid metabolites with marked fold changes (elderly/young subjects ratio >1.5 or <0.67) and $p<0.01$ values between young and elderly age groups in either plasma or serum are plotted for male (A) and female (B) groups. Statistical significance was determined by Mann-Whitney U-test (*$p<0.05$, **$p<0.01$, ***$p<0.001$). PC, phosphatidylcholine; Cer, ceramide; ChE, cholesteryl ester; TG, triacylglycerol; HEPE, hydroxyeicosapentaenoic acid; lysoPC, lysophosphatidylcholine; HETE, hydroxyeicosatetraenoic acid.

A 18:0-18:2 PC

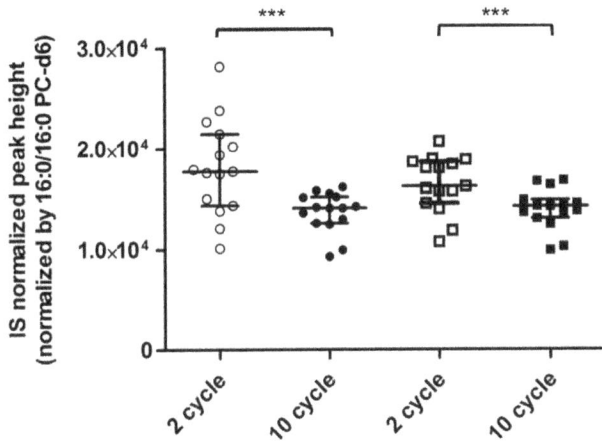

B arachidonic acid

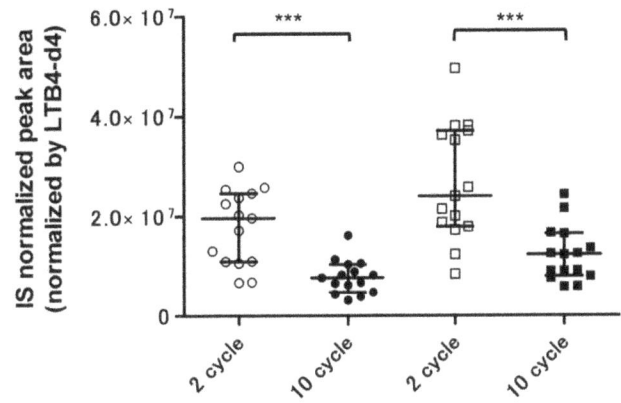

C Ch

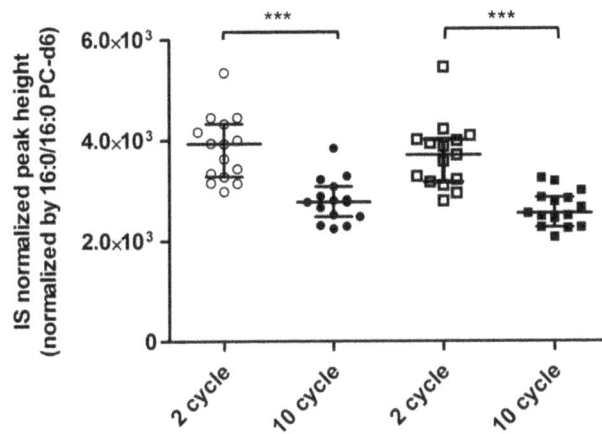

D 20:4 ChE

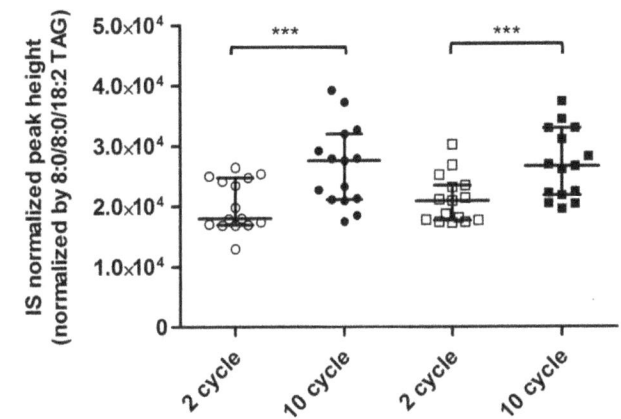

Figure 6. Effects of repeated freeze-thawing on lipid metabolite stability in plasma and serum of young males. Lipid metabolite levels after 10 freeze-thaw cycles were compared with those after two freeze-thaw cycles for plasma and serum samples. (A) 18:0–18:2 PC, (B) arachidonic acid, (C) Ch, (D) 20:4 ChE. The graph shows medians and interquartile ranges. Statistical significance was determined by Wilcoxon matched-pairs signed-rank test (**$p<0.01$, ***$p<0.001$). PC, phosphatidylcholine; Ch, cholesterol; ChE, cholesteryl ester.

(Figures 3 and 5). A recent study suggested that estradiol influence the activity of enzymes that are involved in the synthesis of DHA in rats [34]. Our observations suggest that the change in estrogen secretion following menopause may affect DHA metabolism. Although their mechanisms involved in this potential connection between estrogen levels and DHA metabolism remain to be elucidated, DHA metabolite levels may be influenced by both ageing and genders.

Age-associated differences in lipid metabolite levels were more prevalent in females than in males. Many TGs, in particular, showed age-associated differences in females but not in males (Figure 5). This observation suggests that the decrease in estrogen secretion following menopause [33] affects the metabolism of several lipoproteins, resulting in an increase in the levels of TGs in the elderly. Estrogen may be involved in regulation of the metabolism of lipoproteins such as low-density lipoprotein [35]. As for males, a 5-lipoxygenase (5-LOX) product, 5-HEPE, was significantly higher in elderly than in young subjects, and a similar trend was observed between young and elderly females (Figure 4D). The bioactive lipid 5-HEPE is known to be a potent agonist of

GPR119 and enhances glucose-dependent insulin secretion [36], suggesting the molecule as a possible biomarker candidate for diabetes. Although the mechanisms regulating age-associated increases in 5-HEPE are unknown, Schuchardt et al. reported that serum levels of EPA-derived oxylipins, including 5-HEPE, strongly correlate with the EPA content of the erythrocyte membrane, and thus with the availability of the substrate EPA [37]. Furthermore, results from a dietary intervention study suggest that elderly subjects have a greater capacity to incorporate dietary EPA into plasma phospholipids and cells than younger subjects do [38,39]. Taken together, these results indicate that increased EPA availability in elderly subjects may result in higher 5-HEPE levels in elderly males than in young males. Indeed, EPA-containing lipids such as 16:0–20:5 PC and 20:5 ChE showed age-associated increases in males (Figure 5).

One major ceramide species in plasma, d18:1–22:0 Cer, also exhibited age-associated increases in the serum of male subjects. Plasma ceramide is most likely synthesized in the liver by ceramide synthases from sphinganine and acyl coenzyme A, after which it is incorporated into lipoproteins and transported into plasma [40].

The augmented levels of this species in elderly males may be linked with the biological function of ceramides as signaling molecules capable of regulating vital cellular functions, including apoptosis, cell growth, differentiation, senescence, diabetes, insulin resistance, inflammation, neurodegenerative disorders, and atherosclerosis [41]. It is interesting to note that two metabolites with opposing roles, 5-HEPE (insulin secretion) and d18:1–22:0 Cer (insulin resistance), are concurrently increased with age in healthy males.

Based on the results of the lipidomic analysis of serum and plasma samples subjected to repeated freeze-thaw cycles compared to normal samples, we recommend that plasma and serum samples should be aliquoted after their collection, and repeated freeze-thaw cycles should be avoided. Most lipid metabolite levels were significantly decreased by repeated freeze-thaw cycles in both plasma and serum (Tables S6 and S7, Figure 5A–C), suggesting that lipid metabolites are degraded or metabolized by various lipases such as phospholipase A_2 (PLA_2) during the process of freeze-thawing. However, PUFA-esterified ChEs were significantly increased by repeated freeze-thaw cycles (Figure 5D). Because lecithin-cholesterol acyltransferase (LCAT) is primarily responsible for producing PUFA-esterified ChEs such as 20:4 ChE and 22:6 ChE [42], these results suggest that the LCAT-mediated reaction in which the acyl chain at the sn-2 position is transferred from PLs to Ch is activated during the freeze-thaw process in each matrix.

Conclusions

Lipidomics could be a promising new approach for the identification of new biomarkers for monitoring or predicting disease states and/or drug responsiveness. In this study, we obtained fundamental information on sample selection and handling procedures applied to measuring lipid metabolites in blood, which is important in ensuring the quality of biomarker discovery and qualification processes. Our results suggest that plasma (rather than serum), subjected to a minimum number of freeze-thaw cycles, is suitable for obtaining reliable measurements of lipid levels reflecting basal, physiological levels. Notably, repeated freeze-thawing is a detrimental procedure resulting in a reduction in the levels of the majority of lipid metabolites, consequently leading to erroneous data. In general, ideal metabolomic biomarkers are those whose levels are drastically modulated by disease or drugs so as to overcome any variations in subjects' background levels. Otherwise, we should keep in mind that age and gender are confounding factors when measuring several metabolites. Our current data showing gender- and age-

associated differences in each lipid metabolite are key pieces of information to be considered when selecting candidate biomarkers for further validation. It is conceivable that lifestyle (e.g., drinking, smoking and dietary habits) and ethnicity may also affect metabolite levels in blood, and further study is needed to examine these effects.

Supporting Information

Table S1   Dataset of internal standard (IS)-normalized peak hight of phosphoglycerolipids, sphingolipids and neutral lipids in human blood.

Table S2   Dataset of internal standard (IS)-normalized peak area of polyunsaturated fatty acids and their oxidative metabolites.

Table S3   Matrix-associated differences (fold changes and statistical analysis) in the levels of lipid metabolites.

Table S4   Gender-associated differences (fold changes and statistical analysis) in the levels of lipid metabolites.

Table S5   Age-associated differences (fold changes and statistical analysis) in the levels of lipid metabolites.

Table S6   Effects of freeze-thawing on phosphoglycerolipids, sphingolipids and neutral lipids stability.

Table S7   Effects of freeze-thawing on polyunsaturated fatty acids and their oxidative metabolites stability.

Acknowledgments

We thank Ms. Chie Sudo for secretarial assistance and support.

Author Contributions

Conceived and designed the experiments: KM YK Y. Saito. Performed the experiments: MI KM KS Y. Senoo MU MM YT. Analyzed the data: MI KM KS YK Y. Saito. Wrote the paper: MI KM KS Y. Saito.

References

1. Liu JY, Li N, Yang J, Li N, Qiu H, et al. (2010) Metabolic profiling of murine plasma reveals an unexpected biomarker in rofecoxib-mediated cardiovascular events. Proc Natl Acad Sci U S A 107: 17017–17022.
2. Vouk K, Hevir N, Ribić-Pucelj M, Haarpaintner G, Scherb H, et al. (2012) Discovery of phosphatidylcholines and sphingomyelins as biomarkers for ovarian endometriosis. Hum Reprod 27: 2955–2965.
3. Sato Y, Suzuki I, Nakamura T, Bernier F, Aoshima K, et al. (2012) Identification of a new plasma biomarker of Alzheimer's disease using metabolomics technology. J Lipid Res 53: 567–576.
4. Seijo S, Lozano JJ, Alonso C, Reverter E, Miquel R, et al. (2013) Metabolomics discloses potential biomarkers for the noninvasive diagnosis of idiopathic portal hypertension. Am J Gastroenterol 108: 926–932.
5. Soga T, Baran R, Suematsu M, Ueno Y, Ikeda S, et al (2006). Differential metabolomics reveals ophthalmic acid as an oxidative stress biomarker indicating hepatic glutathione consumption. J Biol Chem 281: 16768–16776.
6. Beger RD, Sun J, Schnackenberg LK (2010) Metabolomics approaches for discovering biomarkers of drug-induced hepatotoxicity and nephrotoxicity. Toxicol Appl Pharmacol 243: 154–166.
7. Parman T, Bunin DI, Ng HH, McDunn JE, Wulff JE, et al. (2011) Toxicogenomics and metabolomics of pentamethylchromanol (PMCol)-induced hepatotoxicity. Toxicol Sci 124: 487–501.
8. McClay JL, Adkins DE, Vunck SA, Batman AM, Vann RE, et al. (2013) Large-scale neurochemical metabolomics analysis identifies multiple compounds associated with methamphetamine exposure. Metabolomics 9: 392–402.
9. Han X, Gross RW (2005) Shotgun lipidomics: electrospray ionization mass spectrometric analysis and quantitation of cellular lipidomes directly from crude extracts of biological samples. Mass Spectrom Rev 24: 367–412.
10. Taguchi R, Nishijima M, Shimizu T (2007) Basic analytical systems for lipidomics by mass spectrometry in Japan. Methods Enzymol 432: 185–211.
11. Spiegel S, Milstien S (2003) Sphingosine-1-phosphate: an enigmatic signaling lipid. Nat Rev Mol Cell Biol 4: 397–407.
12. Serhan CN, Chiang N, Van Dyke TE (2008) Resolving inflammation: dual anti-inflammatory and pro-resolution lipid mediators. Nat Rev Immunol 8: 349–361.
13. Sparvero LJ, Amoscato AA, Kochanek PM, Pitt BR, Kagan VE, et al. (2010) Mass-spectrometry based oxidative lipidomics and lipid imaging: applications in traumatic brain injury. J Neurochem 115: 1322–1336.
14. Han X, Rozen S, Boyle SH, Hellegers C, Cheng H, et al. (2011) Metabolomics in Early Alzheimer's Disease: Identification of Altered Plasma Sphingolipidome Using Shotgun Lipidomics. PLoS One 6: e21643.
15. Wang-Sattler R, Yu Z, Herder C, Messias AC, Floegel A, et al. (2012) Novel biomarkers for pre-diabetes identified by metabolomics. Mol Syst Biol 8: 615.
16. Demirkan A, Isaacs A, Ugocsai P, Liebisch G, Struchalin M, et al. (2013) Plasma phosphatidylcholine and sphingomyelin concentrations are associated with

depression and anxiety symptoms in a Dutch family-based lipidomics study. J Psychiatr Res 47: 357–62.
17. Wishart DS, Jewison T, Guo AC, Wilson M, Knox C, et al. (2013) HMDB 3.0-The Human Metabolome Database in 2013. Nucleic Acid Res 41 (Database issue): D801–7.
18. Liu L, Aa J, Wang G, Yan B, Zhang Y, et al. (2010) Differences in metabolite profile between blood plasma and serum. Anal Biochem 406: 105–112.
19. Denery JR, Nunes AA, Dickerson TJ (2010) Characterization of differences between blood sample matrices in untargeted metabolomics. Anal Chem 83: 1040–1047.
20. Yu Z, Kastenmüller G, He Y, Belcredi P, Möller G, et al. (2011) Differences between human plasma and serum metabolite profiles. PLoS One 6: e21230.
21. Ishikawa M, Tajima Y, Murayama M, Senoo Y, Maekawa K, et al. (2013) Plasma and serum from nonfasting men and women differ in their lipidomic profiles. Biol Pharm Bull 36: 682–685.
22. Hodson L, Skeaff CM, Fielding BA (2008) Fatty acid composition of adipose tissue and blood in humans and its use as a biomarker of dietary intake. Prog Lipid Res 47: 348–380.
23. Nikkilä J, Sysi-Aho M, Ermolov A, Seppänen-Laakso T, Simell O, et al. (2008) Gender-dependent progression of systemic metabolic states in early childhood. Mol Syst Biol 4: 197.
24. Mittelstrass K, Ried JS, Yu Z, Krumsiek J, Gieger C, et al. (2011) Discovery of sexual dimorphisms in metabolic and genetic biomarkers. PLoS Genet 7: e1002215.
25. Bligh EG, Dyer WJ (1959) A rapid method of total lipid extraction and purification. Can J Biochem Physiol 37: 911–917.
26. Taguchi R, Ishikawa M (2010) Precise and global identification of phospholipid molecular species by an Orbitrap mass spectrometer and automated search engine Lipid Search. J chromatogr A 1217: 4229–4239.
27. Ono M, Shitashige M, Honda K, Isobe T, Kuwabara H, et al. (2006) Label-free quantitative proteomics using large peptide data sets generated by nanoflow liquid chromatography and mass spectrometry. Mol Cell Proteomics 5: 1338–1347.
28. Sangkuhl K, Shuldiner AR, Klein TE, Altman RB (2011) Platelet aggregation pathway. Pharmacogenet Genomics 21: 516–521.
29. Lagarde M, Chen P, Véricel E, Guichardant M (2010) Fatty acid-derived lipid mediators and blood platelet aggregation. Prostaglandins Leukot Essent Fatty Acids 82: 227–230.
30. Mas E, Croft KD, Zahra P, Barden A, Mori TA (2012) Resolvins D1, D2, and other mediators of self-limited resolution of inflammation in human blood following n-3 fatty acid supplementation. Clin Chem 58: 1476–1484.
31. Ben-Amor N, Redondo PC, Bartegi A, Pariente JA, Salido GM, et al. (2005) A role for 5,6-epoxyeicosatrienoic acid in calcium entry by de novo conformational coupling in human platelets. J Physiol 570: 309–323.
32. Merrill AH Jr, Wang E, Innis WS, Mullins R (1985) Increases in serum sphingomyelin by 17 beta-estradiol. Lipids 20: 252–254.
33. Bjørnerem A, Straume B, Midtby M, Fønnebø V, Sundsfjord J, et al. (2004) Endogenous sex hormones in relation to age, sex, lifestyle factors, and chronic diseases in a general population: the Tromsø Study. J Clin Endocrinol Metab 89: 6039–6047.
34. Alessandri JM, Extier A, Al-Gubory KH, Harbeby E, Lallemand MS, et al. (2012) Influence of gender on DHA synthesis: the response of rat liver to low dietary α-linolenic acid evidences higher ω3 Δ4-desaturation index in females. Eur J Nutr 51: 199–209.
35. Knopp RH, Zhu X, Bonet B (1994) Effects of estrogens on lipoprotein metabolism and cardiovascular disease in women. Atherosclerosis Suppl: S83–91.
36. Kogure R, Toyama K, Hiyamuta S, Kojima I, Takeda S (2011) 5-Hydroxy-eicosapentaenoic acid is an endogenous GPR119 agonist and enhances glucose-dependent insulin secretion. Biochem Biophys Res Commun 416: 58–63.
37. Schuchardt JP, Schmidt S, Kressel G, Dong H, Willenberg I, et al. (2013) Comparison of free serum oxylipin concentrations in hyper- vs. normolipidemic men. Prostaglandins Leukot Essent Fatty Acids 89: 19–29.
38. Rees D, Miles EA, Banerjee T, Wells SJ, Roynette CE, et al. (2006) Dose-related effects of eicosapentaenoic acid on innate immune function in healthy humans: a comparison of young and older men. Am J Clin Nutr 83: 331–342.
39. Walker CG, Browning LM, Mander AP, Madden J, West AL, et al. (2013) Age and sex differences in the incorporation of EPA and DHA into plasma fractions, cells and adipose tissue in humans. Br J Nutr: 1–11.
40. Jiang H, Hsu FF, Farmer MS, Peterson LR, Schaffer JE, et al. (2013) Development and validation of LC-MS/MS method for determination of very long acyl chain (C22:0 and C24:0) ceramides in human plasma. Anal Bioanal Chem 405: 7357–7365.
41. Arana L, Gangoiti P, Ouro A, Trueba M, Gomez-Munoz A (2010) Ceramide and ceramide 1-phosphate in health and disease. Lipids Health Dis 9: 15.
42. Subbaiah PV, Jiang XC, Belikova NA, Aizezi B, Huang ZH, et al. (2012) Regulation of plasma cholesterol esterification by sphingomyelin: effect of physiological variations of plasma sphingomyelin on lecithin-cholesterol acyltransferase activity. Biochim Biophys Acta 1821: 908–913.

4

Visualization and Analysis of Hepatitis C Virus Structural Proteins at Lipid Droplets by Super-Resolution Microscopy

Dennis Eggert[1,2]◑, **Kathrin Rösch**[1]◑, **Rudolph Reimer**[1], **Eva Herker**[1]*

1 Heinrich-Pette-Institute, Leibniz Institute for Experimental Virology, University of Hamburg, Hamburg, Germany, **2** Institute of Physical Chemistry, University of Hamburg, Hamburg, Germany

Abstract

Cytosolic lipid droplets are central organelles in the Hepatitis C Virus (HCV) life cycle. The viral capsid protein core localizes to lipid droplets and initiates the production of viral particles at lipid droplet–associated ER membranes. Core is thought to encapsidate newly synthesized viral RNA and, through interaction with the two envelope proteins E1 and E2, bud into the ER lumen. Here, we visualized the spatial distribution of HCV structural proteins core and E2 in vicinity of small lipid droplets by three-color 3D super-resolution microscopy. We observed and analyzed small areas of colocalization between the two structural proteins in HCV-infected cells with a diameter of approximately 100 nm that might represent putative viral assembly sites.

Editor: MinKyung Yi, University of Texas Medical Branch at Galveston, United States of America

Funding: The Heinrich Pette Institute, Leibniz Institute for Experimental Virology is supported by the Free and Hanseatic City of Hamburg and the Federal Ministry of Health. The funders had no role in study design, data collection and analysis, decision to publish, or preparation of the manuscript.

Competing Interests: The authors have declared that no competing interests exist.

* Email: eva.herker@hpi.uni-hamburg.de

◑ These authors contributed equally to this work.

Introduction

Hepatitis C virus (HCV) persistently infects ~180 million people worldwide and the associated morbidity and mortality are a major public health concern [1]. Vaccines are not available and albeit recent advances in therapy through development of direct-acting antivirals, current treatment regimens remain very challenging. HCV is a single positive-stranded RNA virus of the *Flaviviridae* family that is translated into a single polyprotein upon entry into host cells. Host and viral proteases cleave this polyprotein, releasing the 10 individual viral proteins. The structural proteins, the viral capsid core and the envelope glycoproteins E1 and E2, are the components of virions while the nonstructural proteins NS3–5B form the viral RNA replication complex. The establishment of fully permissive cell culture systems (HCVcc) [2,3] revealed a close connection between host cell lipids and HCV replication at each step of the viral replication cycle reviewed in [4]. Interestingly, the cellular storage organelles of lipids, lipid droplets, emerged as putative viral assembly sites [5–7].

Two viral proteins localize to lipid droplets in the absence of full viral replication: the capsid protein core and the non-structural protein NS5A [8,9]. Other viral proteins are found in close-proximity of lipid droplets in infected cells, but they lack intrinsic lipid droplet targeting features as they fail to localize to lipid droplets in uninfected cells [10]. Interestingly, both core and NS5A require triglyceride biosynthesis for trafficking to lipid

droplets as inhibitors of diacylglycerol acyltransferase-1 (DGAT1) impair their lipid droplet localization [7,10].

During translation the capsid protein core is released from the polyprotein by two subsequent cleavages, which generate a 179-amino acid mature protein that is believed to migrate via lateral diffusion into ER subcompartments, to mitochondria and onto the surface of lipid droplets [11–13]. Core binds to lipid droplets via an amphipathic helix turn helix motif and mutations that prevent trafficking of the core protein to lipid droplets strongly inhibit virus assembly [5,6,14]. During virus assembly, the core protein must be retrieved from the surface of lipid droplets to the site of viral budding at the opposing ER membrane and interaction between viral NS2 and NS3-4A is essential for this recruitment process [15]. Colocalization of the envelope proteins with the capsid core, the prerequisite for assembly events, were analyzed by confocal laser scanning microscopy and either not detected [16], abundant [17], or limited to areas adjacent to small lipid droplets [15]. However, visualizing the HCV assembly process was largely unsuccessful so far as confocal laser scanning microscopy lacks the resolution required and electron microscopic analysis is hindered by the rarity of HCV assembly processes and the heterogeneity of the viral particles [18].

The major drawback of fluorescence microscopy is its limited resolution. The resolution of a microscope is limited by the diffraction barrier to approximately half of the wavelength of the emitted or diffracted light [19]. In three-dimensional fluorescence microscopy like confocal laser scanning microscopy the axial resolution is even worse than the lateral, resulting in a maximal resolution of 200 nm laterally and 500 nm axially for blue emission light [20]. In recent years several microscopy techniques have been developed that overcome the diffraction limit. The most prominent approaches are stimulated emission depletion STED [21], structured illumination microscopy SIM [22], (fluorescence) photoactivated localization microscopy (f)PALM [23,24], and (direct) stochastic optical reconstruction microscopy (d)STORM [25,26]. The latter, (f)PLAM, STORM, and dSTORM, have in common that their resolution improvement is based on the precise localization of single fluorescent molecules, therefore they are sometimes summed up as single molecule localization microscopy or in short localization microscopy. The maximum resolution in localization microscopy (with commercially available microscopes) is 20 nm laterally and 50 nm axially [27], a tenfold resolution improvement in all three dimensions compared to confocal laser scanning microscopy. Therefore, the volume in which colocalization is detected is up to $1000\times$ smaller in localization microscopy than in confocal microscopy resulting in much more precise colocalization analyses. The improvement in colocalozation precision has recently been shown in 2D [28]. Localization microscopy has been used successfully to get new insights in structural details or infection processes of pathogens like bacteria [29,30], plant infecting fungi [31], and human pathogenic viruses like HIV [32–34].

Here, for the first time simultaneous three-color 3D dSTORM is employed to study a viral infection. We visualize and analyze the spatial distribution of HCV structural proteins core and E2 in vicinity of lipid droplets, the putative viral assembly site.

Results

Lipid droplets can be visualized in dSTORM

To visualize lipid droplets and surrounding structures by super-resolution microscopy (dSTORM) we first analyzed different lipophilic fluorescent dyes (BODIPY, LipidTox Green, LipidTox Red) for their ability to blink upon excitation. Of the dyes tested, LipidTox Red showed the best properties for dSTORM on our microscope. We validated the ON and OFF switch of LipidTox Red signal over time. Shown is the signal intensity of the fluorophore over several minutes (Figure 1A). To verify that the signals detected in dSTORM represent lipid droplets we correlated widefield fluorescence microscopy of LipidTox Red stained Huh7 Lunet cells with the corresponding dSTORM image. The signals detected with each method overlapped almost completely, with the dSTORM signal usually fitting in the middle of larger signals detected by widefield microscopy (Figure 1B). The lipid droplets we observe in dSTORM mode are generally small (<0.5 μm in diameter) as they are only detectable up to 2 μm axial distance from the coverslip. The large lipid droplets additionally observed in widefield images are mainly more than 2 μm above the coverslip (Figure S1), therefore of those large lipid droplets we only detect the edge in dSTORM mode (Figure S1, arrows). Lipid droplets in hepatoma cells can also be detected by staining with antibodies directed against adipose differentiation-related protein (ADRP)/perilipin 2, that is strongly expressed in liver cells and associates with lipid droplets. In two-dimensional dSTORM images ADRP signals either overlapped with lipid droplets stained with LipidTox Red or tightly surrounded them (Figure 1C, inlays). Therefore, LipidTox Red is suitable for staining lipid droplets in dSTORM experiments.

Construction and validation of JFH1[Flag-E2] and Jc1[Flag-E2] infectious HCV clones

To study the subcellular localization of core and E2, we constructed Flag-E2-tagged variants of two infectious HCV clones, the original isolate that replicates in cultured hepatoma cells (JFH1), and a genotype 2a/2a chimera (Jc1) that produces high titers of infectious particles (Figure 2A). To produce infectious HCVcc particles, Huh7.5 cells were transfected with *in vitro* transcribed viral RNA and expression of the viral proteins core and Flag-E2 was verified by western blotting (Figure 2B). Transfected cells produced progeny particles with infectious titers reaching up to 4×10^4 $TCID_{50}$/ml for JFH1[Flag-E2]- and 4×10^7 $TCID_{50}$/ml for Jc1[Flag-E2]-infected Huh7.5 cells (Figure 2C). Culture supernatant was then used to infect Huh7 Lunet cells, a subtype of Huh7 cells that is highly permissive to HCV infection and has a superior flat morphology than Huh7.5 cells for the subsequent microscopy experiments [35]. To verify the suitability of anti-Flag staining to localize Flag-E2 we performed co-immunostaining with Flag and E2 antibodies. Signals for Flag and E2 overlapped almost completely (Pearson Coefficient r_P (Jc1[Flag-E2]) = 0.92; r_P (JFH1[Flag-E2]) = 0.852), indicating the suitability of the Flag antibodies to localize Flag-E2 protein (Figure 2D).

Confocal microscopy of JFH1[Flag-E2]- and Jc1[Flag-E2]-infected cells

Infected cells were then processed for immunostaining with antibodies directed against Flag (E2) and core. Previously it has been suggested that different permeabilization methods might influence lipid droplet localization of select lipid droplet binding proteins [36]. Therefore, we compared two different permeabilization methods, 5 min 0.1% Triton X-100 vs. 5 min 0.5% saponin and addition of 0.1% saponin to the antibody staining solutions. Following the incubation with primary and secondary antibodies, cells were stained with LipidTox Red to visualize lipid droplets and analyzed by confocal microscopy. Signal intensities of core were stronger in JFH1[Flag-E2]- than in Jc1[Flag-E2]-infected cells with core partially localizing to lipid droplets Jc1[Flag-E2]-infected cells and tightly surrounding larger lipid droplets in JFH1[Flag-E2]-infected cells (figure S2). Flag-E2 displayed a predominantly reticular localization and partial colocalization with core at lipid droplets. We measured the degree of colocalization between the two viral proteins according to Manders. The Manders overlap coefficient indicates the portion of the intensity in each channel that overlaps with some intensity in the other channel and is accordingly calculated for each fluorophore. Flag-E2 partially colocalized with core (M1(Jc1[Flag-E2]) = 0.31±0.15, M1(JFH1[Flag-E2]) = 0.45±0.14), while most of core colocalized with Flag-E2 (M2(Jc1[Flag-E2]) = 0.97±0.04, M2(JFH1[Flag-E2]) = 0.87±0.17). We did not observe statistically significant differences in colocalization of core with Flag-E2 between the two viral strains. In our hands both permeabilization methods yielded similar results in confocal microscopy, although permeabilization with Triton X-100 resulted in stronger fluorescence signals than permeabilization with saponin (figure S2).

Super-resolution microscopy of Jc1[Flag-E2]- and JFH1[Flag-E2]-infected cells

Super-resolution datasets were then acquired on a custom modified Nikon N-STORM microscope in dSTORM mode [26]. For 3D dSTORM we used astigmatism imaging as described [27]. Super-resolution images were reconstructed from a series of 10,000–15,000 widefield images per channel using the N-STORM v. 2.0 module of NIS Elements AR v. 4.0. (Nikon). We next

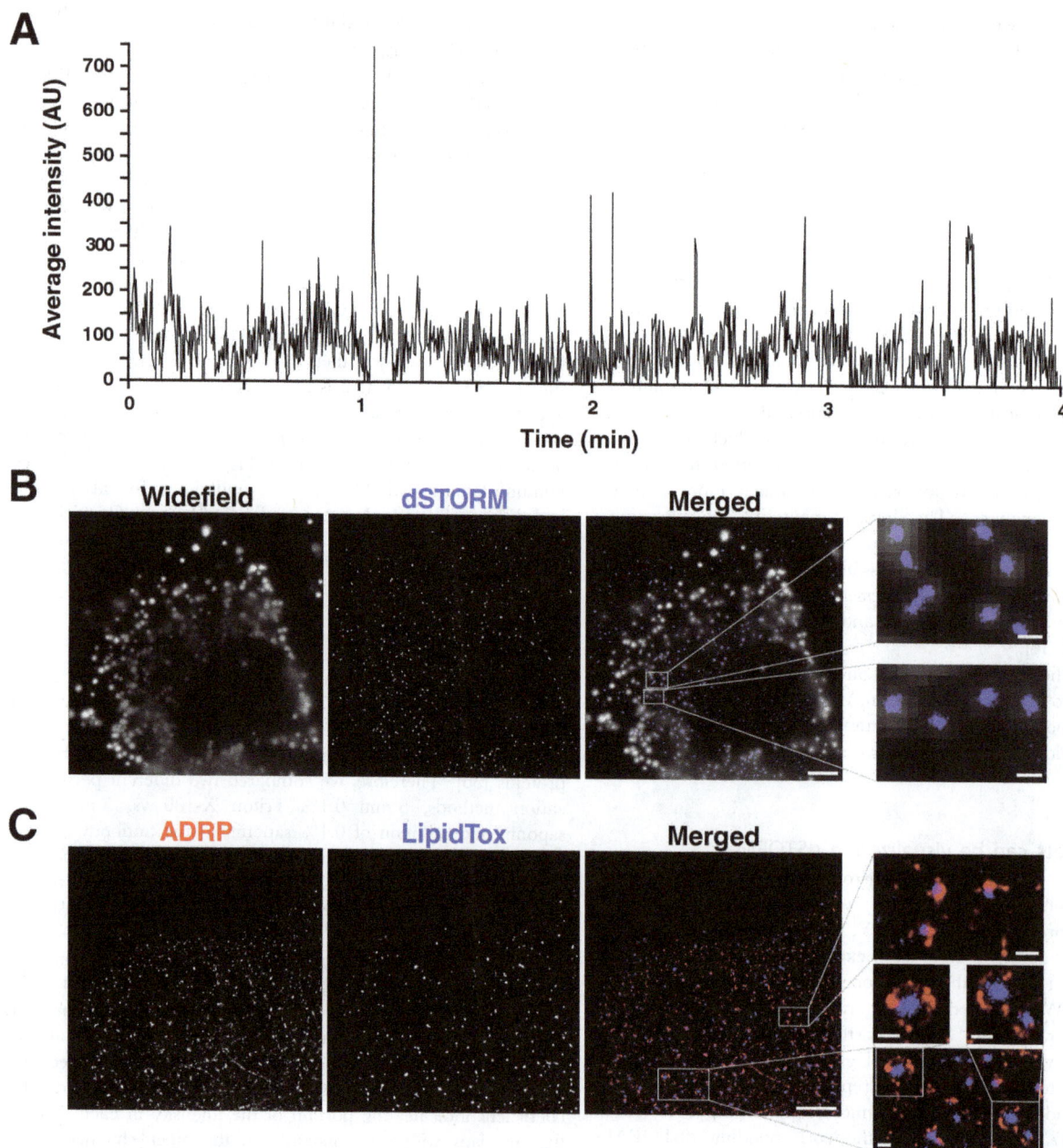

Figure 1. Visualization of lipid droplets in dSTORM. (**A**) Huh7 Lunet cells were incubated with LipidTox Red neutral lipid stain and analyzed by dSTORM. Time-dependent switch between "ON" and "OFF" states of LipidTox Red, shown is the signal of a dSTORM time series. (**B**) Huh7 Lunet cells were incubated with LipidTox Red neutral lipid stain and analyzed by widefield microscopy and dSTORM. Single channels are shown in black and white; the merged image is pseudocolored with widefield in white and dSTORM in blue (scale bar 5 μm, in magnified images 0.5 μm). (**C**) dSTORM of Huh7 Lunet cells after immunostaining with antibodies directed against endogenous ADRP/perilipin 2 and staining of lipid droplets with LipidTox red. Single channels are shown in black and white; the merged image is pseudocolored with ADRP in red and LipidTox in blue (scale bar 5 μm, in magnified images 0.5 μm).

defined cytosolic regions of interests (ROIs) that were reconstructed in 3D with the axial position information encoded according to the color-coded bar. Shown are example ROIs of 3 independent experiments with Triton X-100–permeabilized and one experiment with saponin-permeabilized cells of Huh7 Lunet cells infected either with Jc1$^{Flag-E2}$ or JFH1$^{Flag-E2}$ (Figure 3 and Figures S3, S4). For each ROI the projected 2D image and the merge as well as the color-coded 3D image are shown. Core staining was generally confined to more distinct areas than the Flag-E2 staining (Figure 3 and Figures S3, S4). Lipid droplets appeared as tight

round spots when located close to the coverslip but displayed more scattered signal distribution when further apart from the coverslip, which usually represents the edge of larger droplets (Figure S1). When we compared the two viral strains or the two different permeabilization methods we could not detect any obvious differences. To determine if there are differences in the degree of colocalization of the viral proteins between the two viral strains we calculated the degree of colocalization according to Manders (measuring the portion of the intensity in each channel that coincides with some intensity in the other channel). Overall

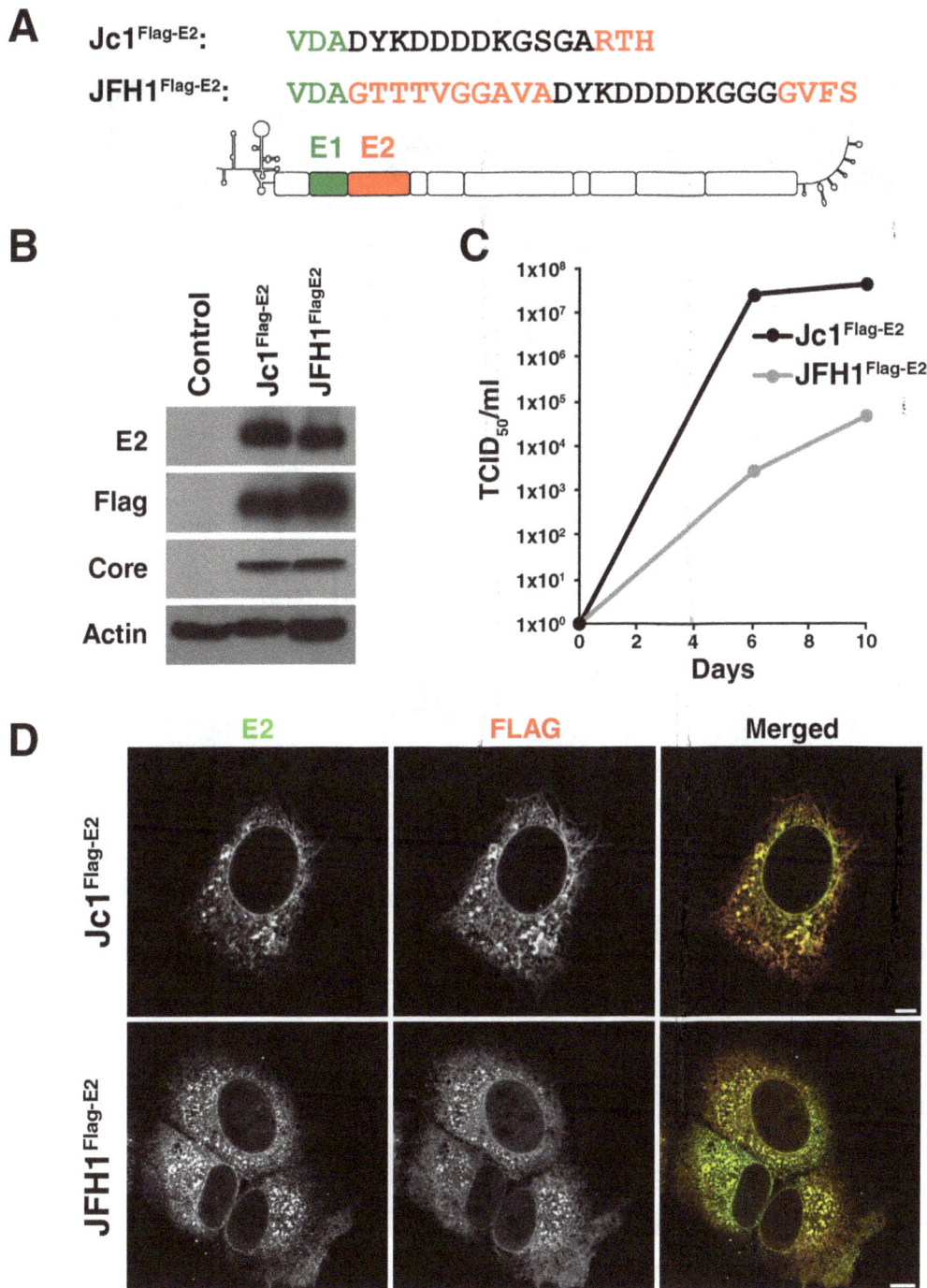

Figure 2. Characterization of Jc1^Flag-E2^ and JFH1^Flag-E2^ infectious HCVcc constructs. (**A**) Scheme of the JFH1^Flag-E2^ and Jc1^Flag-E2^ constructs used in this study. (**B**) Western blot analysis of Huh7.5 cells transfected with *in vitro* transcribed JFH1^Flag-E2^ and Jc1^Flag-E2^ RNA with anti-Flag, anti-E2, anti-core, anti-actin antibodies. (**C**) TCID$_{50}$ of Huh7.5 cells transfected with *in vitro* transcribed JFH1^Flag-E2^ and Jc1^Flag-E2^ RNA. (**D**) Confocal microscopy of Huh7 Lunet cells infected with JFH1^Flag-E2^ and Jc1^Flag-E2^ viral stocks after immunostaining with antibodies recognizing E2 and Flag. Single channels are shown in black and white; the merged image is pseudocolored with ADRP in red and LipidTox in blue (scale bars 5 μm).

colocalization coefficients were very low reflecting the high resolution of the dSTORM images. While the levels of colocalization of core with lipid droplets and colocalization of Flag-E2 with core were similar in both viral strains, Flag-E2 displayed slightly less colocalization with lipid droplets in JFH1^Flag-E2^- versus Jc1^Flag-E2^-infected cells, indicating more efficient recruitment of Flag-E2

protein to lipid droplets in cells infected with a viral strain that produces higher titers of progeny virions (Figure 3B).

Analysis of the 3D distribution and colocalization of core and Flag-E2 at lipid droplets

We then analyzed the areas surrounding lipid droplets in regard to the localization of core and Flag-E2 in 3D. Of note, we only

A

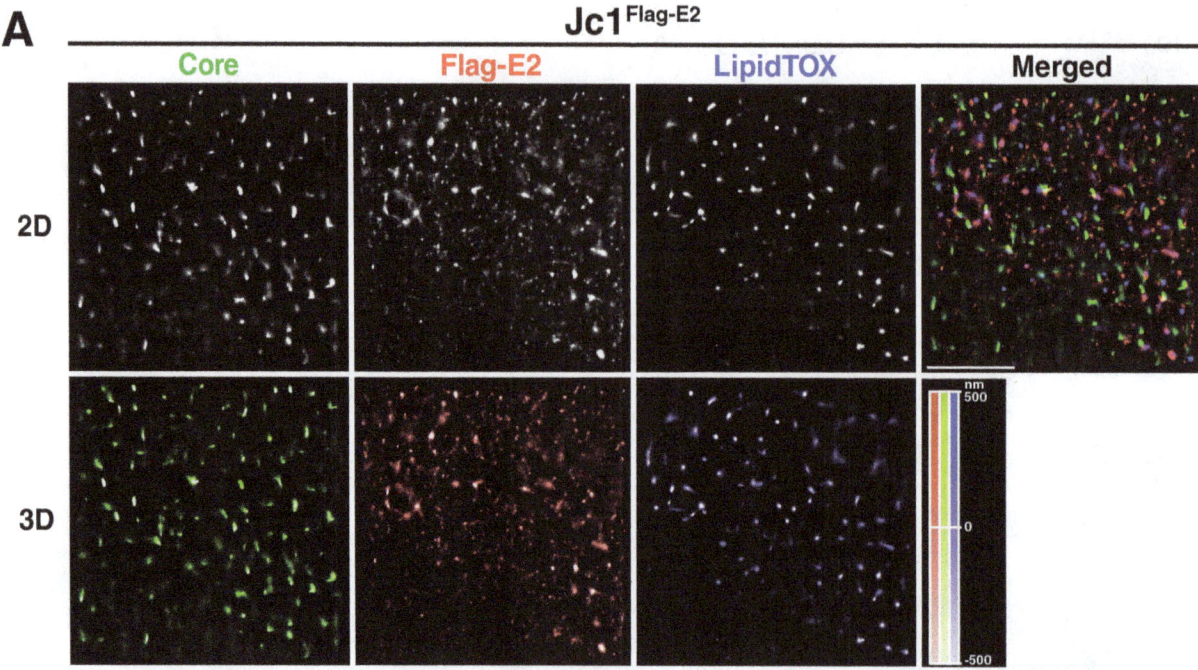

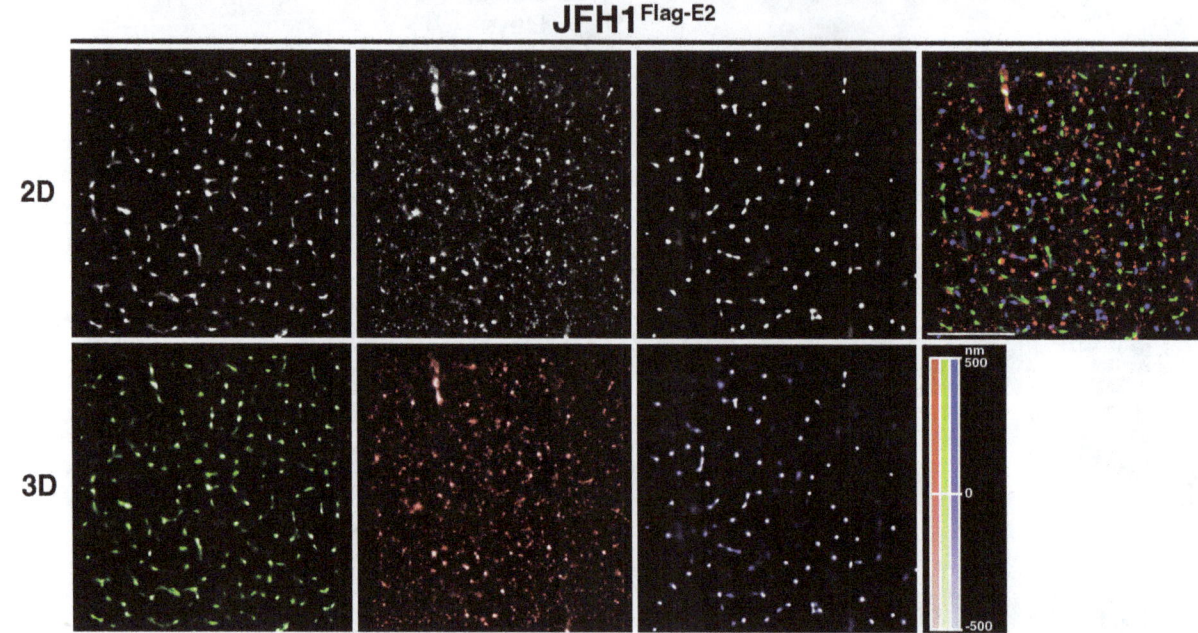

B

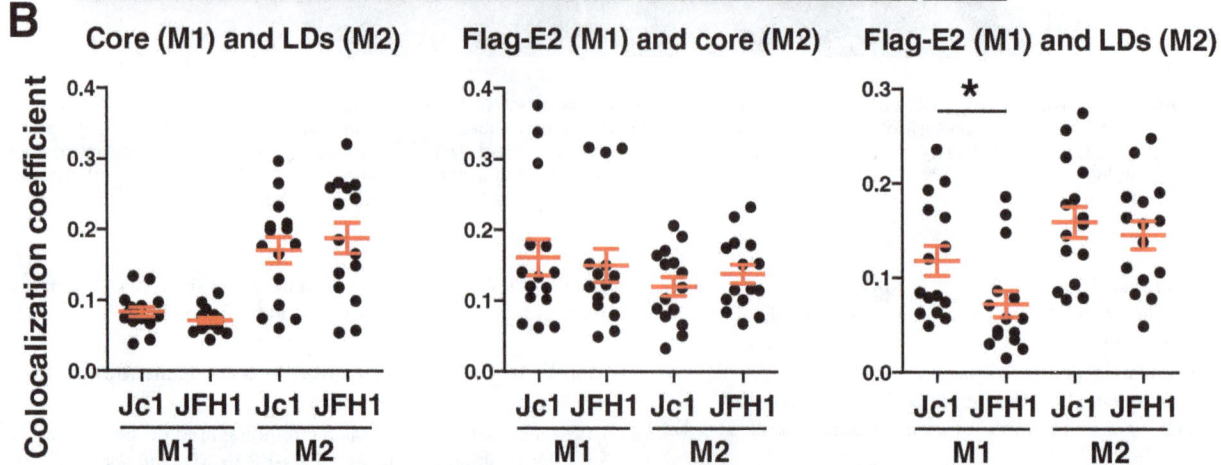

Figure 3. 2D and 3D dSTORM images of core, Flag-E2, and lipid droplets in JFH1$^{Flag-E2}$- and Jc1$^{Flag-E2}$-infected cells. Huh7 Lunet cells were infected with JFH1$^{Flag-E2}$ and Jc1$^{Flag-E2}$ viral stocks and processed for immunofluorescence staining. Cells were permeabilized with Triton X-100 followed by staining with anti-core and anti-Flag antibodies and LipidTox Red. 3D super-resolution datasets were acquired using astigmatism imaging. dSTORM images were reconstructed from a time series of 10,000–15,000 raw images per channel (A) Shown are the single channel 2D projections and the merge and 3D images with the axial position color-coded according to the scale on the right. Scale bars of the x–y image represent 5 μm. (B) dSTORM datasets were analyzed for the degree of colocalization using the JACoP (Just Another Colocalization Plugin) plugin for Image J [42]. We analyzed cytosolic regions of interest in 2D of 4 independent experiments and calculated the degree of colocalization using the Manders colocalization coefficient (mean ± sem, *p<0.05, unpaired two-tailed student's t-test).

analyzed lipid droplets where both viral proteins were detected in close proximity. Lipid droplet localization was scored as follows: A) core and Flag-E2 adjacent to lipid droplets, B) Flag-E2 adjacent to core, core adjacent to lipid droplets, and C) core adjacent to Flag-E2, Flag-E2 adjacent to lipid droplets. Furthermore, the degree of colocalization between the viral proteins was scored as I) complete, II) partial, or III) no colocalization. Analysis of the colocalization between core and Flag-E2 in the area surrounding lipid droplets revealed that in Jc1$^{Flag-E2}$-infected cells there is only at around 30% of the lipid droplets a partial colocalization. In contrast, in JFH1$^{Flag-E2}$-infected cells approximately 70% of the lipid droplets showed colocalization, with ~10% of lipid droplet surrounding areas showing complete colocalization between core and Flag-E2 (Figure 4). Additionally, in both Jc1$^{Flag-E2}$- and JFH1$^{Flag-E2}$-infected cells core was adjacent to lipid droplets in 70–80% of the analyzed lipid droplets, with Flag-E2 co-coating lipid droplets in 50% (Jc1$^{Flag-E2}$) and 80% (JFH1$^{Flag-E2}$) of the cases (Figure 4). So even though more Flag-E2 is recruited to the sites of viral assembly (Figure 3B), when analyzing individual lipid droplets in 3D less lipid droplets are surrounded by colocalizing core and Flag-E2 proteins.

Three-dimensional reconstruction of core and Flag-E2 localization in areas surrounding lipid droplets

Next we visualized the spatial distribution of viral proteins surrounding lipid droplets by calculating the volumes occupied by each protein and of the lipid droplets using Imaris software. Depicted are three examples of Jc1$^{Flag-E2}$- and of JFH1$^{Flag-E2}$-infected cells, each showing the stacks that were used to calculate the localization volumes, then the volumes of all three channels, followed by core or Flag-E2 and the lipid droplet volume and finally the volume of colocalization between core and Flag-E2 (in yellow) and the lipid droplet volume (Figure 5 and 6).

In Jc1$^{Flag-E2}$-infected cells core usually was found in distinct spots adjacent to the lipid droplet volume sometimes partially colocalizing with Flag-E2 or on opposing sites of the lipid droplet (Figure 5). The volumes of colocalization were usually around or below 100 nm in diameter. In contrast, in JFH1$^{Flag-E2}$-infected cells core occupied more space in the areas surrounding lipid droplets and the colocalization volumes with Flag-E2 were larger (Figure 6). Overall Flag-E2 occupied volumes that were more scattered than the ones occupied by core.

Discussion

In this study we performed simultaneous super-resolution microscopy of two viral proteins (core and E2) together with lipid droplets in 3D. We first identified a lipophilic dye suitable for dSTORM imaging and validated the staining with antibodies directed against a lipid droplet-binding protein. To achieve a photoblinking of the fluorophores that was suited for dSTORM the TIRF angle was adjusted to oblique incidence excitation [37]. In this case a thickness of 1–2 μm of the sample is illuminated with maximum light intensity facilitating the blinking of the fluorophores. We observed that within this distance lipid droplets are generally small (0.2–0.5 μm in diameter). In corresponding widefield images larger droplets that are above this area are also

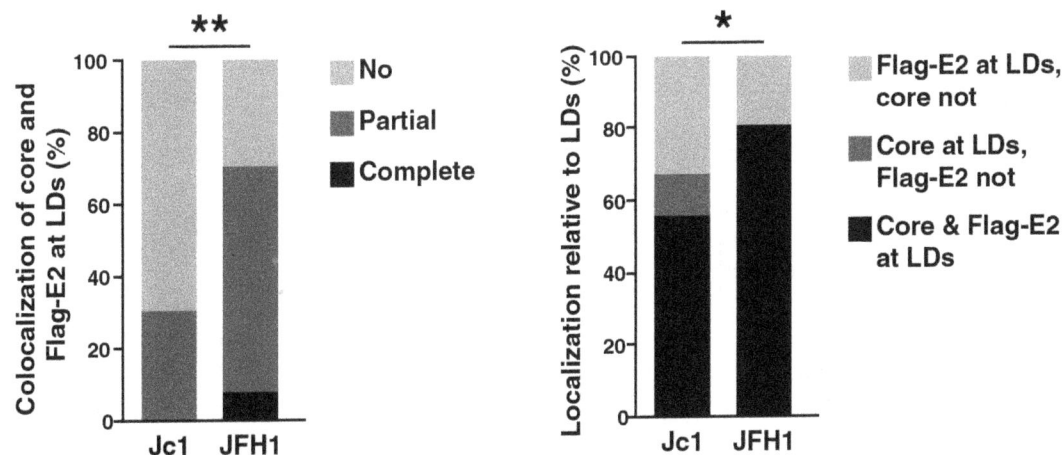

Figure 4. Analysis of the 3D distribution and colocalization of the viral proteins at lipid droplets. dSTORM datasets were analyzed for the distribution and colocalization of core and Flag-E2. Cytosolic regions of interest were reconstructed in 3D and lipid droplets with core and Flag-E2 staining in close proximity were analyzed as follows: lipid droplet localization was scored A) core and Flag-E2 adjacent to lipid droplets, B) Flag-E2 adjacent to core, core adjacent to lipid droplets, and C) core adjacent to Flag-E2, Flag-E2 adjacent to lipid droplets. The degree of colocalization between the viral proteins was scored as I) complete, II) partial, or III) no colocalization. We analyzed reconstructed images from three independent experiments of Triton X-100–permeabilized cells and one experiment of saponin-permeabilized cells with a total amount of 40 lipid droplets in Jc1$^{Flag-E2}$-infected cells and 36 lipid droplets in JFH1$^{Flag-E2}$-infected cells. Colocalization and localization relative to lipid droplets was significantly different between cells infected with the two viral strains (*p<0.05, **p<0.01, chi-square test).

Jc1^Flag-E2 (Core / Flag-E2 / LipidTOX)

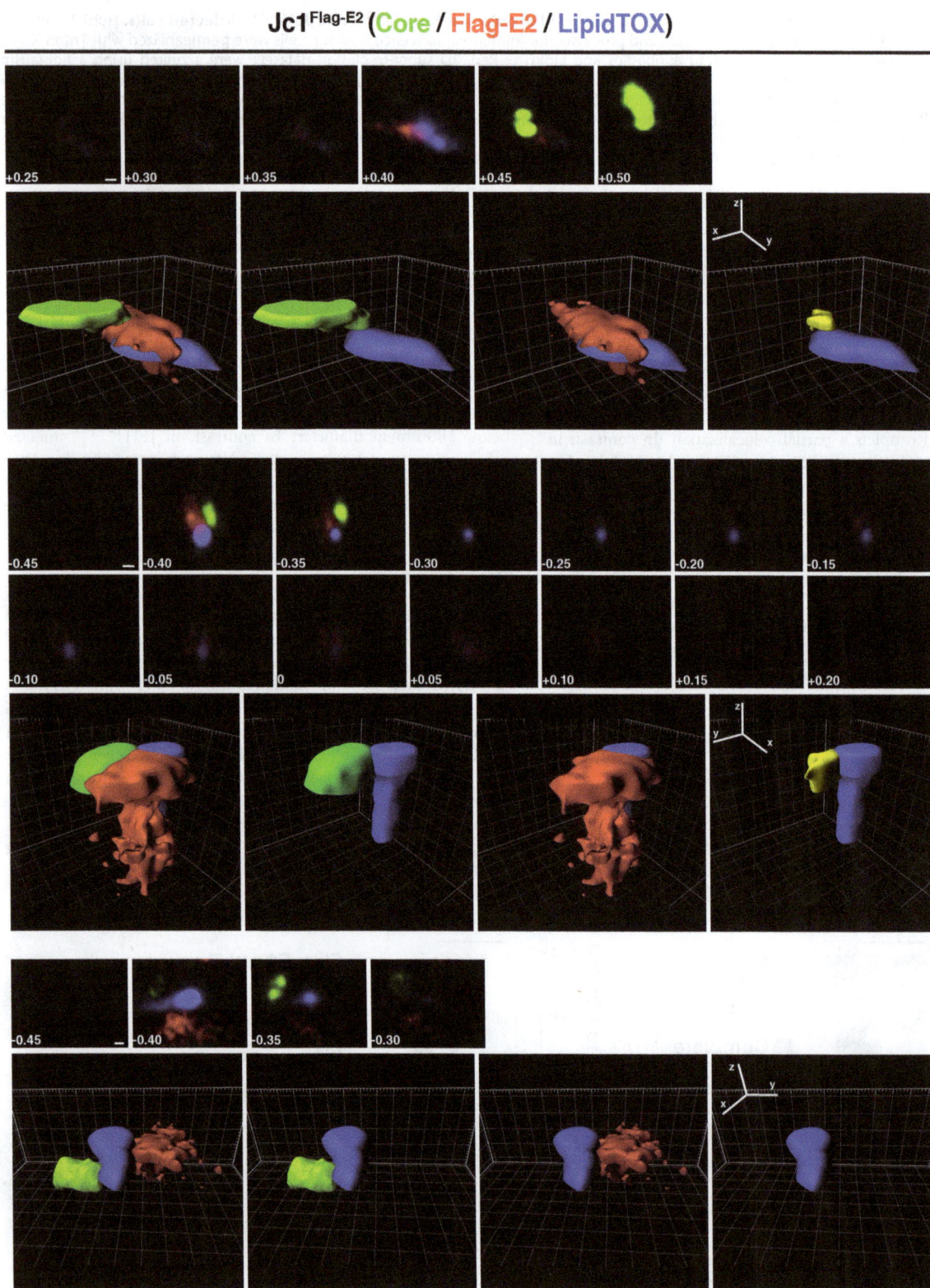

Figure 5. 3D reconstitution of core, Flag-E2, and lipid droplets in Jc1^Flag-E2-infected cells. Representative 3D reconstructions of core and Flag-E2 spatial localization and colocalization around lipid droplets. Shown are three examples of Jc1^Flag-E2-infected cells, first the image stacks used for the 3D reconstruction and the 3D reconstruction of all three channels, then core or Flag-E2 and the lipid droplet volume and finally volume of colocalization between core and Flag-E2 and the lipid droplet volume. Scale bars on the image stacks and ruler hatch marks in the 3D images are 100 nm.

JFH1^Flag-E2 (Core / Flag-E2 / LipidTOX)

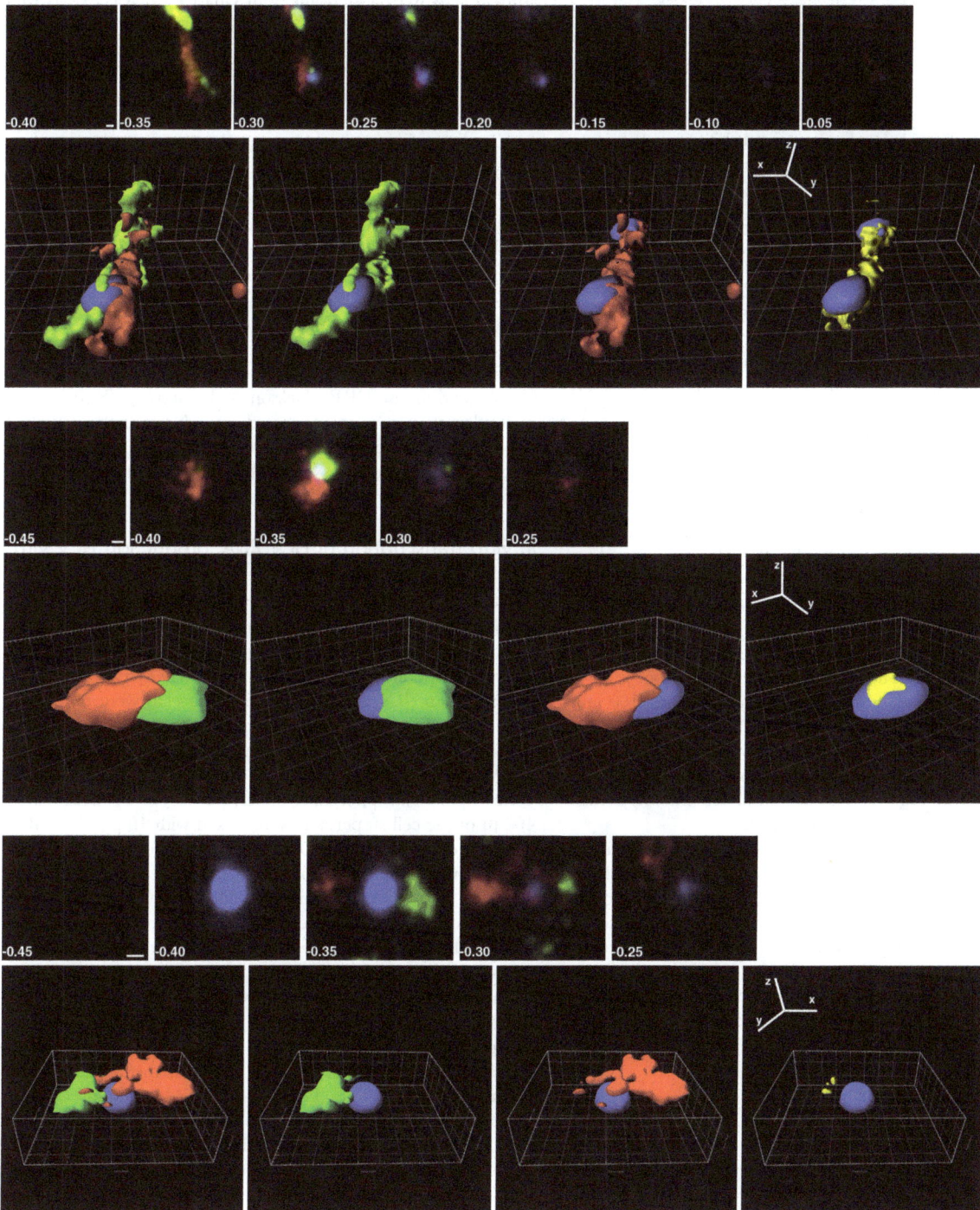

Figure 6. 3D reconstitution of core, Flag-E2, and lipid droplets in JFH1^Flag-E2-infected cells. Representative 3D reconstructions of core and Flag-E2 spatial localization and colocalization around lipid droplets. Shown are three examples of JFH1^Flag-E2-infected cells, first the image stacks used for the 3D reconstruction and the 3D reconstruction of all three channels, then core or Flag-E2 and the lipid droplet volume and finally volume of colocalization between core and Flag-E2 and the lipid droplet volume. Scale bars on the image stacks and ruler hatch marks in the 3D images are 100 nm.

visible. Edges of these larger droplets could be detected in dSTORM mode indicated by a scattered staining pattern just below the maximal distance from the coverslip. Well-resolved lipid droplets generally appeared as round to oval shaped objects.

We then analyzed the spatial distribution of core and Flag-E2 of two viral strains (Jc1^Flag-E2 vs JFH1^Flag-E2) that differ in the amount of progeny virions they produce. As lipid droplets are the putative assembly sites of HCV we hypothesized to observe distinct differences in the distribution of core and Flag-E2 (as structural proteins they are components of the progeny virions). As noted above we could only analyze rather small lipid droplets, but it has been suggested by confocal microscopic analyses and live cell imaging that these small lipid droplets may represent viral assembly sites [15]. We reconstructed and analyzed the 3D distribution in four independent experiments of cells permeabilized with two different methods. Staining patterns and colocalization events were similar when cells were permeabilized with Triton X-100 or saponin. Over all images acquired, slightly more Flag-E2 colocalizes with lipid droplets in Jc1$^{Flag-E2}$-infected cells indicating that the recruitment process of the envelope proteins to the putative sites of assembly is more efficient in cells infected with a strain capable of efficient progeny virus production. All other colocalization analyses did not reveal any significant differences.

We next focused our analysis on lipid droplets that had both viral proteins present in close proximity. As noted above colocalization events in projected 2D images were rather infrequent. It has recently been shown that the amount of colocalization decreases in dSTORM compared to confocal microscopy due to the improved resolution [28]. In line with these results, when we analyzed the spatial distribution in 3D only very small areas of colocalization volumes were observed. In addition, the frequency of such core Flag-E2 colocalization events in cells was very low, which could reflect the low number of assembly events in HCV-infected cells. We observed even less colocalization of the capsid protein core with the envelope E2 in cells that produce higher amounts of progeny virions, indicating that assembled virions are transported from the site of assembly or that in the absence of viral assembly, core accumulates and colocalizes with Flag-E2. The volumes of colocalization in Jc1$^{Flag-E2}$-infected cells are generally very tiny, with only around 100 nm in diameter. Within these volumes of colocalization we could not resolve any specific distribution pattern as we are very close to the limit of resolution. In addition, colocalization events were so rare that we could not average the localization of the proteins across colocalization areas of different lipid droplets.

The limitations of our approach are that we can only analyze the space of 2 μm above the coverslip and are therefore unable to analyze the bigger droplets that are, at least in JFH1-infected cells, completely core-coated when analyzed by confocal microscopy. However, the methodology could, combined with correlative electron microscopy, enable finding putative HCV assembly sites in cells by dSTORM, defined through the presence of colocalizing core and Flag-E2 volumes, and thus facilitate analysis of relevant areas in electron microscopy.

Materials and Methods

Plasmids

Flag-tagged E2 expressing JFH1 and Jc1 virus constructs were essentially cloned as described [38,39] by overlap extension PCR using the following primers: BsiWI_J6 sense GGA CAT GAT GAT GAA CTG G, E2_ Flag as CCC TTG TCA TCG TCG TCC TTG TAG TCC GCG TCC ACC CCG GCG GCC, E2_ Flag sense GGA CGA CGA TGA CAA GGG ATC AGG AGC ACG CAC CCA TAC TGT TGG GGG, NS2_JFH1 Not as CCATCGCGGCCGCCGCGCAC and BsiWI_J6 sense GGA CAT GAT GAT GAA CTG G, JFH1 E2_ Flag AS CTT GTC ATC GTC ATC CTT GTA ATC TGC AAC AGC GCC TCC AAC GGT GG, JFH1 E2_ FLAG sense AAG GAT GAC GAT GAC AAG GGA GGC GGT GGC GTG TTC AGC CAT GGC CC, NS2_JFH1 Not as CCA TCG CGG CCG CCG CGC AC (Flag tag underlined).

Cell lines and culture conditions

Huh7 Lunet cells were a kind gift from Ralf Bartenschlager, Huh7.5 cells were a kind gift from Charles M. Rice. Cells were grown under standard cell culture conditions in high glucose DMEM supplemented with 10% FBS (Biochrom Superior). JFH1$^{Flag-E2}$ and Jc1$^{Flag-E2}$ viral stocks were produced in Huh7.5 cells, while Huh7 Lunet were used for the microscopy studies due to their superior properties for immunofluorescence microscopy [35]. Of note, viral RNA replication and infectious virus release are similar in both cell lines, but Huh7.5 express higher levels CD81 than Huh7 Lunet cells [40].

Antibodies and reagents

The following antibodies were obtained commercially: anti-core (clone C7-50, Affinity BioReagents), anti-NS5A (clone 2F6/G11, IBT), anti-Flag (F7425, Sigma), anti-ADRP (ab52355, Abcam), anti beta-actin (Sigma), anti-mouse Alexa 488 (Invitrogen), anti-rabbit Alexa 647 (Invitrogen), anti-rabbit-HRP (Jackson Laboratories), anti-mouse-HRP (Jackson Laboratories). Enzymes for molecular cloning were purchased from New England Biolabs, cell culture reagents from Gibco, and fine chemicals, if not noted otherwise, from Sigma. Neutral lipids were stained with HCS LipidTOX Red (H34476, Invitrogen).

In vitro transcription of HCV RNA, production HCV virus stocks, and titer test

HCV viral stocks were prepared as described [7]. Briefly, plasmids encoding JFH1$^{Flag-E2}$ and Jc1$^{Flag-E2}$ viruses were linearized with SspI and purified by phenol-chloroform extraction. *In vitro* transcription was carried out using the MegaScript T7 kit (Ambion). For RNA transfection, Huh7.5 cells were trypsinized, washed once in Opti-MEM (Invitrogen), and resuspended in Cytomix buffer (120 mM KCl, 5 mM MgCl$_2$, 0.15 mM CaCl$_2$, 2 mM EGTA, 1.9 mM ATP, 4.7 mM GSH, 25 mM HEPES, 10 mM potassium phosphate buffer, pH 7.6) at 10^7 cells per ml. 400 μl of the cell suspension were mixed with 10 μg HCV RNA and pulsed at 260 V and 950 μF using the Gene Pulser II (Biorad). Culture supernatant of Huh7.5 cells transfected with JFH1$^{Flag-E2}$ and Jc1$^{Flag-E2}$ RNA was harvested, filtered and concentrated by polyethylene glycol 8000 precipitation. For infection experiments, naïve Huh7 Lunet cells were incubated with virus preparations for 3 h at 37°C.

Virus titration was performed by seeding Huh 7.5 cells in 96-well plates and infecting them with serial dilutions of culture supernatants. After 3–4 days, cells were washed with phosphate-buffered saline (PBS), fixed for 10–20 min with methanol at −20°C, permeabilized with 0.1% Triton X-100 and blocked for 1 hour in blocking solution (5% BSA, 1% fish skin gelatin, 50 mM Tris in PBS). Endogenous peroxidase was blocked by a 5 min incubation with PBS containing 0.3% (v/v) hydrogen peroxide. After overnight incubation with anti-NS5A antibodies in blocking solution at 4°C, cells were washed and incubated with anti-rabbit-HRP antibodies for 1 h and subsequently stained with diaminobenzidine (DAB, Roche) substrate for 10 min. Nuclei were counterstained with hematoxylin.

Western blotting

For western blot analysis cells were lysed in RIPA buffer (1% NP-40, 0.5% sodium deoxycholate, 0.1% SDS in PBS supplemented with protease inhibitor cocktail (Sigma)) and mixed with Laemmli buffer for SDS-PAGE. For chemiluminescent detection we used ECL Lumilight (Roche) and ECL Hyperfilm (Amersham).

Immunofluorescence and HCS LipidTox Red neutral lipid staining

Cells grown on coverslip or in labtec II chamber slides (Nunc) were fixed in 4% paraformaldehyde for 1 h at room temperature, washed with PBS, and permeabilized in 0.1% Triton X-100 for 5 min or 0.5% saponin for 5 min. After incubation in blocking solution (5% BSA, 1% fish skin gelatin, 50 mM Tris in PBS) for 1 h, cells were incubated with primary antibodies in blocking solution for 1 h, washed and incubated with secondary antibodies for 1 h. When cells were permeabilized with saponin we added 0.1% saponin to the antibody staining solutions. For lipid droplet staining, fixed cells were stained for 20 min with HCS LipidTox Red neutral lipid stain (Invitrogen), diluted 1000-fold in PBS solution. Coverslips were embedded in Mowiol (Calbiochem) mounting medium [41].

Confocal microscopy

CLSM was either performed on a Nikon C2+ confocal laser scanning microscope or on a LSM 510 (Zeiss) confocal laser scanning microscope. The Nikon C2+ microscope was equipped with four lasers: a 405 nm diode laser (100 mW, Coherent Inc.), a 488 nm DPSS Laser (10 mW, Melles Griot GmbH), a 543 nm HeNe laser (5 mW, Melles Griot GmbH), and a 642 nm diode laser (45 mW, Melles Griot GmbH). The intensity of the laserlight was controlled by an acusto optical tunable filter (AOTF). We used a 60× violet corrected oil objective with a NA of 1.4 for imaging (Plan Apo VC 60× H, Nikon). The LSM 510 microscope was equipped with three lasers: a multiline argon laser (458 nm, 477 nm, 488 nm, 514 nm, 30 mW, LASOS Lasertechnik), a 543 nm HeNe laser (1 mW, LASOS Lasertechnik) and a 633 nm HeNe laser (5 mW, LASOS Lasertechnik). The intensity of the laser light was controlled by an acusto optical tunable filter (AOTF). For imaging a 63× objective with a NA of 1.4 (Plan Apochromat 63×, Zeiss) was used.

dSTORM and image analysis

Datasets for dSTORM were acquired on a custom modified Nikon N-STORM microscope equipped with an Apo TIRF 100× oil immersion objective with a numerical aperture of 1.49 (Nikon), an electron multiplying charge-coupled device (EMCCD) camera (iXon+ DU-897, Andor Technology) and a quadband filter composed of a quad line beamsplitter (zt405/488/561/640rpc TIRF, Chroma Technology Corporation) and a quad line emission filter (brightline HC 446, 523, 600, 677, Semrock, Inc.).

For excitation of the fluorophores the following lasers were used: for Alexa Fluor 647 a 647 nm continuous wave visible fiber laser (2RU-VFL-P-300-647 MPB Communications Inc.), for Lipid Tox Red a 150 mW 561 nm optically pumped semiconductor laser (Sapphire 561 LP, Coherent, Inc.), and for Alexa Fluor 488 the 488 nm line of an argon gas laser (35-IMA-840-019, Melles Griot GmbH). A 405 nm diode laser (CUBE 405-100C, Coherent Inc.) was used for switching back the fluorophores from the dark to the fluorescent state. For multicolour imaging the lasers were switched on and off alternately controlled by an acoustooptic optical tunable filter (AOTF). The integration time of the EMCCD camera was set to 50–70 ms per frame with an EM gain of 300.

The TIRF angle was adjusted to oblique incidence excitation as described [37]. This allowed super-resolution imaging 1–2 μm deep into the sample. The focus was kept stable during acquisition using Nikon's perfect focus system.

Super-resolution images were reconstructed from a series of 10,000–15,000 images per channel using the N-STORM analysis module v. 2.0.0.76 of NIS Elements AR v. 4.00.07 (Laboratory imaging s.r.o.). 3D super-resolution microscopy was performed using an astigmatism-based approach as described [27]. For each color

channel a separate 3D calibration curve was used. 3D super-resolution images were visualized using Bitplane Imaris 7.6.1. software.

The imaging was performed in LabTek II chamber slides using an imaging buffer containing 100 mM β-Mercaptoethanolamin (MEA) as described [37] to facilitate a sufficient blinking of the fluorophores.

Supporting Information

Figure S1 In dSTORM mode only small lipid droplets can be visualized. Huh7 Lunet cells infected with Jc1$^{Flag-E2}$ viral stocks were incubated with LipidTox Red neutral lipid stain and analyzed by dSTORM and widefield microscopy. Shown are the single channel 2D and 3D images with the axial position color-coded according to the scale on the right (same image as in Figure 3A upper panel) together with the corresponding widefield image. Scale bar represents 5 μm. Arrows highlight the large lipid droplets observed in widefield microscopy. In dSTORM mode only the lowest edge of those large lipid droplets is visible due to oblique incidence illumination.

Figure S2 Confocal microscopy of JFH1$^{Flag-E2}$- and Jc1$^{Flag-E2}$-infected cells. Huh7 Lunet cells were infected with JFH1$^{Flag-E2}$ and Jc1$^{Flag-E2}$ viral stocks and processed for immunofluorescence staining. Cells were either permeabilized with Triton X-100 (A) or saponin (B) followed by staining with anti-core and anti-Flag antibodies and LipidTox Red. Samples were analyzed by confocal microscopy. Single channels are shown in black and white; the merged image is pseudocolored with core in green, Flag-E2 in red, and LipidTox in blue (scale bars 5 μm). Colocalization was determined using the JACoP Image J plugin.

Figure S3 3D dSTORM images of core, Flag-E2, and lipid droplets in Jc1$^{Flag-E2}$-infected cells. Huh7 Lunet cells were infected with Jc1$^{Flag-E2}$ viral stocks and processed for immunofluorescence staining. Cells were permeabilized with Triton X-100 or saponin followed by staining with anti-core and anti-Flag antibodies and LipidTox Red. Super-resolution datasets were acquired as described above. Shown are the single channel 2D and 3D images with the axial position color-coded according to the scale on the right. Scale bars of the x–y image represent 5 μm.

Figure S4 3D dSTORM images of core, Flag-E2, and lipid droplets in JFH1$^{Flag-E2}$-infected cells. Huh7 Lunet cells were infected with JFH1^{Flag-} viral stocks and processed for immunofluorescence staining. Cells were permeabilized with Triton X-100 or saponin followed by staining with anti-core and anti-Flag antibodies and LipidTox Red. Super-resolution datasets were acquired as described above. Shown are the single channel 2D and 3D images with the axial position color-coded according to the scale on the right. Scale bars of the x–y image represent 5 μm.

Acknowledgments

We thank R. Bartenschlager (University of Heidelberg) for Jc1 constructs and Huh7 Lunet cells, C.M. Rice (Rockefeller University) for Huh7.5 cells, J. McLauchlan (Medical Research Council Virology Unit) for JFH1, T. Wakita (National Institute of Infectious Diseases, Japan) for the JFH1 construct. We thank the Nikon Application Center Northern Germany (Nikon GmbH) for providing the Nikon N-STORM microscope.

Author Contributions

Conceived and designed the experiments: DE KR RR EH. Performed the experiments: DE KR. Analyzed the data: DE KR RR EH. Contributed to the writing of the manuscript: DE KR RR EH.

References

1. Lauer GM, Walker BD (2001) Hepatitis C virus infection. N Engl J Med 345: 41–52.
2. Wakita T, Pietschmann T, Kato T, Date T, Miyamoto M, et al. (2005) Production of infectious hepatitis C virus in tissue culture from a cloned viral genome. Nat Med 11: 791–796.
3. Lindenbach BD, Evans MJ, Syder AJ, Wolk B, Tellinghuisen TL, et al. (2005) Complete replication of hepatitis C virus in cell culture. Science 309: 623–626.
4. Herker E, Ott M (2011) Unique ties between hepatitis C virus replication and intracellular lipids. Trends Endocrinol Metab 22: 241–248.
5. Miyanari Y, Atsuzawa K, Usuda N, Watashi K, Hishiki T, et al. (2007) The lipid droplet is an important organelle for hepatitis C virus production. Nat Cell Biol 9: 1089–1097.
6. Boulant S, Targett-Adams P, McLauchlan J (2007) Disrupting the association of hepatitis C virus core protein with lipid droplets correlates with a loss in production of infectious virus. J Gen Virol 88: 2204–2213.
7. Herker E, Harris C, Hernandez C, Carpentier A, Kaehlcke K, et al. (2010) Efficient hepatitis C virus particle formation requires diacylglycerol acyltransferase-1. Nat Med 16: 1295–1298.
8. Barba G, Harper F, Harada T, Kohara M, Goulinet S, et al. (1997) Hepatitis C virus core protein shows a cytoplasmic localization and associates to cellular lipid storage droplets. Proc Natl Acad Sci U S A 94: 1200–1205.
9. Shi ST, Polyak SJ, Tu H, Taylor DR, Gretch DR, et al. (2002) Hepatitis C virus NS5A colocalizes with the core protein on lipid droplets and interacts with apolipoproteins. Virology 292: 198–210.
10. Camus G, Herker E, Modi AA, Haas JT, Ramage HR, et al. (2013) Diacylglycerol acyltransferase-1 localizes hepatitis C virus NS5A protein to lipid droplets and enhances NS5A interaction with the viral capsid core. J Biol Chem 288: 9915–9923.
11. Hope RG, McLauchlan J (2000) Sequence motifs required for lipid droplet association and protein stability are unique to the hepatitis C virus core protein. J Gen Virol 81: 1913–1925.
12. McLauchlan J, Lemberg MK, Hope G, Martoglio B (2002) Intramembrane proteolysis promotes trafficking of hepatitis C virus core protein to lipid droplets. EMBO J 21: 3980–3988.
13. Schwer B, Ren S, Pietschmann T, Kartenbeck J, Kaehlcke K, et al. (2004) Targeting of hepatitis C virus core protein to mitochondria through a novel C-terminal localization motif. J Virol 78: 7958–7968.
14. Targett-Adams P, Hope G, Boulant S, McLauchlan J (2008) Maturation of hepatitis C virus core protein by signal peptide peptidase is required for virus production. J Biol Chem 283: 16850–16859.
15. Counihan NA, Rawlinson SM, Lindenbach BD (2011) Trafficking of hepatitis C virus core protein during virus particle assembly. PLoS Pathog 7: e1002302.
16. Rouille Y, Helle F, Delgrange D, Roingeard P, Voisset C, et al. (2006) Subcellular localization of hepatitis C virus structural proteins in a cell culture system that efficiently replicates the virus. J Clin Virol 36: S53–S54.
17. Coller KE, Heaton NS, Berger KL, Cooper JD, Saunders JL, et al. (2012) Molecular determinants and dynamics of hepatitis C virus secretion. PLoS Pathog 8: e1002466.
18. Popescu CI, Rouille Y, Dubuisson J (2011) Hepatitis C virus assembly imaging. Viruses 3: 2238–2254.
19. Abbe E (1873) Beiträge zur Theorie des Mikroskops und der mikroskopischen Wahrnehmung. Archiv für mikroskopische Anatomie 9: 413–418.
20. Hell SW (2007) Far-field optical nanoscopy. Science 316: 1153–1158.
21. Hell SW, Wichmann J (1994) Breaking the diffraction resolution limit by stimulated emission: stimulated-emission-depletion fluorescence microscopy. Opt Lett 19: 780–782.
22. Gustafsson MG (2000) Surpassing the lateral resolution limit by a factor of two using structured illumination microscopy. J Microsc 198: 82–87.
23. Betzig E, Patterson GH, Sougrat R, Lindwasser OW, Olenych S, et al. (2006) Imaging intracellular fluorescent proteins at nanometer resolution. Science 313: 1642–1645.
24. Hess ST, Girirajan TP, Mason MD (2006) Ultra-high resolution imaging by fluorescence photoactivation localization microscopy. Biophys J 91: 4258–4272.
25. Rust MJ, Bates M, Zhuang X (2006) Sub-diffraction-limit imaging by stochastic optical reconstruction microscopy (STORM). Nat Methods 3: 793–795.
26. Heilemann M, van de Linde S, Schuttpelz M, Kasper R, Seefeldt B, et al. (2008) Subdiffraction-resolution fluorescence imaging with conventional fluorescent probes. Angew Chem Int Ed Engl 47: 6172–6176.
27. Huang B, Wang W, Bates M, Zhuang X (2008) Three-dimensional super-resolution imaging by stochastic optical reconstruction microscopy. Science 319: 810–813.
28. Flottmann B, Gunkel M, Lisauskas T, Heilemann M, Starkuviene V, et al. (2013) Correlative light microscopy for high-content screening. BioTechniques 55: 243–252.
29. Wang W, Li GW, Chen C, Xie XS, Zhuang X (2011) Chromosome organization by a nucleoid-associated protein in live bacteria. Science 333: 1445–1449.
30. Han JJ, Kunde YA, Hong-Geller E, Werner JH (2014) Actin restructuring during Salmonella typhimurium infection investigated by confocal and super-resolution microscopy. J Biomed Opt 19: 16011.
31. Eggert D, Naumann M, Reimer R, Voigt CA (2014) Nanoscale glucan polymer network causes pathogen resistance. Sci Rep 4: 4159.
32. Pereira CF, Rossy J, Owen DM, Mak J, Gaus K (2012) HIV taken by STORM: super-resolution fluorescence microscopy of a viral infection. Virol J 9: 84.
33. Muranyi W, Malkusch S, Muller B, Heilemann M, Krausslich HG (2013) Super-resolution microscopy reveals specific recruitment of HIV-1 envelope proteins to viral assembly sites dependent on the envelope C-terminal tail. PLoS Pathog 9: e1003198.
34. Van Engelenburg SB, Shtengel G, Sengupta P, Waki K, Jarnik M, et al. (2014) Distribution of ESCRT machinery at HIV assembly sites reveals virus scaffolding of ESCRT subunits. Science 343: 653–656.
35. Shavinskaya A, Boulant S, Penin F, McLauchlan J, Bartenschlager R (2007) The lipid droplet binding domain of hepatitis C virus core protein is a major determinant for efficient virus assembly. J Biol Chem 282: 37158–37169.
36. Ohsaki Y, Maeda T, Fujimoto T (2005) Fixation and permeabilization protocol is critical for the immunolabeling of lipid droplet proteins. HistochemCell Biol 124: 445–452.
37. Dempsey GT, Vaughan JC, Chen KH, Bates M, Zhuang X (2011) Evaluation of fluorophores for optimal performance in localization-based super-resolution imaging. Nat Methods 8: 1027–1036.
38. Merz A, Long G, Hiet MS, Brugger B, Chlanda P, et al. (2011) Biochemical and morphological properties of hepatitis C virus particles and determination of their lipidome. J Biol Chem 286: 3018–3032.
39. Wakita T, Suzuki T, Takahashi H (2009) Production and use of epitope-tagged hepatitis c virus particle. In: Organization WIP, editor.
40. Koutsoudakis G, Herrmann E, Kallis S, Bartenschlager R, Pietschmann T (2007) The level of CD81 cell surface expression is a key determinant for productive entry of hepatitis C virus into host cells. J Virol 81: 588–598.
41. Longin A, Souchier C, Ffrench M, Bryon PA (1993) Comparison of anti-fading agents used in fluorescence microscopy: image analysis and laser confocal microscopy study. J Histochem Cytochem 41: 1833–1840.
42. Bolte S, Cordelieres FP (2006) A guided tour into subcellular colocalization analysis in light microscopy. J Microsc 224: 213–232.

5

A Protective Lipidomic Biosignature Associated with a Balanced Omega-6/Omega-3 Ratio in *fat-1* Transgenic Mice

Giuseppe Astarita[1,2]*[9], Jennifer H. McKenzie[3][9], Bin Wang[3], Katrin Strassburg[4,5], Angela Doneanu[1], Jay Johnson[1], Andrew Baker[1], Thomas Hankemeier[4,5], James Murphy[1], Rob J. Vreeken[4,5], James Langridge[1], Jing X. Kang[3]*

1 Health Sciences, Waters Corporation, Milford, Massachusetts, United States of America, 2 Department of Biochemistry and Molecular & Cellular Biology, Georgetown University, Washington, DC, United States of America, 3 Laboratory for Lipid Medicine and Technology, Department of Medicine, Massachusetts General Hospital and Harvard Medical School, Boston, Massachusetts, United States of America, 4 Analytical Biosciences, Leiden Academic Centre for Drug Research, Leiden University, Leiden, The Netherlands, 5 Netherlands Metabolomics Centre, Leiden University, Leiden, The Netherlands

Abstract

A balanced omega-6/omega-3 polyunsaturated fatty acid (PUFA) ratio has been linked to health benefits and the prevention of many chronic diseases. Current dietary intervention studies with different sources of omega-3 fatty acids (omega-3) lack appropriate control diets and carry many other confounding factors derived from genetic and environmental variability. In our study, we used the *fat-1* transgenic mouse model as a proxy for long-term omega-3 supplementation to determine, in a well-controlled manner, the molecular phenotype associated with a balanced omega-6/omega-3 ratio. The *fat-1* mouse can convert omega-6 to omega-3 PUFAs, which protect against a wide variety of diseases including chronic inflammatory diseases and cancer. Both wild-type (WT) and *fat-1* mice were subjected to an identical diet containing 10% corn oil, which has a high omega-6 content similar to that of the Western diet, for a six-month duration. We used a multi-platform lipidomic approach to compare the plasma lipidome between *fat-1* and WT mice. In fat-1 mice, an unbiased profiling showed a significant increase in the levels of unesterified eicosapentaenoic acid (EPA), EPA-containing cholesteryl ester, and omega-3 lysophospholipids. The increase in omega-3 lipids is accompanied by a significant reduction in omega-6 unesterified docosapentaenoic acid (omega-6 DPA) and DPA-containing cholesteryl ester as well as omega-6 phospholipids and triacylglycerides. Targeted lipidomics profiling highlighted a remarkable increase in EPA-derived diols and epoxides formed via the cytochrome P450 (CYP450) pathway in the plasma of *fat-1* mice compared with WT mice. Integration of the results of untargeted and targeted analyses has identified a lipidomic biosignature that may underlie the healthful phenotype associated with a balanced omega-6/omega-3 ratio, and can potentially be used as a circulating biomarker for monitoring the health status and the efficacy of omega-3 intervention in humans.

Editor: David William Pond, Scottish Association for Marine Science, United Kingdom

Funding: This work was supported by funding from the Fortune Education Foundation and Sansun Life Sciences (to JXK) and the Alzheimer's Association grant (NIRG-11-203674) (to GA). Waters Corporation provided support in the form of salaries for authors GA, AD, JJ, AB, JM and JL. The specific roles of these authors are articulated in the 'author contributions' section. The funders had no role in study design, data collection and analysis, decision to publish, or preparation of the manuscript.

Competing Interests: GA, AD, JJ, AB, JM and JL are employed by Waters, who manufacture and supply mass spectrometers. No other conflicts exist.

* E-mail: giuseppe_astarita@waters.com (GA); kang.jing@mgh.harvard.edu (JXK)

[9] These authors contributed equally to this work.

Introduction

Most Western diets are deficient in omega-3 polyunsaturated fatty acids (PUFAs) and abundant in omega-6 PUFAs [1]. Current nutritional research shows that a diet enriched in omega-3s offers health benefits and anti-inflammatory properties and that an excess of omega-6s might contribute to the pathogenesis of many chronic diseases, including cardiovascular, autoimmune and Alzheimer's diseases [2–10]. The imbalance between omega-6s and omega-3s is largely the result of the traditional reliance of Western diets on vegetables oils such as corn, soybean, safflower, and sunflower. These oils are enriched in omega-6 PUFAs, such as linoleic acid (LA), which can be metabolized in animals and

humans to form longer chain fatty acids such as di-homo-gamma-linolenic acid (DGLA), docosapentaenoic acid (omega-6 DPA), and arachidonic acid (AA) are (Figure 1). At the same time, Western diets are lacking in leafy green vegetables, which are enriched in the omega-3 fatty acid, alpha-linolenic acid (ALA), and in oily fish, which contain the longer-chain omega-3 PUFAs such as eicosapentaenoic acid (EPA), omega-3 DPA, and docosahexaenoic acid (DHA) (Figure 1).

The human body cannot synthesize PUFAs de novo and must rely entirely on dietary intake for these essential nutrients. Also, it cannot interconvert between omega-6 and omega-3 fatty acids. The PUFAs, once absorbed in the intestines, are then transported,

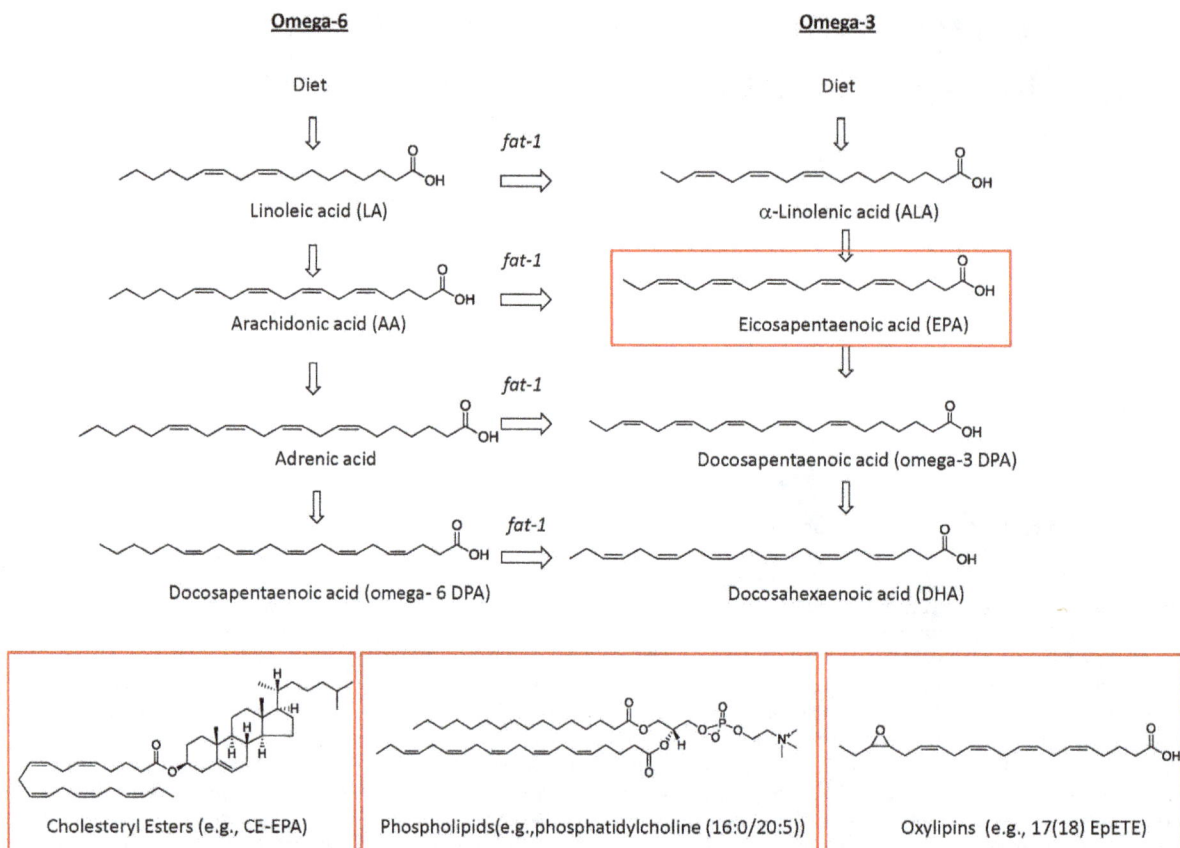

Figure 1. Overview of the omega-6 and omega-3 PUFAs metabolism. Diet-derived omega-6 linoleic acid (LA) and omega-3 alpha-linolenic acid (ALA) are transformed into longer chains PUFAs by the sequential action of desaturases and elongases. PUFAs can be found in blood as unesterified fatty acids, esterified to more complex lipids such as cholesteryl esters and phospholipids, or converted into the oxygenated metabolites oxylipins. The figure shows chemical structures of fatty acids and derivatives highlighted in our study.

via the bloodstream, to all tissues. They can be found as unesterified fatty acids or esterified to complex lipids (e.g., phospholipids, cholesteryl esters and triacylglycerols) and can be metabolized into bioactive species (e.g., oxylipins) (Figure 1). Hundreds of small molecules have been identified as metabolites of these few omega-3 and omega-6 precursors in human tissue. Yet, it is the overall balance between omega-3s and omega-6s that seems to modulate many biological processes including the relaxation and contraction of smooth muscle tissue, blood coagulation, and – significantly – inflammation [11–13].

Although much research demonstrates a potentially important relationship between PUFA intake and the risk of disease, it remains challenging in current dietary intervention studies to accurately evaluate the impact of increased intake of omega-3s by food or supplementation. A frequently confounding factor is the variability inherent in studies of control diets. A different nutritional value may accompany such diets compared with a diet enriched in omega-3– a different composition of fatty acids, for example. Other frequent issues relate to the chemical nature, source, and dose of the omega-3 used in the dietary intervention studies. These issues include the mixed use of the different types of omega-3s, such as EPA and DHA, or the different forms of omega-3s, such as triacylglycerols, phospholipids, or ethyl esters. Furthermore, dietary intervention studies in humans are often associated with high individual genetic and environmental variability [14]. All of these factors militate against an accurate

evaluation of the biological effects of omega-3s, and no molecular markers of omega-3 intake currently exist.

In 2004, the *fat-1* transgenic mouse model was developed to eliminate many of the confounders inherent in omega-6/omega-3 research [15]. The mouse was engineered to carry the *C. elegans fat-1* gene, which can add a double bond into an unsaturated fatty-acid hydrocarbon chain, thus converting omega-6 to omega-3 fatty acids (Figure 1) [15]. Though the mice are not exposed to an omega-3 diet, this conversion results in an abundance of omega-3 and a reduction in omega-6 fatty acids in their organs. The resulting omega-6/omega-3 fatty acid profile has also been shown to be comparable to those obtained by dietary supplementation [16]. The animals therefore provide a controlled approach for evaluating the effects of a balanced omega-6/omega-3 ratio, one that does not introduce the confounding factors that result from enforcing different test diets. To date, the *fat-1* transgenic mouse model has been widely used, and has demonstrated that balancing the omega-6/omega-3 ratio can protect against a wide variety of diseases, including chronic inflammatory diseases and cancer [3]. However, the molecular mechanisms underlying these beneficial effects remain to be fully elucidated.

In the present study, we used a multi-platform lipidomic approach to compare the molecular phenotype of *fat-1* and WT mice exposed for six months to an identical high-omega-6 diet in order to identify the biosignature involved in the health benefits associated with a balanced omega-6/omega-3 tissue ratio.

Materials and Methods

Materials

All chemicals were purchased from Sigma-Aldrich (Seelze, Germany) and were of analytical grade or higher purity. Lipid standards were purchased from Avanti Polar Lipids (Alabaster, Alabama USA), Cayman Chemical (Ann Arbor, Michigan USA), Biomol (Plymouth Meeting, Pennsylvania USA), and Larodan Fine Chemicals (Malmö, Sweden). For solid-phase extraction, 96-well, Waters Oasis HLB plates (60 mg) were obtained from Waters Corporation (Milford, Massachusetts USA).

Animals and sample collection

Transgenic fat-1 C57BL/6J mice were generated as previously described [15]. Only heterozygous female mice, aged 6 months, were used in this study. The mice were fed an identical diet rich in omega-6 PUFA and low in omega-3 PUFA (Catalog #1812692, Test Diet, Richmond, Indiana USA), a modification of the Test Diet AIN-76A semi-purified diet 58B0, with 10% total corn oil. Both wild type and fat-1 mice were fed the same diet since weaning until the mice were sacrificed. The diet components and fatty acid composition of the diet are listed in Table S1. The animals were phenotyped according to their omega-6 and omega-3 PUFA profiles, as determined by gas chromatography [17]. For the experiments described, WT and fat-1 littermates were used (n = 5). At approximately 6 months of age, all mice underwent 18 hours of fasting and were afterward euthanized by means of carbon dioxide. Total blood volume was immediately collected using cardiac puncture, and the blood was retained on ice. Whole-blood samples were centrifuged at 1,000×g for 15 min at 4°C. Plasma was transferred to a clean polypropylene tube and centrifuged at 10,000×g for 10 min at 4°C. Aliquots of the samples were drawn and stored at −80°C until assayed. All animal procedures were performed in accordance with guidelines prescribed by the Massachusetts General Hospital (MGH) Animal Committee and with Institutional Animal Care and Use Committee (IACUC) for MGH approval (Protocol #2010N000038). All efforts possible were made to minimize animal suffering. Animal investigations conformed to the Public Health Policy on Humane Care and Use of Laboratory Animals.

Lipid extraction from mouse plasma

Total lipids were extracted from 10 μL of plasma after protein precipitation with 490 μL of isopropanol (IPA) spiked with the following internal standard: d_8 arachidonic acid, cholesterol-d_7, C19:0-cholesteryl ester, trinonadecenoin, 1,2-dimyristoyl-sn-glycero-3-phosphoethanolamine, 1,2-dimyristoyl-sn-glycero-3-phosphocholine, 1-heptadecenoyl-2-hydroxy-sn-glycero-3-phosphocholine. Samples were vortex mixed, incubated at 4°C for 30 min, and centrifuged at 10,000×g and 4°C for 10 min. Supernatant was then transferred to mass-spectrometry vials for further analysis. Extraction of oxylipins was performed as previously described [18]. Briefly, after thawing on ice, 50 μL of plasma samples were treated immediately with antioxidants (0.2 mg butylated hydroxytoluene (BHT/EDTA)) and spiked with the following internal standards: 6k-PGF$_{1\alpha}$-d$_4$, TXB$_2$-d$_4$, PGF$_{2\alpha}$-d$_4$, PGE2-d$_4$, PGD$_2$-d$_4$, LTE$_4$-d$_3$, LTB$_4$-d$_4$, 12,13 diHOME-d$_4$, 9,10-diHOME-d$_4$, 14,15-DiHETrE-d11, 15-deoxy-δ-12,14-PGJ$_2$-d$_4$, 20-HETE-d$_6$, 9S-HODE-d$_4$, 12S-HETE-d$_8$, 5S-HETE-d$_8$, 5-oxo-ETE-d$_7$. Plasma samples were loaded onto the solid-phase extraction plate and then eluted with 1.5 mL ethyl acetate, after wetting the plate wells with 0.5 mL methanol. The eluent was reduced under nitrogen and subsequently reconstituted in a 50-μL solution of methanol and acetonitrile (1:1) containing 100 nM 1-cyclohexyluriedo-3-dodecanoic acid, a quality marker for the analysis.

Liquid Chromatography Mass Spectrometry

Total lipid analysis was performed with IonKey/MS system comprised of an ACQUITY UPLC M-Class, the ionKey source and an iKey CSH C18 130 Å, 1.7 μm particle size, 150 μm ×100 mm column (Waters Corporation, Milford, Massachusetts, USA) coupled to a Synapt G2-S or Xevo TQ-S mass spectrometer (Waters Corporation, Manchester, UK). The capillary voltage was 2.8 kV and the source temperature 110°C. Injections were 0.5 μL using partial-loop mode with a column temperature of 55°C and flow rate of 3 μL/min. Mobile phase A consisted of acetonitrile/water (60/40) with 10 mM ammonium formate +0.1% formic acid. Mobile phase B consisted of IPA/acetonitrile (90/10) with 10 mM ammonium formate +0.1% formic acid. The gradient was programmed as follows: 0.0–2.0 min from 40% B to 43% B, 2.0–2.1 min to 50% B, 2.1–12.0 min to 99% B, 12.0–12.1 min to 40% B, and 12.1–14.0 min at 40% B. For the untargeted analyses, a mass range of 50–1500 m/z was selected in both positive and negative electrospray ionization.

Oxylipins were analyzed using UPLC (Waters Corporation, Milford, Massachusetts USA) coupled to electrospray ionization on a Xevo TQ-S mass spectrometer (Waters Corporation, Manchester, UK). The auto sampler was cooled to 10°C. For each analysis, 3 μL of sample was injected on a BEH C18 1.7 μm, 2.1×100 mm column (Waters Corporation, Milford, Massachusetts, USA) using a flow rate of 0.6 mL/min at 40°C. Mobile phases were composed as follows: A = 0.1% acetic acid, and B = 90:10 v/v acetonitrile/IPA (0.00–1.00 min from 25% B to 33% B, 1.00–8.00 min B to 95%, 8.00–8.50 min to 95% B, 8.51–10.00 min to 25% B). Electrospray ionization was performed in the negative-ion mode applying a capillary voltage of 2 kV, a source temperature of 120°C, a desolvation gas temperature of 350°C, a desolvation gas flow of 650 L/h, a cone-gas flow of 150 L/h, a collision voltage of 15 V, and a cone voltage of 35 V. Oxylipins were detected using MRM transitions optimized using synthetic standards (Table S2). Limits of quantification were in the lower pM range.

Data Processing and Analysis

For the untargeted lipidomic approach, a combination of analysis of the variance (ANOVA) and multivariate statistics, including principal component analysis (PCA) and partial, least-squares discriminant analysis (PLS-DA), identified lipids most responsible for differences between fat-1 and WT sample groups. Compounds were identified by database searches as performed by Human Metabolome Database (HMDB) [19] and METLIN [20] as well as by fragmentation patterns, retention times and ion-mobility-derived collision cross sections versus commercially available reference standards, when available. For the targeted approach, mean concentrations and SEM values (n = 5) for each group (fat-1 and WT) were calculated using appropriate internal standards to normalize for variations in sample preparation and MS detection. Univariate analyses (Student's t-tests) were conducted to assess for significance (p-value) and false-discovery rate (FDR) was used to control for multiple comparison. A volcano plot was produced to compare the fold-change versus the p-value for the targeted lipids.

Data processing and analysis was conducted using Progenesis QI Informatics (Nonlinear Dynamics, Newcastle, UK) and MetaboAnalyst 2.0 [21]. Data quantification was performed using TargetLynx software (Waters Corporation, Milford, Massachusetts USA).

Results

To characterize the molecular phenotype associated with a balanced omega-6/omega-3 ratio, a multi-platform lipidomic approach was used (Figure 2). Plasma samples from the fat-1 and WT groups were divided into two aliquots, one for an untargeted lipidomic approach to screen for major lipid alterations, and the other for a multiplexed targeted lipidomic approach to quantify low-abundance, bioactive lipid species.

To identify the most abundant lipid alterations resulting from a balanced omega-6/omega-3 tissue ratio, we adopted an untargeted lipidomic approach. Using electrospray-mass spectrometry in both positive and negative ionization mode, we detected thousands of molecular features in plasma samples. Significant changes were identified using ANOVA (Table S3). Multivariate

statistics, including PCA (Figure S1) and PLS-DA, highlighted the molecular features contributing most to the variance between the two groups of mice (Figure 3A and B). A fold-change analysis showed a marked increase in the levels of EPA and EPA-containing cholesteryl ester (CE-EPA) (Figure 3C and D), whereas a corresponding, significant decrease in the levels of omega-6 DPA and DPA-containing CE (CE-omega-6-DPA) in *fat-1* mice compared to WT littermates (Figure 3C). Slight but significant increases were also observed for unesterified DHA and CE-DHA, accompanied by a modest decrease for unesterified AA and CE-AA (Tables 1 and 2).

The plasma of *fat-1* mice exhibited a balanced omega-6/omega-3 ratio for unesterified PUFA, differing from the plasma of the WT littermates, for which the ratio shifted toward the omega-6 PUFAs

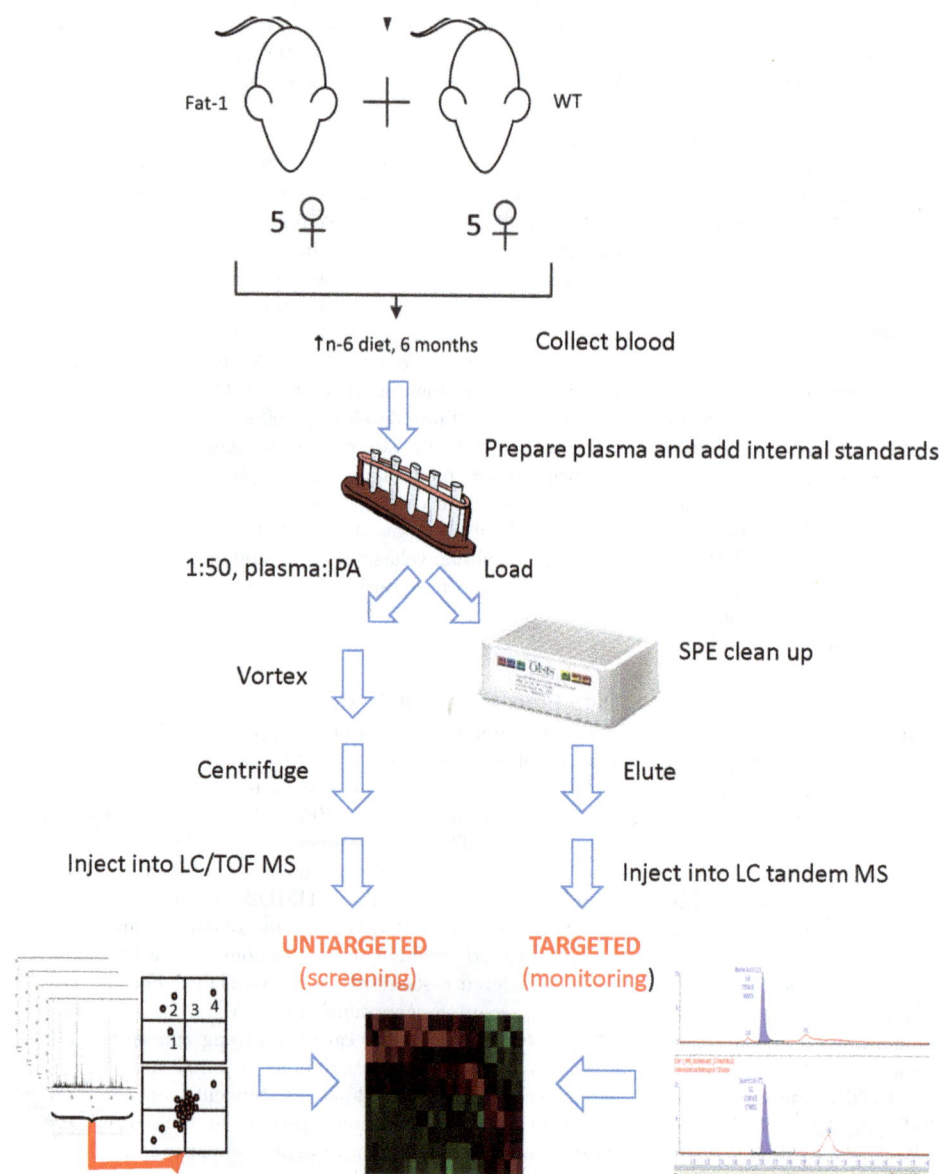

Figure 2. Study design and workflow for the lipidomic analyses. Heterozygous female *fat-1* and WT mice were fed a 6-month-long diet containing 10% corn oil, which is particularly enriched in omega-6 PUFAs. Blood was collected and plasma samples were prepared and divided into two aliquots. Before extraction, a mixture of internal standards was added to the plasma to normalize for variations in sample preparation or MS detection. Complementary untargeted and targeted lipidomic analyses were conducted, and the results were integrated for the generation of a unique lipidomic biosignature characteristic of the balanced omega-6/omega-3 ratio found in *fat-1* mice.

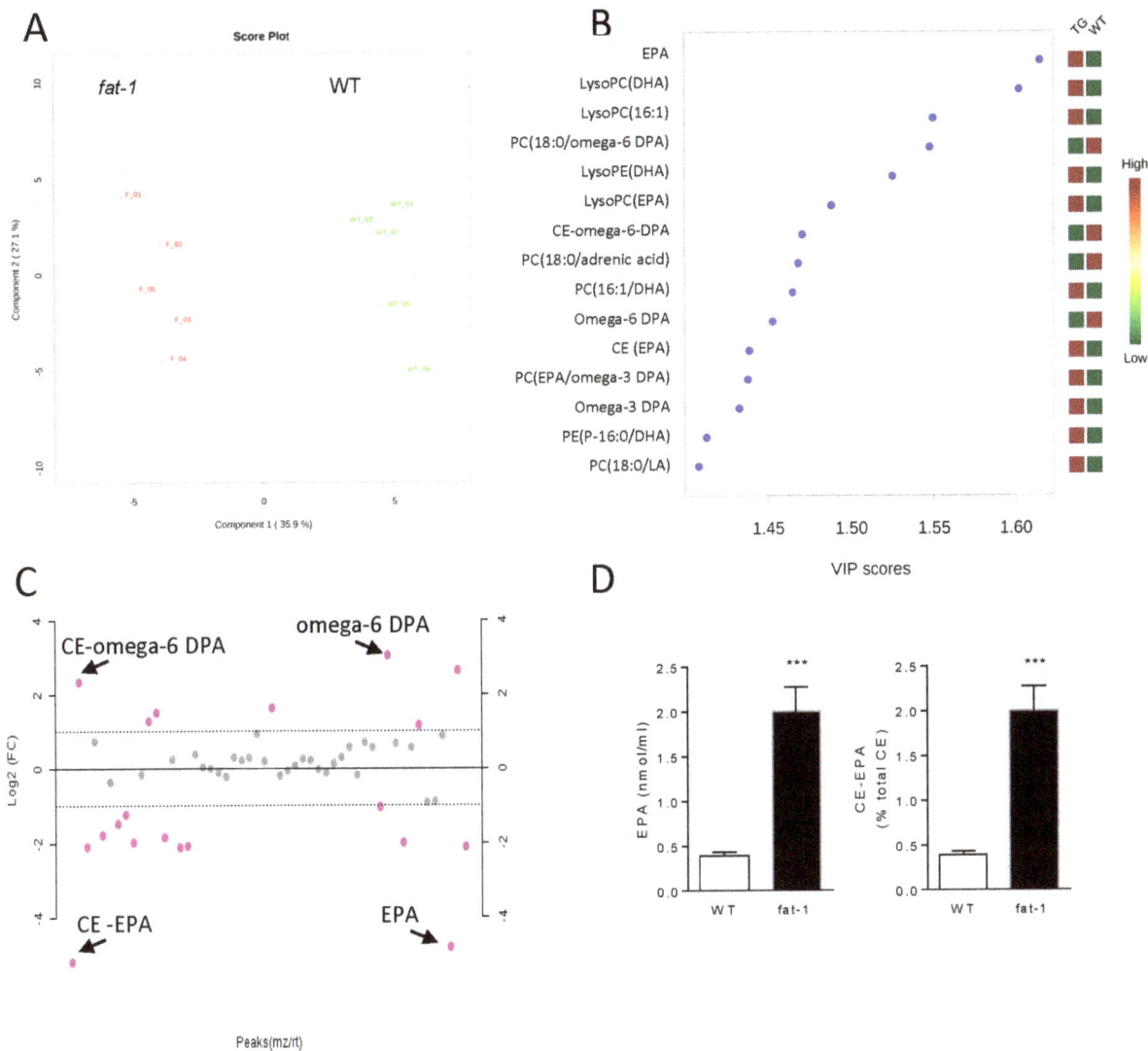

Figure 3. Untargeted lipidomic analysis. A, PLS-DA analysis showed a marked separation of plasma samples belonging to WT and *fat-1* mice, highlighting the features that contributed most to the variance between the two groups. B, Important features identified by PLS-DA. The colored boxes on the right indicate the relative concentrations of the corresponding metabolite in each group under study. Variable Importance in Projection (VIP) is a weighted sum of squares of the PLS loadings taking into account the amount of explained Y-variation in each dimension. C, Important features selected by fold-change analysis (relative to WT) with threshold 2. The red circles represent features above the threshold. Note the values are on log scale, so that both up-regulated and downregulated features can be plotted in a symmetrical way. D, Levels of EPA and percent composition of CE-EPA in WT and *fat-1* mice (n = 5, Student's t test; ***, p<0.001). The data present the mean ± SEM.

(Table 1). Other significant differences between *fat-1* and WT mice include the levels of phosphatidylcholines (PCs) and phosphatidylethanolamines (PEs) (Figure 3B and Table S3). Most notably, we observed a marked increase in the levels of various lysophospholipids, including DHA-containing lyso PC and lyso PE, in *fat-1* mice (Figure 3B). Finally, the levels of various TG species containing omega-6 fatty acids, as well as cholesterol and cholesterol sulfate, were also decreased in *fat-1* mice (Table S3).

PUFAs can be converted into hundreds of bioactive oxygenated metabolites (i.e., oxylipins) by means of enzymatic and nonenzymatic reactions. To measure the levels of such usually low-abundance metabolites, we used a multiplexed-targeted lipidomic approach (Figure 2). We measured the levels of more than 100 common oxylipins using tandem MS (multiple reaction monitoring). A volcano plot analysis using Student t-test p<0.01 and a fold change >2 as cut-off parameters highlighted a marked increase in

the levels of EPA epoxides and diols in the plasma of *fat-1* versus WT mice, including 17(18)-epoxy-eicosatetraenoic acid (17(18)-EpETE) and 17,18-dihydroxy-eicosatetraenoic acid (17,18-Di-HETE) (Figure 4B). Such differences were found to be significant after multiple comparison correction (Figure 4C and Table S1). Minor increases were also observed for other EPA- or DHA-derived metabolites (Figure 4C), whereas decreases were observed in the levels of DGLA- and AA-derived metabolites in *fat-1* mice (Figure 4C and Table 1).

To determine the overall differences in the molecular phenotypes of *fat-1* and WT mice, we combined the results derived from both the untargeted and targeted analyses. Correlation analysis revealed a marked association of many features with EPA and EPA metabolites (Figures 5A and B). Thes lipidomic biosignature for *fat-1* mice was underscored by a significant molecular contrast between the two groups of mice (Figure 5C).

Table 1. Levels of unesterified fatty acids (nmol/ml) in WT and *fat-1* mice (n = 5).

Fatty acid species	WT	*fat-1*	p-value	FDR
18:3 n-3 (ALA)	24.33±1.49	24.49±1.96	0.9484737	0.63176
20:5 n-3 (EPA)	2.19±0.16	17.04±0.36	2.558E-10	0.001165
22:5 n-3 (omega-3 DPA)	0.96±0.09	3.39±0.19	2.473E-06	0.009725
22:6 n-3 (DHA)	22.92±1.68	27.10±0.91	0.0017493	0.81954
Total n-3	50.40±2.41	72.02±2.54	0.0002642	
18:2 n-6 (LA)	64.64±7.42	50.06±3.51	0.1135015	0.15112
20:3 n-6 (DGLA)	8.32±0.74	7.06±0.19	0.1382162	0.25896
20:4 n-6 (AA)	48.72±6.21	34.63±1.62	0.0596187	0.10569
22:4 n-6 (adrenic acid)	2.10±0.13	1.53±0.2	0.0440865	0.15112
22:5 n-6 (omega-6 DPA)	8.16±0.6	3.18±0.33	8.443E-05	0.009361
Total n-6	131.93±13.35	96.45±4.96	0.0374117	
Omega-6/omega-3 ratio	2.61±0.21	1.34±0.04	0.0003226	

P- values derived from Student's t test. FDR, false discovery rate.

Discussion

In this study, we compared the plasma lipidome of transgenic *fat-1* and WT mice after exposing the animals for six months to a diet high in omega-6 content. The *fat-1* mice are able to endogenously convert omega-6s into omega-3s, serving as a good model of a balanced omega-6/omega-3 ratio [3,15,16]. The design of this comparative study allowed an investigation of the effects of long-term omega-3 PUFA enrichment without the need for two different diets, thereby eliminating confounders derived from differences in calories, flavor, or nutrients. Our approach also avoided other confounders associated with individual genetic and environmental variability, which can be found in nutritional intervention studies involving human subjects [14,22].

Previous studies have determined the total fatty acid composition of various tissues from *fat-1* mice using gas chromatography (GC), showing that the omega-6/omega-3 PUFA ratio can range from about 1 to 5 depending on the tissue, while wild type mice usually range from 20–50 (with the exception of the brain which is about 4) [3,15,16,22]. Such studies, however, aimed to measure the total fatty acid content without detailing information on the levels of unesterified fatty acids, their product of oxygenation (oxylipins), and complex lipid species carrying PUFA, which may

be particularly enriched in the plasma lipoproteins of fasted animals. By using two complementary analytical approaches, untargeted and multiplexed targeted lipidomic analyses, we were able to compare, in a more comprehensive fashion, the molecular phenotypes of the plasma from *fat-1* and WT mice. Indeed, our study provided information about the availability of unesterified fatty acids that might serve as precursors for the formation of oxylipins. Furthermore, our study described the rearrangement of fatty acyl chains in complex lipid structures, which might determine 1) the physicochemical properties of the membranes and lipoproteins containing such lipids and 2) the ability to transport PUFAs to various organs [15,22].

The most obvious observation deriving from the untargeted lipidomic analysis is a rearrangement of the fatty acyl chains of CEs and phospholipids, which largely reflected the availability of omega-3 and omega-6 PUFAs. Such results are consistent with previous animal and human studies showing that blood lipid composition reflects the intake of PUFAs [14,23–28]. We observed, however, specific alterations of PUFA-containing lipids that appear to be independent from the PUFA availability. In particular, we observed a marked increase for CE-EPA in the plasma of the *fat-1* mice, as compared to CE-DHA (Table 2). This observation generally aligns with previous reports showing a

Table 2. Composition (% of the total) of cholesteryl esters in WT and *fat-1* mice (n = 5).

CE species	WT	*fat-1*	p-value	FDR
18:3 n-3	1.17±0.1	1.16±0.03	0.8664	0.74976
20:5 n-3	0.39±0.04	2.08±0.27	0.0002	0.009725
22:5 n-3	1.00±0.13	1.36±0.11	0.039	0.008543
22:6 n-3	9.66±0.63	12.73±0.54	0.0054	0.58567
18:2 n-6	24.9±1.08	25.72±1.08	0.4847	0.86083
20:3 n-6	7.96±0.44	7.75±0.29	0.6161	0.63176
20:4 n-6	50.94±0.58	45.81±1.49	0.009	0.42803
22:5 n-6	0.82±0.07	0.21±0.04	0.0001	0.008543
18:1 n-9	3.18±0.6	3.18±0.27	0.9351	0.77652

P-values derived from Student's t test.

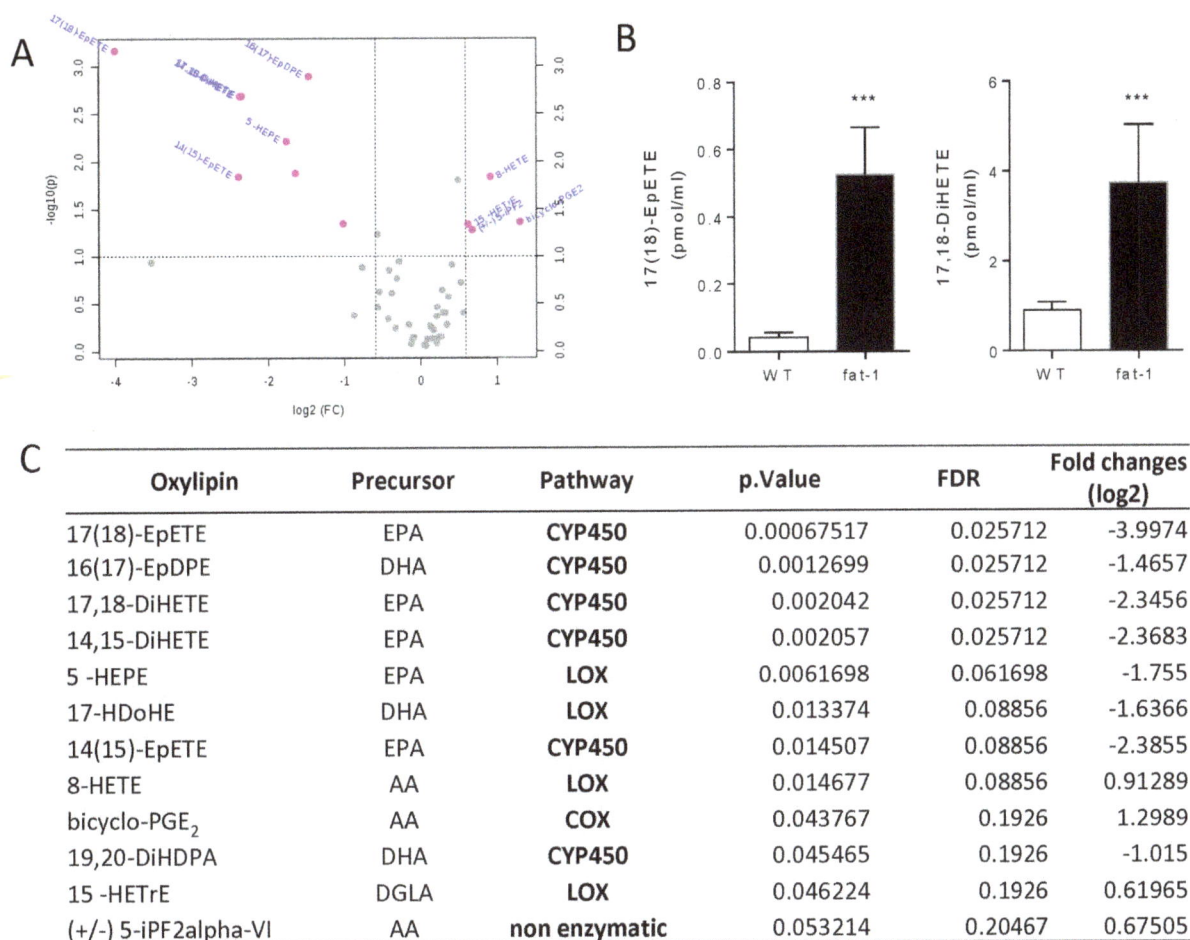

Oxylipin	Precursor	Pathway	p.Value	FDR	Fold changes (log2)
17(18)-EpETE	EPA	CYP450	0.00067517	0.025712	-3.9974
16(17)-EpDPE	DHA	CYP450	0.0012699	0.025712	-1.4657
17,18-DiHETE	EPA	CYP450	0.002042	0.025712	-2.3456
14,15-DiHETE	EPA	CYP450	0.002057	0.025712	-2.3683
5 -HEPE	EPA	LOX	0.0061698	0.061698	-1.755
17-HDoHE	DHA	LOX	0.013374	0.08856	-1.6366
14(15)-EpETE	EPA	CYP450	0.014507	0.08856	-2.3855
8-HETE	AA	LOX	0.014677	0.08856	0.91289
bicyclo-PGE$_2$	AA	COX	0.043767	0.1926	1.2989
19,20-DiHDPA	DHA	CYP450	0.045465	0.1926	-1.015
15 -HETrE	DGLA	LOX	0.046224	0.1926	0.61965
(+/-) 5-iPF2alpha-VI	AA	non enzymatic	0.053214	0.20467	0.67505

Figure 4. Targeted lipidomic analyses. A, Important features selected by volcano plot with fold change threshold (x) 2 and t-tests threshold (y) 0.1. The red circles represent features above the threshold. Note both fold changes (relative to WT) and p values are log transformed. The further its position away from the (0,0), the more significant the feature is. B, Levels of representative EPA-derived epoxide and diol species in WT and *fat-1* mice (n = 5, Student's t test; ***, p<0.001). The data present the mean ± SEM. C, Most important oxylipins features as determined by volcano plot; also reported: Student's t test p-value, false discovery rate (FDR) and fold-change.

limited effect on CE-DHA concentration following supplementation with high doses of DHA in humans [29,30]. Although some have speculated that a retroconversion of DHA to EPA could contribute to the limited biosynthesis of CE-DHA [31–37], this biosynthesis could also come about because of differential specificity of enzymes involved in the metabolism of PC species carrying DHA and EPA. Previous studies indeed showed that the CE composition is largely determined by the specificity of the enzyme lecithin-cholesterol acyl transferase (LCAT). The modest changes observed for CE-DHA may therefore arise because DHA is a poor substrate for LCAT, compared with EPA [26,38–41]. Although the literature included relatively little evidence regarding the role of different cholesteryl esters, the observed increase in CE-EPA might affect the size or action of lipoproteins [42,43].

Determining the plasma profile of unesterified fatty acids is normally problematic due to the direct influence of diet on plasma fatty acid content. Using a single diet for both WT and *fat-1* animals made it possible for plasma content analysis to be used as a reliable indicator of endogenous PUFA metabolism. Notably, the diet does not contain longer chain PUFAs (such as AA, EPA, and DHA) and any circulating lipids longer than 18 carbons must come from endogenous metabolism of shorter PUFAs. Our study highlighted a significant difference in the omega-6/omega-3 ratio

between unesterified plasma lipids and other previously reported tissues [3,15,16]. In particular, our results showed a marked increase for unesterified EPA compared to unesterified DHA in the plasma of *fat-1* mice (Table 1), which could be explained, in part, by a retroconversion of DHA to EPA [31–37]. The concomitant elevation of the DHA-containing phospholipids, however, suggests that EPA might be a preferred substrate for the hydrolytic activity of phospholipase A$_2$ (PLA$_2$) in plasma, rather than DHA [44]. These observations point to a differential metabolism for EPA and DHA, which could explain the diverse physiological effects previously reported for these two omega-3 PUFAs [45–47]. On the other hand, DHA-containing phospholipids and lysophospholipids may play roles in the cellular membrane properties and in the transport of DHA to other tissues [40,42]. Lysophospholipids containing DHA do, in fact, appear to be best suited as carriers of this essential fatty acid to eye and brain tissues, where it modulates the membrane fluidity of synaptic vesicles [48–50] and displays neuroprotective properties [51]. Notably, our untargeted analysis highlights a wide increase in other lysophospholipid species, which comports with a previous report showing that dietary supplementation with omega-3s can dynamically regulate plasma lysoPC [52].

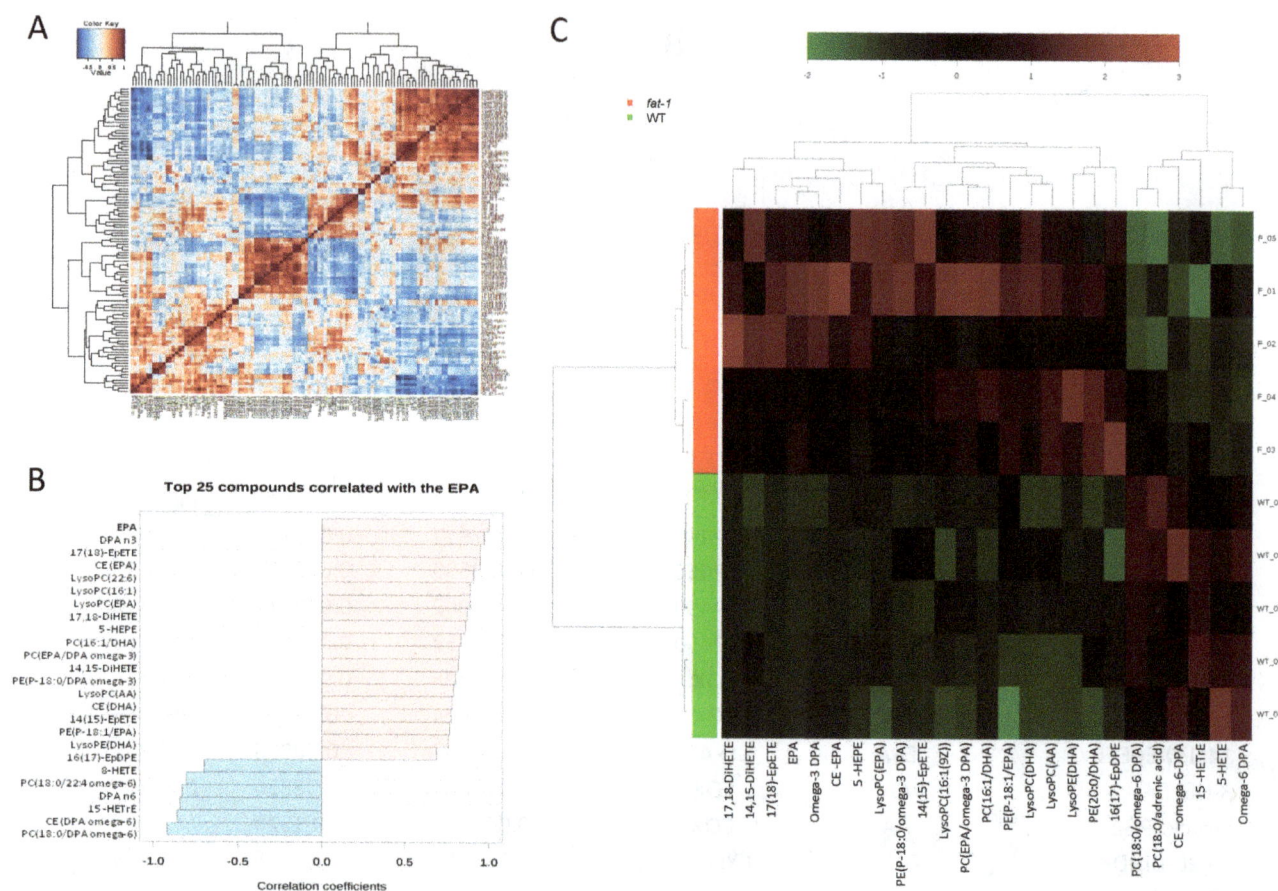

Figure 5. Lipidomic biosignature of *fat-1* mice. A, Correlation analysis was used to visualize the overall relationships between different features and (B) to identify which features are correlated with EPA. C, Clustering result shown as heatmap (distance measure using Pearson, and clustering algorithm using ward), providing an intuitive visualization of the characteristic lipidomic biosignature found in *fat-1* mice versus WT mice. Each colored cell on the map corresponds to a concentration value, with samples in rows and features/compounds in columns. Displayed are the top 25 lipids ranked by t-tests.

The untargeted lipidomic analysis also showed a significant decrease in triacylglycerols and cholesterol. Such a decrease is in accordance with the fact that the *fat-1* phenotype is resistant to metabolic syndrome, obesity, and liver steatosis [53–56]. Significantly, the unexpected decrease in cholesterol-sulfate in the plasma of *fat-1* mice may also be linked with some of the protective effect of omega-3s. Cholesterol sulfate is present in lipoproteins, and it has been found in atherosclerotic lesions of the human aorta, where it plays a role in platelet adhesion, possibly determining the prothrombotic potential of atherosclerotic lesions [57]. Cholesterol sulfate is also found to be particularly enriched in DHA-rich cellular membranes, where it seems to modulate the lipid raft formation [58,59]. One might reasonably speculate that the observed decrease in the circulating levels of cholesterol sulfate might indicate the effect of possible sequestration caused by the DHA-rich cellular membranes in *fat-1* mice. To establish the validity of such a hypothesis is beyond the scope of this paper.

The multiplexed targeted lipidomics experiments highlighted the marked separation of the panel of oxylipins in the plasma of WT and *fat-1* mice. These lipid mediators are currently the focus of considerable interest, for they are also key messengers for cellular homeostasis, inflammation, platelet aggregation, and vascularization [11–13]. Oxylipins are produced via enzymatic or nonenzymatic oxygenation of both omega-6 and omega-3 PUFAs. Three major enzymatic pathways are involved in their

generation: cyclooxygenase (COX), lipoxygenase (LOX) and cytochrome P450 (CYP) (Figures 6A and 6B). These pathways are important drug targets for various diseases. The ability to control such pathways with dietary interventions (i.e., omega-3 supplementation) could offset many of the side effects linked to pharmacological treatments. Our study indicated that the availability of unesterified omega-3 PUFA precursors correlated with the increase in the levels of the corresponding omega-3 oxylipins while decreasing the omega-6 oxylipins. Such results suggest that the higher levels of omega-3s compete with omega-6s for enzymatic and nonenzymatic activities to ultimately reduce the pro-inflammatory lipid mediators (HETEs, bicyclo-PGE$_2$ and $(+/-)$5-iPF2alpha-VI) while increasing the levels of anti-inflammatory mediators (5-HEPE, 15-HETrE, and 17-HDoHE).

Notably, the most abundant oxylipins alterations in *fat-1* mice were related to the metabolism of the CYP450 pathway [37] (Figure 6). The CYP450 family of enzymes can produce epoxides from PUFAs, which are subsequently metabolized by the soluble epoxide hydrolase (sEH) to the corresponding vicinal diols, dihydroxyeicosatrienoic acids [60]. In human and animal studies, the CYP-dependent metabolite profiles were generally reflective of the PUFA composition [14,37,61–63], suggesting that most of the CYP-epoxygenases accept omega-3s and omega-6s as equally efficient substrates [64]. Recent evidence shows that DHA intake increases the levels of the EPA-derived vicinal diol 17,18-DiHETE

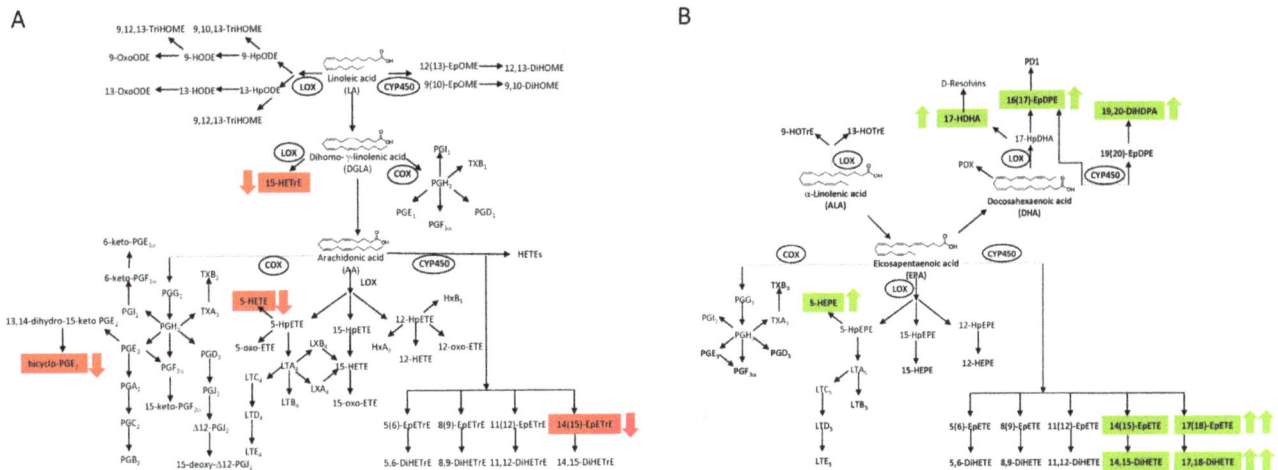

Figure 6. Pathway analysis. The activities of COX, LOX and CYP450 enzymes catalyze the formation of hundreds of oxylipins species with different biological activities starting from the omega-6 PUFAs precursors (*panel A*) and the omega-3 PUFAs (*panel B*). The *fat-1* mice had marked alterations in the CYP450 pathway and minor alterations in the LOX/COX pathways resulting in the increase of omega-3 oxylipins (green) and decrease of omega-6 oxylipins (red).

metabolized in the CYP/sEH pathway in plasma from piglets, suggesting that DHA retroconversion to EPA may occur to some extent [37].

Although the physiologic properties of AA-derived metabolites of CYP have been studied extensively [65–67], the study of DHA and EPA-derived metabolites of CYP450 and their physiologic properties has only recently begun. Omega-3 CYP-metabolites have been described as possessing anti-inflammatory [68] and analgesic properties [69], as inhibitors of platelet aggregation [70], and as pulmonary, smooth-muscle relaxants [71]. It has been suggested that some of the beneficial effects of fish-oil-enhanced diets on cardiovascular function may be mediated by the levels of these metabolites [72]. Our results support the hypothesis that CYP-450-mediated omega-3 metabolism might represent a major physiological pathway underlying the reduced disease risk and health benefits observed across numerous *fat-1* mouse studies.

Finally, the integration of untargeted and targeted lipidomic results provided a detailed molecular signature for a balanced omega-6/omega-3 tissue ratio. Overall, EPA levels were the most remarkable molecular change observed in the plasma of *fat-1* mice. Although the use of the *fat-1* transgenic mouse model allowed us to eliminate confounding factors of the diet, further work would be needed to establish the validity of these molecular changes in humans.

In conclusion, this study demonstrates a protective lipidomic biosignature in *fat-1* mice that may contribute to the low risk of disease and health benefits associated with a balanced omega-6/omega-3 tissue ratio. Such a lipidomic biosignature could be used as a potential circulating biomarker for monitoring the "health

status" or the efficacy of nutritional intervention with omega-3s in humans.

Supporting Information

Figure S1 PCA plot have been applied to the ions that were found to have statistically significant alterations in *fat-1* mice compared to WT mice. The separation between clusters of the samples from *fat-1* transgenic mice (TG in red) and WT mice (in green) is indicative of the potential discriminating power of the statistically significant lipid identified.

Table S1 Nutritional and fatty acid composition of the 10% corn oil diet.

Table S2 Levels of oxylipins expressed as mean concentration and SEM values (n = 5) for each group (*fat-1* and WT).

Table S3 Tentative identification of the most significant lipid alterations obtained from the unbiased lipidomic analysis.

Author Contributions

Conceived and designed the experiments: GA JL JXK. Performed the experiments: GA JHM BW KS. Analyzed the data: GA JHM. Contributed reagents/materials/analysis tools: AD JJ AB JM TH RJV. Wrote the paper: GA JXK.

References

1. Blasbalg TL, Hibbeln JR, Ramsden CE, Majchrzak SF, Rawlings RR (2011) Changes in consumption of omega-3 and omega-6 fatty acids in the United States during the 20th century. Am J Clin Nutr 93: 950–962.
2. James MJ, Gibson RA, Cleland LG (2000) Dietary polyunsaturated fatty acids and inflammatory mediator production. Am J Clin Nutr 71: 343S–348S.
3. Kang JX (2011) The omega-6/omega-3 fatty acid ratio in chronic diseases: animal models and molecular aspects. World Rev Nutr Diet 102: 22–29.
4. Loef M, Walach H (2013) The omega-6/omega-3 ratio and dementia or cognitive decline: a systematic review on human studies and biological evidence. J Nutr Gerontol Geriatr 32: 1–23.
5. Simopoulos AP (2008) The importance of the omega-6/omega-3 fatty acid ratio in cardiovascular disease and other chronic diseases. Exp Biol Med (Maywood) 233: 674–688.
6. Simopoulos AP (2006) Evolutionary aspects of diet, the omega-6/omega-3 ratio and genetic variation: nutritional implications for chronic diseases. Biomed Pharmacother 60: 502–507.
7. Wang D, Dubois RN (2010) Eicosanoids and cancer. Nat Rev Cancer 10: 181-193.
8. Zhang L, Geng Y, Yin M, Mao L, Zhang S, et al. (2010) Low omega–6/omega–3 polyunsaturated fatty acid ratios reduce hepatic C-reactive protein expression in apolipoprotein E-null mice. Nutrition 26: 829–834.

9. de Lorgeril M, Salen P (2012) New insights into the health effects of dietary saturated and omega-6 and omega-3 polyunsaturated fatty acids. BMC Med 10: 50.

10. Hooijmans CR, Pasker-de Jong PC, de Vries RB, Ritskes-Hoitinga M (2012) The effects of long-term omega-3 fatty acid supplementation on cognition and Alzheimer's pathology in animal models of Alzheimer's disease: a systematic review and meta-analysis. J Alzheimers Dis 28: 191–209.

11. Dalli J, Colas RA, Serhan CN (2013) Novel n-3 immunoresolvents: structures and actions. Sci Rep 3: 1940.

12. Serhan CN (2008) Systems approach with inflammatory exudates uncovers novel anti-inflammatory and pro-resolving mediators. Prostaglandins Leukot Essent Fatty Acids 79: 157–163.

13. Serhan CN, Samuelsson B (1988) Lipoxins: a new series of eicosanoids (biosynthesis, stereochemistry, and biological activities). Adv Exp Med Biol 229: 1–14.

14. Nording ML, Yang J, Georgi K, Hegedus Karbowski C, German JB, et al. (2013) Individual variation in lipidomic profiles of healthy subjects in response to omega-3 Fatty acids. PLoS One 8: e76575.

15. Kang JX, Wang J, Wu L, Kang ZB (2004) Transgenic mice: fat-1 mice convert n-6 to n-3 fatty acids. Nature 427: 504.

16. Orr SK, Tong JY, Kang JX, Ma DW, Bazinet RP (2010) The fat-1 mouse has brain docosahexaenoic acid levels achievable through fish oil feeding. Neurochem Res 35: 811–819.

17. Kang JX, Wang J (2005) A simplified method for analysis of polyunsaturated fatty acids. BMC Biochem 6: 5.

18. Strassburg K, Huijbrechts AM, Kortekaas KA, Lindeman JH, Pedersen TL, et al. (2012) Quantitative profiling of oxylipins through comprehensive LC-MS/MS analysis: application in cardiac surgery. Anal Bioanal Chem 404: 1413–1426.

19. Wishart DS, Jewison T, Guo AC, Wilson M, Knox C, et al. (2013) HMDB 3.0–The Human Metabolome Database in 2013. Nucleic Acids Res 41: D801–807.

20. Smith CA, O'Maille G, Want EJ, Qin C, Trauger SA, et al. (2005) METLIN: a metabolite mass spectral database. Ther Drug Monit 27: 747–751.

21. Xia J, Mandal R, Sinelnikov IV, Broadhurst D, Wishart DS (2012) MetaboAnalyst 2.0– a comprehensive server for metabolomic data analysis. Nucleic Acids Res 40: W127–133.

22. Kang JX (2007) Fat-1 transgenic mice: a new model for omega-3 research. Prostaglandins Leukot Essent Fatty Acids 77: 263–267.

23. Balogun KA, Albert CJ, Ford DA, Brown RJ, Cheema SK (2013) Dietary omega-3 polyunsaturated Fatty acids alter the Fatty Acid composition of hepatic and plasma bioactive lipids in C57BL/6 mice: a lipidomic approach. PLoS One 8: e82399.

24. Zock PL, Mensink RP, Harryvan J, de Vries JH, Katan MB (1997) Fatty acids in serum cholesteryl esters as quantitative biomarkers of dietary intake in humans. Am J Epidemiol 145: 1114–1122.

25. Katan MB, Deslypere JP, van Birgelen AP, Penders M, Zegwaard M (1997) Kinetics of the incorporation of dietary fatty acids into serum cholesteryl esters, erythrocyte membranes, and adipose tissue: an 18-month controlled study. J Lipid Res 38: 2012–2022.

26. Subbaiah PV, Kaufman D, Bagdade JD (1993) Incorporation of dietary n-3 fatty acids into molecular species of phosphatidyl choline and cholesteryl ester in normal human plasma. Am J Clin Nutr 58: 360–368.

27. Ma J, Folsom AR, Lewis L, Eckfeldt JH (1997) Relation of plasma phospholipid and cholesterol ester fatty acid composition to carotid artery intima-media thickness: the Atherosclerosis Risk in Communities (ARIC) Study. Am J Clin Nutr 65: 551–559.

28. Ottestad I, Hassani S, Borge GI, Kohler A, Vogt G, et al. (2012) Fish oil supplementation alters the plasma lipidomic profile and increases long-chain PUFAs of phospholipids and triglycerides in healthy subjects. PLoS One 7: e42550.

29. Bronsgeest-Schoute HC, van Gent CM, Luten JB, Ruiter A (1981) The effect of various intakes of omega 3 fatty acids on the blood lipid composition in healthy human subjects. Am J Clin Nutr 34: 1752–1757.

30. von Lossonczy TO, Ruiter A, Bronsgeest-Schoute HC, van Gent CM, Hermus RJ (1978) The effect of a fish diet on serum lipids in healthy human subjects. Am J Clin Nutr 31: 1340–1346.

31. Brossard N, Croset M, Pachiaudi C, Riou JP, Tayot JL, et al. (1996) Retroconversion and metabolism of [13C]22:6n-3 in humans and rats after intake of a single dose of [13C]22:6n-3-triacylglycerols. Am J Clin Nutr 64: 577–586.

32. Conquer JA, Holub BJ (1996) Supplementation with an algae source of docosahexaenoic acid increases (n-3) fatty acid status and alters selected risk factors for heart disease in vegetarian subjects. J Nutr 126: 3032–3039.

33. Conquer JA, Holub BJ (1997) Dietary docosahexaenoic acid as a source of eicosapentaenoic acid in vegetarians and omnivores. Lipids 32: 341–345.

34. Gronn M, Christensen E, Hagve TA, Christophersen BO (1991) Peroxisomal retroconversion of docosahexaenoic acid (22:6(n-3)) to eicosapentaenoic acid (20:5(n-3)) studied in isolated rat liver cells. Biochim Biophys Acta 1081: 85–91.

35. Stark KD, Holub BJ (2004) Differential eicosapentaenoic acid elevations and altered cardiovascular disease risk factor responses after supplementation with docosahexaenoic acid in postmenopausal women receiving and not receiving hormone replacement therapy. Am J Clin Nutr 79: 765–773.

36. von Schacky C, Weber PC (1985) Metabolism and effects on platelet function of the purified eicosapentaenoic and docosahexaenoic acids in humans. J Clin Invest 76: 2446–2450.

37. Bruins MJ, Dane AD, Strassburg K, Vreeken RJ, Newman JW, et al. (2013) Plasma oxylipin profiling identifies polyunsaturated vicinal diols as responsive to arachidonic acid and docosahexaenoic acid intake in growing piglets. J Lipid Res 54: 1598–1607.

38. Parks JS, Bullock BC, Rudel LL (1989) The reactivity of plasma phospholipids with lecithin:cholesterol acyltransferase is decreased in fish oil-fed monkeys. J Biol Chem 264: 2545–2551.

39. Thornburg JT, Parks JS, Rudel LL (1995) Dietary fatty acid modification of HDL phospholipid molecular species alters lecithin: cholesterol acyltransferase reactivity in cynomolgus monkeys. J Lipid Res 36: 277–289.

40. Holub BJ, Bakker DJ, Skeaff CM (1987) Alterations in molecular species of cholesterol esters formed via plasma lecithin-cholesterol acyltransferase in human subjects consuming fish oil. Atherosclerosis 66: 11–18.

41. Parks JS, Thuren TY, Schmitt JD (1992) Inhibition of lecithin:cholesterol acyltransferase activity by synthetic phosphatidylcholine species containing eicosapentaenoic acid or docosahexaenoic acid in the sn-2 position. J Lipid Res 33: 879–887.

42. Nozaki S, Matsuzawa Y, Hirano K, Sakai N, Kubo M, et al. (1992) Effects of purified eicosapentaenoic acid ethyl ester on plasma lipoproteins in primary hypercholesterolemia. Int J Vitam Nutr Res 62: 256–260.

43. Satoh N, Shimatsu A, Kotani K, Himeno A, Majima T, et al. (2009) Highly purified eicosapentaenoic acid reduces cardio-ankle vascular index in association with decreased serum amyloid A-LDL in metabolic syndrome. Hypertens Res 32: 1004–1008.

44. Shikano M, Masuzawa Y, Yazawa K, Takayama K, Kudo I, et al. (1994) Complete discrimination of docosahexaenoate from arachidonate by 85 kDa cytosolic phospholipase A2 during the hydrolysis of diacyl- and alkenylacylglycerophosphoethanolamine. Biochim Biophys Acta 1212: 211–216.

45. Mori TA, Burke V, Puddey IB, Watts GF, O'Neal DN, et al. (2000) Purified eicosapentaenoic and docosahexaenoic acids have differential effects on serum lipids and lipoproteins, LDL particle size, glucose, and insulin in mildly hyperlipidemic men. Am J Clin Nutr 71: 1085–1094.

46. Kobatake Y, Kuroda K, Jinnouchi H, Nishide E, Innami S (1984) Differential effects of dietary eicosapentaenoic and docosahexaenoic fatty acids on lowering of triglyceride and cholesterol levels in the serum of rats on hypercholesterolemic diet. J Nutr Sci Vitaminol (Tokyo) 30: 357–372.

47. Jacobson TA, Glickstein SB, Rowe JD, Soni PN (2012) Effects of eicosapentaenoic acid and docosahexaenoic acid on low-density lipoprotein cholesterol and other lipids: a review. J Clin Lipidol 6: 5–18.

48. Stillwell W, Wassall SR (2003) Docosahexaenoic acid: membrane properties of a unique fatty acid. Chem Phys Lipids 126: 1–27.

49. Scott BL, Bazan NG (1989) Membrane docosahexaenoate is supplied to the developing brain and retina by the liver. Proc Natl Acad Sci U S A 86: 2903–2907.

50. Lagarde M, Bernoud N, Brossard N, Lemaitre-Delaunay D, Thies F, et al. (2001) Lysophosphatidylcholine as a preferred carrier form of docosahexaenoic acid to the brain. J Mol Neurosci 16: 201–204; discussion 215–221.

51. Bazan NG, Calandria JM, Gordon WC (2013) Docosahexaenoic acid and its derivative neuroprotectin D1 display neuroprotective properties in the retina, brain and central nervous system. Nestle Nutr Inst Workshop Ser 77: 121–131.

52. Block RC, Duff R, Lawrence P, Kakinami L, Brenna JT, et al. (2010) The effects of EPA, DHA, and aspirin ingestion on plasma lysophospholipids and autotaxin. Prostaglandins Leukot Essent Fatty Acids 82: 87–95.

53. White PJ, Arita M, Taguchi R, Kang JX, Marette A (2010) Transgenic restoration of long-chain n-3 fatty acids in insulin target tissues improves resolution capacity and alleviates obesity-linked inflammation and insulin resistance in high-fat-fed mice. Diabetes 59: 3066–3073.

54. Romanatto T, Fiamoncini J, Wang B, Curi R, Kang JX (2013) Elevated tissue omega-3 fatty acid status prevents age-related glucose intolerance in fat-1 transgenic mice. Biochim Biophys Acta 1842: 186–191.

55. Bellenger J, Bellenger S, Bataille A, Massey KA, Nicolaou A, et al. (2011) High pancreatic n-3 fatty acids prevent STZ-induced diabetes in fat-1 mice: inflammatory pathway inhibition. Diabetes 60: 1090–1099.

56. Lopez-Vicario C, Gonzalez-Periz A, Rius B, Moran-Salvador E, Garcia-Alonso V, et al. (2014) Molecular interplay between Delta5/Delta6 desaturases and long-chain fatty acids in the pathogenesis of non-alcoholic steatohepatitis. Gut 63: 344–355.

57. Merten M, Dong JF, Lopez JA, Thiagarajan P (2001) Cholesterol sulfate: a new adhesive molecule for platelets. Circulation 103: 2032–2034.

58. Wassall SR, Stillwell W (2008) Docosahexaenoic acid domains: the ultimate non-raft membrane domain. Chem Phys Lipids 153: 57–63.

59. Schofield M, Jenski LJ, Dumaual AC, Stillwell W (1998) Cholesterol versus cholesterol sulfate: effects on properties of phospholipid bilayers containing docosahexaenoic acid. Chem Phys Lipids 95: 23–36.

60. Morisseau C, Hammock BD (2013) Impact of soluble epoxide hydrolase and epoxyeicosanoids on human health. Annu Rev Pharmacol Toxicol 53: 37–58.

61. Shearer GC, Harris WS, Pedersen TL, Newman JW (2010) Detection of omega-3 oxylipins in human plasma and response to treatment with omega-3 acid ethyl esters. J Lipid Res 51: 2074–2081.

62. Newman JW, Stok JE, Vidal JD, Corbin CJ, Huang Q, et al. (2004) Cytochrome p450-dependent lipid metabolism in preovulatory follicles. Endocrinology 145: 5097–5105.

63. Keenan AH, Pedersen TL, Fillaus K, Larson MK, Shearer GC, et al. (2012) Basal omega-3 fatty acid status affects fatty acid and oxylipin responses to high-dose n3-HUFA in healthy volunteers. J Lipid Res 53: 1662–1669.

64. Arnold C, Markovic M, Blossey K, Wallukat G, Fischer R, et al. (2010) Arachidonic acid-metabolizing cytochrome P450 enzymes are targets of {omega}-3 fatty acids. J Biol Chem 285: 32720–32733.

65. Capdevila JH, Falck JR (2002) Biochemical and molecular properties of the cytochrome P450 arachidonic acid monooxygenases. Prostaglandins Other Lipid Mediat 68–69: 325–344.

66. Spector AA (2009) Arachidonic acid cytochrome P450 epoxygenase pathway. J Lipid Res 50 Suppl: S52–56.

67. Zeldin DC (2001) Epoxygenase pathways of arachidonic acid metabolism. J Biol Chem 276: 36059–36062.

68. Morin C, Sirois M, Echave V, Albadine R, Rousseau E (2010) 17,18-epoxyeicosatetraenoic acid targets PPARgamma and p38 mitogen-activated protein kinase to mediate its anti-inflammatory effects in the lung: role of soluble epoxide hydrolase. Am J Respir Cell Mol Biol 43: 564–575.

69. Morisseau C, Inceoglu B, Schmelzer K, Tsai HJ, Jinks SL, et al. (2010) Naturally occurring monoepoxides of eicosapentaenoic acid and docosahexaenoic acid are bioactive antihyperalgesic lipids. J Lipid Res 51: 3481–3490.

70. VanRollins M (1995) Epoxygenase metabolites of docosahexaenoic and eicosapentaenoic acids inhibit platelet aggregation at concentrations below those affecting thromboxane synthesis. J Pharmacol Exp Ther 274: 798–804.

71. Morin C, Sirois M, Echave V, Rizcallah E, Rousseau E (2009) Relaxing effects of 17(18)-EpETE on arterial and airway smooth muscles in human lung. Am J Physiol Lung Cell Mol Physiol 296: L130–139.

72. Arnold C, Konkel A, Fischer R, Schunck WH (2010) Cytochrome P450-dependent metabolism of omega-6 and omega-3 long-chain polyunsaturated fatty acids. Pharmacol Rep 62: 536–547.

6

Roles of Raft-Anchored Adaptor Cbp/PAG1 in Spatial Regulation of c-Src Kinase

Takashi Saitou[1]*[⑨], Kentaro Kajiwara[2]*[⑨], Chitose Oneyama[2], Takashi Suzuki[3,4], Masato Okada[2]

1 Department of Molecular Medicine for Pathogenesis, Graduate School of Medicine, Ehime University, Shitsukawa, Toon, Ehime, Japan, **2** Department of Oncogene Research, Research Institute for Microbial Diseases, Osaka University, Suita, Osaka, Japan, **3** Division of Mathematical Science, Department of Systems Innovation, Graduate School of Engineering Science, Osaka University, Toyonaka, Osaka, Japan, **4** JST, CREST, Chiyoda-ku, Tokyo, Japan

Abstract

The tyrosine kinase c-Src is upregulated in numerous human cancers, implying a role for c-Src in cancer progression. Previously, we have shown that sequestration of activated c-Src into lipid rafts via a transmembrane adaptor, Cbp/PAG1, efficiently suppresses c-Src-induced cell transformation in Csk-deficient cells, suggesting that the transforming activity of c-Src is spatially regulated via Cbp in lipid rafts. To dissect the molecular mechanisms of the Cbp-mediated regulation of c-Src, a combined analysis was performed that included mathematical modeling and *in vitro* experiments in a c-Src- or Cbp-inducible system. c-Src activity was first determined as a function of c-Src or Cbp levels, using focal adhesion kinase (FAK) as a crucial c-Src substrate. Based on these experimental data, two mathematical models were constructed, the sequestration model and the ternary model. The computational analysis showed that both models supported our proposal that raft localization of Cbp is crucial for the suppression of c-Src function, but the ternary model, which includes a ternary complex consisting of Cbp, c-Src, and FAK, also predicted that c-Src function is dependent on the lipid-raft volume. Experimental analysis revealed that c-Src activity is elevated when lipid rafts are disrupted and the ternary complex forms in non-raft membranes, indicating that the ternary model accurately represents the system. Moreover, the ternary model predicted that, if Cbp enhances the interaction between c-Src and FAK, Cbp could promote c-Src function when lipid rafts are disrupted. These findings underscore the crucial role of lipid rafts in the Cbp-mediated negative regulation of c-Src-transforming activity, and explain the positive role of Cbp in c-Src regulation under particular conditions where lipid rafts are perturbed.

Editor: Mikel Garcia-Marcos, Boston University School of Medicine, United States of America

Funding: This work was supported by the Japan Science Technology Agency (JST), Core Research for Evolutional Science and Technology (CREST). The funders had no role in study design, data collection and analysis, decision to publish, or preparation of the manuscript.

Competing Interests: The authors have declared that no competing interests exist.

* E-mail: kajiwara@biken.osaka-u.ac.jp (KK); t-saito@m.ehime-u.ac.jp (TS)

⑨ These authors contributed equally to this work.

Introduction

The first identified proto-oncogene product, c-Src [1], is a membrane-associated non-receptor tyrosine kinase that plays pivotal roles in coordinating a broad range of cellular responses such as differentiation, proliferation, adhesion, and migration [2]. The kinase activity of c-Src is enhanced by autophosphorylation at tyrosine 418 (Y418) in response to extracellular stimuli such as growth factors and extracellular matrices, while c-Src activity is negatively regulated by phosphorylation of its regulatory tyrosine 529 (Y529), which is catalyzed by the C-terminal Src kinase (Csk) [3–5]. The regulatory mechanism for c-Src function has been extensively analyzed by molecular studies [6] as well as theoretical studies [7–10], but c-Src signaling dynamics and their roles in cell physiology and diseases such as cancer are not yet fully understood.

c-Src is frequently overexpressed and activated in a wide variety of human cancers [6,11,12], despite the fact that Csk is normally expressed and the *c-src* gene is not mutated [13,14]. These observations suggest that other components in the c-Src regulatory system may be perturbed during cancer progression, although the underlying mechanisms remain unclear. Upregulation of c-Src has

been implicated in cancer invasion and metastasis, which are associated with the activation of cell-migration machinery [15]. Cell migration is mediated by the formation and disassembly of focal adhesions [16], which is controlled by c-Src-mediated phosphorylation of focal adhesion components such as focal adhesion kinase (FAK) and cortactin [17,18]. Constitutive activation of FAK promotes not only focal adhesion turnover, but also cell growth and survival signaling, thereby promoting tumor progression [16]. Upon activation of c-Src, c-Src and FAK tightly interact to phosphorylate and activate each other, but the mechanism through which activated c-Src efficiently accesses FAK remains elusive.

c-Src is anchored to the membrane via its myristoylated N-terminus, whereas Csk is a cytoplasmic protein; thus Csk requires membrane-anchor proteins to efficiently access c-Src. A transmembrane phosphoprotein, Csk-binding protein (Cbp) [19], also called PAG1 [20] but hereafter referred to as Cbp, has been identified as such a membrane anchor for Csk. Cbp is exclusively localized to lipid rafts, membrane microdomains enriched in cholesterol and sphingolipids [21], by palmitoylation anchoring. Lipid rafts have been regarded as signaling platforms that harbor various signaling molecules and positively transduce cell signaling, although the specific function of lipid rafts is still under debate [22]. In normal cells, Cbp in lipid rafts plays a scaffolding role in the Csk-dependent negative regulation of c-Src [19,23]. We analyzed the role of Cbp in the regulation of the transforming activity of c-Src using Csk-deficient cells as a model system [24]. This system enabled us to dissect the initial events following c-Src activation, and to mimic the activity status of c-Src in cancer cells, in which the proportion of c-Src present in the active form is increased despite the expression of Csk [24,25]. Because Csk-deficient cells can be transformed by expression of a limited amount of wild-type (WT) c-Src, the cells were also used to identify c-Src targets required for cell transformation. In this system, we found that phosphorylated Cbp binds to c-Src via its Src homology 2 (SH2) domain and recruits active c-Src to lipid rafts, and that sequestration of c-Src in lipid rafts is sufficient to suppress c-Src-mediated transformation (Figure 1) [25,26]. The tumor-suppressive role of Cbp was verified in v-Src-transformed cells and in human cancer cells, both of which express Csk [26].

We also showed that the distribution of Src family kinases (SFKs) to lipid rafts varies, depending upon their N-terminal fatty-acylation status [26]. c-Src and Blk, which have a single myristoyl moiety, have a low affinity for lipid rafts and mostly distribute to non-raft membranes when Cbp is downregulated. By contrast, other SFK members, such as Lyn and Fyn, which contain additional palmitoyl moieties, have a higher affinity for lipid rafts, and can therefore distribute to lipid rafts independently of Cbp. It was reported that Cbp could function positively in SFK-mediated signal transduction when bound to raft-localized SFKs, such as Lyn and Fyn, in lipid rafts [27–29]. These findings suggest that Cbp positively supports the function of SFKs that are intrinsically localized in lipid rafts, while it negatively regulates non-raft c-Src/Blk by spatially controlling their raft localization. However, the mechanisms by which Cbp exerts such reciprocal functions remain unknown.

In this study, we addressed the molecular mechanisms for the Cbp-mediated spatial regulation of the transforming activity of c-Src. For this purpose, a mathematical modeling analysis was combined with in vitro experiments using a c-Src or Cbp inducible system in Csk-deficient cells. This system is suitable for analyzing the initial processes in cell transformation or its suppression [30]. The combined analysis reveals that in addition to Cbp-Src and Src-FAK complexes, a ternary complex consisting of Cbp, c-Src

and FAK is required for the Cbp-mediated regulation of c-Src via lipid rafts. Moreover, the proposed model predicts that, if Cbp enhances the interaction between c-Src and FAK, Cbp can further promote c-Src function when lipid rafts are perturbed. These findings underscore the crucial role of lipid rafts in the Cbp-mediated regulation of c-Src, and suggest that Cbp can positively regulate c-Src function under particular conditions, such as where lipid rafts are perturbed or c-Src substrates are present in lipid rafts.

Results

In vitro analysis of c-Src activity regulated by Cbp and lipid rafts

To analyze the regulation of c-Src activity by Cbp and lipid rafts, two cell lines were used, as follows: Csk-deficient mouse embryonic fibroblasts (MEFs) harboring a doxycycline (Dox)-inducible c-Src expression system (Csk$^{-/-}$ MEF/pBKT2-c-Src), and c-Src-expressing Csk$^{-/-}$ MEFs harboring a Dox-inducible Cbp expression system (Csk$^{-/-}$ MEF/c-Src/pBKT2-Cbp) [30]. Dox-induced c-Src expression can induce cell transformation, and the c-Src-induced transformation is efficiently suppressed by Dox-induced Cbp expression as observed previously [30].

The expression and phosphorylation levels of c-Src were first determined. Analysis of whole-cell lysates showed that c-Src protein expression and phosphorylation at tyrosine 418 (pY418) were induced by Dox treatment in a time-dependent manner (Figure 2A). During this period, Cbp levels were unchanged, although there were band shifts due to phosphorylation by c-Src. The raft localization of activated c-Src was next assessed by separating detergent-resistant membrane (DRM) (lipid rafts) and non-DRM fractions (non-raft membranes). The DRM-separation assay showed that activated c-Src was predominantly localized to non-DRM fractions (Figure 2G), where FAK and cortactin were also localized [25,31]. Following c-Src induction, phosphorylation of FAK at tyrosine 576 (pY576), which enhances its activity [16], was dramatically elevated (Figure 2A) in non-DRM fractions (Figure 2G). The relative intensity of FAK phosphorylation (p-FAK/FAK) increased as c-Src levels increased (Figure 2B). Likewise, c-Src induction increased the phosphorylation of cortactin at tyrosine 421 (pY421) (Figure 2A and 2C) in non-DRM fractions (Figure 2G). These observations demonstrate that c-Src activation induces concurrent phosphorylation of its

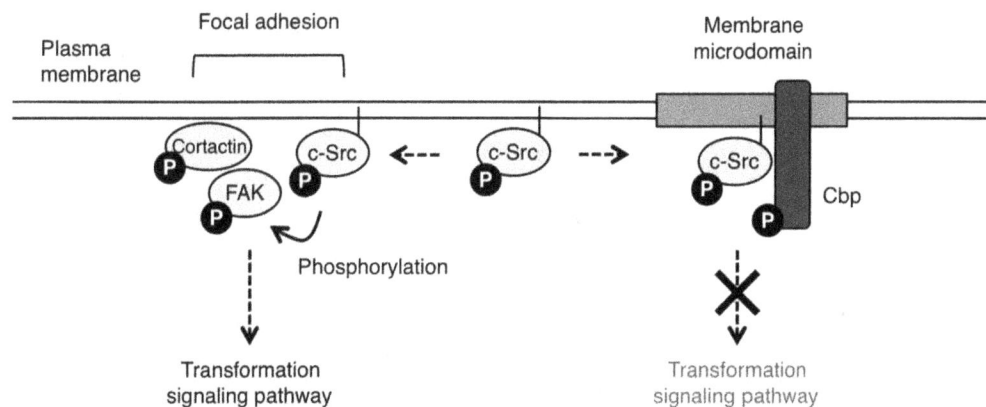

Figure 1. Proposed hypothesis for the spatial control of c-Src phosphorylation. c-Src is anchored to the plasma membrane via its N-terminal myristoyl modification. In focal adhesions, activated c-Src phosphorylates its substrates, FAK and cortactin. When Cbp is expressed, activated c-Src is retained in lipid rafts by Cbp. This results in the sequestration of c-Src from Src substrates, thereby suppressing the phosphorylation of Src substrates.

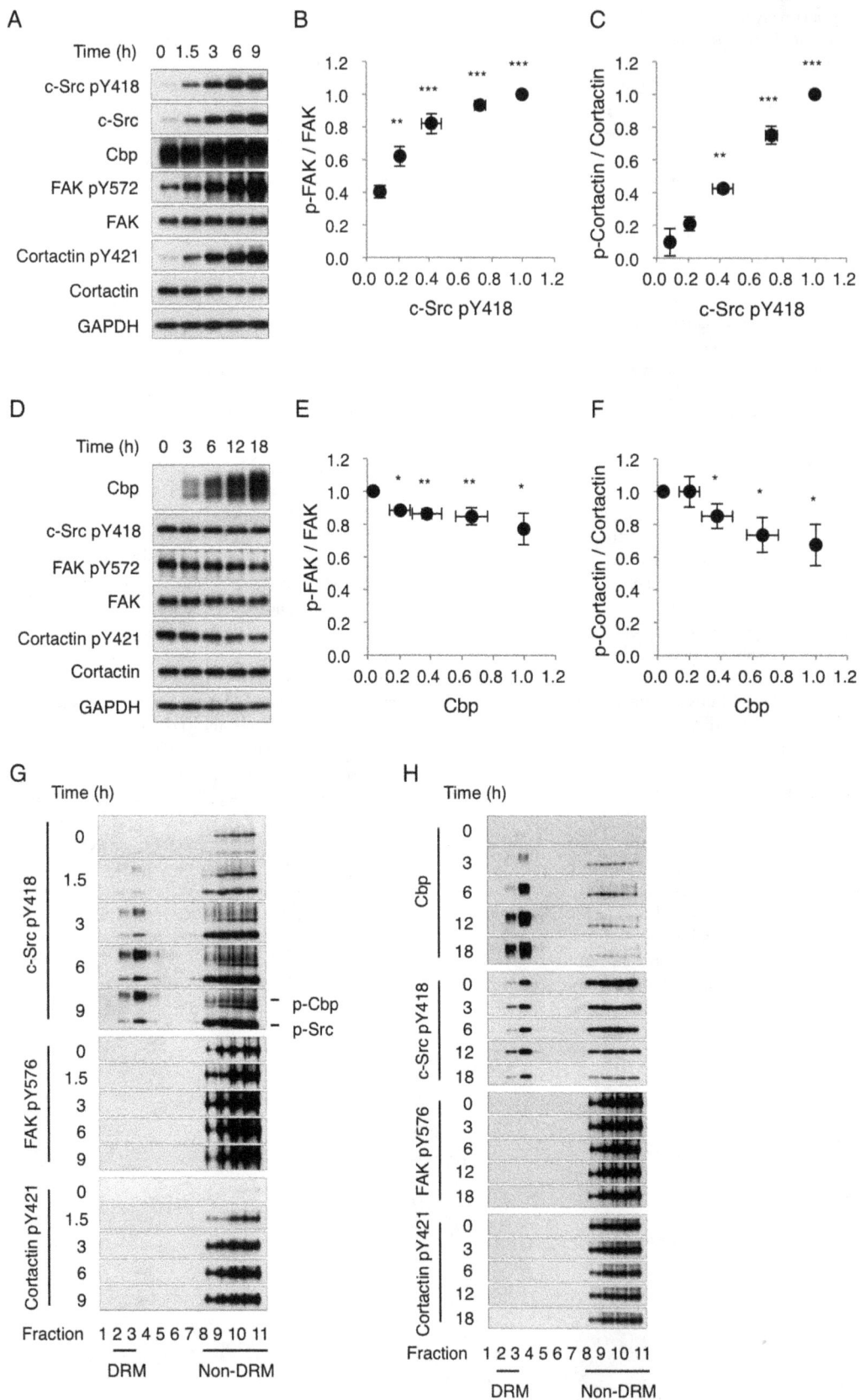

Figure 2. Phosphorylation status of FAK and cortactin following expression of c-Src and Cbp. (A and D) Csk$^{-/-}$ MEFs harboring pBKT2-c-Src (Csk$^{-/-}$ MEF/p-BKT2-c-Src cells, A) and c-Src-expressing Csk$^{-/-}$ MEFs harboring pBKT2-Cbp (Csk$^{-/-}$ MEF/c-Src/p-BKT2-Cbp cells, D) were incubated with or without 1 µg/ml Dox for the indicated time periods. Cell lysates from these cells were subjected to immunoblot analysis with the indicated antibodies. (B, C, E, and F) Relative intensity of phosphorylated FAK and cortactin as a function of c-Src and Cbp. The band intensities of c-

Src pY418 and Cbp were normalized to the GAPDH band intensities, and then the relative intensities were calculated by dividing the band intensities of 9 h treated Csk$^{-/-}$ MEF/p-BKT2-c-Src cells (B and C) and 18 h treated Csk$^{-/-}$ MEF/c-Src/p-BKT2-Cbp cells (E and F). The band intensities of phosphorylated FAK and cortactin were normalized as p-FAK/FAK and p-cortactin/cortactin, respectively. The relative intensities were calculated in the same way as employed for the c-Src and Cbp intensity calculation. Means ± standard deviations (SD) were obtained from three independent experiments. (G and H) Csk$^{-/-}$ MEF/p-BKT2-c-Src cells (G) and Csk$^{-/-}$ MEF/c-Src/p-BKT2-Cbp cells (H) were incubated with or without 1 µg/ml Dox for the indicated time periods. DRM and non-DRM fractions of treated cells were separated on a sucrose density gradient. Aliquots of the fractions were immunoblotted with the indicated antibodies.

substrates, FAK and cortactin, in non-raft membranes, a process associated with cell transformation [30].

In the Cbp-inducible cells, expression of Cbp was also induced by Dox treatment in a time-dependent manner (Figure 2D). However, c-Src activation (pY418) did not change during this time period (Figure 2D), indicating that Cbp overexpression does not inhibit total c-Src kinase activity. The DRM-separation assay showed that Cbp is exclusively distributed to DRM fractions, and that the levels of activated c-Src in non-DRM fractions decreased as Cbp levels increased (Figure 2H). This indicates that activated c-Src is sequestered into lipid rafts by Cbp overexpression. Consistent with the reduction of c-Src in non-DRM fractions, phosphorylation of FAK (p-FAK/FAK) and cortactin (p-cortactin/cortactin) was decreased in a Cbp-dependent manner (Figure 2D-2F and 2H). These finding support our previous observation that Cbp serves as a suppressor of activated c-Src by sequestering it into lipid rafts [25,26,31].

Mathematical modeling

Based on the *in vitro* experimental data, mathematical models were next developed for the interactions between lipid rafts, c-Src, Cbp, and FAK. To construct possible models, we postulated that c-Src signaling is carried out through a series of tyrosine phosphorylation events in two cellular compartments, lipid rafts and non-raft membranes. All reactions were assumed to be the same in both lipid rafts and non-raft membranes. Two possible reaction schemes can be considered based on the literature. We previously showed that c-Src binds to Cbp through its SH2 domain and forms a complex, Cbp-Src [19,25]. c-Src also binds to FAK through its SH2 domain [16], and possibly through its SH3 domain [32]. Therefore, by assuming that phosphorylation of FAK by c-Src occurs through a Michaelis-Menten-type reaction and that its dephosphorylation is simply described by first-order kinetics, we developed a simple model, the sequestration model (Figure 3A). However, the possibility remains that the Cbp-Src complex is able to bind to Src substrates (SS) and form a ternary complex, Cbp-Src-FAK, to phosphorylate FAK, which led us to develop an alternate model, the ternary model (Figure 3A). We then considered these two models to determine which one best reflects the system. The sequestration model consists of five biochemical reactions with three elementary species: activated c-Src, Cbp, and FAK, in two compartments. On the other hand, the ternary model consists of seven biochemical reactions with three elementary species: activated c-Src, Cbp, and FAK, in two compartments. In these two models, all biochemical reactions are modeled by a system of ordinary differential equations (ODEs). A schematic representation of the models is shown in Figure 3A. The full sets of equations are shown in Materials and Methods. The parameters are assigned as $ks1$: Src and FAK binding constant, k_s1: Src and FAK dissociation constant, $ks2$: Cbp-Src and FAK binding constant, k_s2: Cbp-Src and FAK dissociation constant, kc: Cbp and Src binding rate constant, k_c: Cbp and Src dissociation rate constant, $kp1$: FAK phosphorylation rate constant through the Src-FAK complex, $kp2$: FAK phosphory-

lation rate constant through Cbp-Src-FAK complex, and kd: FAK dephosphorylation rate constant.

Shuttling behavior into and out of lipid rafts is described by first order kinetics, according to a previous report [33]. Thus, molecular localization in lipid rafts and non-raft membranes can be described by adjusting the rate constants. The experimental data show that Cbp is predominantly located to lipid rafts independently of c-Src-expression levels (Figure 2G), and that when Cbp expression is induced, c-Src is relocated to lipid rafts (Figure 2H). These results indicate that the Cbp-Src complex is distributed to lipid rafts. In contrast, FAK is mostly located to non-raft membranes (Figure 2G) [25,31], enabling us to assume that the Src-FAK complex is formed in non-raft membranes and that the Cbp-Src-FAK complex can also be located in non-raft membranes. The parameters assigned are k_{cin}: import rate (into raft fractions) for Cbp and Cbp-Src, k_{cout}: export rate (from raft fractions) for Cbp and Cbp-Src, k_{sin}: import rate for Src, k_{sout}: export rate for Src, k_{ssin}: import rate for Src-FAK and Cbp-Src-FAK, and k_{ssout}: export rate for Src-FAK and Cbp-Src-FAK.

Computational analysis revealed that only the ternary model includes the raft-volume dependence on FAK phosphorylation

Parameters were searched by random sampling (Text S1 and S2, Figure S1, Table S1), and the FAK phosphorylation curve obtained using the estimated parameter values was superimposed on the experimental data (Figure 3B and S2, Table S2 and S3). The properties of the steady-state solution of the two models were then compared. The dependence of raft volume on c-Src function was first examined by modulating Vr, which represents the ratio of the raft volume to the total membrane volume, beginning with $Vr = 0.1$. In the sequestration model, decreasing Vr did not affect the phosphorylation level of FAK at any concentration of c-Src (Figure 4A), while in the ternary model, as the Vr decreased, the phosphorylation of FAK increased at all concentrations of c-Src (Figure 4B). Even when c-Src activity was analyzed as a function of the Cbp level, a decrease in Vr had no effect on the phosphorylation of FAK in the sequestration model (Figure 4C and 4E), while a decrease in Vr attenuated the inhibitory activity of Cbp on the phosphorylation of FAK in the ternary model (Figure 4D and 4F). Interestingly, when Vr was decreased to 0.01 (a condition where a large amount of Cbp is distributed to non-raft fractions) Cbp loses its inhibitory activity on c-Src function, irrespective of Cbp concentration (Figure 4D). We further investigated whether these properties depended on the initial choice of raft-volume ratio, Vr. The raft-volume ratio was therefore set to $Vr = 0.05$, and the parameters were re-estimated by random sampling, as was done previously. The results showed that the raft-volume dependence is independent of the initial choice of Vr (Figure S3). These results support our proposal that raft localization of Cbp is required to exert an inhibitory effect on c-Src function. Since Cbp is exclusively localized to lipid rafts under normal conditions, the model reflects our *in vitro* observations and underscores the importance of lipid rafts in the negative regulation of c-Src.

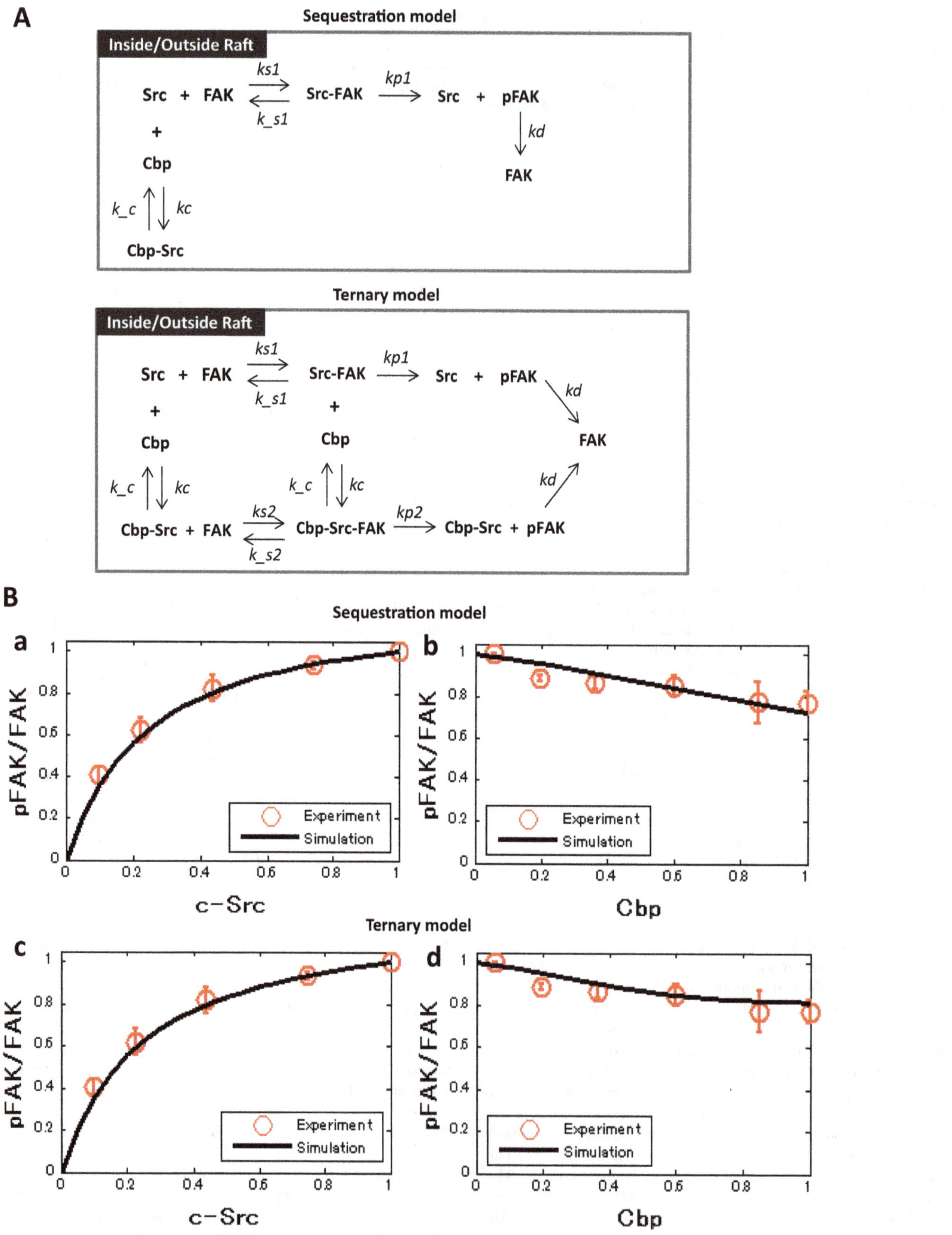

A

B

Figure 3. Schematic representation of mathematical models. (A) Two possible mechanisms are proposed based on the literature. c-Src binds to Cbp through its SH2 domain and forms a complex, Cbp-Src. c-Src also binds to FAK through its SH3 domain and phosphorylates it by a Michaelis-Menten mechanism. In the sequestration model, Cbp inhibits c-Src activity in a competitive way by sequestering c-Src. The ternary model was developed by adding the ternary complex consisting of Cbp, c-Src, and FAK to the sequestration model. Reaction schemes are assumed to be similar in raft and non-raft compartments. (B) Simulated phosphorylation curve of FAK. Simulated data are shown in black lines. Experimental data points depicted in red circles with bars show the mean ± SD (n=3). (a) FAK phosphorylation ratio as a function of the total Src concentration for the sequestration model. (b) FAK phosphorylation ratio as a function of the total Cbp concentration for the sequestration model. (c) FAK phosphorylation ratio as a function of the total Src concentration for the ternary model. (d) FAK phosphorylation ratio as a function of the total Cbp concentration for the ternary model.

Experimental validation of the raft volume dependence on FAK phosphorylation

To determine which model is more appropriate, the raft-volume dependence on FAK phosphorylation was examined in *in vitro* experiments. Based on the ternary model, when Cbp is excluded from lipid rafts, the active Cbp-Src complex can efficiently phosphorylate non-raft substrates. To validate this proposal, the effect of forced exclusion of Cbp from lipid rafts on c-Src function

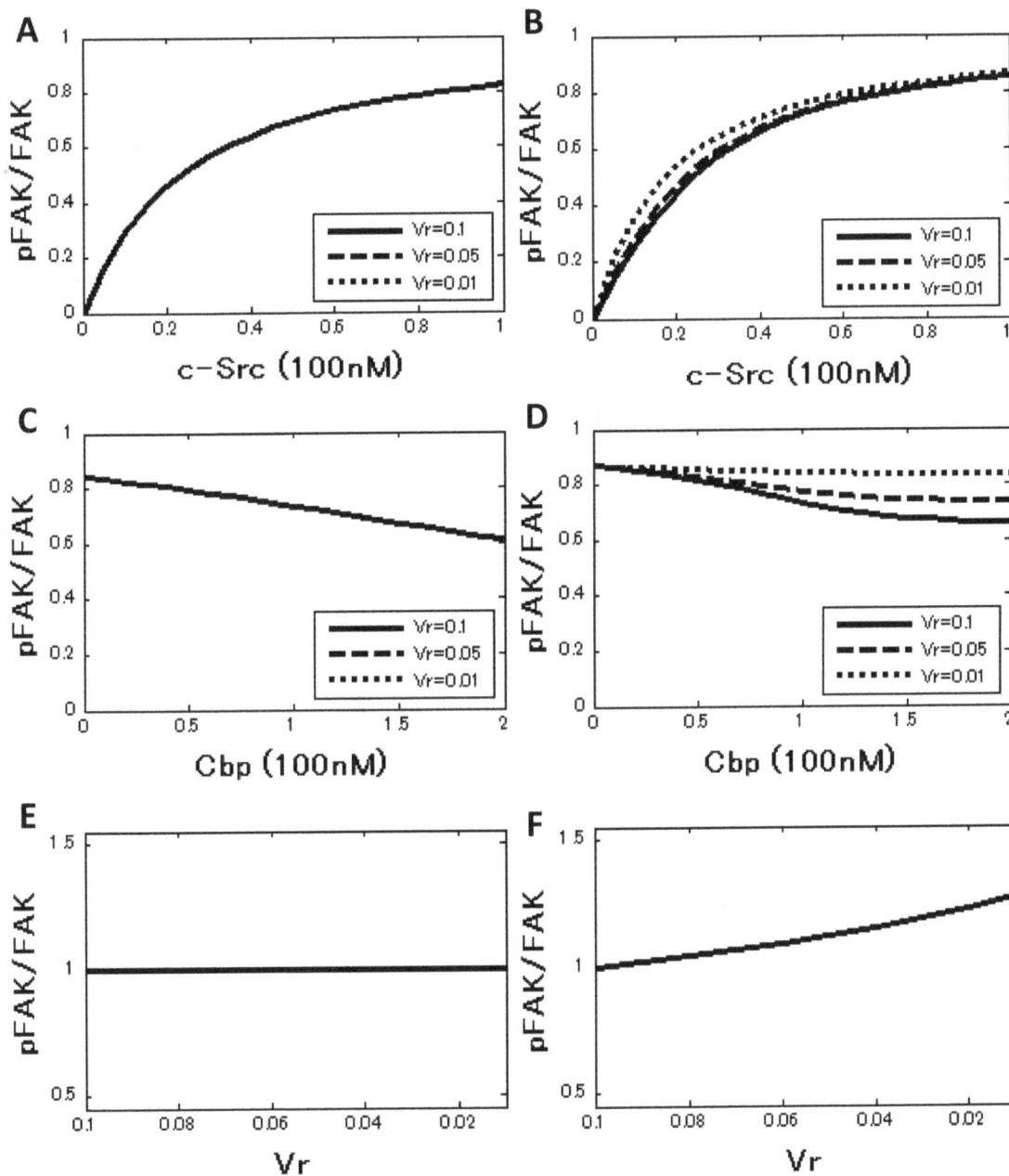

Figure 4. Simulated phosphorylation curve for the sequestration model and the ternary model. (A and B) FAK phosphorylation ratio as a function of the total c-Src concentration for three different raft-volume ratios in the sequestration model (A) and the ternary model (B). The black line denotes the raft-volume ratio *Vr*=0.1 (control), the dashed line denotes *Vr*=0.05, and the pointed line denotes *Vr*=0.01. (C and D) FAK phosphorylation ratio as a function of the total Cbp concentration for three different raft-volume ratios in the sequestration model (C) and the ternary model (D). (E and F) The fold change in FAK phosphorylation level with respect to the raft-volume ratio *Vr*. The values are normalized by *Vr*=0.1.

was examined. For this, lipid rafts were disrupted by depleting cholesterol with methyl-β-cyclodextrin (MβCD) in Csk-deficient cells. Treatment with MβCD reduced cholesterol levels (data not shown) and induced the relocation of the raft marker proteins flotillin-1 and caveolin-1 from DRM fractions to non-DRM fractions (Figure 5A), indicating that lipid rafts were successfully disrupted. Under these conditions, active c-Src and phosphorylated Cbp were relocated from DRM fractions to non-DRM fractions (Figure 5A). An immunoprecipitation assay revealed that c-Src and Cbp interacted with each other, and that the Cbp-Src complex was relocated to non-DRM fractions in a manner dependent on MβCD concentration (Figure 5B). Furthermore, association of FAK with this complex was detected only in non-DRM fractions (Figure 5B). Consistent with a previous report [26], FAK phosphorylation was significantly enhanced by MβCD treatment (Figure 5C and 5D). Phosphorylation of ERK, a downstream component of FAK signaling, was also significantly increased, confirming the activation of the FAK signaling pathway. These observations demonstrate that the ternary complex, Cbp-Src-FAK, is indeed formed in non-raft membranes and that the phosphorylation of FAK increases when raft volume is reduced. These *in vitro* experimental results are in agreement with the assumptions and simulation results of the ternary model (Figure 4B, 4D and 4F).

Computational analysis revealed that enhancement of the association of c-Src with FAK by Cbp is required for the positive regulatory role of Cbp

To evaluate the potential positive role of Cbp in c-Src regulation, computational analysis was again performed by adopting the ternary model. We previously found that phosphorylated Cbp binds Csk tyrosine kinase at its SH2 domain and enhances its catalytic activity [23]. Thus, the contribution of c-Src activation by Cbp was input into the ternary model. The rate constants were first set as $ks2 = ks1$, $k\_s2 = k\_s1$, and $kp2 = \alpha \cdot kp1$ ($\alpha = 2$), where α is defined as the activating potential of the rate of FAK phosphorylation, and then random parameter sets were re-sampled and the parameter values were re-estimated. As in the previous result (Figure 4D), the phosphorylation of FAK decreased as Cbp increased, while Cbp lost its inhibitory activity on c-Src function irrespective of Cbp concentration when Vr was decreased to 0.01 (Figure 6A). Setting the activating potential $\alpha = 10$ resulted in no significant difference from the $\alpha = 2$ results (Figure S4A), indicating that sequestration of c-Src into lipid rafts by binding to Cbp can suppress c-Src function even though c-Src catalytic activity is enhanced at least 10-fold. The rate constants were next set as $ks2 = \beta \cdot ks1$ ($\beta = 2$), $k\_s2 = k\_s1$, and $kp2 = kp1$, where β is defined as the activating potential of the Src/FAK

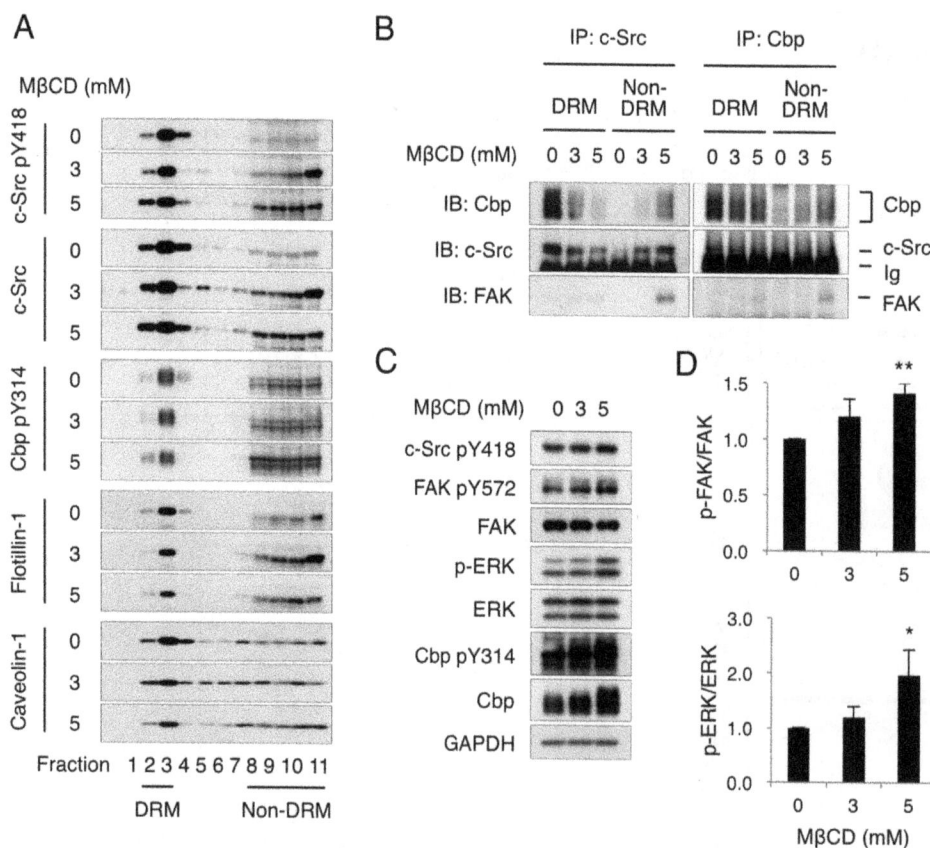

Figure 5. Experimental validation and prediction for the ternary model. (A) Csk$^{-/-}$ MEFs were incubated with MβCD at the indicated concentrations for 1 h. DRM and non-DRM fractions of treated cells were separated on sucrose density gradient. Aliquots of the fractions were immunoblotted with the indicated antibodies. (B) DRM and non-DRM fractions of the MβCD-treated Csk$^{-/-}$ MEFs were subjected to immunoprecipitation (IP) with anti-Src or anti-Cbp antibodies, followed by immunoblotting with the indicated antibodies. Ig, immunoglobulin heavy chain. (C) Total cell lysates from the MβCD-treated Csk$^{-/-}$ MEFs were subjected to immunoblot analysis with the indicated antibodies. (D) The phosphorylation status of FAK (p-FAK/FAK) and ERK (p-ERK/ERK) was obtained from the band intensities of panel C and calculated by defining the value for non-treated controls as 1.0. The relative specific activities were calculated by defining the value for non-treated controls as 1.0. Means ± SD were obtained from three independent experiments. $^*P < 0.05$, $^{**}P < 0.01$, Student's t-test.

binding rate. When $Vr = 0.1$, the phosphorylation of FAK decreased as Cbp increased. However, when Vr was set to 0.01, the phosphorylation of FAK was slightly elevated relative to the levels observed before Cbp induction (Figure 6B). When α and β were set to 2 to investigate the synergy of the activating potentials α and β, enhancement of FAK phosphorylation was also observed, but no significant difference was observed compared to the case when $\alpha = 1$ and $\beta = 2$ (Figure S4B). The effect of the initial choice of raft-volume ratio, Vr, on these properties was next investigated. The results showed that the raft-volume dependence was independent of the initial choice of Vr (Figure S5), suggesting that enhancement of affinity for substrates ($ks2$) can contribute to the positive effect of Cbp on c-Src function. Therefore, these observations imply that if Cbp directly induces the association of c-Src with FAK, Cbp can exert a positive effect on c-Src function when the Cbp-Src complex is delivered to non-raft membranes, where it can encounter substrates.

Discussion

This study addressed molecular mechanisms of the spatial regulation of c-Src by combining mathematical modeling and *in vitro* experimental approaches. Based on initial experimental data, two mathematical models, the sequestration model and the ternary model, were developed, and their properties were investigated by random sampling approaches. Both models support our proposal that raft localization of Cbp is crucial for suppression of c-Src function, but the ternary model predicts the dependence of c-Src function on the lipid-raft volume. Raft disruption experiments using MβCD showed that c-Src function depends on raft volume and that a ternary complex consisting of Cbp, c-Src and FAK is formed in non-raft fractions, indicating that the ternary model best describes the system. To address the potential positive role of Cbp, computational analysis was also performed by setting the ternary model to include the effect of Cbp binding on c-Src catalytic activity. Analyses using an increased binding constant for the c-Src and FAK complex suggest that if Cbp is excluded from lipid rafts, it can positively regulate c-Src activity. This prediction was supported by the previous observation that disruption of lipid rafts in Csk-deficient cells induced c-Src-mediated cell transformation [26], although detailed experimental analysis will be necessary to elucidate the mode of c-Src activation. These findings not only underscore the crucial role of lipid rafts in the Cbp-mediated negative regulation of c-Src transforming activity, but may also

explain the positive regulation of c-Src by Cbp under certain conditions where lipid rafts are perturbed.

To analyze the spatial regulation of the transforming activity of c-Src, Csk-deficient cells were used as a model system. Cbp serves as an initial substrate of c-Src, and phosphorylated Cbp recruits cytosolic Csk to lipid rafts to allow close contact with c-Src [19]. Since this negative feedback mechanism occurs constitutively, it has been difficult to analyze the initial events of c-Src activation in normal fibroblasts. In some cancer cells, however, a substantial population of c-Src is present as an active form even though Csk is expressed [25]. Therefore, to dissect the initial molecular events following c-Src activation, and to mimic the activity status of c-Src in cancer cells, we used Csk-deficient cells. This system was successfully used to detect the direct association between activated c-Src and Cbp, and characterize the role of raft-anchored Cbp as a negative regulator of c-Src-mediated transformation. This tumor-suppressive function of Cbp has been verified in v-Src-transformed cells and some human cancer cells, both of which express Csk [26]. We previously found in this system that disruption of lipid rafts was sufficient to induce robust cell transformation [26]. Analysis of this phenomenon showed that liberation of c-Src from lipid rafts allows c-Src to make contact with non-raft substrates, such as FAK, which triggers cell transformation. To validate this mechanism and gain new insights into the mechanisms regulating cell transformation, we conducted a simulation study using a c-Src- or Cbp-inducible system as an experimental model [30]. However, since Csk is ubiquitously expressed even in cancer cells [34], Csk function will be incorporated into these models in follow-up studies, particularly those investigating regulatory mechanisms for stimulus-dependent c-Src function.

c-Src activity is regulated by intramolecular domain interaction [6]. When the negative regulatory tyrosine in the C-terminus is phosphorylated, its own SH2 domain interacts with this site to adopt a closed conformation that cannot access substrates. As phosphorylated Cbp directly interacts with c-Src SH2 domain [19,26], it is likely that Cbp can induce a conformation change in c-Src, resulting in an increase in its catalytic activity as well as its affinity for substrates. Computational analysis revealed that if Cbp activates the association of c-Src with FAK, Cbp could exert a positive effect on c-Src function. Since c-Src interacts with FAK through its SH2 and/or SH3 domain [32,35], it is possible that the binding of Cbp to the c-Src SH2 domain may promote the

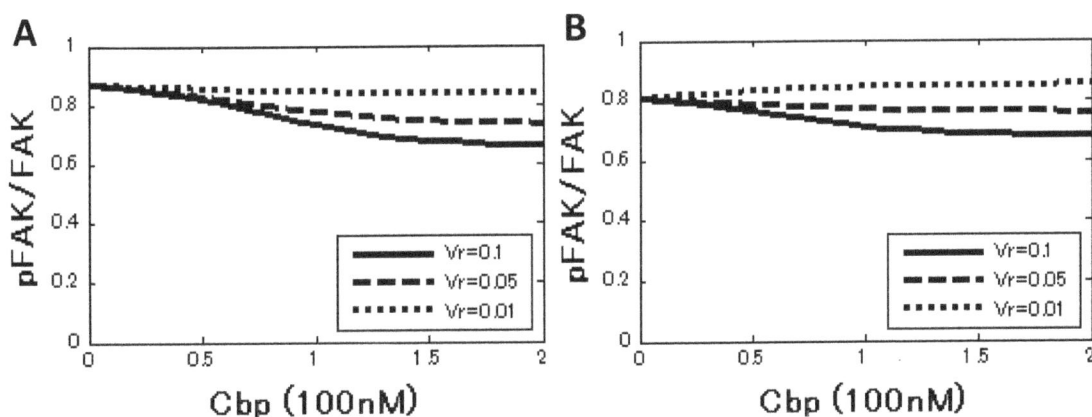

Figure 6. Simulated phosphorylation curve for the effects of Cbp on c-Src function. (A) FAK phosphorylation ratio as a function of the total Cbp concentration. The phosphorylation rate constant is set as $kp2 = 2*kp1$ ($\alpha = 2$). (B) FAK phosphorylation ratio as a function of the total Cbp concentration. The Src-FAK binding rate constant is set as $ks2 = 2*ks1$ ($\beta = 2$).

interaction of FAK with the c-Src SH3 domain. Although the complete mechanisms of this interaction must await future study, the association of FAK with the Cbp-Src complex would further enhance the mutual activation between c-Src and FAK, promoting cell transformation, as observed in the previous study [25].

Scaffold proteins play key roles in regulating signal transduction [36–39], physically assembling pathway components and sequestering them into subcellular compartments. By binding and spatiotemporally organizing pathway components, scaffold proteins promote cell signaling along a particular pathway. Our analysis showed that Cbp plays a scaffolding role in c-Src signaling by sequestering c-Src and its substrates. Because c-Src serves as a signaling hub in response to diverse extracellular stimuli, Cbp may regulate the multiple functions of c-Src by spatiotemporally controlling c-Src activity. Therefore, it would be interesting to investigate how Cbp is involved in controlling stimulus-dependent c-Src signaling.

The SFKs Fyn and Lyn also interact with Cbp through their SH2 and SH3 domains [28,40,41], suggesting that Cbp is involved in the regulation of all SFKs. The function of Fyn and Lyn in early events in B cell antigen-receptor (BCR) signaling was computer-simulated based on a model incorporating site-specific protein-protein interactions among BCR, Lyn, Fyn, Syk, Csk, and Cbp/PAG1 [42]. By incorporating the Cbp/PAG1-Csk complex as a negative regulator of SFKs, the analysis successfully reconstituted the function of Fyn and Lyn in BCR signaling, although the subcellular localization of signaling molecules was not considered in that study. On the other hand, it has been reported in a lymphoma cell line that the Cbp-Lyn complex phosphorylates signal transducer and activator of transcription 3 (STAT3) in lipid rafts, resulting in the efficient activation of transforming signals [29]. This finding demonstrates that Cbp can positively regulate SFKs when SFK substrates are present in lipid rafts. Because SFKs other than c-Src and Blk have a high affinity for lipid rafts and their raft localization is independent of Cbp [26], it is likely that intrinsic substrates for these SFKs are located in lipid rafts. Therefore, Cbp may function as a positive regulator specifically for these raft resident SFKs. By contrast, raft localization of c-Src is highly dependent on Cbp [26] and c-Src substrates, e.g., FAK and cortactin, being distributed in non-raft membranes. Due to these unique features of c-Src, Cbp can specifically act as a suppressor of c-Src by sequestering it into lipid rafts to limit its access to its substrates. However, as predicted by computational analysis, it is also possible that Cbp positively regulates c-Src when c-Src substrates are accumulated in lipid rafts or when the Cbp-Src complex is delivered to non-raft membranes under certain conditions where lipid rafts are perturbed. Cell transformation induced by MβCD in Csk-deficient cells might be mediated by Cbp acting in this fashion in non-raft membranes.

The ternary model showed that c-Src function is dependent on the lipid-raft volume. To validate this prediction, lipid rafts were disrupted by artificial treatment with MβCD [26]. However, we found that exclusion of Cbp from lipid rafts is an early event in cell transformation that occurs because of alterations in lipid-raft cholesterol and sphingolipid content [43]. This observation suggests that Cbp has the potential to serve as an activator of c-Src during cancer development when lipid metabolism is dysregulated. It is also possible that other pathological conditions that disturb lipid metabolism, such as aging, starvation, diabetes mellitus, or metabolic disorders, would allow Cbp to activate c-Src in non-raft membranes. Another possible mechanism is depalmitoylation or fatty acid remodeling of Cbp [44]. Because a mutant Cbp lacking a palmitoylation site loses its raft localization [25], perturbation of palmitoylation metabolism under some conditions

may also contribute to the positive effect of Cbp on c-Src function. By contrast, we also observed that Cbp is downregulated in fully transformed cells [25] and that Cbp expression is silenced in cancer cells potentially through an epigenetic mechanism [45]. Re-expression of Cbp in c-Src transformed cells and some cancer cells successfully suppressed malignant progression of these cells by sequestering active c-Src into lipid rafts [25]. These findings clearly demonstrate that Cbp serves as a suppressor of Src-mediated tumor progression when it is localized to lipid rafts. Overall, it is possible that the function of Cbp is dependent on its localization, exhibiting tumor-suppressive activity in lipid rafts and tumor-promoting activity in non-raft membranes. The ternary model clearly predicted these reciprocal functions of Cbp. Therefore, given that appropriate parameters in particular cancer cells are available, the mathematical models constructed in this study would be applicable to predict the behavior of Src-activated cancer cells [46–48].

In conclusion, our combination analysis unveiled opposing roles for Cbp in the regulation of transforming c-Src. Using computational models, we were able to identify a positive regulatory role for Cbp localized to non-raft membranes, an these models highlighted the importance of the subcellular localization of c-Src substrates. The modeling study also produced new interesting questions that must be addressed, plus the need for the molecular basis for the strict barrier function of lipid rafts against non-raft membranes to be elucidated, the physiological relevance of raft perturbation to c-Src activation to be widely investigated, and the identification of lipid-raft-residing c-Src substrates which would shed new light on the normal function of lipid-raft-anchored c-Src. The establishment of a mathematical model that fully represents the c-Src regulatory system would be useful for *in silico* screening of therapeutic targets in human cancers in which c-Src is upregulated.

Materials and Methods

Cell and cell culture

Csk-deficient ($Csk^{-/-}$) MEFs were a kind gift from Dr. A Imamoto [49]. $Csk^{-/-}$ MEFs transfected with pBKT2-c-Src (Csk$^{-/-}$/p-BKT2-c-Src) and $Csk^{-/-}$ MEFs transfected with both c-Src and pBKT2-Cbp (Csk$^{-/-}$/c-Src/p-BKT2-Cbp) were kindly donated by Dr. F Imamoto [30]. Cells were cultured in Dulbecco's modified Eagle's medium (DMEM) supplemented with 10% fetal bovine serum (FBS). For induction of c-Src and Cbp expression, cells were incubated with 1 mg/ml doxycycline (Dox, Sigma).

Western blot analysis

Cells were washed with PBS and lysed with n-octyl-β-D-glucoside (ODG) buffer [20 mM Tris-HCl (pH 7.4), 150 mM sodium chloride, 1 mM EDTA, 1 mM sodium orthovanadate, 20 mM sodium fluoride, 1% Nonidet P-40, 5% glycerol, 2% ODG, and a protease inhibitor cocktail], and immunoblotting was carried out as described previously [26]. The following primary antibodies were used: anti-Src (Ab-1; Calbiochem), anti-Src pY418 (Invitrogen), anti-phosphotyrosine (4G10; Millipore), anti-cortactin (Sigma), anti-cortactin pY421 (Sigma), anti-FAK (Santa Cruz), anti-FAK pY576 (Cell Signaling), anti-ERK pT202/Y204 (Cell Signaling), anti-ERK (Cell Signaling), anti-GAPDH (Sigma), anti-Flottilin-1 (BD) and anti-Caveolin-1 (BD). Anti-Cbp and anti-Cbp pY314 were prepared as described previously [19]. GAPDH was used as a loading control. Horseradish peroxidase (HRP)–conjugated anti-mouse or anti-rabbit IgG (Zymed) was used as the secondary antibody. All blots were visualized and quantitated using a LAS-4000 luminescent image analyzer (GE Healthcare).

DRM fractionation

Fractionation of membrane compartments on a sucrose gradient was performed as described previously [26,50]. Cells were washed with PBS, lysed with homogenization buffer [50 mM Tris-HCl (pH 7.4), 150 mM sodium chloride, 1 mM EDTA, 1 mM sodium orthovanadate, 20 mM sodium fluoride, 0.25% Triton X-100 and a protease inhibitor cocktail] and separated on a discontinuous sucrose gradient (5-35-40%) by ultracentrifugation at 150,000×g for 12 h at 4°C using an Optima L-100XP with a SW55Ti (Beckman Coulter). Eleven fractions were collected from the top of the sucrose gradient.

Immunoprecipitation

Fractions of membrane compartments were solubilized with 2% ODG and 1% Triton X-100 for 30 min at 4°C. The solubilized fractions were incubated with the indicated antibodies for 1 h at 4°C, and then further incubated with protein A-sepharose beads (GE healthcare) for 1 h at 4°C. The immunoprecipitates were washed with ODG buffer and analyzed by Western blotting.

Kinetic modeling and numerical computation

The model is described by a system of ODEs. The details of the model are provided below. Numerical simulations of ODEs were implemented by MATLAB software (The Mathworks Inc.) using the ode23 function. In all simulations, the time evolution was performed up to the time $T = 1,000$ sec. Parameter estimation was performed by random sampling. The essence of the analysis is minimization of the target function, which was chosen as the residual sum of squares (RSS) between the experimental data and the simulation data,

RSS = (Σ Src data+Σ Cbp data){Experimental data – Simulated data}^{2},

where "Src data" and "Cbp data" show data sequences as the Src and Cbp concentrations change. For details, see Text S1 and S2, Table S1, S2 and S3.

Ordinary differential equations

Here, we present a system of ODEs for the sequestration model and the ternary model. The differences between the sequestration model and the ternary model are the existence of a ternary complex, Cbp-Src-FAK, and its related kinetic reactions. In the sequestration model, the system of ordinary differential equations is composed of 12 variables. The equations are given by:

$$\frac{d[\text{Src}_{\text{in}}]}{dt} = -kc[\text{Cbp}_{\text{in}}][\text{Src}_{\text{in}}] + k\_c[\text{Cbp} - \text{Src}_{\text{in}}]$$
$$- ks1[\text{Src}_{\text{in}}][\text{SS}_{\text{in}}] + k\_s1[\text{Src} - \text{SS}_{\text{in}}] + kp1[\text{Src} - \text{SS}_{\text{in}}]$$
$$- ks_{out}[\text{Src}_{\text{in}}] + ks_{in}[\text{Src}_{\text{out}}],$$

$$\frac{d[\text{Cbp}_{\text{in}}]}{dt} = -kc[\text{Cbp}_{\text{in}}][\text{Src}_{\text{in}}] + k\_c[\text{Cbp} - \text{Src}_{\text{in}}]$$
$$- kc_{out}[\text{Cbp}_{\text{in}}] + kc_{in}[\text{Cbp}_{\text{out}}],$$

$$\frac{d[\text{SS}_{\text{in}}]}{dt} = -ks1[\text{Src}_{\text{in}}][\text{SS}_{\text{in}}] + k\_s1[\text{Src} - \text{SS}_{\text{in}}] + kd[\text{pSS}_{\text{in}}]$$
$$- kss_{out}[\text{SS}_{\text{in}}] + kss_{in}[\text{SS}_{\text{out}}],$$

$$\frac{d[\text{Cbp} - \text{Src}_{\text{in}}]}{dt} = kc[\text{Cbp}_{\text{in}}][\text{Src}_{\text{in}}] - k\_s[\text{Cbp} - \text{Src}_{\text{in}}]$$
$$- ks[\text{Cbp} - \text{Src}_{\text{in}}][\text{SS}_{\text{in}}] + k\_s[\text{Cbp} - \text{Src} - \text{SS}_{\text{in}}]$$
$$+ kp[\text{Cbp} - \text{Src} - \text{SS}_{\text{in}}] - kc_{out}[\text{Cbp} - \text{Src}_{\text{in}}]$$
$$+ kc_{in}[\text{Cbp} - \text{Src}_{\text{out}}],$$

$$\frac{d[\text{Src} - \text{SS}_{\text{in}}]}{dt} = ks1[\text{Src}_{\text{in}}][\text{SS}_{\text{in}}] - k\_s1[\text{Src} - \text{SS}_{\text{in}}]$$
$$- kp1[\text{Src} - \text{SS}_{\text{in}}] - kss_{out}[\text{Src} - \text{SS}_{\text{in}}]$$
$$+ kss_{in}[\text{Src} - \text{SS}_{\text{out}}]$$

$$\frac{d[\text{pSS}_{\text{in}}]}{dt} = kp1[\text{Src} - \text{SS}_{\text{in}}] - kd[\text{pSS}_{\text{in}}] - kss_{out}[\text{pSS}_{\text{in}}]$$
$$+ kss_{in}[\text{pSS}_{\text{out}}]$$

$$\frac{d[\text{Src}_{\text{out}}]}{dt} = -kc[\text{Cbp}_{\text{out}}][\text{Src}_{\text{out}}] + k\_c[\text{Cbp} - \text{Src}_{\text{out}}]$$
$$- ks1[\text{Src}_{\text{out}}][\text{SS}_{\text{out}}] + k\_s1[\text{Src} - \text{SS}_{\text{out}}] + kp1[\text{Src} - \text{SS}_{\text{out}}]$$
$$+ ks_{out}(\text{Vr}/\text{Vn})[\text{Src}_{\text{in}}] - ks_{in}(\text{Vr}/\text{Vn})[\text{Src}_{\text{out}}],$$

$$\frac{d[\text{Cbp}_{\text{out}}]}{dt} = -kc[\text{Cbp}_{\text{out}}][\text{Src}_{\text{out}}] + k\_c[\text{Cbp} - \text{Src}_{\text{out}}]$$
$$+ kc_{out}(\text{Vr}/\text{Vn})[\text{Cbp}_{\text{in}}] - kc_{in}(\text{Vr}/\text{Vn})[\text{Cbp}_{\text{out}}],$$

$$\frac{d[\text{SS}_{\text{out}}]}{dt} = -ks1[\text{Src}_{\text{out}}][\text{SS}_{\text{out}}] + k\_s1[\text{Src} - \text{SS}_{\text{out}}] + kd[\text{pSS}_{\text{out}}]$$
$$+ kd[\text{pSS}_{\text{out}}] + kss_{out}(\text{Vr}/\text{Vn})[\text{SS}_{\text{in}}]$$
$$- kss_{in}(\text{Vr}/\text{Vn})[\text{SS}_{\text{out}}],$$

$$\frac{d[\text{Cbp} - \text{Src}_{\text{out}}]}{dt} = kc[\text{Cbp}_{\text{out}}][\text{Src}_{\text{out}}] - k\_c[\text{Cbp} - \text{Src}_{\text{out}}]$$
$$+ kc_{out}(\text{Vr}/\text{Vn})[\text{Cbp} - \text{Src}_{\text{in}}]$$
$$- kc_{in}(\text{Vr}/\text{Vn})[\text{Cbp} - \text{Src}_{\text{out}}],$$

$$\frac{d[\text{Src} - \text{SS}_{\text{out}}]}{dt} = ks1[\text{Src}_{\text{out}}][\text{SS}_{\text{out}}] - k\_s1[\text{Src} - \text{SS}_{\text{out}}]$$
$$- kp1[\text{Src} - \text{SS}_{\text{out}}] + kss_{out}(\text{Vr}/\text{Vn})[\text{Src} - \text{SS}_{\text{in}}]$$
$$- kss_{in}(\text{Vr}/\text{Vn})[\text{Src} - \text{SS}_{\text{out}}],$$

$$\frac{d[\mathrm{pSS_{out}}]}{dt} = kp1[\mathrm{Src - SS_{out}}] - kd[\mathrm{pSS_{out}}]$$
$$+ kss_{out}(\mathrm{Vr/Vn})[\mathrm{pSS_{in}}] - kss_{in}(\mathrm{Vr/Vn})[\mathrm{pSS_{out}}],$$

In the ternary model, the system of ordinary differential equations is composed of 14 variables. The equations are given by:

$$\frac{d[\mathrm{Src_{in}}]}{dt} = -kc[\mathrm{Cbp_{in}}][\mathrm{Src_{in}}] + k\_c[\mathrm{Cbp - Src_{in}}]$$
$$- ks1[\mathrm{Src_{in}}][\mathrm{SS_{in}}] + k\_s1[\mathrm{Src - SS_{in}}] + kp1[\mathrm{Src - SS_{in}}]$$
$$- ks_{out}[\mathrm{Src_{in}}] + ks_{in}[\mathrm{Src_{out}}],$$

$$\frac{d[\mathrm{Cbp_{in}}]}{dt} = -kc[\mathrm{Cbp_{in}}][\mathrm{Src_{in}}] + k\_c[\mathrm{Cbp - Src_{in}}]$$
$$- kc[\mathrm{Cbp_{in}}][\mathrm{Src - SS_{in}}] + k\_c[\mathrm{Cbp - Src - SS_{in}}]$$
$$- kc_{out}[\mathrm{Cbp_{in}}] + kc_{in}[\mathrm{Cbp_{out}}],$$

$$\frac{d[\mathrm{SS_{in}}]}{dt} = -ks1[\mathrm{Src_{in}}][\mathrm{SS_{in}}] + k\_s1[\mathrm{Src - SS_{in}}] + kd[\mathrm{pSS_{in}}]$$
$$- ks2[\mathrm{Cbp - Src_{in}}][\mathrm{SS_{in}}] + k\_s2[\mathrm{Cbp - Src - SS_{in}}]$$
$$- kss_{out}[\mathrm{SS_{in}}] + kss_{in}[\mathrm{SS_{out}}],$$

$$\frac{d[\mathrm{Cbp - Src_{in}}]}{dt} = kc[\mathrm{Cbp_{in}}][\mathrm{Src_{in}}] - k\_s[\mathrm{Cbp - Src_{in}}]$$
$$- ks[\mathrm{Cbp - Src_{in}}][\mathrm{SS_{in}}] + k\_s[\mathrm{Cbp - Src - SS_{in}}]$$
$$+ kp[\mathrm{Cbp - Src - SS_{in}}] - kc_{out}[\mathrm{Cbp - Src_{in}}]$$
$$+ kc_{in}[\mathrm{Cbp - Src_{out}}],$$

$$\frac{d[\mathrm{Src - SS_{in}}]}{dt} = ks1[\mathrm{Src_{in}}][\mathrm{SS_{in}}] - k\_s1[\mathrm{Src - SS_{in}}]$$
$$- kc[\mathrm{Cbp_{in}}][\mathrm{Src - SS_{in}}] + k\_c[\mathrm{Cbp - Src - SS_{in}}]$$
$$- kp1[\mathrm{Src - SS_{in}}] - kss_{out}[\mathrm{Src - SS_{in}}]$$
$$+ kss_{in}[\mathrm{Src - SS_{out}}]$$

$$\frac{d[\mathrm{Cbp - Src - SS_{in}}]}{dt} = ks2[\mathrm{Cbp - Src_{in}}][\mathrm{SS_{in}}]$$
$$- k\_s2[\mathrm{Cbp - Src - SS_{in}}] + kc[\mathrm{Cbp_{in}}][\mathrm{Src - SS_{in}}]$$
$$- k\_c[\mathrm{Cbp - Src - SS_{in}}] - kp2[\mathrm{Cbp - Src - SS_{in}}]$$
$$- kss_{out}[\mathrm{Cbp - Src - SS_{in}}] + kss_{in}[\mathrm{Cbp - Src - SS_{out}}],$$

$$\frac{d[\mathrm{pSS_{in}}]}{dt} = kp1[\mathrm{Src - SS_{in}}] + kp2[\mathrm{Cbp - Src - SS_{in}}] - kd[\mathrm{pSS_{in}}]$$
$$- kss_{out}[\mathrm{pSS_{in}}] + kss_{in}[\mathrm{pSS_{out}}]$$

$$\frac{d[\mathrm{Src_{out}}]}{dt} = -kc[\mathrm{Cbp_{out}}][\mathrm{Src_{out}}] + k\_c[\mathrm{Cbp - Src_{out}}]$$
$$- ks1[\mathrm{Src_{out}}][\mathrm{SS_{out}}] + k\_s1[\mathrm{Src - SS_{out}}]$$
$$+ kp1[\mathrm{Src - SS_{out}}] + ks_{out}(\mathrm{Vr/Vn})[\mathrm{Src_{in}}]$$
$$- ks_{in}(\mathrm{Vr/Vn})[\mathrm{Src_{out}}],$$

$$\frac{d[\mathrm{Cbp_{out}}]}{dt} = -kc[\mathrm{Cbp_{out}}][\mathrm{Src_{out}}] + k\_c[\mathrm{Cbp - Src_{out}}]$$
$$- kc[\mathrm{Cbp_{out}}][\mathrm{Src - SS_{out}}] + k\_c[\mathrm{Cbp - Src - SS_{out}}]$$
$$+ kc_{out}(\mathrm{Vr/Vn})[\mathrm{Cbp_{in}}] - kc_{in}(\mathrm{Vr/Vn})[\mathrm{Cbp_{out}}],$$

$$\frac{d[\mathrm{SS_{out}}]}{dt} = -ks1[\mathrm{Src_{out}}][\mathrm{SS_{out}}] + k\_s1[\mathrm{Src - SS_{out}}] + kd[\mathrm{pSS_{out}}]$$
$$- ks2[\mathrm{Cbp - Src_{out}}][\mathrm{SS_{out}}] + k\_s2[\mathrm{Cbp - Src - SS_{out}}]$$
$$+ kss_{out}(\mathrm{Vr/Vn})[\mathrm{SS_{in}}] - kss_{in}(\mathrm{Vr/Vn})[\mathrm{SS_{out}}],$$

$$\frac{d[\mathrm{Cbp - Src_{out}}]}{dt} = kc[\mathrm{Cbp_{out}}][\mathrm{Src_{out}}] - k\_c[\mathrm{Cbp - Src_{out}}]$$
$$- ks2[\mathrm{Cbp - Src_{out}}][\mathrm{SS_{out}}] + k\_s2[\mathrm{Cbp - Src - SS_{out}}]$$
$$+ kp2[\mathrm{Cbp - Src - SS_{out}}] + kc_{out}(\mathrm{Vr/Vn})[\mathrm{Cbp - Src_{in}}]$$
$$- kc_{in}(\mathrm{Vr/Vn})[\mathrm{Cbp - Src_{out}}],$$

$$\frac{d[\mathrm{Src - SS_{out}}]}{dt} = ks1[\mathrm{Src_{out}}][\mathrm{SS_{out}}] - k\_s1[\mathrm{Src - SS_{out}}]$$
$$- kc[\mathrm{Cbp_{out}}][\mathrm{Src - SS_{out}}] + k\_c[\mathrm{Cbp - Src - SS_{out}}]$$
$$- kp1[\mathrm{Src - SS_{out}}] + kss_{out}(\mathrm{Vr/Vn})[\mathrm{Src - SS_{in}}]$$
$$- kss_{in}(\mathrm{Vr/Vn})[\mathrm{Src - SS_{out}}],$$

$$\frac{d[\mathrm{Cbp - Src - SS_{out}}]}{dt} = ks2[\mathrm{Cbp - Src_{out}}][\mathrm{SS_{out}}]$$
$$- k\_s2[\mathrm{Cbp - Src - SS_{out}}]$$
$$+ kc[\mathrm{Cbp_{out}}][\mathrm{Src - SS_{out}}]$$
$$- k\_c[\mathrm{Cbp - Src - SS_{out}}] - kp2[\mathrm{Cbp - Src - SS_{out}}]$$
$$+ kss_{out}(\mathrm{Vr/Vn})[\mathrm{Cbp - Src - SS_{in}}]$$
$$- kss_{in}(\mathrm{Vr/Vn})[\mathrm{Cbp - Src - SS_{out}}],$$

$$\frac{d[\mathrm{pSS_{out}}]}{dt} = kp1[\mathrm{Src\_SS_{out}}] + kp2[\mathrm{Cbp - Src - SS_{out}}]$$
$$- kd[\mathrm{pSS_{out}}] + kss_{out}(\mathrm{Vr/Vn})[\mathrm{pSS_{in}}] - kss_{in}(\mathrm{Vr/Vn})[\mathrm{pSS_{out}}],$$

Supporting Information

Figure S1 Schematic representation of the subsystems. (A) Michaelis-Menten-type reaction of c-Src mediated phosphorylation of SS. (B) c-Src and Cbp binding and dissociation reaction. (C) Import into and export from membrane microdomains.

Figure S2 Results of random sampling for parameter estimation. Two-dimensional scatter plots in the K_M and K_{dep} plane for the sequestration model (A) and the ternary model (B). Green dots indicate parameters satisfying RSS>0.03, red dots indicate parameters satisfying 0.015<RSS<0.03, and blue dots indicate parameters satisfying RSS<0.015.

Figure S3 Simulated phosphorylation curve for the sequestration model and the ternary model with $Vr = 0.05$. (A and B) FAK phosphorylation ratio as a function of the total Src concentration for three different raft volume ratios in the sequestration model (A) and the ternary model (B). The black line denotes the raft volume ratio $Vr = 0.05$ (control), the dashed line denotes $Vr = 0.01$, and the pointed line denotes $Vr = 0.001$. (C and D) FAK phosphorylation ratio as a function of the total Cbp concentration for three different raft volume ratios in the sequestration model (C) and the ternary model (D).

Figure S4 Simulated phosphorylation curve for allosteric effects. (A) FAK phosphorylation ratio as a function of the total Cbp concentration. The phosphorylation rate constant is set as $kp2 = 10*kp1$ ($\alpha = 10$). (B) FAK phosphorylation ratio as a function of the total Cbp concentration. The phosphorylation rate constant is set as $kp2 = 2*kp1$ ($\alpha = 2$) and the Src-FAK binding rate constant is set as $ks2 = 2*ks1$ ($\beta = 2$).

Figure S5 Simulated phosphorylation curve for allosteric effects with $Vr = 0.05$. (A) FAK phosphorylation ratio as a function of the total Cbp concentration. The phosphorylation rate constant is set as $kp2 = 2*kp1$ ($\alpha = 2$). (B) FAK phosphorylation ratio as a function of the total Cbp concentration. The Src-FAK binding rate constant is set as $ks2 = 2*ks1$ ($\beta = 2$).

Table S1 Search range for parameter estimation.

Table S2 Parameter values for the sequestration model with $Vr = 0.1$.

Table S3 Parameter values for the ternary model with $Vr = 0.1$.

Text S1 Analysis of subsystems.

Text S2 Random sampling for parameter searches.

Acknowledgments

We thank Drs. A. Imamoto and F. Imamoto for providing cells, and Ms. Y. Yoshikawa for helpful discussions. We also thank Prof. T. Imamura and T. Iimura for useful discussions and comments.

Author Contributions

Conceived and designed the experiments: KK CO MO. Performed the experiments: KK. Analyzed the data: T. Saitou KK. Contributed reagents/materials/analysis tools: T. Saitou KK. Wrote the paper: T. Saitou KK CO T. Suzuki MO.

References

1. Jove R, Hanafusa H (1987) Cell transformation by the viral src oncogene. Annu Rev Cell Biol 3: 31–56.
2. Brown MT, Cooper JA (1996) Regulation, substrates and functions of src. Biochem Biophys Acta 1278: 121–149.
3. Nada S, Okada M, MacAuley A, Cooper JA, Nakagawa H (1991) Cloning of a complementary DNA for a protein-tyrosine kinase that specifically phosphorylates a negative regulatory site of p60c-src. Nature 351: 69–72.
4. Nada S, Yagi T, Takeda H, Tokunaga T, Nakagawa H, et al. (1993) Constitutive activation of Src family kinases in mouse embryos that lack Csk. Cell 73.
5. Okada M (2012) Regulation of the SRC family kinases by Csk. Int J Biol Sci 8: 1385–1397.
6. Yeatman TJ (2004) A renaissance for SRC. Nat Rev Cancer 4: 470–480.
7. Fuss H, Dubitzky W, Downes S, Kurth MJ (2006) Bistable switching and excitable behaviour in the activation of Src at mitosis. Bioinformatics 22: e158–165.
8. Fuss H, Dubitzky W, Downes CS, Kurth MJ (2007) Deactivation of Src family kinases: hypothesis testing using a Monte Carlo sensitivity analysis of systems-level properties. J Comput Biol 14: 1185–1200.
9. Fuss H, Dubitzky W, Downes CS, Kurth MJ (2008) SRC family kinases and receptors: analysis of three activation mechanisms by dynamic systems modeling. Biophys J 94: 1995–2006.
10. Kaimachnikov NP, Kholodenko BN (2009) Toggle switches, pulses and oscillations are intrinsic properties of the Src activation/deactivation cycle. FEBS J 276: 4102–4118.
11. Frame MC (2002) Src in cancer: deregulation and consequences for cell behaviour. Biochem Biophys Acta 1602: 114–130.
12. Ishizawar R, Parsons SJ (2004) c-Src and cooperating partners in human cancer. Cancer Cell 6: 209–214.
13. Irby RB, Mao Y, Coppola D, Kang J, Loubeau JM, et al. (1999) Activating SRC mutation in a subset of advanced human colon cancers. Nat Genet 21: 187–190.
14. Irby RB, Yeatman TJ (2000) Role of Src expression and activation in human cancer. Oncogene 19: 5636–5642.
15. Hanahan D, Weinberg RA (2011) Hallmarks of cancer: the next generation. Cell 144: 646–674.
16. Mitra SK, Schlaepfer DD (2006) Integrin-regulated FAK-Src signaling in normal and cancer cells. Curr Opin Cell Biol 18: 516–523.
17. Wang W, Liu Y, Liao K (2011) Tyrosine phosphorylation of cortactin by the FAK-Src complex at focal adhesions regulates cell motility. BMC Cell Biol 12: 49.
18. Tomar A, Lawson C, Ghassemian M, Schlaepfer DD (2012) Cortactin as a target for FAK in the regulation of focal adhesion dynamics. PLoS One 7: e44041.
19. Kawabuchi M, Satomi Y, Takao T, Shimonishi Y, Nada S, et al. (2000) Transmembrane phosphoprotein Cbp regulates the activities of Src-family tyrosine kinases. Nature 404: 999–1003.
20. Brdicka T, Pavlistová D, Leo A, Bruyns E, Korínek V, et al. (2000) Phosphoprotein Associated with Glycosphingolipid-Enriched Microdomains (Pag), a Novel Ubiquitously Expressed Transmembrane Adaptor Protein, Binds the Protein Tyrosine Kinase Csk and Is Involved in Regulation of T Cell Activation. J Exp Med 191: 1591–1604.
21. Simons K, Toomre D (2000) Lipid rafts and signal transduction. Nat Rev Mol Cell Biol 1.
22. Munro S (2003) Lipid Rafts: Elusive or Illusive? Cell 115: 377–388.
23. Takeuchi S, Takayama Y, Ogawa A, Tamura K, Okada M (2000) Transmembrane phosphoprotein Cbp positively regulates the activity of the carboxyl-terminal Src kinase, Csk. J Biol Chem 275: 29183–29186.
24. Oneyama C, Hikita T, Nada S, Okada M (2008) Functional dissection of transformation by c-Src and v-Src. Genes Cells 13: 1–12.
25. Oneyama C, Hikita T, Enya K, Dobenecker MW, Saito K, et al. (2008) The lipid raft-anchored adaptor protein Cbp controls the oncogenic potential of c-Src. Mol Cell 30: 426–436.
26. Oneyama C, Iino T, Saito K, Suzuki K, Ogawa A, et al. (2009) Transforming potential of Src family kinases is limited by the cholesterol-enriched membrane microdomain. Mol Cell Biol 29: 6462–6472.
27. Shima T, Nada S, Okada M (2003) Transmembrane phosphoprotein Cbp senses cell adhesion signaling mediated by Src family kinase in lipid rafts. Proc Natl Acad Sci U S A 100: 14897–14902.
28. Solheim SA, Torgersen KM, Tasken K, Berge T (2008) Regulation of FynT function by dual domain docking on PAG/Cbp. J Biol Chem 283: 2773–2783.

29. Tauzin S, Ding H, Khatib K, Ahmad I, Burdevet D, et al. (2008) Oncogenic association of the Cbp/PAG adaptor protein with the Lyn tyrosine kinase in human B-NHL rafts. Blood 111: 2310–2320.

30. Inoue K, Sone T, Oneyama C, Nishiumi F, Kishine H, et al. (2009) A versatile nonviral vector system for tetracycline-dependent one-step conditional induction of transgene expression. Gene Ther 16: 1383–1394.

31. Kanou T, Oneyama C, Kawahara K, Okimura A, Ohta M, et al. (2011) The transmembrane adaptor Cbp/PAG1 controls the malignant potential of human non-small cell lung cancers that have c-src upregulation. Mol Cancer Res 9: 103–114.

32. Arold ST (2011) How focal adhesion kinase achieves regulation by linking ligand binding, localization and action. Curr Opin Struct Biol 21: 808–813.

33. Lipniacki T, Paszek P, Brasier AR, Luxon B, Kimmel M (2004) Mathematical model of NF-kappaB regulatory module. J Theor Biol 228: 195–215.

34. Okada M, Nada S, Yamanashi Y, Yamamoto T, Nakagawa H (1991) CSK: a protein-tyrosine kinase involved in regulation of src family kinases. J Biol Chem 266: 24249–24252.

35. Thomas JW (1998) SH2- and SH3-mediated Interactions between Focal Adhesion Kinase and Src. Journal of Biological Chemistry 273: 577–583.

36. Levchenko A, Bruck J, Sternberg PW (2000) Scaffold proteins may biphasically affect the levels of mitogen-activated protein kinase signaling and reduce its threshold properties. Proc Natl Acad Sci U S A 97: 5818–5823.

37. Burack WR, Shaw AS (2000) Signal transduction: hanging on a scaffold. Curr Opin Cell Biol 12: 211–216.

38. Kolch W (2005) Coordinating ERK/MAPK signalling through scaffolds and inhibitors. Nat Rev Mol Cell Biol 6: 827–837.

39. Good MC, Zalatan JG, Lim WA (2011) Scaffold proteins: hubs for controlling the flow of cellular information. Science 332: 680–686.

40. Solheim SA, Petsalaki E, Stokka AJ, Russell RB, Tasken K, et al. (2008) Interactions between the Fyn SH3-domain and adaptor protein Cbp/PAG derived ligands, effects on kinase activity and affinity. FEBS J 275: 4863–4874.

41. Ingley E, Schneider JR, Payne CJ, McCarthy DJ, Harder KW, et al. (2006) Csk-binding protein mediates sequential enzymatic down-regulation and degradation of Lyn in erythropoietin-stimulated cells. J Biol Chem 281: 31920–31929.

42. Barua D, Hlavacek WS, Lipniacki T (2012) A computational model for early events in B cell antigen receptor signaling: analysis of the roles of Lyn and Fyn. J Immunol 189: 646–658.

43. Kajiwara K, Yamada T, Bamba T, Fukusaki E, Imamoto F, et al. (2013) c-Src-induced activation of ceramide metabolism impairs membrane microdomains and promotes malignant progression by facilitating the translocation of c-Src to focal adhesions. Biochemical J 458: 81–93.

44. Resh MD (2006) Trafficking and signaling by fatty-acylated and prenylated proteins. Nat Chem Biol 2: 584–590.

45. Suzuki K, Oneyama C, Kimura H, Tajima S, Okada M (2011) Down-regulation of the tumor suppressor C-terminal Src kinase (Csk)-binding protein (Cbp)/PAG1 is mediated by epigenetic histone modifications via the mitogen-activated protein kinase (MAPK)/phosphatidylinositol 3-kinase (PI3K) pathway. J Biol Chem 286: 15698–15706.

46. Fitzgerald JB, Schoeberl B, Nielsen UB, Sorger PK (2006) Systems biology and combination therapy in the quest for clinical efficacy. Nat Chem Biol 2: 458–466.

47. Stites EC, Trampont PC, Ma Z, Ravichandran KS (2007) Network analysis of oncogenic Ras activation in cancer. Science 318: 463–467.

48. Berger SI, Iyengar R (2009) Network analyses in systems pharmacology. Bioinformatics 25: 2466–2472.

49. Imamoto A, Soriano P (1993) Disruption of the csk gene, encoding a negative regulator of Src family tyrosine kinases, leads to neural tube defects and embryonic lethality in mice. Cell 73: 1117–1124.

50. Lingwood D, Simons K (2007) Detergent resistance as a tool in membrane research. Nat Protoc 2: 2159–2165.

7

Comparative Analyses of Three *Chlorella* Species in Response to Light and Sugar Reveal Distinctive Lipid Accumulation Patterns in the Microalga *C. sorokiniana*

Julian N. Rosenberg[1,2☉], Naoko Kobayashi[3☉], Austin Barnes[3], Eric A. Noel[3], Michael J. Betenbaugh[1], George A. Oyler[1,2,3*]

1 Department of Chemical & Biomolecular Engineering, Johns Hopkins University, Baltimore, Maryland, United States of America, 2 Synaptic Research LLC, Baltimore, Maryland, United States of America, 3 Department of Biochemistry, University of Nebraska–Lincoln, Lincoln, Nebraska, United States of America

Abstract

While photosynthetic microalgae, such as *Chlorella*, serve as feedstocks for nutritional oils and biofuels, heterotrophic cultivation can augment growth rates, support high cell densities, and increase triacylglycerol (TAG) lipid content. However, these species differ significantly in their photoautotrophic and heterotrophic characteristics. In this study, the phylogeny of thirty *Chlorella* strains was determined in order to inform bioprospecting efforts and detailed physiological assessment of three species. The growth kinetics and lipid biochemistry of *C. protothecoides* UTEX 411, *C. vulgaris* UTEX 265, and *C. sorokiniana* UTEX 1230 were quantified during photoautotrophy in Bold's basal medium (BBM) and heterotrophy in BBM supplemented with glucose (10 g L^{-1}). Heterotrophic growth rates of UTEX 411, 265, and 1230 were found to be 1.5-, 3.7-, and 5-fold higher than their respective autotrophic rates. With a rapid nine-hour heterotrophic doubling time, *Chlorella sorokiniana* UTEX 1230 maximally accumulated 39% total lipids by dry weight during heterotrophy compared to 18% autotrophically. Furthermore, the discrete fatty acid composition of each strain was examined in order to elucidate lipid accumulation patterns under the two trophic conditions. In both modes of growth, UTEX 411 and 265 produced 18:1 as the principal fatty acid while UTEX 1230 exhibited a 2.5-fold enrichment in 18:2 relative to 18:1. Although the total lipid content was highest in UTEX 411 during heterotrophy, UTEX 1230 demonstrated a two-fold increase in its heterotrophic TAG fraction at a rate of 28.9 mg L^{-1} d^{-1} to reach 22% of the biomass, corresponding to as much as 90% of its total lipids. Interestingly, UTEX 1230 growth was restricted during mixotrophy and its TAG production rate was suppressed to 18.2 mg L^{-1} d^{-1}. This constraint on carbon flow raises intriguing questions about the impact of sugar and light on the metabolic regulation of microalgal lipid biosynthesis.

Editor: Rafael Vazquez Duhalt, Center for Nanosciences and Nanotechnology, Mexico

Funding: This work was supported in part by funds from the U.S. Department of Energy (http://www.doe.gov/) under grant numbers DE-SC0004521 and DE-EE0003373 to GAO. Partial support was also provided by grant number NSF-EFRI-1302862 to MJB from the National Science Foundation (http://www.nsf.gov/) and a fellowship to JNR from the Johns Hopkins Environment, Energy, Sustainability & Health Institute (E²SHI, http://e2shi.jhu.edu/). Publication of this article was funded in part by the Open Access Promotion Fund of the Johns Hopkins University Libraries. The funders had no role in study design, data collection and analysis, decision to publish, or preparation of the manuscript.

Competing Interests: GAO is the president and founder of Synaptic Research. JNR is a part-time employee of Synaptic Research.

* E-mail: george@synapticresearch.com

☉ These authors contributed equally to this work.

Introduction

While many benefits of microalgae production are inherent to photosynthetic carbon dioxide assimilation, heterotrophic growth can circumvent certain limitations of photoautotrophic cultivation, such as ineffective light transfer in saturated cultures and low photosynthetic efficiencies [1,2]. During heterotrophy, the presence of a fixed carbon source (*i.e.*, sugar) can promote rapid growth, support high cell densities, and augment lipid accumulation [3]. As such, this mode of growth has been exploited for the industrial production of polyunsaturated fatty acids and bioactive pigments to serve nutritional markets [4,5]. In recent years, however, incentives to grow oleaginous microalgae have shifted toward biofuels as an end product [6–8]. Triacylglycerol (TAG) is a preferred renewable oil because it possesses a high molar ratio of

hydrogen to carbon, can be easily extracted from biomass and directly converted to drop-in biofuels by transesterification, hydrotreating, or pyrolysis [9,10]. If heterotrophic substrates can be sustainably sourced for relatively low cost (*e.g.*, derived from cellulosic biomass, wastewater, or directly produced by other photoautotrophs) [11], the bioconversion of organic compounds to lipids may be a viable model for algal biofuel production [12,13].

For the commercial cultivation of microalgae, nitrogen limitation is a well studied environmental stressor known to induce lipid accumulation [14–17]. Detailed analyses of heterotrophic algal oil production have also been reported and recently reviewed [18–23]. Additionally, mixotrophic cultivation with sugar and light can stimulate high oil yields [24]. While we have previously assessed the biomass and lipid productivities of *Nannochloropsis*, *Dunaliella*, and *Chlorella* species grown on a wide variety of sugars [25], the

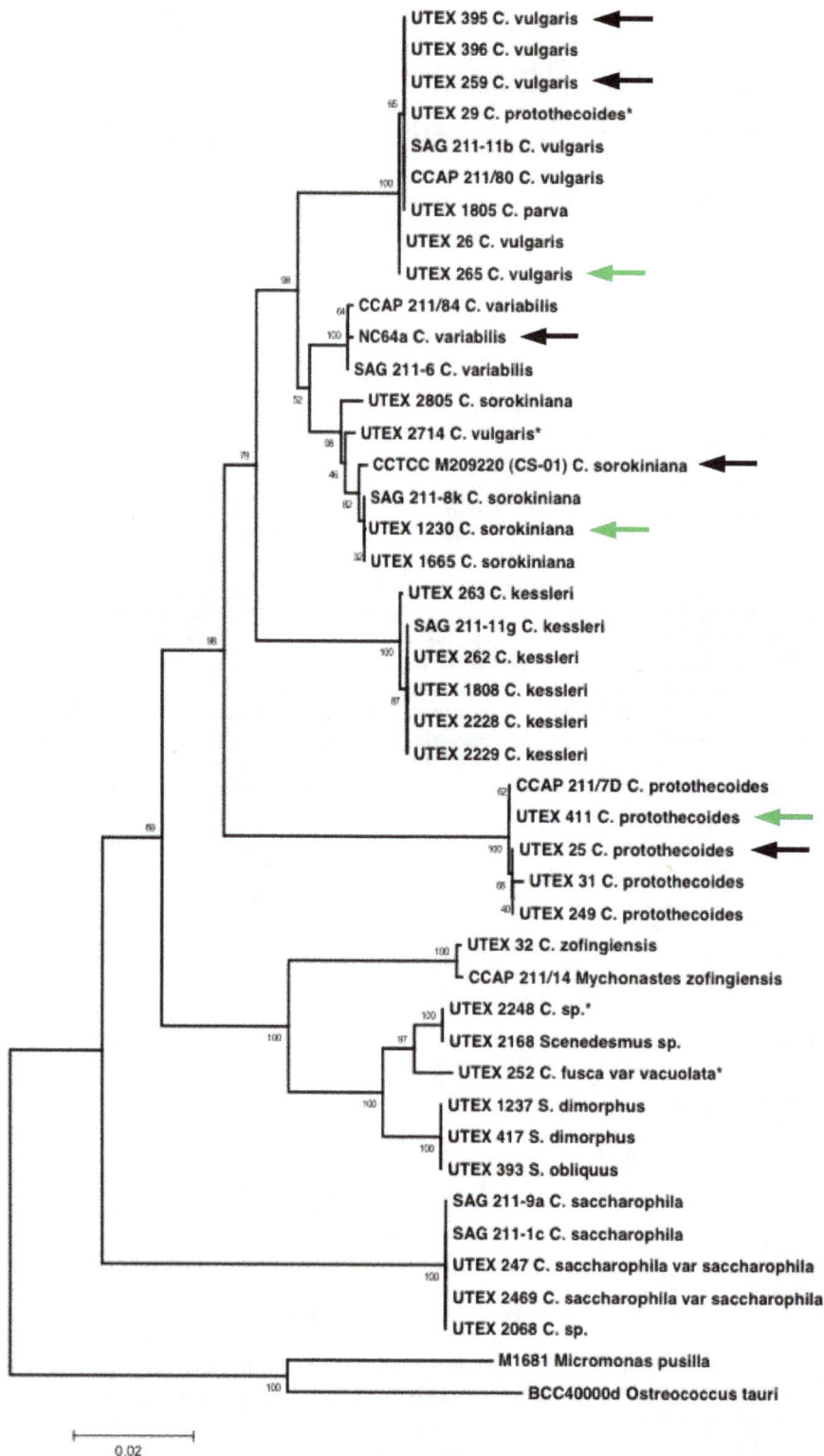

Figure 1. Proposed phylogenetic tree of *Chlorella* species of interest. Green arrows point to branch placement of the three candidate strains examined in this study (*C. vulgaris* UTEX 265, *C. sorokiniana* UTEX 1230, and *C. protothecoides* UTEX 411); black arrows indicate strains with active genome projects. Asterisks denote strains found to align more closely with a species other than their original labeled speciation. The scale bar represents a 2% difference between distinct ITS region sequences.

present study focuses on the *Chlorella* genus of green algae *(Chlorophyceae)*, most commonly known for its nutritional benefits when consumed in whole cell dietary supplements [26–28]. Various aspects of plant physiology and metabolism have also been studied in *Chlorella* for nearly a century [29,30]. In the early 1950s, when open pond cultivation systems first became prevalent, *Chlorella* species were some of the first microalgae to be produced in mass quantities and an investigation of scale-up "from laboratory to pilot plant" was reported by Burlew [31].

Despite the long history of the *Chlorella* genus, currently the only species with a fully sequenced, annotated, and publicly available genome is *Chlorella variabilis* NC64A [32]. This unique strain is both the host to large DNA chloroviruses and can be an endosymbiont of *Paramecium bursaria*, surviving through mixotrophic nutrient exchange [33–35]. When NC64A is not harbored by *P. bursaria*, it is susceptible to viral infection and requires nutrient supplementation to grow [36–38]. While its genome remains a valuable resource for understanding the balance of carbon metabolism, symbiosis, and pathogen-host interactions, NC64A is not a likely candidate for biomass production due to its low yields. Instead, other *Chlorella* species have been cultivated under mixo– and heterotrophic conditions for the production of lutein and astaxanthin (antioxidants) and have served as the basis for mathematical models of sugar-based growth [39–44]. Recent metabolic flux analyses and transcriptomic studies performed under different trophic conditions also provide compelling information about shifts in *Chlorella* lipid metabolism [21,45–47].

In order to fulfill an ongoing search for production organisms and model algal systems, the present study assesses the biodiversity of *Chlorella* species based on biofuel production qualities of heterotrophic growth and TAG accumulation when supplemented with glucose at 10 g L^{-1}. After phylogenetic sequencing of thirty strains from culture repositories, *C. sorokiniana* UTEX 1230, *C. vulgaris* UTEX 265, and *C. protothecoides* UTEX 411 were selected for comparative analyses based on growth rates, biomass yield, and lipid productivities in photoautotrophic and heterotrophic culture. The influence of heterotrophy and mixotrophy on lipid biochemistry was also investigated through examination of the abundance, composition, and distribution of total oils as membrane-associated lipids, TAG neutral lipids, or accessory lipophilic molecules. Finally, discrete lipid profiles were determined using gas chromatography–mass spectrometry (GC-MS) to evaluate the dynamics of fatty acid chain length and degree of saturation during the course of cultivation. As a result of this comprehensive *Chlorella* species screening, the occurrence of differential lipid compositions led to further consideration of *C. sorokiniana* as a potential platform for bioenergy and biotechnology.

Results

Phylogenetic analysis of *Chlorella* species and strain pedigree

For our initial species selection, a "genetic fingerprint" based on the 18S ribosomal RNA's internal transcribed spacer (ITS) regions was established for each isolate from a collection of over thirty *Chlorella* strains and used to construct a phylogenetic tree (Figure 1). The annotated 18S ITS sequences for these organisms have been made available online through the GenBank database. During this survey, we encountered some *Chlorella* strains that had been designated as related species (*e.g.*, UTEX 29, 2714) and some strains that group within a separate genus entirely (UTEX 2248, 252). Sequencing of the ITS regions also revealed close phylogenetic relationships between *Chlorella* species in our collection and other strains with active genome projects or available

transcriptomes, including NC64A, CS-01, UTEX 395, 259, and 25 [45,46].

In order to compare autotrophic *Chlorella* species' propensity to grow on glucose and accumulate lipids under uniform heterotrophic conditions, *C. sorokiniana* UTEX 1230, *C. vulgaris* UTEX 265, and *C. protothecoides* UTEX 411 were selected based on vigorous growth attributes during preliminary examination in minimal nutrient Bold's basal medium (BBM) (unpublished data). These three strains were found to be phylogenetically distinguishable within their species groupings and each possesses unique physiological traits. The first in this group, *Chlorella sorokiniana* UTEX 1230, has been studied extensively for its carbohydrate, lipid, protein, and biopolymer content [48–50] as well as hydrogen photoproduction [51], glucose transport [52], production of biomolecules by fermentation [53,54], and bioremediation capabilities [55–58]. This organism was isolated by Sorokin and Meyers in the early 1950s [59] and maintained in many culture collections under different strain numbers (originally classified as *C. pyrenoidosa*) [60–64].

Chlorella vulgaris is a closely related algal species often used as a health supplement [65]. While some annotated genomic sequences have been made available for this species [66], the *C. vulgaris* strain UTEX 265 examined in the present study has not been fully characterized in this manner. Nonetheless, the population dynamics of UTEX 265 have been studied using genetic probes [67] and this particular strain demonstrates heterotrophic growth in medium containing 0.1–0.6% glucose [68]. A related *C. vulgaris* strain (UTEX 395) is the subject of intense transcriptomic and proteomic analyses of lipid biosynthetic pathways [45,69].

Chlorella (Auxenochlorella) protothecoides is a well established heterotrophic alga that can utilize glucose as an organic carbon source [40]. Furthermore, *C. protothecoides* UTEX 411 can grow on sucrose and glycerol [70,71] and has been exploited for lutein production [40]. Other *C. protothecoides* strains have been cultivated heterotrophically for lipid production— achieving over 55% lipids by dry weight [8] and scale-up to 10,000-L bioreactors [9]. This demonstration has reinforced the potential of heterotrophic *Chlorella* species to generate substantial quantities of lipids for biofuel production.

Growth and biomass yields of *C. sorokiniana* (UTEX 1230), *C. vulgaris* (UTEX 265), and *C. protothecoides* (UTEX 411)

The three selected *Chlorella* species were cultivated using previously reported growth methods in separate photoautotrophic and heterotrophic batches for quantitative growth and biomass assessment [72]. The heterotrophic specific growth rates of UTEX 411, 265, and 1230 (0.48, 0.84, and 1.77 d^{-1}, respectively) were found to be 1.5-, 3.7-, and 5-fold higher than their autotrophic rates (Table 1). A comparison of auto– and heterotrophic growth curves for *C. sorokiniana*, *C. vulgaris*, and *C. protothecoides* cultures inoculated with 1×10^6 cells ml^{-1} is shown in Figure 2A. Although heterotrophic growth has been reported for each algal strain [9,52,68,70,71], *C. sorokiniana* UTEX 1230 demonstrated the most significant heterotrophic advantage in Bold's basal medium with glucose (10 g L^{-1}). Within one week, heterotrophic *C. sorokiniana* UTEX 1230 reached the highest cell density of all strains (113×10^6 cells ml^{-1}), while the final heterotrophic cell densities of *C. vulgaris* and *C. protothecoides* cultures were at least 9-fold lower ($10–2 \times 10^6$ cells ml^{-1}). Under autotrophic conditions, UTEX 265, 1230, and 411 achieved cell densities of roughly 90, 40, and 10 million cells ml^{-1} at their respective stationary phases with comparable photoautotrophic specific growth rates ($\mu = 0.23–0.36$ d^{-1}).

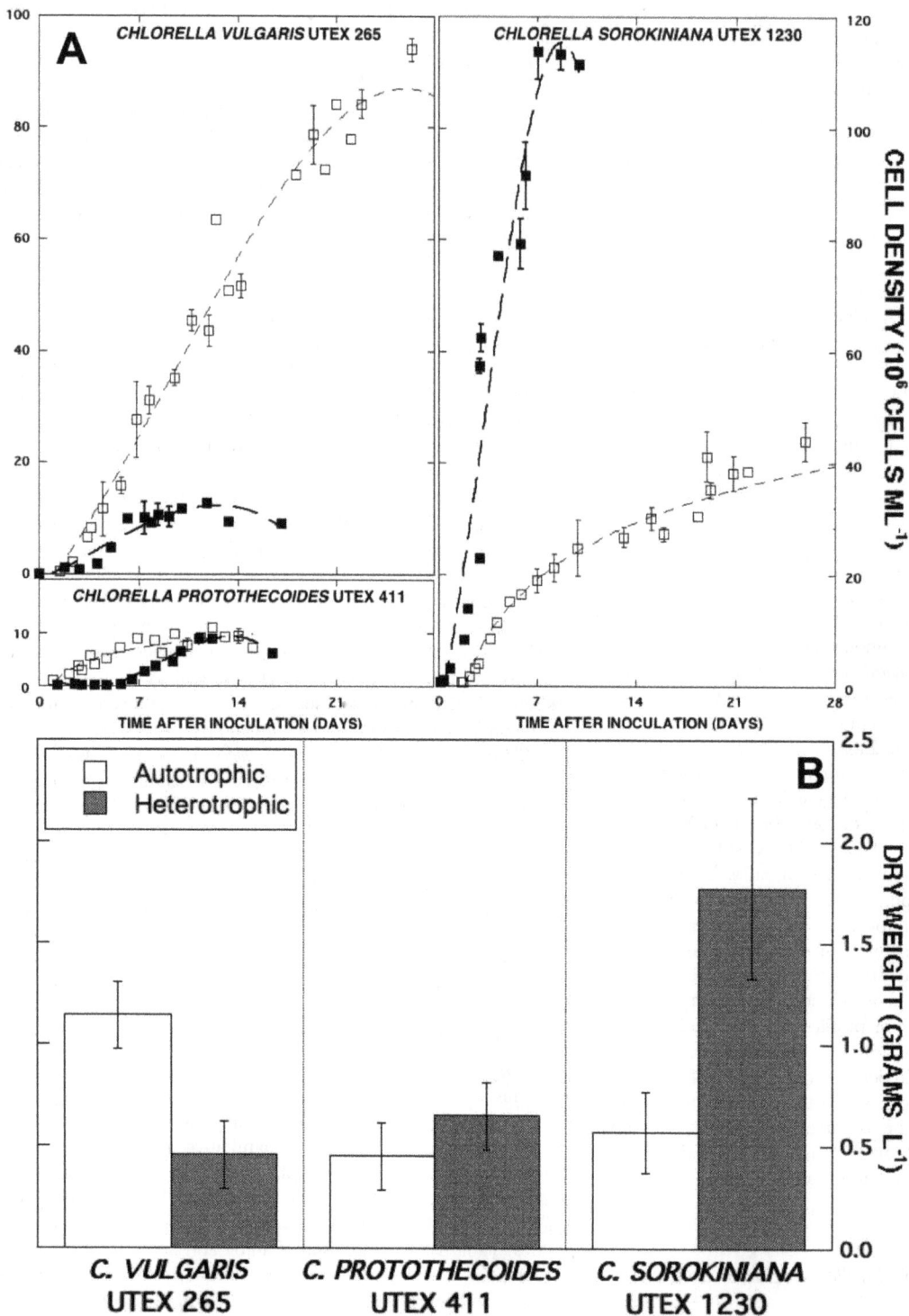

Figure 2. Growth curves and volumetric biomass yields of *C. protothecoides,* *C. vulgaris,* **and** *C. sorokiniana.* (**A**) Photoautotrophic (□) and heterotrophic (■) cultures were followed over the course of four weeks or until the population reached stationary phase. Data points are representative of biological replicates and error bars denote standard deviations greater than 1×10^6 cells ml^{-1}. (**B**) *C. protothecoides, C. vulgaris,* and *C. sorokiniana* final biomass concentrations in 1.5-L culture during auto– (white) and heterotrophy (gray) are compared. Error bars designate standard deviation from the average of three technical replicates.

At the termination of each culture, the dry biomass weights of *C. sorokiniana, C. vulgaris,* and *C. protothecoides* were compared (Figure 2B). Not surprisingly, heterotrophic *C. sorokiniana* UTEX 1230 amassed the greatest final biomass (1.8 g L^{-1}), while the dry

weights of *C. vulgaris* and *C. protothecoides* heterotrophic biomass were both close to 0.5 g L^{-1}. When grown autotrophically, *Chlorella vulgaris* UTEX 265 reached a final biomass density of 1.1 g L^{-1} (94×10^6 cells ml^{-1}) while *C. sorokiniana* UTEX 1230

Table 1. Growth characteristics of *C. protothecoides* UTEX 411, *C. vulgaris* UTEX 265, *C. sorokiniana* UTEX 1230.

Chlorella **Species**	C. protothecoides	C. vulgaris	C. sorokiniana
Strain	UTEX 411	UTEX 265	UTEX 1230
Specific Growth Rate, μ (d^{-1})	**0.48±0.01**	**0.84±0.09**	**1.77±0.04**
	0.32±0.05	0.23±0.01	0.36±0.05
Doubling Time (hr)	**35±1**	**20±2**	**9±1**
	52±4	72±3	46±3

The major heterotrophic and photoautotrophic growth parameters for each *Chlorella* strain were calculated using data points from the exponential phase of batch culture. Bold values correspond to heterotrophic populations; regular typeface represents autotrophic rates. All kinetic quantities are reported as the average of biological replicates ± standard deviations.

reached a maximal cell density of 44×10^6 cells ml^{-1} at 0.6 g L^{-1} dry weight, similar to the 0.5 g L^{-1} biomass yield of UTEX 411 from only 10×10^6 cells ml^{-1}. Interestingly, *C. protothecoides* UTEX 411 may possess more massive cells based on dry weight although it was unable to reach cell densities above 10 million cells ml^{-1} during either auto– or heterotrophic modes of cultivation in the present study. Both *C. protothecoides* UTEX 411 and *C. vulgaris* UTEX 265 exhibited an extended lag phase during heterotrophy before expansion of the population, suggesting a lengthy acclimation period for growth on simple sugar. Based on its rapid growth and high biomass yield under heterotrophic conditions, *C. sorokiniana* UTEX 1230 was taken into consideration as a lead candidate for an in depth investigation of oil production from heterotrophically-grown cells.

Lipid content and composition of *Chlorella* during auto– and heterotrophic cultivation

In order to evaluate oil production during auto– and heterotrophy, total lipids were extracted from the three *Chlorella* strains followed by transesterification to form fatty acid methyl esters (FAME) for GC-MS analysis. While autotrophic UTEX 1230 lipid extracts remained deeply green due to chlorophyll, heterotrophic lipid extracts exhibited a yellow hue (Figure 3A), indicative of inhibited chlorophyll synthesis. The distribution of each strains' lipid extract is categorized in Figure 3B according to its major components: total lipids measured gravimetrically; total fatty acids measured as FAME; and TAG neutral lipids. In this figure, the inferred difference between FAME (gray bars) and TAG lipids (black bars) in each case is assumed to approximate the amount of membrane lipids [+]. In a similar manner, comparing FAME lipids quantified by GC-MS to total lipid extracts (white bars) leads to the recognition of an unaccounted fraction [*], which represents other lipophilic biomolecules that can be extracted by the organic solvent system, but not measured by GC-MS analysis of the FAME component. This residual fraction is likely to be constituted by chlorophylls, lipid-soluble metabolites, and carotenoids such as lutein. Since many of these accessory "lipids" can be molecules associated with photosynthesis, carbon fluxes are likely to be directed toward these biomolecules rather than TAG during photoautotrophy. For example, UTEX 265 and 411 maintained roughly 60% of their total autotrophic lipids as accessory pigments and metabolites compared to only 16% in UTEX 1230. Consequently, UTEX 1230 partitioned more of its total autotrophic biomass as TAG (11% of the biomass; 60% of total lipids) compared to the markedly lower autotrophic TAG

levels in UTEX 265 and 411. The photoautotrophic biomass of *C. sorokiniana* UTEX 1230 was comprised of 18% total lipids, 15% fatty acids, and 11% TAG by dry weight, while *C. vulgaris* UTEX 265 accumulated equivalent amounts of total lipids (18%), but considerably lower levels of fatty acids (7.6%) and TAG (6.6%). *Chlorella protothecoides* UTEX 411 biomass contained the lowest lipid contents during autotrophy with 12% total lipids, 4.5% fatty acids, and 2.3% TAG.

As evidenced by the early onset of chlorosis in *C. sorokiniana* UTEX 1230, accessory pigments can be dramatically reduced during heterotrophy (Figure 3A). Accordingly, this unique phenomenon is manifested in the similarity between the total lipid content and total fatty acid content in UTEX 1230 with marginal amounts of accessory metabolites and minimal membrane lipids [×] (Figure 3B). Since the absence of chlorophyll observed during heterotrophic cultivation of UTEX 265 and 411 was not as severe as UTEX 1230, there is not nearly as much parity between their total lipids and total FAME contents. During heterotrophy, UTEX 1230 total lipid, fatty acid, and TAG contents increased to 24%, 23%, and 21%, respectively, amassing the highest TAG level among all strains examined, which was also two-fold higher than the TAG content of UTEX 1230 during autotrophy. The total lipid, FAME, and TAG contents of heterotrophic UTEX 265 exhibited proportional increases to 26%, 16%, and 13%, respectively. Compared to the other strains, UTEX 411 accumulated the highest total lipid content of 31% with 24% fatty acids during heterotrophy. However, a smaller portion of these lipids was stored as TAG (12% of the biomass; 40% of total lipids). Despite the respectable oil induction in UTEX 265 and 411 with 1.5–2.5 times higher total lipid contents during heterotrophy relative to autotrophy, these two strains possessed slow growth rates and low heterotrophic biomass yields under the current cultivation conditions. Alternatively, UTEX 1230 demonstrated rapid growth during heterotrophy and accrued an overwhelming majority of its total lipid content as TAG.

The quantitative results of neutral lipid induction gathered from UTEX 1230 lipid extracts during heterotrophy were also confirmed with an *in situ* method using Nile Red (NR) to stain whole cells [73]. As conveyed by fluorescent emission curves (Figure S1), heterotrophic UTEX 1230 exhibited a clear maximum of 5.8×10^4 fluorescent intensity units (FIU) at its expected emission wavelength of 580 nm compared to 2×10^4 FIU autotrophically. The signatures of these NR profiles indicate that UTEX 1230 accumulated more hydrophobic biomolecules under heterotrophic conditions compared to autotrophy. In addition to different carbon sources, corresponding pH changes in auto– and heterotrophic media during algal growth can also contribute to lipid accumulation and influence fatty acid speciation [74]. While autotrophic cultures of *C. sorokiniana* UTEX 1230 became alkaline with pH rising from 7.3 to 10, heterotrophic cultures experienced slight acidification from pH 6 to 4.8 (Figure S2).

Distribution of fatty acid chain length and degree of saturation in *Chlorella* species

In order to further characterize the lipid metabolic profiles of these three *Chlorella* species, detailed fatty acid compositions were elucidated by GC-MS for auto– and heterotrophy (Figure 4). The typical distribution of sixteen- to eighteen-carbon fatty acids with no more than three degrees of unsaturation (16–18 : $\leq$3) characteristic of many freshwater green algae was confirmed [74,75]. While all species exhibited increased total lipid and TAG levels during heterotrophy, Figure 4 highlights the species specificity of fatty acid accumulation with respect to lipid chain length. The discernable differences between the three strains are

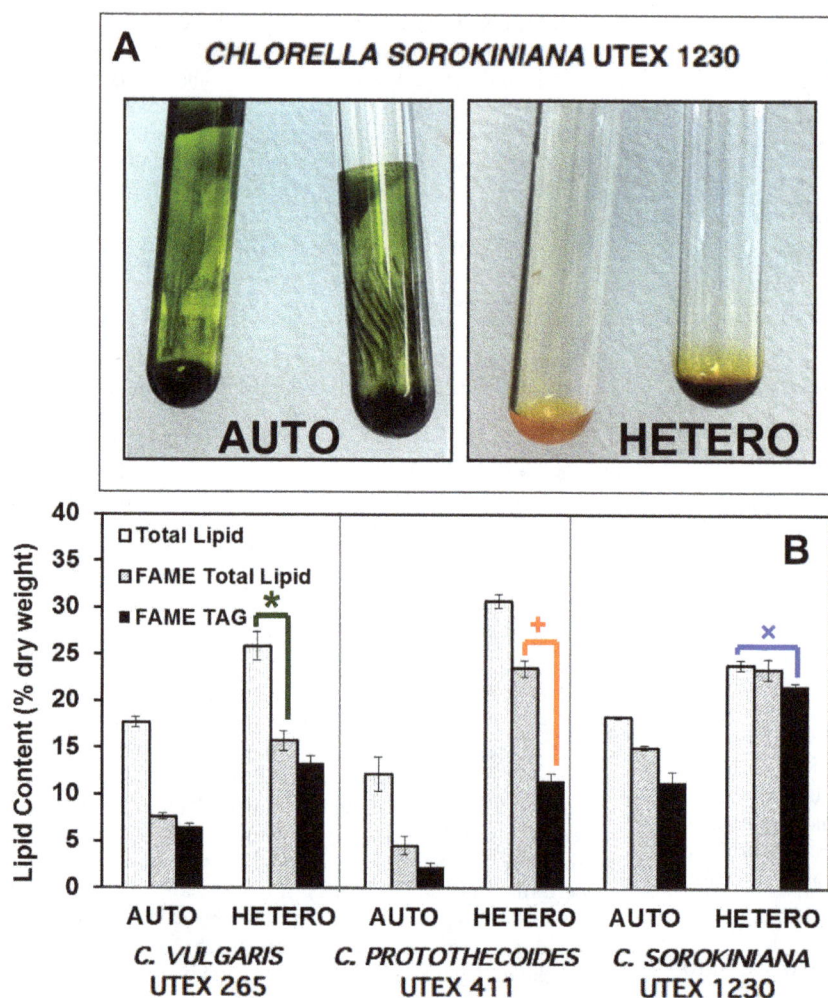

Figure 3. Distribution of total lipid extracts as fatty acids and TAG in three *Chlorella* **strains. (A)** *C. sorokiniana* UTEX 1230 exhibits more prominent chlorosis during heterotrophy than UTEX 265 and 411, which is readily apparent in the pigmentation of total lipid extracts (in color online). **(B)** Total lipids (white) in *C. protothecoides* UTEX 411, *C. vulgaris* UTEX 265, and *C. sorokiniana* UTEX 1230 are classified as FAME (gray) and TAG storage lipids (black). By comparing these values, the relative amounts of supposed accessory metabolites [*] and membrane lipids [+] emerge, while UTEX 1230 appears to be devoid of accessory pigmentation with minimal membrane lipids [×]. Error bars represent one standard deviation from the average of three technical replicates.

the ratios of fatty acids accumulated as 16:0, 18:1, and 18:2 as well as the amount of TAG, which is consistent with Figure 3B. During both auto– and heterotrophic conditions, UTEX 1230 showed a relatively even distribution of palmitic (16:0) and linoleic (18:2) acids, both between 5–10% by dry weight. In contrast, UTEX 265 and 411 produced oleic acid (18:1) as the principal fatty acid, reaching heterotrophic maxima of 7.5% and 14%, respectively, with secondary storage lipids of 16:0 and 18:2 each at less than 5% of the biomass. Furthermore, UTEX 1230 exhibited greater than 2.5-fold enrichment in 18:2 relative to 18:1, which is in direct opposition with the relative levels of oleic and linoleic acids prevalent in the other strains. This polyunsaturation in UTEX 1230 may indicate higher activity of lipid desaturases [76], particularly during heterotrophy [77].

Time-course comparison of lipid production during hetero– and mixotrophy in UTEX 1230

In order to compare the dynamics of lipid production in *C. sorokiniana* UTEX 1230 during hetero– and mixotrophy, cultures were grown in BBM with glucose (10 g L^{-1}) either in the dark or

together with artificial illumination. Separate 8-L bioreactors were sampled over two weeks for cell density, biomass concentration, and total gravimetric lipid measurements followed by GC-MS analysis. Despite the equivalent amounts of initial glucose, mixotrophic cultures displayed significantly reduced biomass and lipid yields compared to heterotrophic cultures (Figure 5). During the first three to five days of cultivation, both populations expanded at similar specific growth rates; however, clear differences arose later in cultivation (Figure 5A). While heterotrophic growth of UTEX 1230 peaked at 130×10^6 cells ml^{-1}, the mixotrophic culture plateaued at only 70×10^6 cells ml^{-1}, which is more similar to the UTEX 1230 growth pattern during photoautotrophy. The respective hetero– and mixotrophic biomass yields of 2 and 1.7 g L^{-1} were more similar than would be expected from the cell densities (Figure 5B), suggesting a possible photoinhibitory or pH-induced effect on cell cycle progression during mixotrophy, which is also reflected in diminished total lipid content (Figure 5C). Mixotrophic UTEX 1230 biomass contained $31.8 \pm 0.7\%$ total lipids after two weeks compared to $38.7 \pm 0.9\%$ in the heterotrophic biomass, but experienced a two-fold reduction

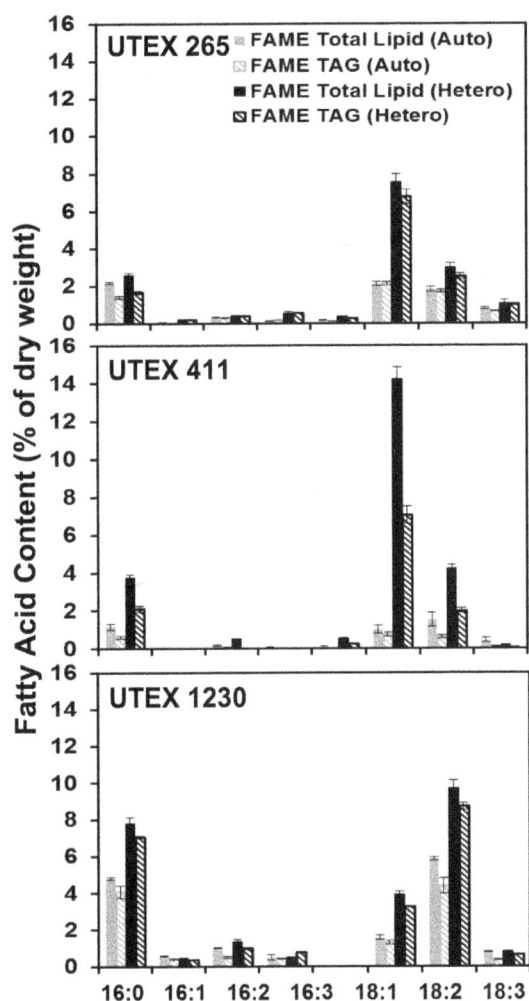

Figure 4. Fatty acid composition of three *Chlorella* strains during auto- and heterotrophy. The distribution of total fatty acids measured as FAME and TAG are plotted as solid and diagonal bars, respectively, for *C. protothecoides* UTEX 411, *C. vulgaris* UTEX 265, and *C. sorokiniana* UTEX 1230. Compared to autotrophy (gray), heterotrophy (black) induces remarkable changes in FAME and TAG lipid profiles. Error bars represent the standard deviation from the average of three biological replicates.

in TAG accumulation relative to heterotrophy. While mixotrophy engendered a final TAG content of 11.8±0.8% in UTEX 1230, heterotrophy achieved 22.1±0.7% TAG (Figure 5D), corresponding to volumetric TAG productivities of 18.2 and 28.9 mg L^{-1} d^{-1}, respectively.

Progressive induction of polyunsaturated fatty acids in UTEX 1230 during heterotrophy

Lipid profiles from this time-course evaluation of UTEX 1230 revealed that the kinetics of fatty acid biosynthesis in heterotrophy are accelerated relative to mixotrophy, especially with regard to the 16:0–18:1–18:2 triad (Figure 6). Time dependent induction of unsaturation led to the biogenesis of 18:2 (linoleic acid) from precursor 18:1 (oleic acid) and persisted in late-stage heterotrophic cultures (Figure 6A). In this case, both 18:1 and 18:2 showed a steady increase in the first ten days of heterotrophic culture from less than 2% of the biomass (<0.5% TAG) to nearly 5% and 8%, respectively (5–6% TAG). While 18:1 exhibited an insignificant

increase between days ten and thirteen, 18:2 levels reached 10% of the total dry biomass with 7.7% TAG by the end of heterotrophic culture. Conversely, Figure 6B illustrates that 18:1 barely exceeded 2% during mixotrophy and 18:2 reached a peak at approximately 5% dry weight. While total 18:1 content remained unchanged during the final three days of mixotrophic culture, 18:1 TAG levels actually dropped while 18:2 TAG only increased slightly. Interestingly, 16:0, 18:1, and 18:2 TAG fractions had higher initial levels during mixotrophy (closer to 1%), but did not exhibit the high rates of TAG accumulation that occurred during heterotrophy and subsisted at less than 4% of the total biomass by the end of two-week cultivation. Furthermore, the increase in unsaturated 18:1 and 18:2 in UTEX 1230 may have been balanced by an enrichment in 16:0 to decrease membrane fluidity as pH became more acidic during heterotrophy [78].

Discussion and Conclusions

For the production of microalgal oils, the interdependence of total lipid content, fatty acid composition, and biomass productivity throughout cultivation is critical [79,80]. Novel approaches to maximize carbon storage as neutral lipids can lead to significant benefits for algae biotechnology, biomanufacturing, and bioenergy [81]. The accumulation of different lipids during heterotrophy and autotrophy can be especially important for the interrogation of lipid biosynthetic pathways in commercially relevant *Chlorella* species. Toward this end, the current study has revealed distinctive auto– and heterotrophic capabilities of three phylogenetically classified *Chlorella* strains with respect to the trade-offs between oil content and growth rate.

Molecular phylogeny and genomic resources

While microalgae can be generally classified based on cellular features [82], species of the same genus are nearly morphologically identical and can only be definitively differentiated by genotypic analysis [83,84]. Therefore, the foundation of our species characterization relied on determining the molecular phylogeny of the conserved 18S sequence and species-specific ITS regions for each *Chlorella* strain [85–87] employing a rapid colony PCR method [88]. Confirming an organism's true phylogeny is important for understanding what adaptive traits have been collected through natural selection and how they may be useful for biotechnology. In the course of mapping the branches of the phylogenetic tree in this study (Figure 1), the occurrence of mismatched species (*e.g.*, UTEX 29, 252, 2248, 2714) was more prevalent than anticipated. Indeed, another case of species misidentification had previously been identified during genome sequencing of NC64A in tandem with the putative *C. vulgaris* C-169 that is resistant to the *Chlorella* virus [89]. Despite physiological similarities to *C. variabilis*, the C-169 strain was actually determined to be *Coccomyxa subellipsoidea* by comparative genomics [90,91].

While genetic resources for the *Chlorella* genus are improving with genome sequencing projects underway for many common species [92], the 46-Mb *C. variabilis* NC64A genome is currently the only publicly available *Chlorella* genome [32]. Although peculiar in its growth habits, this organism holds information about the genetic-basis of sugar transport in green algae during symbiosis, which is of particular relevance to the present study [35]. Sequencing of *Chlorella* ITS regions fulfills the need to relate macroscopic growth characteristics of potential production organisms to molecular genetic traits [93]. Thus, the ribosomal ITS-based phylogenetic tree constructed during the present investigation can indicate a strain's proximity to *C. variabilis* NC64A or other organisms with available genetic tools. In addition, this map

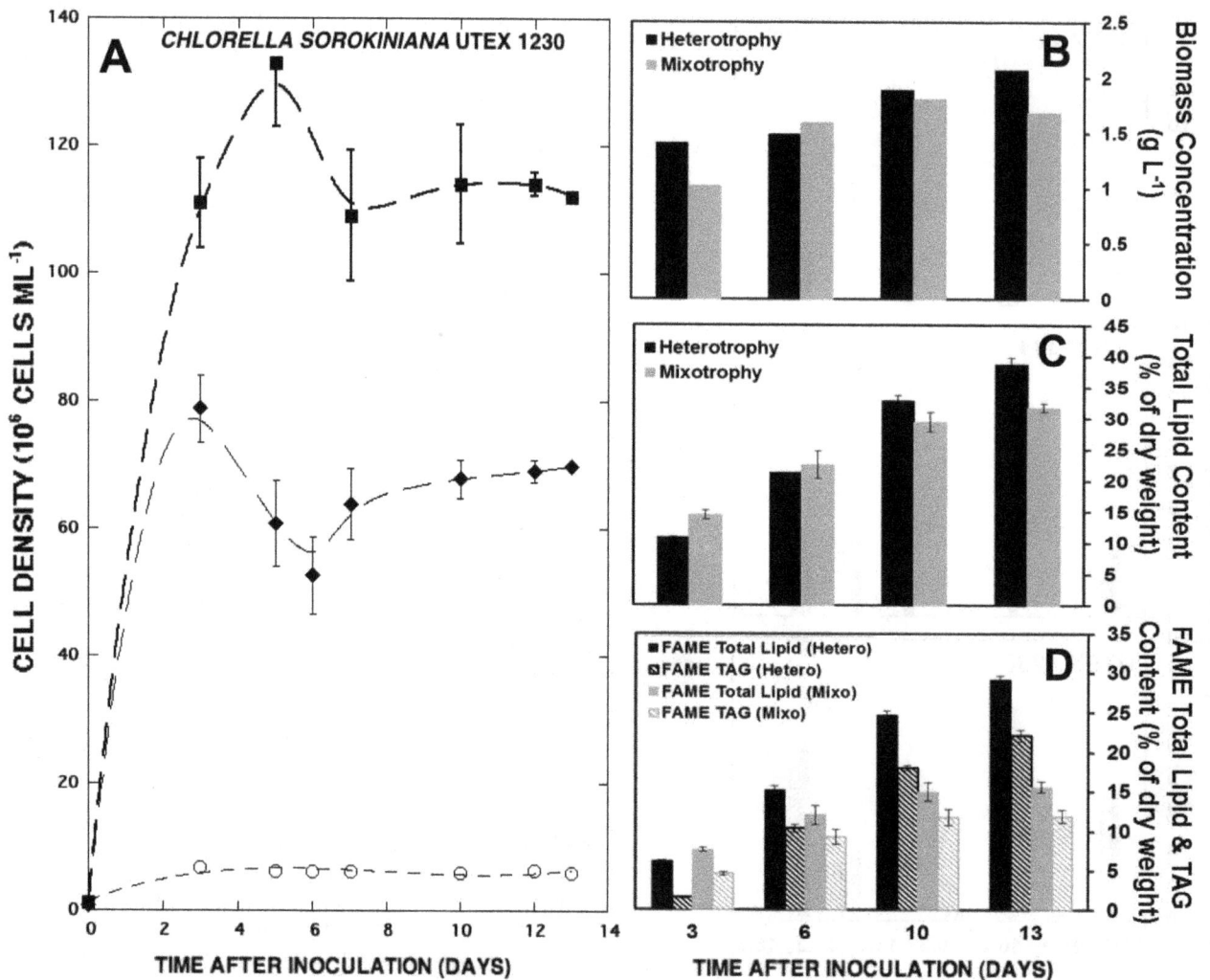

Figure 5. Growth curves, biomass yield, and lipid accumulation of *C. sorokiniana* **UTEX 1230 throughout mixo- and heterotrophy in 8-L bioreactors.** (**A**) Heterotrophic (■) and mixotrophic (◆) cell densities are compared to UTEX 1230 grown in BBM without sugar or light (○), which served as a negative control. (**B**) The corresponding dry biomass, (**C**) total lipid content, and (**D**) total FAME and TAG percentages at days 3, 6, 10, and 13 of mixo- (gray) and heterotrophic (black) cultivation are plotted. FAME and TAG contents for each lipid type are plotted as solid and diagonal bars, respectively. Error bars represent the standard deviation from the average of three technical replicates.

of phylogeny makes it possible to identify *Chlorella* isolates with suitable biomass and lipid productivities that are both close in genetic relation, yet distinct in physiological characteristics.

Physiological and biochemical survey of three *Chlorella* species

While the biomass productivity of *Chlorella* cultures has been extensively studied, there remains little consensus between the research community and commercial endeavors regarding preferred *Chlorella* species or growth methodologies to maximize oil yields. Since *C. vulgaris*, *C. prototohecoides*, and *C. sorokiniana* are the primary organisms used for process development and strain improvement [94,95], this work focused on representatives of these species.

In measuring the growth characteristics of *C. sorokiniana* UTEX 1230, *C. vulgaris* UTEX 265 and *C. prototohecoides* UTEX 411 (Table 1), UTEX 1230 produced the greatest amount of heterotrophic biomass (Figure 2) in general agreement with previous literature supporting the heterotrophic cultivation of

Chlorella [96–100]. However, *C. prototohecoides* UTEX 411 growth did not compare well with the other *Chlorellas* during either hetero- or photoautotrophic modes in our experiments. Compositional differences in media or the effect of pH stress on cell cycle may have accounted for this observed growth restriction, cellular enlargement, and reduced lipid productivity [101]. Compared to BBM used in this study, prior heterotrophic *Chlorella* growth media have sometimes replaced sodium nitrate with glycine and vitamin B_1 [8,102], which implicates a potential vitamin requirement or preferred nitrogen source for UTEX 411 [80]. However, we decided to maintain the medium as consistent as possible to limit environmental factors in this comparison. From a technoeconomic perspective, we recognize that the most cost-effective sources and optimal amounts of fixed carbon, nitrogen, and micronutrients may vary for different *Chlorella* species. Furthermore, microalgal strains that can prosper in relatively low nitrogen and carbon concentrations can be desirable for large-scale biomass production by reducing overall nutrient costs and lowering the impact of these resources on bioprocessing costs.

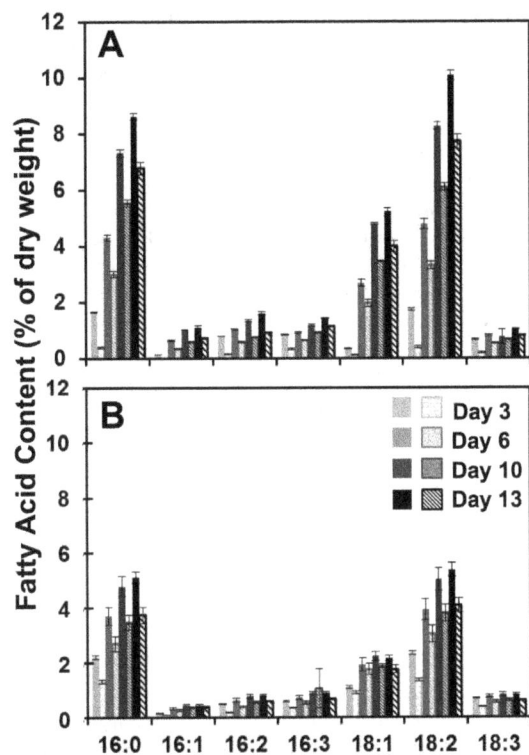

Figure 6. Comparison of total fatty acid and TAG composition of UTEX 1230 during mixo– and heterotrophic cultivation. The discrete lipid profiles of *C. sorokiniana* UTEX 1230 under (**A**) heterotrophic and (**B**) mixotrophic conditions from samples taken on days 3, 6, 10 and 13 of cultivation show fatty acid accumulation rates and patterns. Total FAME and TAG contents for each lipid type are plotted as solid and diagonal bars, respectively. Error bars represent the standard deviation from the average of three biological replicates.

Lipid allocation between membranes, metabolism, and energy storage in *Chlorella*

One of the main goals of this study was to quantify and compare the differential apportioning of lipid types in the three *Chlorella* strains during auto– and heterotrophy. The divergence in lipid generation between these algal species was clearly evident from the distribution of total lipids, total fatty acids measured as FAME, and TAG neutral lipids. In Figure 3B, heterotrophic *C. vulgaris* UTEX 265 exhibited roughly 10% accessory pigments and lipid-soluble metabolites by dry weight [*], which can serve as high-value nutritional coproducts (*e.g.*, chlorophylls, carotenoids, phenols, tocopherols, sterols). This may indicate that the cellular metabolism of UTEX 265 remains more active than lipid storage during heterotrophy, as observed previously with *C. sorokiniana* strains UTEX 2714 and CS-01 as well as *Chlamydomonas reinhardtii* [56,103]. In a complementary case, *C. protothecoides* UTEX 411 produced approximately 10% membrane lipids during heterotrophy [+] (*e.g.*, phospho-, galacto-, and sulpholipids). In addition to these detractions from TAG accumulation due to the competing cellular demand for lipid metabolites and membrane structures, both UTEX 265 and 411 exhibited less vigorous heterotrophic growth capacities (Figure 2A). Alternatively, *C. sorokiniana* UTEX 1230 benefited from rapid heterotrophic growth and contained 24% total lipids with close to 90% of these total lipids allocated as TAG [×] (Figure 3B). With energy stores accounting for the majority of UTEX 1230 total lipids, this leaves a limited amount of fatty acids available for membranes during heterotrophy, which

may arise from a substantial reduction in the number of chloroplast and thylakoid membranes. Combined with the higher cell densities supported during heterotrophy, UTEX 1230 displayed a 3.8-fold increase in lipids per liter at the end of batch culture relative to autotrophy with an abrupt shift from non-fatty acid lipid production to TAG storage. The increase in pH during autotrophy may be primarily responsible for TAG accumulation due to cell cycle arrest [74]. Alternatively, during heterotrophic cultivation with minor changes in pH, the ultimate lipid yields of UTEX 1230 increased from 0.14 to 0.53 g L^{-1} perhaps due to the up-regulation of stearoyl-ACP desaturase in the presence of glucose, leading to an abundance of 18:1 precursor fatty acid [77].

Further mixo– and heterotrophic bioreactor optimization at the 8-L scale enabled UTEX 1230 to accumulate 30–40% of its cell mass as lipids, possibly as a result of increased mixing and enhanced gas exchange. Additionally, nutrient utilization trends in previous *Chlorella* studies using BBM demonstrate that nitrate and nitrite levels were reduced to 5–10% of their initial concentrations, indicating that BBM may not be completely deprived of nitrogen at stationary phase [56,103]. This residual nitrogen may, therefore, sustain the presence of chlorophyll and basal metabolic activity. Furthermore, TAG storage in *Chlorella* can exhibit differential responses to reduced nitrogen levels compared to other algae. While *C. reinhardtii* may require complete nitrogen depletion to induce TAG accumulation [104], *Chlorella* species can store moderate amounts of TAG in nitrogen replete media. This connection between nutrient limitation, cell growth, and lipid biosynthesis offers additional insight into *Chlorella* metabolism.

Bioenergetics of mixo– and heterotrophic growth and its effect on fatty acid distribution

Considering that the breakdown of sugar by glycolysis may take precedence over carbon dioxide utilization through photosynthesis [72], our study of *C. sorokiniana* UTEX 1230 indeed found no apparent metabolic benefit from light input in addition to glucose. Furthermore, this strain appeared to be potentially maladapted for oil accumulation during mixotrophy. Restricted mixotrophic growth suggests that glucose uptake may be inhibited by light or the energy balance of mixotrophic cells is unfavorably altered (Figure 5A). In extreme cases, pH can also negatively affect sugar uptake via the hexose-proton symporter, although the pH drift encountered in the present study fell within the optimal range for glucose uptake by *Chlorella* (Figure S2) [105]. Interestingly, incongruencies between biomass and cell density in mixo– and heterotrophic comparisons (Figure 5B) also suggest that mixotrophic cells may be larger due to the simultaneous activity of glycolytic lipid accumulation in storage vesicles and photosynthetic carbon fixation, which requires multiple chloroplasts. While these findings contradict a recent report of high mixotrophic productivities from a different *C. sorokiniana* strain, this discrepancy simply highlights the strain-specific metabolic portraits drawn for different *Chlorella* isolates under investigation [106]. Previous studies have indeed found evidence for the inhibition of organic carbon uptake by other *Chlorellas* in the presence of light [107,108]. In addition, a recent evaluation of *C. sorokiniana* CS-01 (closely related to UTEX 1230) revealed that cytosolic acetyl-coA carboxylase (ACCase) is overexpressed compared to chloroplast ACCase during mixotrophy, suggesting that fatty acid precursors contributing to TAG synthesis are derived from glycolysis of exogenous sugars rather than photosynthetically fixed carbon [72]. This underscores the importance of understanding the bifurcation of carbon utilization during these potentially competing modes of growth.

From the distribution of lipid classes under heterotrophic conditions, it is clear that TAG can constitute a significant portion of lipids in all three of these *Chlorella* strains with 18:1 and 18:2 as the primary fatty acids, which are more suitable for biofuel applications than polyunsaturated lipids. From a bioenergetics standpoint, more cellular energy is expended to generate polyunsaturated fatty acids compared to saturated lipids. Since oxidation also has a negative impact on fuel stability, unsaturated lipids must be catalytically hydrogenated prior to fuel blending, which requires additional capital and operating expenses. While the absence of long-chain polyunsaturated fatty acids is not uncommon in *Chlorella* species [109], all strains in the present study exhibited surprisingly low levels of 18:3 (linolenic acid). Other studies have demonstrated that UTEX 1230 is capable of 18:3 biosynthesis under autotrophic conditions with nitrogen deprivation [103,110] and anaerobic digester effluent [56]. In contrast, UTEX 1230 can accumulate higher levels of 18:1 when supplemented with 3% carbon dioxide (unpublished data).

As a whole, these results support the *Chlorella* genus as a vital source of commercially relevant production organisms with unique lipid metabolic properties, which can vary significantly depending on the presence or absence of sugars and light. While glucose was used in these controlled growth studies, low-cost organic carbon sources are likely to be required for commodity scale biofuel production [56–58,111]. This study's phylogenetic and physiological characterizations of *C. vulgaris* UTEX 265, *C. protothecoides* UTEX 411, and *C. sorokiniana* UTEX 1230 offer insight into the diverse carbon partitioning strategies even within a single genus. In one exemplary case, *C. sorokiniana* UTEX 1230 may serve as a versatile model organism for photosynthetic carbon utilization and heterotrophic lipid biosynthesis with implications for bioenergy and bioprocessing.

Materials and Methods

Microalgal cultivation

Stock samples of microalgal strains were obtained from the Culture Collection of Algae at University of Texas at Austin (http://web.biosci.utexas.edu/utex/) and maintained on sterile agar plates (1.5% w:v) containing Bold's basal medium (BBM) [112]. Cells were cultivated in 1.5-L glass Fernbach flasks or 8-L glass Bellco bioreactors (New Jersey, USA) at 27°C ($\pm$1) using BBM. Each axenic batch culture was inoculated with exponentially growing cells at a density of 1×10^6 cells ml^{-1}, constantly stirred, and bubbled with sterile air and monitored with a calibrated M240 digital pH meter (Corning, Inc., New York, USA). Photoautotrophic cultures were illuminated with cool-white fluorescent light at an intensity of 100 μE m^{-2} s^{-1}. Heterotrophic and mixotrophic cultures were supplemented with glucose at a concentration of 10 g L^{-1}. While heterotrophic cultures were grown in complete darkness, mixotrophic cultures were grown with a light intensity of 100 μE m^{-2} s^{-1}. Cell densities were measured by hemocytometer with an Axiovert 100 inverted light microscope (Carl Zeiss, Göttingen, Germany). Measurements were taken in duplicate and experiments were repeated at least three times. Liquid cultures were harvested using a high-speed centrifuge (Sorvall RC-5B, Delaware, USA) at 6,000$\times$ *g* for 20 minutes.

Species identification by genetic sequencing and phylogeny

Species-specific genetic fingerprints were determined by sequencing the internal transcribed spacer (ITS) regions of ribosomal DNA [85–87]. Nucleic acids were extracted from clonal algal populations using a 5% Chelex-100 solution as described previously [88]. After boiling at 100°C for 15 minutes,

samples were centrifuged at 16,000$\times$ *g* for 2 minutes and DNA recovery was quantified using a NanoDrop 2000 spectrophotometer (Thermo Fisher Scientific, Delaware, USA). The primer pair designed for this study (5'-ACTCCGCCGGCACCTTATGAG-3'; 5'-CCGGTTCGCTCGCCGTTACTA-3') was used to amplify the ITS regions with the Top Taq Master Mix Kit (Qiagen, California, USA) according to the manufacturer's protocol employing thermocycler conditions with an initial melting at 95°C for 2 minutes, followed by 35 cycles of [94°C for 30 sec, 60°C for 30 sec, 72°C for 2 min] and a final elongation at 72°C for 10 minutes. Amplified fragments were separated by electrophoresis on an acrylamide (1% w:v) tris-borate-EDTA gel. Molecular weights were determined with a GeneRuler 1 kb Plus DNA Ladder (Fermentas, Delaware, USA) and ultimately purified using the GenCatch™ PCR Extraction Kit (Epoch). The resulting nucleotide sequences of 18S ITS regions (Eurofins MWG Operon, Ebersberg, Germany) can be found as annotated entries in the GenBank database (http://www.ncbi.nlm.nih.gov/genbank/) and were aligned using version five of the Molecular Evolutionary Genetics Analysis (MEGA) software suite employing the MUSCLE alignment feature and neighbor joining method to produce phylogenetic trees [113].

Measurement of algal biomass dry weight and total lipid content

In order to measure dry biomass yield, algal cultures were harvested as described previously, dried overnight in a vacuum oven at 60°C (Precision Scientific Model 5831, NAPCO, Virginia, USA) and weighed in triplicate. For total lipid extraction, washed cell pellets were freeze-dried with a lyophilizer (Model 25 SRC, Virtis, New York, USA) to preserve the integrity of fatty acids and homogenized with a mortar and pestle in liquid nitrogen. For each sample, 100 mg of homogenized biomass was extracted in 6 ml of 1:2 (v:v) chloroform:methanol containing 0.01% butylhydroxyltoluene and 500 μg of tripentadecanoin (15:0 TAG, Nu-Check Prep, Minnesota, USA) was added as an internal standard [114]. To the 6 ml total extraction volume, 1 ml of 0.7 mm diameter zirconia beads (BioSpec Products, Oklahoma, USA) was added and vortexed at room temperature for 30 minutes. After centrifugation at 1,500$\times$ *g* for 5 minutes, the chloroform phase was collected and the water phase was re-extracted using 5 ml of chloroform; the re-extraction was repeated three times. The pooled chloroform phases were evaporated to dryness under a stream of nitrogen. To confirm the methodological validity of manual oil separation [115], total lipid extractions were performed in parallel using an automated Accelerated Solvent Extraction 150 system, which employs elevated temperature and pressure (120°C, 1,500 psi, 4 reflux cycles) to accomplish rigorous oil separation and is generally recognized as the most effective oil extraction equipment (Dionex, Thermo Fischer, Delaware, USA) [116,117]. The closely comparable total lipid contents of manual and automated extractions were determined in triplicate by gravimetric methods.

Fatty acid methyl ester (FAME) analysis

Extracted lipids were transmethylated to FAME as previously described [15] and TAG separation was accomplished by thin layer chromatography. FAME was analyzed using an Agilent 6890 Series Gas Chromatography System with an Agilent 5973 Network Mass Selective Detector (Agilent Technologies, Delaware, USA). Chromatography was carried out using a 200 m$\times$250 μm$\times$0.25 μm Varian GC capillary column (Varian Inc., California, USA) with the inlet held at 270°C while 1 μl of the sample was injected using helium as the carrier gas. The oven temperature was programmed for 130°C (10 min) to 160°C

(7 min); 160°C to 190°C (7 min), from 190°C to 220°C (22 min) and from 220°C to 250°C (17 min) at a rate of $10°C\,min^{-1}$ for each step. The total analysis time was 75 minutes using 70 eV electron impact ionization and data was evaluated with total ion count.

Relative lipid content analysis *in situ* by Nile Red fluorescence

Nile Red (AnaSpec, Inc., California, USA) was dissolved in acetone to yield a 250× stock solution following previous protocols [73]. For each *Chlorella* sample, 10 μl of Nile Red stock was mixed with 2.5 ml of algal culture diluted to 4×10^7 cells ml^{-1} in a glass fluorometer cuvette (Starna Cells, Inc., California, USA). A PTI spectrofluorometer (Photon Technology International, Inc., New Jersey, USA) comprised of the SC-500 Shutter Control, MD-5020 Motor Driver, LPS-220B Lamp Power Supply, and 814 Photo-multiplier Detection System was used to excite samples at 486 nm and collect fluorescent emission spectra over a 500–660 nm range using the accompanying PTI FeliX32 software.

Supporting Information

Figure S1 Nile Red fluorescent emission curves of *C. sorokiniana* UTEX 1230. The shifts in emission peaks from UTEX 1230 cultivated under (**A**) autotrophic and (**B**) heterotrophic conditions implicate more significant lipid accumulation during heterotrophy.

Figure S2 Change in media pH during auto– and heterotrophic growth of *C. sorokiniana*. The pH of auto– (□) and heterotrophic (■) UTEX 1230 cultures was monitored during exponential growth and remained stable throughout stationary phase. The resulting pH curves demonstrate the interdependence of cell metabolism and the surrounding environment. The standard deviation for each data point is less than 0.5 pH unit (error bars not visible).

Acknowledgments

The authors thank Adithya Balasubramanian and Gunjan Andlay at Synaptic Research for their assistance with sample preparation and appreciate the fruitful discussions with Dr. Marc Donohue, Mr. Thomas Fekete, Dr. Minxi Wan, and Dr. Scott Williams at Johns Hopkins University. The authors also wish to thank Drs. Paul N. Black and Concetta C. DiRusso and members of their laboratories at the University of Nebraska-Lincoln for assistance with lipid preparation and analyses using GC-MS.

Author Contributions

Conceived and designed the experiments: JNR NK GAO. Performed the experiments: JNR NK AB EAN. Analyzed the data: JNR NK GAO MJB. Contributed reagents/materials/analysis tools: GAO MJB. Wrote the paper: JNR NK GAO MJB.

References

1. Blankenship RE, Tiede DM, Barber J, Brudvig GW, Fleming G, et al. (2011) Comparing photosynthetic and photovoltaic efficiencies and recognizing the potential for improvement. Science 332: 805–809.
2. Zhu X-G, Long SP, Ort DR (2008) What is the maximum efficiency with which photosynthesis can convert solar energy into biomass? Curr Opin Biotech 19: 153–159.
3. Bumbak F, Cook S, Zachleder V, Hauser S, Kovar K (2011) Best practices in heterotrophic high-cell-density microalgal processes: achievements, potential and possible limitations. Appl Microbiol Biot 91: 31–46.
4. Senanayake SPJN, Ahmed N, Fichtali J (2010) Chapter 37. Nutraceuticals and Bioactives from Marine Algae. In: C . Alasavar, F . Shahidi, K . Miyashita and U . Wanasundara, editors. Handbook of Seafood Quality, Safety and Health Applications. Oxford, UK: Blackwell Publishing Ltd. pp. 455–462.
5. Patnaik S, Samocha TM, Davis DA, Bullis RA, Browdy CL (2006) The use of HUFA-rich algal meals in diets for *Litopenaeus vannamei*. Aquacult Nutr 12: 395–401.
6. Davis R, Fishman D, Frank ED, Wigmosta MS, Aden A, et al. (2012) Renewable Diesel from Algal Lipids: An Integrated Baseline for Cost, Emissions, and Resource Potential from a Harmonized Model. Technical Report Numbers: ANL/ESD/12-4; NREL/TP-5100-55431; PNNL-21437. Argonne, IL: Argonne National Laboratory; Golden, CO: National Renewable Energy Laboratory; Richland, WA: Pacific Northwest National Laboratory.
7. Schenk PM, Thomas-Hall SR, Stephens E, Marx UC, Mussgnug JH, et al. (2008) Second generation biofuels: high-efficiency microalgae for biodiesel production. Bioenerg Res 1: 20–43.
8. Wu Q, Miao X (2006) Biodiesel production from heterotrophic microalgal oil. Bioresource Technol 97: 841–846.
9. Li X, Xu H, Wu Q (2007) Large-scale biodiesel production from microalga *Chlorella prototheocoides* through heterotrophic cultivation in bioreactors. Biotechnol Bioeng 98: 764–771.
10. Harwood JL, Guschina IA (2009) The versatility of algae and their lipid metabolism. Biochimie 91: 679–684.
11. Niederholtmeyer H, Wolfstadter BT, Savage DF, Silver PA, Way JC (2010) Engineering cyanobacteria to synthesize and export hydrophilic products. Appl Environ Microb 76: 3462–3466.
12. National Research Council (2012) Sustainable Development of Algal Biofuels in the United States. In: L.G . Coplin, editor. Washington DC: The National Academies Press. Available: http://www.nap.edu/catalog.php?record_id=13437. Accessed 2012 Oct 25.
13. Vasudevan V, Stratton RW, Pearlson MN, Jersey GR, Beyene AG, et al. (2012) Environmental performance of algal biofuel technology options. Environ Sci Technol 46: 2451–2459.
14. Converti A, Casazza AA, Ortiz EY, Perego P, Borghi MD (2009) Effect of temperature and nitrogen concentration on the growth and lipid content of *Nannochloropsis oculata* and *Chlorella vulgaris* for biodiesel production. Chem Eng Process 48: 1146–1151.
15. Msanne J, Xu D, Konda AR, Casas-Mollano JA, Awada T, et al. (2011) Metabolic and gene expression changes triggered by nitrogen deprivation in the photoautotrophically grown microalgae *Chlamydomonas reinhardtii* and *Coccomyxa* sp. C-169. Phytochemistry 75: 50–59.
16. Sheehan J, Dunahay T, Benemann J, Roessler P (1998) A Look Back at the U.S. Department of Energy's Aquatic Species Program: Biodiesel from Microalgae. Report Number: TP-580-24190. Golden, Colorado: National Renewable Energy Lab.
17. Stephenson AL, Dennis JS, Howe CJ, Scott SA, Smith AG (2010) Influence of nitrogen-limitation regime on the production by *Chlorella vulgaris* of lipids for biodiesel feedstocks. Biofuels 1: 47–58.
18. Spoehr HA, Milner HW (1949) The chemical composition of *Chlorella*; effect of environmental conditions. Plant Physiol 24: 120–149.
19. Perez-Garcia O, Escalante FME, de-Bashan LE, Bashan Y (2011) Heterotrophic cultures of microalgae: metabolism and potential products. Water Res 45: 11–36.
20. Samejima H, Meyers J (1958) On the heterotrophic growth of *Chlorella pyrenoidosa*. J Gen Microbiol 18: 107–117.
21. Xiong W, Liu L, Wu C, Yang C, Wu Q (2010) ^{13}C-tracer and gas chromatography-mass spectrometry analyses reveal metabolic flux distribution in the oleaginous microalga *Chlorella protothecoides*. Plant Physiol 154: 1001–1011.
22. Fan J, Huang J, Li Y, Han F, Wang J, et al. (2012) Sequential heterotrophy-dilution-photoinduction cultivation for efficient microalgal biomass and lipid production. Bioresource Technol 112: 206–211.
23. Zheng Y, Chi Z, Lucker B, Chen S (2012) Two-stage heterotrophic and phototrophic culture strategy for algal biomass and lipid production. Bioresource Technol 103: 484–488.
24. Lee Y-K (2006) Ch. 7 Algal Nutrition: Heterotrophic Carbon Nutrition. In: A . Richmond, editor. Handbook of Microalgal Culture: Biotechnology and Applied Phycology. Oxford, UK: Backwell Science Ltd. pp. 116–124.
25. Wan M-X, Wang R-M, Xia J-L, Rosenberg JN, Nie Z-Y, et al. (2012) Physiological evaluation of a new *Chlorella sorokiniana* isolate for its biomass production and lipid accumulation in photoautotrophic and heterotrophic cultures. Biotechnol Bioeng 109: 1958–1964.
26. Kiron V, Phromkunthong W, Huntley M, Archibald I, Scheemaker GD (2012) Marine microalgae from biorefinery as a potential feed protein source for Atlantic salmon, common carp and whiteleg shrimp. Aquacult Nutr 18: 521–531.
27. Tredici MR, Biondi N, Ponis E, Rodolfi L (2009) Chapter 20: Advances in microalgal culture for aquaculture feed and other uses. In: G . Burnell, G . Allan, editors. New technologies in aquaculture production efficiency, quality and environmental management. Boca Raton, FL: CRC Press. pp. 610–676.

28. Gouveia L, Batista AP, Sousa, Raymundo A, Bandarra NM (2008) Microalgae in novel food products. In: N . Konstantinos, P.P . Papadopoulos, editors. Food Chemistry Research Developments. New York, NY: Nova Science Publishers.

29. Warburg O, Negelein E (1920) Uber die Reduktion der Salpetersaure in griinen Zellen. Biochem Z 110: 66–115.

30. Kessler E (1976) Comparative physiology, biochemistry, and the taxonomy of *Chlorella* (Chlorophyceae). Plant Syst Evol 125: 129–138.

31. Burlew JS (1953) Algal Culture from Laboratory to Pilot Plant. Washington DC: Carnegie Institution of Washington.

32. Blanc G, Duncan G, Agarkova I, Borodovsky M, Gurnon J, et al. (2010) The *Chlorella variabilis* NC64A genome reveals adaptation to photosymbiosis, coevolution with viruses, and cryptic sex. Plant Cell 22: 2943–2955.

33. McAuley PJ (1986) Glucose uptake by symbiotic *Chlorella* in the green-hydra symbiosis. Planta 168: 523–529.

34. Muscatine L, Karakashian SJ, Karakashian MW (1967) Soluble extracellular products of algae symbiotic with a ciliate, a sponge, and a mutant hydra. Comp Biochem Physiol 20: 1–12.

35. Ziesenisz E, Reisser W, Wiessner W (1981) Evidence of *de novo* synthesis of maltose excreted by the endosymbiotic *Chlorella* from *Paramecium bursaria*. Planta 153: 481–485.

36. Dunigan DD, Cerny RL, Bauman AT, Roach JC, Lane LC, et al. (2012) *Paramecium bursaria Chlorella* virus 1 proteome reveals novel architectural and regulatory features of a giant virus. J Virol 86: 8821–8834.

37. Van Etten JL (2003) Unusual life style of *Chlorella* viruses. Annu Rev Genet 37: 153–195.

38. Yamada T, Onimatsu H, Van Etten JL (2006) *Chlorella* viruses. Adv Virus Res 66: 293–336.

39. Shi X, Wu Z, Chen F (2006) Kinetic modeling of lutein production by heterotrophic *Chlorella* at various pH and temperatures. Mol Nutr Food Res 50: 763–768.

40. Shi X-M, Zhang X-W, Chen F (2000) Heterotrophic production of biomass and lutein by *Chlorella protothecoides* on various nitrogen sources. Enzyme Microb Technol 27: 312–318.

41. Ip P-F, Chen F (2005) Production of astaxanthin by the green microalga *Chlorella zofingiensis* in the dark. Process Biochem 40: 733–738.

42. Sun N, Wang Y, Li Y-T, Huang J-C, Chen F (2008) Sugar-based growth, astaxanthin accumulation and carotenogenic transcription of heterotrophic *Chlorella zofingiensis* (Chlorophyta). Process Biochem 43: 1288–1292.

43. Feng P, Deng Z, Fan L, Hu Z (2012) Lipid accumulation and growth characteristics of *Chlorella zofingiensis* under different nitrate and phosphate concentrations. J Biosci Bioeng 114: 405–410.

44. Matsukawa R, Hotta M, Masuda Y, Chihara M, Karube I (2000) Antioxidants from carbon dioxide fixing *Chlorella sorokiniana*. J Appl Phycol 12: 263–267.

45. Guarnieri MT, Nag A, Smolinski SL, Darzins A, Seibert M, et al. (2011) Examination of triacylglycerol biosynthetic pathways via *de novo* transcriptomic and proteomic analyses in an unsequenced microalga. PLoS ONE (6:10): e25851.

46. Wan M, Faruq J, Rosenberg JN, Xia J, Oyler GA, et al. (2012) Achieving high throughput sequencing of cDNA library utilizing an alternative protocol for the bench top next-generation sequencing system. J Microbiol Meth 92: 122–126.

47. Fan J, Cui Y, Huang J, Wang W, Yin W, et al. (2012) Suppression subtractive hybridization reveals transcript profiling of *Chlorella* under heterotrophy to photoautotrophy transition. PLoS ONE 7: e50414.

48. Takeda H (1988) Classification of *Chlorella* strains by cell wall sugar composition. Phytochemistry 27: 3823–3826.

49. Banskota AH, Stefanova R, Gallant P, Osbornw JA, Melanson R, et al. (2013) Nitric oxide inhibitory activity of monogalactosylmonoacylglycerols from a freshwater microalgae *Chlorella sorokiniana*. Nat Prod Res 27: 1028–1031.

50. Kodner RB, Summons RE, Knoll AH (2009) Phylogenetic investigation of the aliphatic, nonhydrolyzable biopolymer algaenan, with a focus on green algae. Org Geochem 40: 854–862.

51. Brand JJ, Wright JN, Lien S (1989) Hydrogen production by eukaryotic algae. Biotechnol Bioeng 33: 1482–1488.

52. Komor E, Cho B-H, Kraus W (1988) The occurrence of the glucose-inducible transport systems for glucose, proline, and arginine in different species of *Chlorella*. Biotanica Acta 101: 321–326.

53. Lee YK, Ding SY, Hoe CH, Low CS (1996) Mixotrophic growth of *Chlorella sorokiniana* in outdoor enclosed photobioreactor. J Appl Phycol 8: 163–169.

54. Running JA, Huss RJ, Olson PT (1996) Heterotrophic production of ascorbic acid by microalgae. J Appl Phycol 6: 99–104.

55. Jeong ML, Gillis JM, Hwang J-Y (2003) Carbon dioxide mitigation by microalgal photosynthesis. Bull Korean Chem Soc 24: 1763–1766.

56. Kobayashi N, Noel EA, Barnes A, Watson A, Rosenberg JN, et al. (2013) Characterization of *Chlorella sorokiniana* strains in anaerobic digested effluent from cattle manure. Bioresource Technol 150: 377–386.

57. Li Y, Zhou W, Hu B, Min M, Chen P, et al. (2011) Integration of algae cultivation as biodiesel production feedstock with municipal wastewater treatment: Strains screening and significance evaluation of environmental factors. Bioresource Technol 102: 10861–10867.

58. Polakovičová G, Kušnír P, Nagyová S, Mikulec J (2012) Process integration of algae production and anaerobic digestion. Chem Eng Trans 29: 1129–1134.

59. Sorokin C, Meyers J (1953) A high-temperature strain of *Chlorella*. Science 117: 330–331.

60. Fahey R, Buschbacher R, Newton G (1987) The evolution of glutathione metabolism in phototrophic microorganisms. J Mol Evol 25: 81–88.

61. Haehnel W, Nairn JA, Reisberg P, Sauer K (1982) Picosecond fluorescence kinetics and energy transfer in chloroplasts and algae. Biochim Biophys Acta 680: 161–173.

62. Rosen BH, Berliner MD, Petro MJ (1985) Protoplast induction in C. *pyrenoidosa*. Plant Sci 41: 23–30.

63. Sakai N, Sakamoto Y, Kishimoto N, Chihara M, Karube I (1995) *Chlorella* strains from hot springs tolerant to high temperature and high CO_2. Energ Convers Manage 36: 693–696.

64. Zelibor JL, Romankiw L, Hatcher PG, Colwell RR (1988) Comparative analysis of the chemical composition of mixed and pure cultures of green algae and their decomposed residues by ^{13}C nuclear magnetic resonance spectroscopy. Appl Environ Microb 54: 1051–1060.

65. Tanaka K, Tomita Y, Tsuruta M, Konishi F, Okuda M, et al. (1990) Oral administration of *Chlorella vulgaris* augments concomitant antitumor immunity. Immunopharmacol Immunotoxicol 12: 277–291.

66. Wakasugi T, Nagai T, Kapoor M, Sugita M, Ito M, et al. (1997) Complete nucleotide sequence of the chloroplast genome from the green alga *Chlorella vulgaris*: the existence of genes possibly involved in chloroplast division. Proc Natl Acad Sci USA 94: 5967–5972.

67. Meyer JR, Ellner SP, Hairston NG, Jones LE, Yoshida T (2006) Prey evolution on the time scale of predator–prey dynamics revealed by allele-specific quantitative PCR. Proc Natl Acad Sci USA 103: 10690–10695.

68. Yong-Ha P, Seonggi Y, Mi P, Hoon MS (1997) Livestock waste handling and a high degree of functional biological fertilizer. In: O. . Huimok, editor: Korean Institute of Science and Technology: Biotechnology Institute. Available: http://attfile.konetic.or.kr/konetic/xml/use/31A1D0205268.pdf. Accessed 2014 Mar 1.

69. Guarnieri MT, Nag A, Yang S, Pienkos PT (2013) Proteomic analysis of *Chlorella vulgaris*: Potential targets for enhanced lipid accumulation. J Proteomics 93: 245–253.

70. Dillon HF, Day AG, Trimbur DE, Im C-S, Franklin S, et al. (2009) Sucrose feedstock utilization for oil-based fuel manufacturing. Patent Number: US8476059. Solazyme, Inc.

71. Dillon HF, Day AG, Trimbur DE, Im C-S, Franklin S, et al. (2009) Glycerol feedstock utilization for oil-based fuel manufacturing. Patent Application Number: US12/194,389. Solazyme, Inc.

72. Wan M, Li P, Xia J, Rosenberg JN, Oyler GA, et al. (2011) The effect of mixotrophy on microalgal growth, lipid content, and expression levels of three pathway genes in *Chlorella sorokiniana*. Appl Microbiol Biot 91: 835–844.

73. Cooksey KE, Guckert JB, Williams SA, Calli PR (1987) Fluorometric determination of the neutral lipid content of microalgal cells using Nile Red. J Microbiol Meth 6: 333–345.

74. Guckert JB, Cooksey KE (1990) Triglyceride accumulation and fatty acid profile changes in *Chlorella* (Chlorophyta) during high pH-induced cell cycle inhibition. J Phycol 26: 72–79.

75. Hu Q, Sommerfeld M, Jarvis E, Ghirardi M, Posewitz M, et al. (2008) Microalgal triacylglycerols as feedstocks for biofuel production: perspectives and advances. Plant J 54: 621–639.

76. Nguyen HM, Cuiné S, Beyly-Adriano A, Légeret B, Billon E, et al. (2013) The green microalga *Chlamydomonas reinhardtii* has a single ω-3 fatty acid desaturase which localizes to the chloroplast and impacts both plastidic and extraplastidic membrane lipids. Plant Physiol 113: 223941.

77. Liu J, Sun Z, Zhong Y, Huang J, Hu Q, et al. (2012) Stearoyl-acyl carrier protein desaturase gene from the oleaginous microalga *Chlorella zofingiensis*: cloning, characterization and transcriptional analysis. Planta 236: 1665–1676.

78. Poerschmann J, Spijkerman E, Langer U (2004) Fatty acid patterns in *Chlamydomonas* sp. as a marker for nutritional regimes and temperature under extremely acidic conditions. Microb Ecol 48: 78–89.

79. Stephens E, Ross IL, King Z, Mussgnug JH, Kruse O, et al. (2010) An economic and technical evaluation of microalgal biofuels. Nat Biotechnol 28: 126–128.

80. Rogers JN, Rosenberg JN, Guzman BJ, Oh VH, Mimbela LE, et al. (2014) A critical analysis of paddlewheel-driven raceway ponds for algal biofuel production at commercial scales. Algal Res *In press*.

81. Fan J, Yan C, Zhang X, Xu C (2013) Dual role for phospholipid:diacylglycerol acyltransferase: enhancing fatty acid synthesis and diverting fatty acids from membrane lipids to triacylglycerol in *Arabidopsis* leaves. Plant Cell *Online advance*.

82. Prescott GW (1954) How to know the fresh-water algae. Dubuque, IA: C. Brown Company.

83. Bock C, Krienitz L, Pröschold T (2011) Taxonomic reassessment of the genus *Chlorella* (Trebouxiophyceae) using molecular signatures (barcodes), including description of seven new species. Fottea 11: 293–312.

84. Luo W, Flugmacher SP, Pröschold N, Walz N, Krienitz L (2006) Genotype versus phenotype variability in *Chlorella* and *Micractinium* (Chlorophyta, Trebouxiophyceae). Protist 157: 315–333.

85. An SS, Friedl T, Hegewald E (1999) Phylogenetic relationships of *Scenedesmus* and *Scenedesmus*-like coccoid green algae as inferred from ITS-2 rDNA sequence comparisons. Plant Biol 1: 418–428.

86. Hoshina R, Iwataki M, Imamura N (2010) *Chlorella variabilis* and *Micractinium reisseri* sp. nov. (Chlorellaceae, Trebouxiophyceae): Redescription of the endosymbiotic green algae of *Paramecium bursaria* (Peniculia, Oligohymenophorea) in the 120th year. Phycol Res 58: 188–201.

87. Huss VAR, Frank C, Hartmann EC, Hirmer M, Kloboucek A, et al. (1999) Biochemical taxonomy and molecular phylogeny of the genus *Chlorella* sensu lato (Chlorophyta). J Phycol 35: 587–598.

88. Wan M, Rosenberg JN, Faruq J, Betenbaugh MJ, Xia J (2011) An improved colony PCR procedure for genetic screening of *Chlorella* and related microalgae. Biotechnol Lett 33: 1615–1619.

89. Noutoshi Y, Ito Y, Kanetani S, Fujie M, Usami S, et al. (1998) Molecular anatomy of a small chromosome in the green alga *Chlorella vulgaris*. Nucleic Acids Res 26: 3900–3907.

90. Blanc G, Agarkova I, Grinwood J, Kuo A, Brueggeman A, et al. (2012) The genome of the polar eukaryotic microalga *Coccomyxa subellipsoidea* reveals traits of cold adaptation. Genome Biol 13: R39.

91. Joint Genome Institute (2011) *Coccomyxa subellipsoidea* C-169 v2.0. U.S. Department of Energy. Available: http://genome.jgi.doe.gov/Coc_C169_1. Accessed 2013 Nov 28.

92. Tirichine L, Bowler C (2011) Decoding algal genomes: tracing back the history of photosynthetic life on Earth. Plant J 66: 45–67.

93. Rosenberg JN, Oyler GA, Wilkinson L, Betenbaugh MJ (2008) A green light for engineered algae: redirecting metabolism to fuel a biotechnology revolution. Curr Opin Biotech 19: 430–436.

94. Campenni L, Nobre BP, Santos CA, Oliveira AC, Aires-Barros MR, et al. (2013) Carotenoid and lipid production by the autotrophic microalga *Chlorella protothecoides* under nutritional, salinity, and luminosity stress conditions. Appl Microbiol Biot 97: 1383–1393.

95. He PJ, Mao B, Shen CM, Shao LM, Lee DJ, et al. (2013) Cultivation of *Chlorella vulgaris* on wastewater that contains high levels of ammonia for biodiesel production. Bioresource Technol 129: 177–181.

96. Heredia-Arroyo T, Wei W, Hu B (2010) Oil accumulation via heterotrophic/ mixotrophic *Chlorella protothecoides*. Appl Biochem Biotechnol 162: 1978–1995.

97. Liu ZY, Wang GC, Zhou BC (2008) Effect of iron on growth and lipid accumulation in *Chlorella vulgaris*. Bioresource Technol 99: 4717–4722.

98. Lu S, Wang J, Niu Y, Yang J, Zhou J, et al. (2012) Metabolic profiling reveals growth related FAME productivity and quality of *Chlorella sorokiniana* with different inoculum sizes. Biotechnol Bioeng 109: 1651–1662.

99. Santos CA, Ferreira ME, de Silva TL, Gouveia L, Novais JM, et al. (2011) A symbiotic gas exchange between bioreactors enhances microalgal biomass and lipid productivities: taking advantage of complementary nutritional modes. J Ind Microbiol Biotechnol 38: 909–917.

100. Vigeolas H, Duby F, Kaymak E, Niessen G, Motte P, et al. (2012) Isolation and partial characterization of mutants with elevated lipid content in *Chlorella sorokiniana* and *Scenedesmus obliquus*. J Biotechnol 162: 3–12.

101. Shen Y, Yuan W, Pei Z, Mao E (2010) Heterotrophic culture of *Chlorella protothecoides* in various nitrogen sources for lipid production. Appl Biochem Biotechnol 160: 1674–1684.

102. Wu QY, Yin S, Sheng GY, Fu JM (1992) A comparative study of gases generated from stimulant thermal degradation of autotrophic and heterotrophic *Chlorella*. Prog Nat Sci (in Chinese) 3: 435–440.

103. Kobayashi N, Noel EA, Barnes A, Rosenberg J, DiRusso C, et al. (2013) Rapid detection and quantification of triacylglycerol by HPLC–ELSD in *Chlamydomonas reinhardtii* and *Chlorella* strains. Lipids 48: 1035–1049.

104. Wase N, Black PN, Stanley BA, DiRusso CC (2014) Integrated quantitative analysis of nitrogen stress response in *Chlamydomonas reinhardtii* using metabolite and protein profiling. J Proteome Res, *In press*.

105. Komor E, Tanner W (1974) The hexose-proton cotransport system of *Chlorella*: pH-dependent change in K_m values and translocation constants of the uptake system. J Gen Physiol 64: 568–581.

106. Ngangkhama M, Rathaa SK, Prasannaa R, Kumarb R, Babua S, et al. (2013) Substrate amendment mediated enhancement of the valorization potential of microalgal lipids. Biocatal Agric Biotechnol 2: 240–246.

107. Haass D, Tanner W (1974) Regulation of hexose transport in *Chlorella vulgaris*. Plant Physiol 53: 14–20.

108. Kamiya A, Kowallik W (1987) Photoinhibition of glucose uptake in *Chlorella*. Plant Cell Physiol 28: 617–679.

109. Kim DG, Hur SB (2013) Growth and fatty acid composition of three heterotrophic *Chlorella* species. Algae 28: 101–109.

110. Reddy HK, Muppaneni T, Rastegary J, Shirazi SA, Ghassemi A, et al. (2013) Hydrothermal extraction and characterization of bio-crude oils from wet *Chlorella sorokiniana* and *Dunaliella tertiolecta*. Environ Prog Sust Energy 32: 910–915.

111. Mitra D, van Leeuwen J, Lamsal B (2012) Heterotrophic/mixotrophic cultivation of oleaginous *Chlorella vulgaris* on industrial co-products. Algal Res 1: 40–48.

112. Nichols HW, Bold HC (1965) *Trichosarcina polymorpha* Gen. et Sp. Nov. J Phycol 1: 34–38.

113. Tamura K, Peterson D, Peterson N, Stecher G, Nei M, et al. (2011) MEGA5: molecular evolutionary genetics analysis using maximum likelihood, evolutionary distance, and maximum parsimony methods. Mol Biol Evol 28: 2731–2739.

114. Bligh EG, Dyer WJ (1959) A rapid method for total lipid extraction and purification. Can J Biochem Phys 37: 911–917.

115. Laurens LML, Dempster TA, Jones HDT, Wolfrum EJ, Wychen SV, et al. (2012) Algal biomass constituent analysis: method uncertainties and investigation of the underlying measuring chemistries. Anal Chem 84: 1879–1889.

116. Luthria D, Vinjamoori D, Noel K, Ezzell J (2004) Chapter 3: Accelerated Solvent Extraction. In: D.L . Luthria, editor. Oil Extraction and Analysis: Critical Issues and Comparative Studies. Champaign, IL: American Oil Chemists' Society Press. pp. 25–38.

117. Mulbry W, Kondrad S, Buyer J, Luthria DL (2009) Optimization of an oil extraction process for algae from the treatment of manure effluent. J Am Oil Chem Soc 86: 909–915.

8

Oxidative Stress Is a Mediator for Increased Lipid Accumulation in a Newly Isolated *Dunaliella salina* Strain

Kaan Yilancioglu[1], Murat Cokol[1,2], Inanc Pastirmaci[1], Batu Erman[1], Selim Cetiner[1]*

1 Faculty of Engineering and Natural Sciences, Sabanci University, Orhanlı, Istanbul, Turkey, 2 Sabanci University Nanotechnology Research and Application Center, Orhanlı, Istanbul, Turkey

Abstract

Green algae offer sustainable, clean and eco-friendly energy resource. However, production efficiency needs to be improved. Increasing cellular lipid levels by nitrogen depletion is one of the most studied strategies. Despite this, the underlying physiological and biochemical mechanisms of this response have not been well defined. Algae species adapted to hypersaline conditions can be cultivated in salty waters which are not useful for agriculture or consumption. Due to their inherent extreme cultivation conditions, use of hypersaline algae species is better suited for avoiding culture contamination issues. In this study, we identified a new halophilic *Dunaliella salina* strain by using 18S ribosomal RNA gene sequencing. We found that growth and biomass productivities of this strain were directly related to nitrogen levels, as the highest biomass concentration under 0.05 mM or 5 mM nitrogen regimes were 495 mg/l and 1409 mg/l, respectively. We also confirmed that nitrogen limitation increased cellular lipid content up to 35% under 0.05 mM nitrogen concentration. In order to gain insight into the mechanisms of this phenomenon, we applied fluorometric, flow cytometric and spectrophotometric methods to measure oxidative stress and enzymatic defence mechanisms. Under nitrogen depleted cultivation conditions, we observed increased lipid peroxidation by measuring an important oxidative stress marker, malondialdehyde and enhanced activation of catalase, ascorbate peroxidase and superoxide dismutase antioxidant enzymes. These observations indicated that oxidative stress is accompanied by increased lipid content in the green alga. In addition, we also showed that at optimum cultivation conditions, inducing oxidative stress by application of exogenous H_2O_2 leads to increased cellular lipid content up to 44% when compared with non-treated control groups. Our results support that oxidative stress and lipid overproduction are linked. Importantly, these results also suggest that oxidative stress mediates lipid accumulation. Understanding such relationships may provide guidance for efficient production of algal biodiesels.

Editor: Douglas Andrew Campbell, Mount Allison University, Canada

Funding: This work was financially supported by research fund of Sabanci University. The funders had no role in study design, data collection and analysis, decision to publish, or preparation of the manuscript.

Competing Interests: The authors have declared that no competing interests exist.

* E-mail: cetiner@sabanciuniv.edu

Introduction

The idea of using biofuels has gained prominence, since they provide a cleaner alternative to the currently used fossil fuels. It has recently been estimated that utilization of biofuels will result in a 30% decrease in CO_2 emissions in the United States. Biofuels can be derived from different kinds of resources including microalgae, animal fats, soybeans, corns and other oil crops. While none of these options currently has the efficiency to produce the required amounts of biofuel [1], microalgae are considered the most promising venue of biofuel production due to their ease of cultivation, sustainability, and compliance in altering their lipid content resulting in higher biofuel production.

High lipid accumulation and biomass productivity are the two manifestly desired phenotypes in algae for biodiesel production. However, various studies conducted under nutrient depleted conditions have demonstrated that biomass productivity and lipid accumulation are negatively related [2]. These studies have established that stress conditions, which by definition reduce the biomass production, increase lipid content of algae. This problem was addressed by using a two-stage reactor where algal species such as *Oocysti sp.* and *amphora sp.* are grown in optimal conditions

for maximum biomass, followed by stress conditions for maximum lipid accumulation [3]. Within this context, nitrogen depletion can be still considered as a strategy for increasing lipid accumulation since it has been still defined as one of the best lipid accumulator stress condition in algae to date. However the mechanistic insights of this phenomenon are still needed.

Nitrogen deprivation as a stress condition is known to maximize the lipid content up to 90% [4]. However, underlying mechanisms have not been well described in terms of its physiological and molecular aspects. Despite the fact that oxygen itself is not harmful for cells, the presence of reactive oxygen species (ROS) may lead to oxidative damage to the cellular environment, ultimately leading to toxicity resulting from excessive reactive oxygen stress [5]. Redox reactions of the reactive forms of oxygen including hydrogen peroxide (H_2O_2), superoxide (O_2^-) or hydroxyl (OH^-) radicals with cellular lipids, proteins, and DNA result in oxidative stress [6]. A previous study showed that nitrogen depletion results in the co-occurrence of ROS species and lipid accumulation in diatoms [7]. Association of increased reactive oxygen species levels and cellular lipid accumulation under different environmental stress conditions was also shown in green microalgae [8] ROS is known to be an important factor in cellular response and it is well

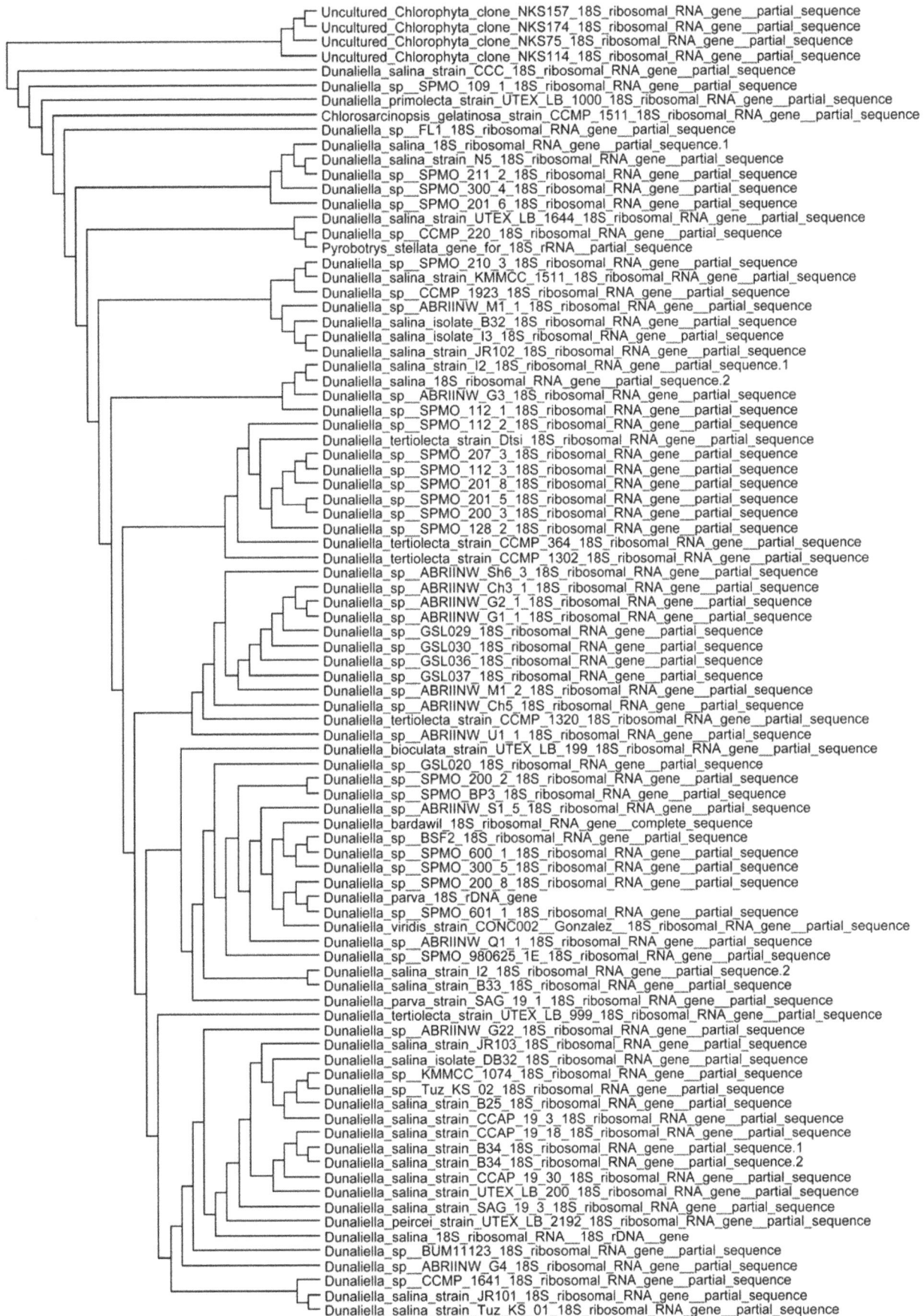

Figure 1. Phylogenetic analysis of *Dunaliella salina* **strain Tuz_KS_01 (GeneBank accession no. JX880083).** Dendrogram was generated using the neighbor-joining analysis based on 18S rDNA gene sequences. The phylogenetic tree shows the position of *Dunaliella salina* strain *Tuz_KS_01* (GeneBank accession no. **JX880083**) relative to other species and strains of *Dunaliella* deposited in NCBI GeneBank.

established that ROS increases when microalgae are exposed to various stresses. However, a mechanistic understanding of the connection between ROS increase and increased lipid accumulation in algae species requires further investigation [9].

Nitrogen depleted conditions trigger reactive oxygen species accumulation, increased cellular lipid content and protein production impairment. However, the temporal order and the causal links between these events are yet to be explored. Here, we aimed at finding the relationship between oxidative stress and increased cellular lipid content under nitrogen depleted conditions in a hypersaline green alga in order to have a better understanding of this phenomenon.

Dunaliella genus [10] is one of the microalgae genus that has been considered for lipid production. *Dunaliella* species are particularly attractive due to their strong resistance characteristics to various unfavourable environmental conditions such as high salinity. Obtaining such strong algal species for lipid production under conditions that are otherwise not useful is an important economical consideration in terms of biodiesel production. In addition, cultivation of algae species in freshwaters may not be feasible due to the limited supply and population expansion. Understanding the mechanisms behind increased lipid accumulation in halophilic *Dunaliella* in response to different stress conditions, especially nitrogen depletion, is crucial to enable key manipulations at the genetic, biochemical and physiological level for decreasing biodiesel production costs.

Materials and Methods

Organism and Culture Conditions

Dunaliella tertiolecta (*D.t.* #LB999) and *Chlamydomonas reinhardtii* (*C.r.* #90) were obtained from UTEX, Collection Culture of Algae, USA and cultivated artificial sea water medium and soil extract medium as instructed by the culture collection protocols respectively. The alga used in this study was isolated from the hypersaline lake "Tuz", which is located in Middle Anatolia, Turkey with the research permission of Republic of Turkey Ministry of Food, Agriculture and Livestock (Permit issue: B.12.0.TAG.404.03.10.03.03–1607). Field studies did not involve endangered or protected species. Collection of water samples was done and isolation location was recorded using a GPS device as 39°4′23.97″K - 33°24′33.11″E at the southern east part of Sereflikochisar province. 10 ml water samples were collected and enriched with the same volume of Bold's Basal Medium (BBM) modified by addition of 5% NaCl. BBM was pH 7.4 and consisted of 5 mM NaNO$_3$ along with CaCl$_2$·2H$_2$O 0.17 mM, MgSO$_4$·7H$_2$O 0.3 mM, K$_2$HPO$_4$ 0.43 mM, KH$_2$PO$_4$ 1.29 mM, Na$_2$EDTA·2H$_2$O 2 mM, FeCl$_3$·6H$_2$O 0.36 mM, MnCl$_2$·4H$_2$O 0.21 mM, ZnCl$_2$ 0.037 mM, CoCl$_2$·6H$_2$O 0.0084 mM, Na$_2$MoO$_4$·2H$_2$O 0.017 mM, Vitamin B$_{12}$ 0.1 mM and 5 mM NaCO$_3$ was supplied as carbon source.

Water samples were plated on petri dishes with modified BBM 5% NaCl and 1% bacteriological agar, while same water samples were also subjected to dilutions with fresh modified BBM 5% NaCl in 48-well plates (1:2, 1:4, 1:8, 1:16, 1:32, 1:64) to obtain monocultures. After 2 and 4 weeks cultivation periods of 48-well plate liquid cultures and petri dishes, respectively, clones were isolated. These clones were transferred to fresh mediums in 100 ml canonical flasks in a final volume of 25 ml for obtaining cell stocks

at 25°C under continuous shaking and photon irradiance of 80 rpm and 150 μEm^{-2}s^{-1}.

All experiments were carried out under same cultivation conditions in 250 ml batch cultures but different nitrogen concentrations, 0.05, 0.5 and 5 mM NaNO$_3$ were used for low, medium and high nitrogen concentrations, respectively. Concentrations ranging between 200 uM and 8 mM were used for H$_2$O$_2$ experiments.

Isolation and Purification of DNA and Amplification of 18S rRNA Encoding Gene

DNA isolation was done by using DNeasy Plant Mini Kit (Qiagen) as instructed by the manufacturer. Quantification of the genomic DNA obtained and assessment of its purity was done on a Nanodrop Spectrophotometer ND-1000 (Thermo Scientific) and on 1% agarose gel elecrophoresis. MA1 [5′-CGGGATCCG-TAGTCATATGCTTGTCTC-3′] and MA2 [5-GGAATT-CCTTCTGCAGGTTCACC-3′] were designed from 18S rDNA genes and were previously reported by Olmos et al [11]. PCR reactions were carried out in a total volume of 50 μl containing 50 ng of chromosomal DNA in dH$_2$O and 200 ng MA1 and MA2 conserved primers. The amplification was carried out using 30 cycles in a MJ Mini Personal Thermal Cycler (BioRad), with a T$_m$ of 52°C for all reactions. One cycle consisted of 1 minute at 95°C, 1 minute at 52°C and 2 minutes at 72°C.

Sequencing and Phylogenetic Analysis

MA1–MA2 PCR products were utilized to carry out sequencing reactions after purification with a QIAquick PCR purification kit (Qiagen). The sequencing reactions were run by MCLAB (San Francisco, CA), employing primers MA1–MA2 in both reverse and forward directions. DNA sequences were imported to BLAST for identification and to search for phylogenetic relationship correlations between other 18S rDNA gene sequences of *Dunaliella* species/strains deposited in NCBI Gene Bank. Dendrogram data generated by BLAST was converted into newick format and submitted to Phyfi [12] for generating phylogenetic tree.

Growth Analysis

Specific growth rate and biomass productivity was calculated according to the equation; $K' = Ln(N2/N1)/(t2-t1)$ where $N1$ and $N2$, biomass at time1 ($t1$) and time2 ($t2$) respectively ($t2 > t1$) [13]. Divisions per day and the generation (doubling) time were calculated according to the equations below:

$$Div \times day^{-1} = K'/Ln2$$

$$Gen't = 1/Div \times day^{-1}$$

Extraction and Measurements of Lipid Contents and Fluorescence Microscopy

Lipid was extracted according to Bligh and Dyer wet extraction method. Briefly, to a 15 ml glass vial containing 100 mg dried algal biomass, 2 ml methanol and 1 ml chloroform were added

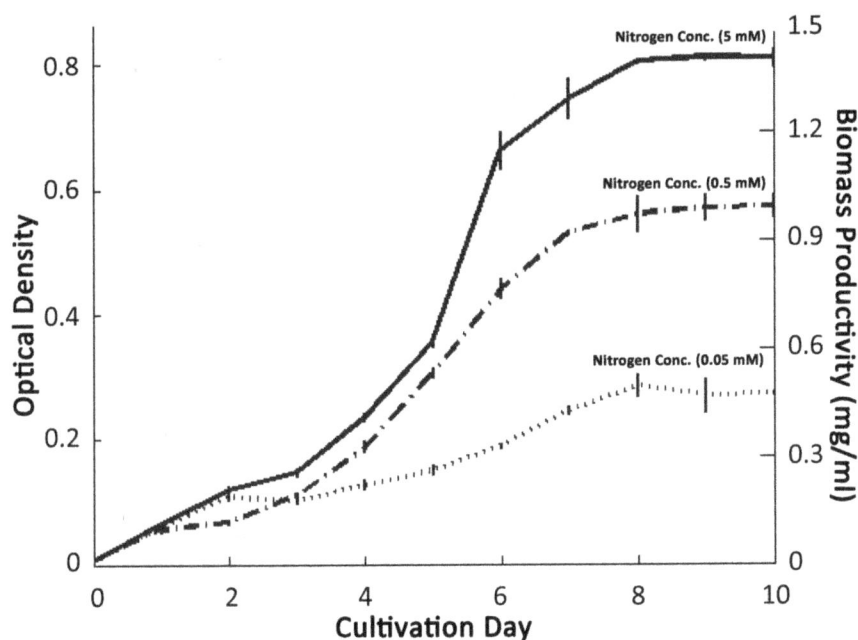

Figure 2. Effects of different nitrogen concentrations on the growth and biomass productivity of *Dunaliella salina* **strain** *Tuz_KS_01* **(GeneBank accession no. JX880083).** 0.05 mM, 0.5 mM and 5 mM NaNO₃ are referred as low, medium and high nitrogen concentrations respectively. Shown OD₆₀₀ optical density and biomass values are the means of three replicates. The error bars correspond to ±1 SD of triplicate optical density measurements.

and kept for 24 h at 25°C. The mixture was then vortexed and sonicated for 10 minutes. One milliliter of chloroform was again added, and the mixture shaken vigorously for 1 min. Subsequently, 1.8 ml of distilled water was added and the mixture vortexed again for 2 min. The aqueous and organic phases were separated by centrifugation for 15 min at 2,000 rpm. The lower (organic) phase was transferred into a previously weighed clean vial (V1). Evaporation occurred in a thermo-block at 95°C, and the residue was further dried at 104°C for 30 min. The weight of the vial was again recorded (V2). Lipid content was calculated by subtracting V1 from V2, and expressed as dcw %. The correspondence between Nile Red fluorescence intensity and % lipid content was determined by plotting relative fluorescence units against % cellular lipid content obtained from triplicate samples. For fluorescence microscopy analysis of Nile Red, cells were stained with 5 μl 0.5 mg/mL Nile Red (Sigma, USA) stock solution after fixing cells with 5% paraformaldehyde and imaged by epifluorescence microscopy with a Leica DMR microscope (Leica Microsystems).

Spectrofluorometric Microplate Analysis for Determination of Lipid and Reactive Oxygen Species Accumulation

Nile red, 9-diethylamino-5H-benzo[alpha]phenoxazine-5-one was first reported that is an excellent vital stain for the detection of intracellular lipid droplets by fluorescence microscopy and flow cytoflourometry [14] and it has been widely used for measuring and comparing cellular lipid content in various organisms in numerous studies [15–19]. A stock solution of Nile Red (NR) (Sigma, 72485) was prepared by adding 5 mg of NR to 10 ml of acetone. The solution was kept in a dark colored bottle and stored in the dark at −20°C. 1 ml of algal cells from a culture of 250 ml glass erlenmeyer flasks containing 100 ml growth media with

different nitrogen concentrations were transferred to 1.5 ml eppendorf tubes for 5 min centrifugation at 5,000 rpm, washed twice with fresh medium, and measured in a spectrophotometer at 600 nm. Each sample was adjusted to an OD₆₀₀ of 0.3 in a 1 ml final volume by dilution with fresh medium. 5 μl of Nile Red solution was added to each tube and mixed well, followed by 20 min incubation in the dark. Finally, cellular neutral lipids were quantified using a 96-well microplate spectrofluorometry (SpectraMAX GEMINI XS) with an excitation wavelength of 485 nm and an emission wavelength of 612 nm.

Dichloro-dihydro-fluorescein diacetate (DCFH-DA) is the most widely used fluorometric probe for detecting intracellular oxidative stress. This probe is cell-permeable and is hydrolyzed intracellularly to the DCFH carboxylate anion. Two-electron oxidation of DCFH results in the formation of a dichlorofluorescein (DCF) as a fluorescent product. The amount of this fluorescent product is highly correlated with the cellular oxidation/oxidative stress level. Investigators have routinely used DCFH-DA to measure intracellular generation of reactive oxidants in cells in response to intra- or extracellular activation with oxidative stimulus [20]. Determination of ROS production related to oxidative stress was done by using DCFH-DA (Sigma, USA). Vital staining for determination of cell survival under H₂O₂ treatment was done by using fluorescein diacetate (FDA) (Sigma, USA). FDA is incorporated into live cells and it is converted into fluorescein by cellular hydrolysis [21]. 0.5 mg/ml stock solutions for both stains were prepared in acetone and the same protocol was used for FDA (Sigma, USA) and DCFH-DA (Sigma, USA) staining as described above for Nile Red staining. Data were recorded as relative fluorescence units [22] for all spectrofluorometric staining experiments.

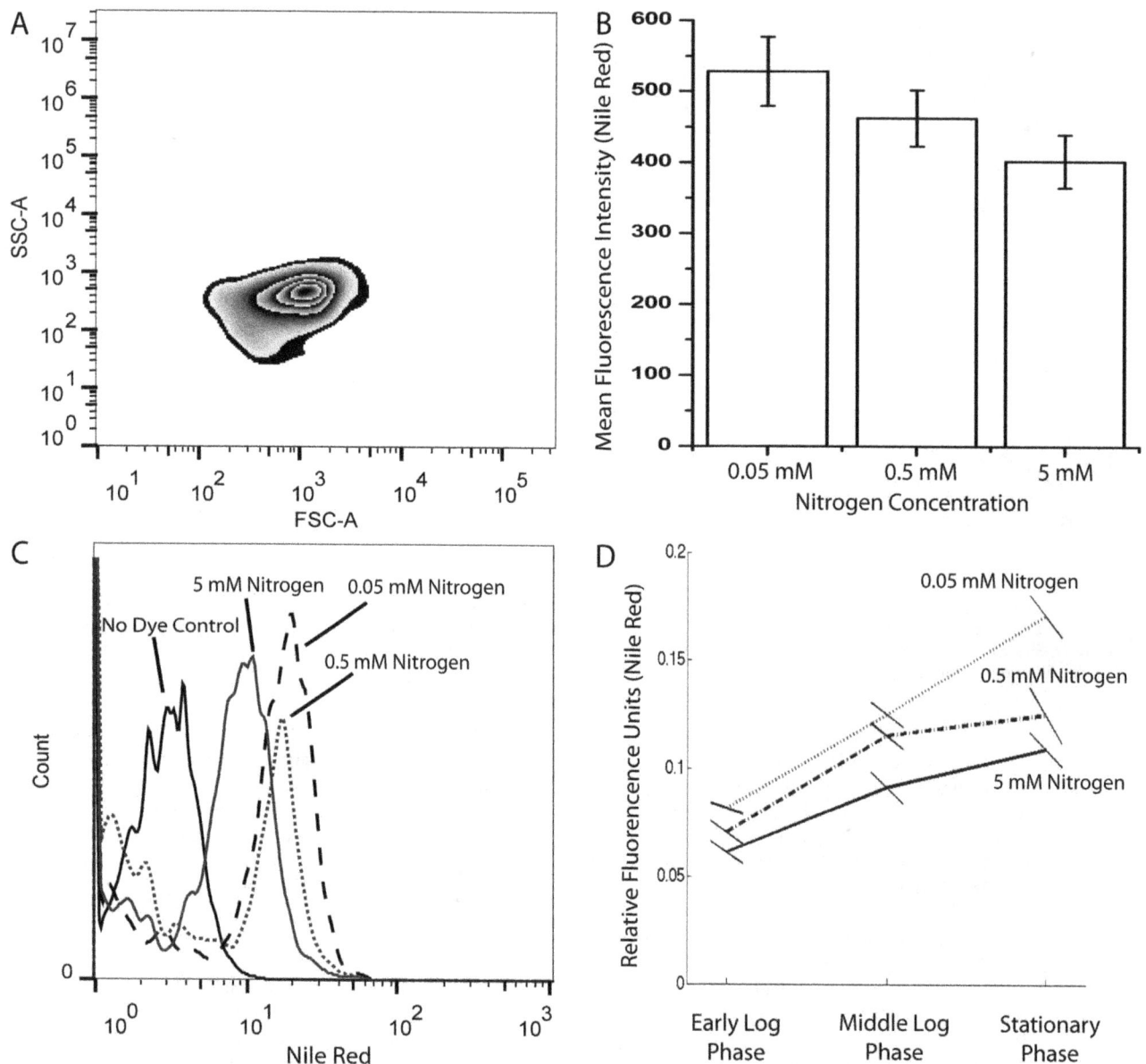

Figure 3. Lipid content analysis of *Dunaliella salina* **strain** *Tuz_KS_01* **under different nitrogen conditions. A)** Zebra-plot of SSC (Side Scatter) and FSC (Forward Scatter) expressing cellular granulation and cellular size, respectively, under a representative nitrogen depletion condition (0.05 mM). **B)** A representative histogram of flow-cytometric analysis of lipid contents under different nitrogen concentrations **C)** Mean fluorescent intensities (MFI) of flow cytometric analysis. **D)** Fluorometric microplate Nile Red analysis of early logarithmic, late logarithmic and stationary growth phases. Data represent the mean values of triplicates. Standard error for each triplicate is shown as tilted lines for clarity, where the minimum and maximum y values of each line corresponds to ±1 SE.

Flow Cytometric Analysis for Determination of Lipid and Reactive Oxygen Species Accumulation

5 μl of Nile Red (Sigma, USA) from stock solution (0.5 mg/mL) was added to 1 ml of a cell suspension at an OD_{600} of 0.3 after washing cells twice with fresh medium. This mixture was gently vortexed and incubated for 20 minutes at room temperature in dark. Nile Red uptake was determined using a BD-FACS Canto flow cytometer (Becton Dickinson Instruments) equipped with a 488 nm argon laser. Upon excitation by a 488 nm argon laser, NR exhibits intense yellow-gold fluorescence when dissolved in neutral lipids. The optical system used in the FACS Canto collects yellow and orange light (560–640 nm, corresponding to neutral

lipids). Approximately 10,000 cells were analysed using a log amplification of the fluorescent signal. Non-stained cells were used as an autofluorescence control. Nile Red fluorescence was measured using a 488 nm laser and a 556 LP+585/42 band pass filter set on a FACS Canto Flow Cytometer. Data were recorded as mean fluorescence intensity (MFI).

DCFH-DA (Sigma, USA) from stock solution (0.5 mg/ml) was used as described above for flow-cytometric microplate Nile Red staining method in order to analyze cellular oxidative stress status. Briefly, 1 ml of algal cells in different nitrogen concentrations were transferred into 1.5 ml eppendorf tubes, centrifuged 5 min at 5,000 rpm, washed twice with fresh medium, and measured in a spectrophotometer at 600 nm. Each sample was adjusted to an

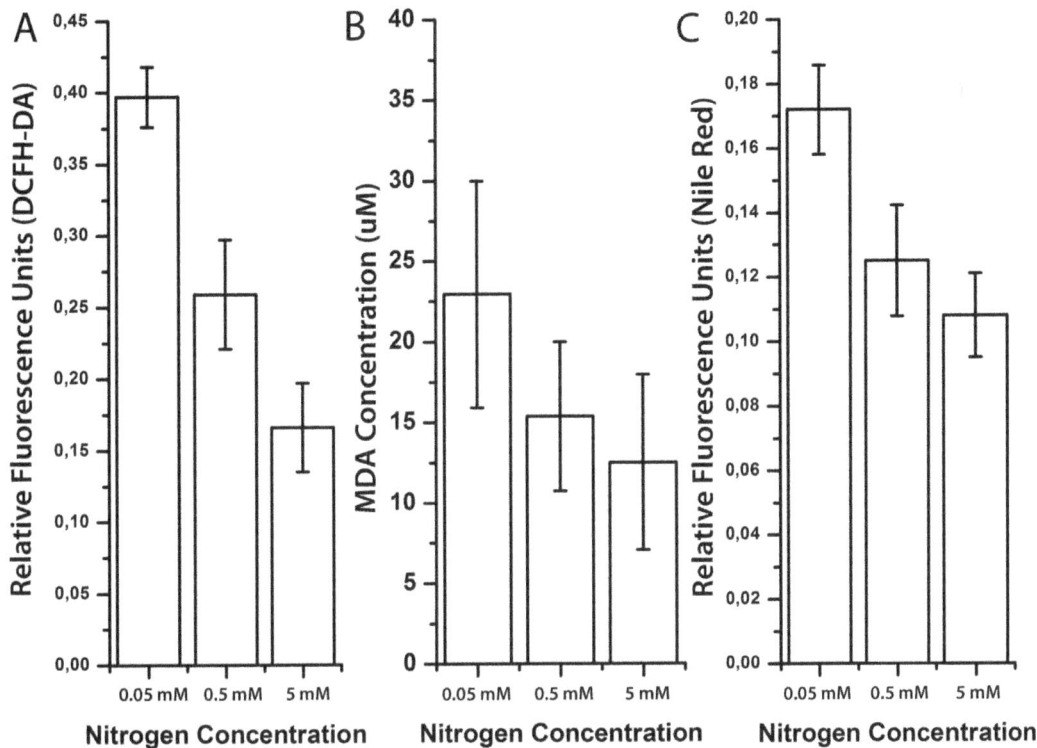

Figure 4. Oxidative stress and lipid accumulation under different nitrogen concentrations. A) Fluorometric microplate DCFH-DA analysis under different nitrogen concentrations. **B)** TBARS analysis for lipid peroxidation under different nitrogen concentrations. **C)** Fluorometric microplate Nile-Red analysis under different nitrogen concentrations. Data represent the mean values of triplicates ± 1 SE.

OD_{600} of 0.3 in a 1 ml final volume by dilution with fresh medium. 5 µl of dye solution was added to each tube and mixed well, followed by 20 min incubation in the dark and analyzed. Nonfluorescent DCFH-DA is taken up and converted into diclorodihydrofluorescein (DCF) by the action of cellular esterases. Fluorescence from oxidized DCF was measured using a 488 nm laser and a 556/Long Pass filter set on a BD-FACS Canto Flow Cytometer. Data were expressed as mean fluorescence intensity (MFI). The data of both staining methods were evaluated using Flowjo Ver. 7.6.1 (Tree Star, Inc.).

Protein, Chlorophyll, Carotenoid, TBARS Analyses and Enzymatic Assays

Total protein isolation was done according to Barbarino et al. [23]. Protein content was determined following the method described by Bradford [24]. Cellular chlorophyll and carotenoid isolations were done using the methanol extraction method. Calculation of chlorophyll and carotenoid contents were carried out using the formula for methanol extraction described by Wellburn [25]. Lipid peroxidation analysis upon different nitrogen concentration regimes was analysed by using thiobarbituric acid reactive substances (TBARS) method described by Sabatini et al [26]. For superoxide dismutase (SOD) analysis 50 mg biomass was homogenized in 2 ml 0.5 M phosphate buffer (pH 7.5) and centrifuged at 13,000 rpm for 10 min at 4°C. SOD activity was determined in the supernatant by inhibition of nitroblue tetrazolium (NBT) using a reaction mixture of 1.5 ml Na_2CO_3 (1 M), 200 µl methionine (200 mM), 100 µl NBT (2.25 mM), 100 µl EDTA (3 mM), 100 µl riboflavin (60 ?M) and 1.5 ml phosphate buffer (pH 7.8, 0.1 M). The absorbance was recorded at 560 nm. One unit of SOD per gram of protein was defined as the amount

causing 50% inhibition of photochemical reduction of NBT [27]. For catalase (CAT) analysis, 50 mg biomass was homogenized in 2 ml phosphate buffer (0.5 M, pH 7.5), centrifuged at 12,000 rpm at 4°C for 30 min and supernatant was taken for CAT activity. A reaction mixture containing 1.6 ml phosphate buffer (pH 7.3), 100 µl EDTA (3 mM), 200 µl H_2O_2 (0.3%) and 100 µl supernatant (containing enzyme extract) was taken in a cuvette and CAT activity in supernatant was determined by monitoring the disappearance of H_2O_2, by measuring a decrease in absorbance at 240 nm against a blank of same reaction mixture without 0.3% H_2O_2 [28]. For analysis of ascorbate peroxidase (APX), 50 mg biomass was homogenized in 2 ml phosphate buffer (0.5 M, pH 7.5) and centrifuged at 12,000 rpm at 4°C for 30 min. The supernatant was taken for APX activity. A reaction mixture containing 1 ml phosphate buffer (pH 7.3), 100 µl EDTA (3 mM), 1 ml ascorbate (5 mM), 200 µl H_2O_2 (0.3%) and 100 µl supernatant (containing enzyme extract) was prepared in a cuvette. The reaction was followed for 3 min at a wavelength of 290 nm against a blank of same reaction mixture without 0.3% H_2O_2 [29].

Results and Discussion

Isolation and Identification of the New *Dunaliella salina* Strain

We isolated a new hypersaline microalga from the biggest salt lake of Turkey located in the Middle Anatolian region in July 2011. The microalga, which is unicellular, biflagellate and tolerant up to 20% NaCl was identified as *Dunaliella salina* based on the morphological characterization prior to molecular identification. We observed that cell color was green under normal conditions and the color changed to red under stress conditions. Morpho-

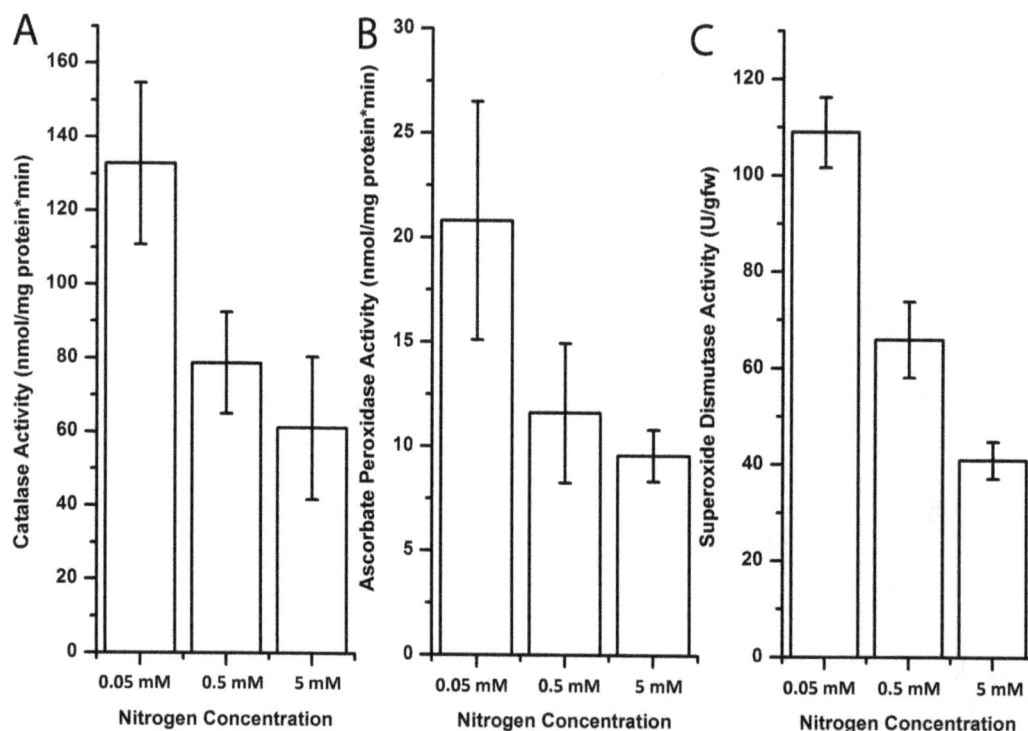

Figure 5. Antioxidant enzyme activities under different nitrogen concentrations. A–C) Spectrophotometric enzymatic assays for catalase (CAT), ascorbate peroxidase (APX) and superoxide dismutase (SOD). Data demonstrate the mean values of triplicates ±1 SE.

logically, ovoid, spherical and cylindrical cell shapes were observed. Stigma was not clearly visible or diffuse. Large pyrenoid with distict amylosphere was observed and refractile granules are

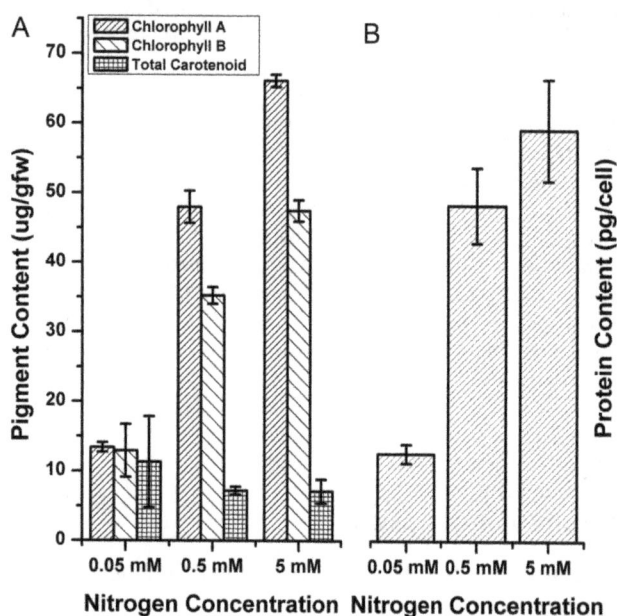

Figure 6. Pigment and protein contents of *Dunaliella salina* strain under different nitrogen concentrations. A) Chlorophyll A, B and total carotenoid contents under different nitrogen concentrations. **B)** Protein contents of *Dunaliella salina* strain Tuz_KS_01 cells cultivated under different nitrogen concentrations Data demonstrate the mean values of triplicates ±1 SE.

absent. Cellular size was measured as ~20 μm. These observed characteristics were consistent with previous morphological characteristics of *Dunaliella salina*.

This strain was deposited in −196°C ultra freeze conditions at Sabanci University algae culture collection as *Dunaliella salina strain Tuz_KS_01*. Molecular identification was based on 18S rDNA gene sequence amplification and sequencing. A single 18S rDNA PCR amplicon was about ~1800 bp in size, in accordance with Olmos et al. [11]. The partial sequence of the 18S rDNA encoding gene was submitted to National Center for Biotechnology Information (NCBI) GeneBank database as *Dunaliella salina strain Tuz_KS_01* (GeneBank accession no. **JX880083**). According to Basic Local Aligment Search Tool (BLAST) analysis, the isolated sequence had very high percentage of identity with other deposited 18S rDNA sequences of *Dunaliella* species shown in the phylogenetic tree plotted in Figure 1. Molecular identification showed a high percentage identity to *Dunaliella salina* strains (Identity 93%, Query Coverage 70%).

Growth Analysis of the New Dunaliella Strain Under Different Nitrogen Concentrations

The newly isolated halotolerant *Dunaliella salina* strain was cultivated under different nitrogen concentrations for growth analysis. 0.05 mM, 0.5 mM and 5 mM NaNO₃ concentrations were considered as low, medium and high nitrogen concentrations. Specific growth rates and/or average biomass productivities (ABP) were found to be 370 mg/dayL, 430 mg/dayL, 520 mg/dayL for low, medium and high nitrogen groups respectively. The highest biomass concentration under low nitrogen regimes was 495 mg/l, compared with 994 mg/l and 1409 mg/l for medium and high nitrogen regimes shown in Figure 2. Other kinetic growth data including doubling time/generation time and division per day were also calculated. Doubling times were calculated as

1.84, 1.59, and 1.32 whereas division of cells per day was calculated as 0.54, 0.62, and 0.75 for low, medium and high nitrogen concentration groups respectively.

Biomass productivity is one of the most important parameters for the feasibility of utilizing algal oil for biodiesel production. Hence, numerous studies have been conducted in various algal species. Tang et al. (2011) demonstrated that *Dunaliella tertiolecta* had a highest biomass concentration of ~400 mg/l [30]. In another study conducted by Ho et al. (2010) the optimal biomass productivity of freshwater alga *Scenedesmus obliquus* were found to be 292.50 mg/dayL [31]. Griffiths et al. (2012) studied 11 different algal species including freshwater, marine and halotolerant species and demonstrated their biomass productivities [32]. Compared to the reported biomass productivities in previous studies, both the highest biomass concentration and average biomass productivity of isolated *Dunaliella salina* strain *Tuz_KS_01* showed very good potential, eventhough small batch cultivation process was utilized instead of using more sophisticated and efficient cultivation systems.

Lipid Accumulation Analysis of the New *Dunaliella salina* Strain Under Different Nitrogen Concentrations

Experimental groups were subjected to flow cytometric Nile red analysis for measuring lipid accumulation at single cell level in stationary growth phase. Cell populations were chosen according to cellular size and granulation as quantified by forward and side scatter values of populations, respectively which is demonstrated in Figure 3A. Figure 3B shows a representative flow cytometric analysis of lipid contents under different nitrogen concentrations, indicating high lipid content under low nitrogen condition. Figure 3C shows the mean and standard error for three biological replicates, clearly exhibiting a negative relationship between lipid accumulation and nitrogen levels. Therefore, cultivation under low nitrogen conditions stimulates lipid accumulation in *Dunaliella salina* strain *Tuz_KS_01*, in agreement with previous studies [33,34].

In addition, we measured lipid accumulation of the green alga under different nitrogen levels using spectrofluorophotometer [35] at early-logarithmic, late-logarithmic and stationary growth phases. While the cellular lipid accumulation in early and late logarithmic phases did not show a drastic change, we observed a marked increase of lipid content in the stationary phase, which was significantly higher under low nitrogen cultivation conditions shown in Figure 3D.

Nitrogen depletion is well known to result in increased lipid accumulation in algal species [36], eventhough its association with other related factors has not been well defined. To demonstrate the oxidative stress under nitrogen limited cultivation conditions, fluorometric DCFH-DA and related TBARS lipid peroxidation assays were utilized. According to these results, increased ROS production (Figure 4A) and lipid peroxidation (Figure 4B) were both observed under nitrogen limited conditions, in association with increased lipid accumulation especially at stationary phase shown in Figure 4C.

Antioxidant Enzyme Activities, Pigment Composition and Protein Analyses Under Different Nitrogen Concentrations

ROS accumulation is prevented by a powerful intrinsic antioxidant system in photosynthetic organisms, involving enzymes such as superoxide dismutase (SOD), catalase (CAT), ascorbate peroxidase (APX) [37]. Next, we analyzed these three oxidative stress indicator antioxidant enzymes to demonstrate the

effect of nitrogen depletion on the intracellular oxidative stress status as also supplementary to DCFH-DA measurements.

Intracellular SOD converts O_2^- to O_2 and H_2O_2, acting as the first line of defence against oxidative stress [38]. We observed that SOD activity of cells increased under low nitrogen concentrations, especially at stationary growth phase. This data suggests that superoxides may be elevated under nitrogen depleted conditions, necessitating increased SOD activity (Figure 5A). CAT is a heme-containing enzyme that catalyzes the conversion of H_2O_2 into oxygen and water [37]. APX is involved in the ascorbate-glutathione cycle occuring in chloroplasts, cytoplasm, mitochondria and perixisomes [39]. We observed elevated levels of CAT and APX activity under nitrogen depleted conditions (Figure 5B, 5C). These data strongly indicate that oxidative stress is induced under nitrogen depletion in *Dunaliella salina* strain *Tuz_KS_01*.

Next, we analyzed the chlorophyll, carotenoid and protein content change in response to nitrogen depletion in *Dunaliella salina* strain *Tuz_KS_01*. Chlorophyll content is an indicator of photosynthetic efficiency, rate and nitrogen [40] status. Carotenoids can perform an essential role in photoprotection by quenching the triplet chlorophyll and scavenging singlet oxygen and other reactive oxygen species [41]. We observed that chlorophyll A and B levels sharply declined in low nitrogen conditions, as expected (Figure 6A). In contrast, we observed no decrease in total carotenoid contents under nitrogen depletion conditions. This observation suggests that cells increase caroteoid content to overcome with oxidative stress induced by singlet oxygens, similar to SOD activity increase discussed above (Figure 6A).

Upon nitrogen starvation, algal cells reduce the synthesis of protein and nucleic acid synthesis and carbon flow is directed from protein synthesis to fatty acid and carbonhydrate synthesis [42]. We found that neutral lipids increase while the protein content of algal cells was reduced under nitrogen depletion conditions (Figure 6B). This indicates that carbon flow direction is to fatty acid synthesis under nitrogen limitation at stationary phase, in accordance with the literature [43].

Reactive Oxygen Species (ROS) Production Induces Lipid Accumulation

Our analysis thus far hints that lipid accumulation might be partially triggered by ROS accumulation and oxidative stress under nitrogen depleted condition (Figure 4). In order to test this hypothesis, we used H_2O_2, a well-known oxidative stress inducer [44]. We treated cells at exponential growth phase in high nitrogen containing media (5 mM) with different concentrations of H_2O_2 and we measured ROS, via using flowcytometric DCFH-DA method, a robust fluorescent assay [45]. In addition, cell viability was measured by using FDA fluorescent assay while using Nile red staining method for determination of lipid contents under different H_2O_2 concentrations. Although cellular size and granularity under H_2O_2 treated conditions were similar to previous experimental conditions (Figure 7A and 3A), cell survival under higher H_2O_2 concentrations were observed significantly reduced due to the oxidative stress induction (Figure 7B–7C). Although reduction of cell survival is to ~20% at 4 mM H_2O_2 concentration, at this high concentration of H_2O_2, new isolate *Dunaliella salina* strain *Tuz_KS_01* showed highly tolerant characteristics to oxidative stress (Figure 7D).

Lipid accumulation of algae is known to increase by various factors including temperature, excessive light, and pH, which may also related to oxidative stress induced by ROS accumulation [33,46]. We used fluorometric Nile red analysis for analysis of lipid accumulation under different H_2O_2 treatments. We observed increased lipid accumulation with increasing concentrations of

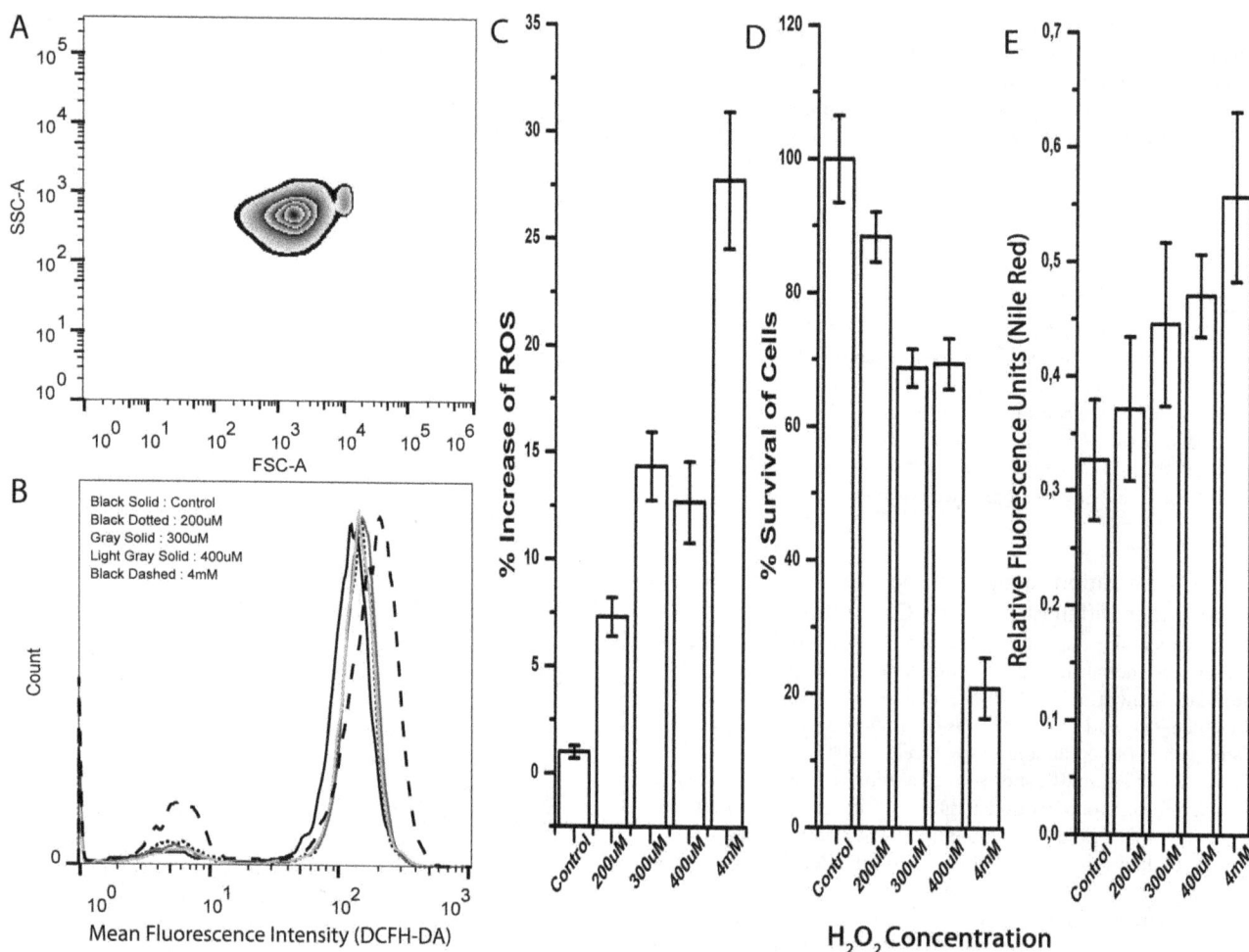

Figure 7. Effect of H₂O₂, a known oxidative stress inducer, on _Dunaliella salina strain Tuz_KS_01_. A) Zebra-plot of SSC (Side Scatter) and FSC (Forward Scatter) expressing cellular granulation and cellular size under a representative H₂O₂ condition (4 mM). **B)** Histogram of flow-cytometric analysis of ROS accumulation under different H₂O₂ concentrations. **C)** Percentage increase of ROS production based on flow-cytometric DCFH-DA analysis. **D)** Fluorometric microplate fluorescent diacetate (FDA) survival analysis cultivated under different H₂O₂ concentrations. **E)** Fluorometric microplate Nile-Red analysis under different H₂O₂ concentrations. Data represent the mean values of triplicates ±1 SE.

H_2O_2 (Figure 7E). This result supports that the lipid accumulation observed under nitrogen depletion conditions is at least partly mediated by oxidative stress. Importantly, this result also suggests a method for obtaining high lipid from green alga, namely by growing cells in high nitrogen (optimal) conditions followed by oxidative stress induction at stationary phase. Such a method may be economically more feasible than two stage reactors described above.

We next wished to visually observe the effect of nitrogen depletion and H_2O_2 treatment on lipid accumulation. After cultivation, we collected and stained cells with Nile red and observed with fluorescence microscopy. We used three different conditions: 1) control, 2) 0.05 mM nitrogen and 3) 4 mM H_2O_2. Microscopy results were in strong agreement with our previous results, as lipid accumulation (increased size and number of cytoplasmic lipid droplets) after nitrogen depletion or H_2O_2 treatment was higher than the control (Figure 8A–8B–8C).

Moreover, in order to validate the lipid increasing effect of nitrogen limitation and H_2O_2 induced oxidative stress, solvent extraction gravimetric lipid analysis in nitrogen depleted (0.05 mM) and H_2O_2 containing media (4 mM) were done. We

found that, nitrogen limitation led to increased lipid accumulation up to 35%, and H_2O_2 induced oxidative stress led to increased lipid accumulation up to 44% as shown in Figure 8D. We also measured the lipid content by Nile red fluorescence analysis for each of these conditions. Consistent with previous studies, we found that gravimetric and fluorometric measurements were correlated ($r^2 = 0.82$) (Figure 8E) [17]. These data suggest that exogenously induced oxidative stress triggered by the application of H_2O_2 resulted to increased lipid accumulation. Therefore, as previously shown by the fluorometric analyses above, oxidative stress and increased lipid accumulation association is coupled under nitrogen limitation conditions. Induction of lipid accumulation by applying exogenous oxidative stress inducers such as H_2O_2 may assist more effective lipid production compared with biomass production lowering nitrogen starvation strategy.

ROS production resulting from various stress factors is known to affect nearly all cellular processes by impairing the structural stability of functional macromolecules including DNA, proteins and structural lipids. Since the green algae life-cycle is reliant on its photosynthetic activity and cellular integrity, it is crucial to protect against oxidative stress. Otherwise cells can not tolerate oxidative

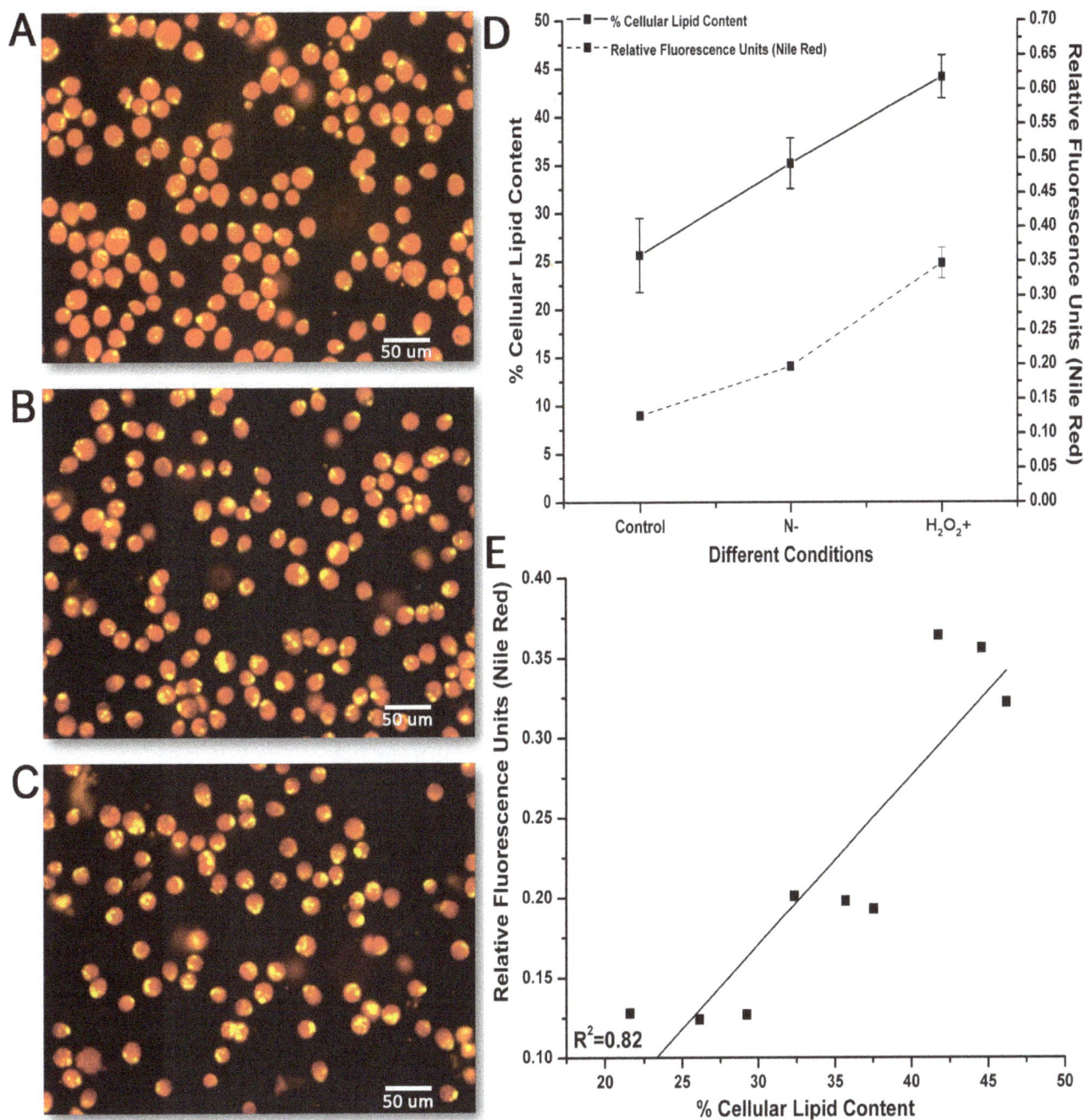

Figure 8. Fluorescence microphotographs of *Dunaliella salina strain Tuz_KS_01* **stained with Nile-Red fluorescence dye and screened under 400X magnification. A)** Control group, 5 mM nitrogen concentration cultivation condition. **B)** Nitrogen limitation group, 0.05 mM nitrogen condentration cultivation condition **C)** Oxidative stress group, 4 mM H_2O_2 cultivation condition. Gravimetric lipid content analysis and Nile-Red fluorescence measurement correlation **D)** Gravimetric and fluorometric Nile-Red lipid content analysis of *Dunaliella salina strain Tuz_KS_01* under 0.05 mM nitrogen and 4 mM H_2O_2 cultivation conditions. Data represent the mean values of triplicates ± 1 SE **E)** Correlation plot of gravimetric and fluorometric Nile-Red lipid content analysis ($r^2 = 0.82$).

damage and eventually die [47–49]. Even though there is limited explanation about the correlation between ROS production and lipid accumulation in algae cultivated under nitrogen depleted conditions. Our results indicate that nitrogen depletion results in ROS accumulation and lipid peroxidation especially at stationary growth phase. This relationship might also be related to survival response of the alga against excessive oxidative stress conditions [7].

Although a few recent studies have suggested an association of ROS levels and cellular lipid accumulation [7,8], underlying mechanistic principles are not clear. ROS is well demonstrated to modify cellular responses against different stressors in corresponding signal transduction pathways. ROS levels increase in microalgae cells when exposed to the different environmental stresses [9]. In a recent study, nitrogen depletion was shown to correlate with increased ROS accumulation and increased cellular

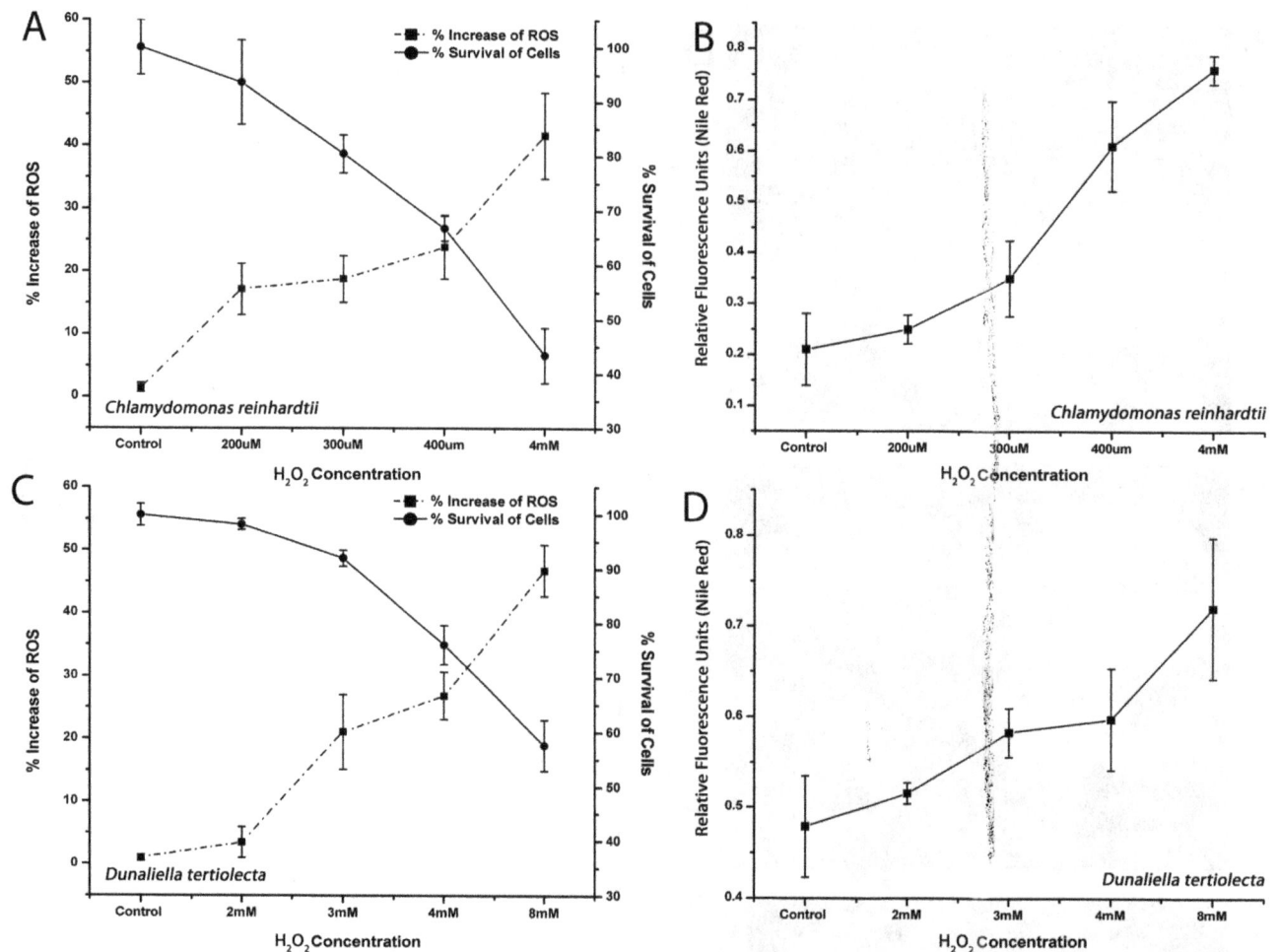

Figure 9. Effect of H₂O₂ induced oxidative stress on cellular lipid accumulation of different algae species. A) Fluorometric DCFH-DA and FDA analyses showing percantage change of ROS production and percentage change of cell survivability of *Chlamydomonas reinhardtii* cells cultivated under different H₂O₂ concentrations. **B)** Fluorometric Nile red analysis showing the effect of different H₂O₂ concentrations on cellular lipid accumulation of *Chlamydomonas reinhardtii* cells **C)** Fluorometric DCFH-DA and FDA analyses showing percantage change of ROS production and percentage change of cell survivability of *Dunaliella tertiolecta* cells cultivated under different H₂O₂ concentrations. **D)** Fluorometric Nile Red analysis showing the effect of different H₂O₂ concentrations on cellular lipid accumulation of *Dunaliella tertiolecta* cells. Data represent the mean values of triplicates ±1 SE.

lipid accumulation in a freshwater algae. This study showed that increased MDA concentration is an indicator of membrane peroxidation which implied increased ROS levels. It was suggested that H_2O_2 induced exogenous oxidative stress was an effective factor for neutral lipid induction in *C. sorokiniana C3* [50].

Microalgae can modify its photosynthetic system under stress, resulting in a decrease in the gene expression of various proteins forming up the photosystem complexes I and II [51]. Such adjustments are thought to occur for minimizing oxidative stress via decreasing photosynthesis rate [52]. Nitrogen deprivation is closely associated with the degradation of ribulose-1,5-bisphosphate carboxylase oxygenase to recycle nitrogen [53]. Degradation of this protein may result in alterations in photosynthesis rate, which is consistent with the observed decrease in chlorophyll content under nitrogen depleted condition [54]. As a result of decreased photosynthesis rate, overall anabolic reaction flux is severely constrained. In this context, algae cells may favor storage of energetic molecules, such as lipids, instead of consumption.

Effects of H₂O₂ Induced Oxidative Stress on Lipid Accumulation in *Dunaliella tertiolecta* and *Chlamydomonas reinhardtii*

Green algae, *Dunaliella tertiolecta* and *Chlamydomonas reinhardtii* were chosen for further investigation and confirmation of the H_2O_2 induced oxidative stress effects on cellular lipid accumulation. We cultivated cells in high nitrogen containing media (5 mM) with different concentrations of H_2O_2 as described for *Dunaliella salina strain Tuz_KS_01*. We measured ROS by using the DCFH-DA method, cell viability by using the FDA fluorescent assay, and lipid content by using the fluorometric Nile Red method, as described above. *Dunaliella tertiolecta* cells were found to be more tolerant to H_2O_2; therefore higher concentrations of H_2O_2 were used to demonstrate the cellular response. *Chlamydomonas reinhardtii* cells were treated within the range from 200 μM to 4 mM H_2O_2 concentrations while *Dunaliella tertiolecta* cells were treated within the range of 2 mM to 8 mM H_2O_2 concentrations. As shown in Figure 9, we found that in response to H_2O_2 in both species ROS accumulation and lipid contents increased, while cell survival

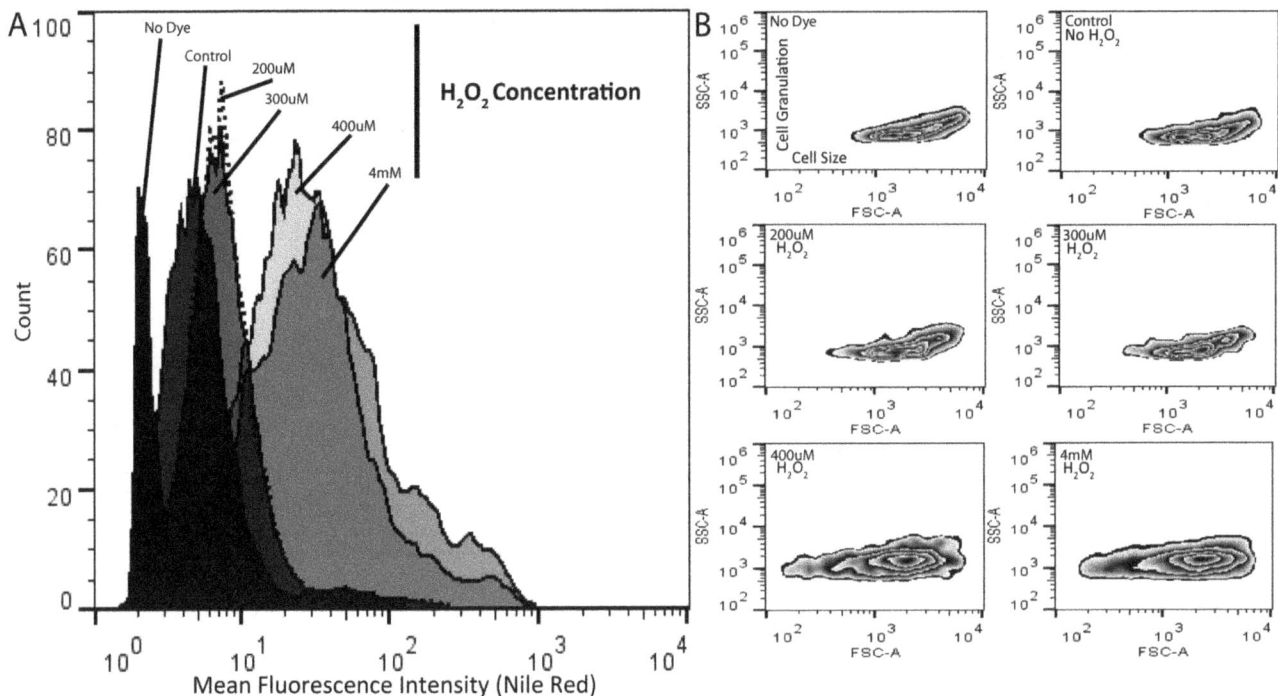

Figure 10. Flow cytometric Nile red, cellular granulation and biovolume analysis of *Chlamydomonas reinhardtii* **cells cultivated under different H₂O₂ concentrations. A)** A representative histogram of flow-cytometric analysis of lipid contents under different H₂O₂ concentrations. **B)** Zebra-plot of SSC (Side Scatter) and FSC (Forward Scatter) expressing cellular granulation and cellular size, respectively, under different H₂O₂ concentrations. Data represent the mean values of triplicates ±1 SE.

dramatically decreased at high H_2O_2 concentrations. In addition, by using flow cytometric analysis, we observed that lipid increase in response to oxidative stress was accompanied with increased granulation and biovolume in *Chlamydomonas* cells (Figure 10). Neither *Dunaliella tertiolecta* nor *Dunaliella salina* strain *Tuz_KS_01* cells showed biovolume increase upon treatment of H_2O_2 at any experimented concentration.

Conclusions

Nitrogen depletion has been shown to be an effective inducer of lipid production of various unicellular green algae by previous studies. But few studies have examined the physiological and biochemical mechanisms underlying this response. In this study, we isolated a new halophilic green alga species. We presented evidence supporting previous observations that nitrogen depletion causes oxidative stress and lipid accumulation. In addition, we

showed that oxidative stress by itself can cause lipid accumulation, suggesting that lipid accumulation under nitrogen depletion is mediated by oxidative stress. These observations are helpful for utilization of green alga for biodiesel production.

Acknowledgments

Authors would like to thank UTEX, The Culture Collection of Algae, for providing kindly of algae species, *Chlamydomonas reinhardtii* and *Dunaliella tertiolecta*. MC was supported by Turkish Academy of Sciences Gebip Programme.

Author Contributions

Conceived and designed the experiments: KY MC SC. Performed the experiments: KY IP. Analyzed the data: KY MC. Contributed reagents/materials/analysis tools: SC MC BE. Wrote the paper: KY MC IP SC.

References

1. Camacho FG, Rodriguez JG, Miron AS, Garcia MC, Belarbi EH, et al. (2007) Biotechnological significance of toxic marine dinoflagellates. Biotechnol Adv 25: 176–194.
2. Li Y, Horsman M, Wang B, Wu N, Lan CQ (2008) Effects of nitrogen sources on cell growth and lipid accumulation of green alga Neochloris oleoabundans. Appl Microbiol Biotechnol 81: 629–636.
3. Csavina JL, Stuart BJ, Riefler RG, Vis ML (2011) Growth optimization of algae for biodiesel production. J Appl Microbiol 111: 312–318.
4. Spolaore P, Joannis-Cassan C, Duran E, Isambert A (2006) Commercial applications of microalgae. J Biosci Bioeng 101: 87–96.
5. Alscher RG, Donahue JL, Cramer CL (1997) Reactive oxygen species and antioxidants: Relationships in green cells. Physiologia Plantarum 100: 224–233.
6. Mallick N, Mohn FH (2000) Reactive oxygen species: response of algal cells. Journal of Plant Physiology 157: 183–193.
7. Liu WH, Huang ZW, Li P, Xia JF, Chen B (2012) Formation of triacylglycerol in Nitzschia closterium f. minutissima under nitrogen limitation and possible

physiological and biochemical mechanisms. Journal of Experimental Marine Biology and Ecology 418: 24–29.
8. Li X, Hu HY, Zhang YP (2011) Growth and lipid accumulation properties of a freshwater microalga Scenedesmus sp. under different cultivation temperature. Bioresour Technol 102: 3098–3102.
9. Hong Y, Hu HY, Li FM (2008) Physiological and biochemical effects of allelochemical ethyl 2-methyl acetoacetate (EMA) on cyanobacterium Microcystis aeruginosa. Ecotoxicol Environ Saf 71: 527–534.
10. Gordillo FL, Goutx M, Figueroa F, Niell FX (1998) Effects of light intensity, CO2 and nitrogen supply on lipid class composition of Dunaliella viridis. Journal of Applied Phycology 10: 135–144.
11. Olmos J, Paniagua J, Contreras R (2000) Molecular identification of Dunaliella sp utilizing the 18S rDNA gene. Letters in Applied Microbiology 30: 80–84.
12. Fredslund J (2006) PHY center dot FI: fast and easy online creation and manipulation of phylogeny color figures. Bmc Bioinformatics 7.

13. Levasseur M, Thompson PA, Harrison PJ (1993) Physiological acclimation of marine-phytoplankton to different nitrogen-sources. Journal of Phycology 29: 587–595.

14. Greenspan P, Mayer EP, Fowler SD (1985) Nile red: a selective fluorescent stain for intracellular lipid droplets. J Cell Biol 100: 965–973.

15. Govender T, Ramanna L, Rawat I, Bux F (2012) BODIPY staining, an alternative to the Nile Red fluorescence method for the evaluation of intracellular lipids in microalgae. Bioresour Technol 114: 507–511.

16. Chen W, Sommerfeld M, Hu Q (2011) Microwave-assisted nile red method for in vivo quantification of neutral lipids in microalgae. Bioresour Technol 102: 135–141.

17. Chen W, Zhang C, Song L, Sommerfeld M, Hu Q (2009) A high throughput Nile red method for quantitative measurement of neutral lipids in microalgae. J Microbiol Methods 77: 41–47.

18. Cole TA, Fok AK, Ueno MS, Allen RD (1990) Use of nile red as a rapid measure of lipid content in ciliates. Eur J Protistol 25: 361–368.

19. Kimura K, Yamaoka M, Kamisaka Y (2004) Rapid estimation of lipids in oleaginous fungi and yeasts using Nile red fluorescence. J Microbiol Methods 56: 331–338.

20. Kalyanaraman B, Darley-Usmar V, Davies KJA, Dennery PA, Forman HJ, et al. (2012) Measuring reactive oxygen and nitrogen species with fluorescent probes: challenges and limitations. Free Radical Biology and Medicine 52: 1–6.

21. Liang ZJ, Ge F, Zeng H, Xu Y, Peng F, et al. (2013) Influence of cetyltrimethyl ammonium bromide on nutrient uptake and cell responses of Chlorella vulgaris. Aquatic Toxicology 138: 81–87.

22. Schonswetter P, Tribsch A, Barfuss M, Niklfeld H (2002) Several Pleistocene refugia detected in the high alpine plant Phyteuma globulariifolium sternb & hoppe (Campanulaceae) in the European Alps. Mol Ecol 11: 2637–2647.

23. Barbarino E, Lourenco SO (2005) An evaluation of methods for extraction and quantification of protein from marine macro- and microalgae. Journal of Applied Phycology 17: 447–460.

24. Bradford MM (1976) A rapid and sensitive method for the quantitation of microgram quantities of protein utilizing the principle of protein-dye binding. Anal Biochem 72: 248–254.

25. Wellburn AR (1994) The spectral determination of chlorophyll-a and chlorophhyll-b, as well as total carotenoids, using various solvents with spectrophotometers of different resolution. Journal of Plant Physiology 144: 307–313.

26. Sabatini SE, Juarez AB, Eppis MR, Bianchi L, Luquet CM, et al. (2009) Oxidative stress and antioxidant defenses in two green microalgae exposed to copper. Ecotoxicology and Environmental Safety 72: 1200–1206.

27. Dhindsa RS, Plumbdhindsa P, Thorpe TA (1981) Leaf senescence - correlated with increased levels of membrane-permeability and lipid-peroxidation, and decreased levels of superoxide-dismutase and catalase. Journal of Experimental Botany 32: 93–101.

28. Aebi H (1984) Catalase invitro. Methods in Enzymology 105: 121–126.

29. Nakano Y, Asada K (1981) Hydrogen-peroxide is scavenged by ascorbate-specific peroxidase in spinach-chloroplasts. Plant and Cell Physiology 22: 867–880.

30. Tang HY, Abunasser N, Garcia MED, Chen M, Ng KYS, et al. (2011) Potential of microalgae oil from Dunaliella tertiolecta as a feedstock for biodiesel. Applied Energy 88: 3324–3330.

31. Ho SH, Chen WM, Chang JS (2010) Scenedesmus obliquus CNW-N as a potential candidate for CO2 mitigation and biodiesel production. Bioresource Technology 101: 8725–8730.

32. Griffiths MJ, van Hille RP, Harrison STL (2012) Lipid productivity, settling potential and fatty acid profile of 11 microalgal species grown under nitrogen replete and limited conditions. Journal of Applied Phycology 24: 989–1001.

33. Chen M, Tang HY, Ma HZ, Holland TC, Ng KYS, et al. (2011) Effect of nutrients on growth and lipid accumulation in the green algae Dunaliella tertiolecta. Bioresource Technology 102: 1649–1655.

34. Sharma KK, Schuhmann H, Schenk PM (2012) High Lipid Induction in Microalgae for Biodiesel Production. Energies 5: 1532–1553.

35. Lee SJ, Yoon BD, Oh HM (1998) Rapid method for the determination of lipid from the green alga Botryococcus braunii. Biotechnology Techniques 12: 553–556.

36. Lin QA, Lin JD (2011) Effects of nitrogen source and concentration on biomass and oil production of a Scenedesmus rubescens like microalga. Bioresource Technology 102: 1615–1621.

37. Bhaduri AM, Fulekar MH (2012) Antioxidant enzyme responses of plants to heavy metal stress. Reviews in Environmental Science and Bio-Technology 11: 55–69.

38. Alscher RG, Erturk N, Heath LS (2002) Role of superoxide dismutases (SODs) in controlling oxidative stress in plants. Journal of Experimental Botany 53: 1331–1341.

39. del Rio LA, Sandalio LM, Corpas FJ, Palma JM, Barroso JB (2006) Reactive oxygen species and reactive nitrogen species in peroxisomes. Production, scavenging, and role in cell signaling. Plant Physiology 141: 330–335.

40. Miller SR, Martin M, Touchton J, Castenholz RW (2002) Effects of nitrogen availability on pigmentation and carbon assimilation in the cyanobacterium Synechococcus sp strain SH-94-5. Archives of Microbiology 177: 392–400.

41. Singh SP, Hader DP, Sinha RP (2010) Cyanobacteria and ultraviolet radiation (UVR) stress: Mitigation strategies. Ageing Research Reviews 9: 79–90.

42. Courchesne NMD, Parisien A, Wang B, Lan CQ (2009) Enhancement of lipid production using biochemical, genetic and transcription factor engineering approaches. Journal of Biotechnology 141: 31–41.

43. Msanne J, Xu D, Konda AR, Casas-Mollano JA, Awada T, et al. (2012) Metabolic and gene expression changes triggered by nitrogen deprivation in the photoautotrophically grown microalgae Chlamydornonas reinhardtii and Coccomyxa sp C-169. Phytochemistry 75: 50–59.

44. Tsukagoshi H (2012) Defective root growth triggered by oxidative stress is controlled through the expression of cell cycle-related genes. Plant Science 197: 30–39.

45. Cash TP, Pan Y, Simon MC (2007) Reactive oxygen species and cellular oxygen sensing. Free Radical Biology and Medicine 43: 1219–1225.

46. Converti A, Casazza AA, Ortiz EY, Perego P, Del Borghi M (2009) Effect of temperature and nitrogen concentration on the growth and lipid content of Nannochloropsis oculata and Chlorella vulgaris for biodiesel production. Chemical Engineering and Processing 48: 1146–1151.

47. Hu Q, Sommerfeld M, Jarvis E, Ghirardi M, Posewitz M, et al. (2008) Microalgal triacylglycerols as feedstocks for biofuel production: perspectives and advances. Plant Journal 54: 621–639.

48. Kobayashi M (2003) Astaxanthin biosynthesis enhanced by reactive oxygen species in the green alga Haematococcus pluvialis. Biotechnology and Bioprocess Engineering 8: 322–330.

49. Pinto E, Sigaud-Kutner TCS, Leitao MAS, Okamoto OK, Morse D, et al. (2003) Heavy metal-induced oxidative stress in algae. Journal of Phycology 39: 1008–1018.

50. Zhang YM, Chen H, He CL, Wang Q (2013) Nitrogen starvation induced oxidative stress in an oil-producing green alga Chlorella sorokiniana C3. PLoS One 8: e69225.

51. Zhang ZD, Shrager J, Jain M, Chang CW, Vallon O, et al. (2004) Insights into the survival of Chlamydomonas reinhardtii during sulfur starvation based on microarray analysis of gene expression. Eukaryotic Cell 3: 1331–1348.

52. Nishiyama Y, Yamamoto H, Allakhverdiev SI, Inaba M, Yokota A, et al. (2001) Oxidative stress inhibits the repair of photodamage to the photosynthetic machinery. EMBO J 20: 5587–5594.

53. Garcia-Ferris C, Moreno J (1993) Redox regulation of enzymatic activity and proteolytic susceptibility of ribulose-1,5-bisphosphate carboxylase/oxygenase fromEuglena gracilis. Photosynth Res 35: 55–66.

54. Cakmak T, Angun P, Ozkan AD, Cakmak Z, Olmez TT, et al. (2012) Nitrogen and sulfur deprivation differentiate lipid accumulation targets of Chlamydomonas reinhardtii. Bioengineered 3: 343–346.

9

Lipid Profiling and Transcriptomic Analysis Reveals a Functional Interplay between Estradiol and Growth Hormone in Liver

Leandro Fernández-Pérez[1,2]*, **Ruymán Santana-Farré**[1], **Mercedes de Mirecki-Garrido**[1], **Irma García**[2,3], **Borja Guerra**[1,2], **Carlos Mateo-Díaz**[1,2], **Diego Iglesias-Gato**[4], **Juan Carlos Díaz-Chico**[1,2], **Amilcar Flores-Morales**[4], **Mario Díaz**[2,3]

1 Department of Clinical Sciences, University of Las Palmas de Gran Canaria - Biomedical and Health Research Institute (IUIBS), Molecular and Translational Endocrinology Group, Las Palmas de Gran Canaria, Spain, 2 Cancer Research Institute of The Canary Islands (ICIC), Las Palmas de Gran Canaria, Canary Islands, Spain, 3 Department of Animal Biology, University of La Laguna, Laboratory of Membrane Physiology and Biophysics, La Laguna, Spain, 4 Molecular Endocrinology group, University of Copenhagen - Novo Nordisk Center for Protein Research, Copenhagen, Denmark

Abstract

17β-estradiol (E2) may interfere with endocrine, metabolic, and gender-differentiated functions in liver in both females and males. Indirect mechanisms play a crucial role because of the E2 influence on the pituitary GH secretion and the GHR-JAK2-STAT5 signaling pathway in the target tissues. E2, through its interaction with the estrogen receptor, exerts direct effects on liver. Hypothyroidism also affects endocrine and metabolic functions of the liver, rendering a metabolic phenotype with features that mimic deficiencies in E2 or GH. In this work, we combined the lipid and transcriptomic analysis to obtain comprehensive information on the molecular mechanisms of E2 effects, alone and in combination with GH, to regulate liver functions in males. We used the adult hypothyroid-orchidectomized rat model to minimize the influence of internal hormones on E2 treatment and to explore its role in male-differentiated functions. E2 influenced genes involved in metabolism of lipids and endo-xenobiotics, and the GH-regulated endocrine, metabolic, immune, and male-specific responses. E2 induced a female-pattern of gene expression and inhibited GH-regulated STAT5b targeted genes. E2 did not prevent the inhibitory effects of GH on urea and amino acid metabolism-related genes. The combination of E2 and GH decreased transcriptional immune responses. E2 decreased the hepatic content of saturated fatty acids and induced a transcriptional program that seems to be mediated by the activation of PPARα. In contrast, GH inhibited fatty acid oxidation. Both E2 and GH replacements reduced hepatic CHO levels and increased the formation of cholesterol esters and triacylglycerols. Notably, the hepatic lipid profiles were endowed with singular fingerprints that may be used to segregate the effects of different hormonal replacements. In summary, we provide *in vivo* evidence that E2 has a significant impact on lipid content and transcriptome in male liver and that E2 exerts a marked influence on GH physiology, with implications in human therapy.

Editor: Toshi Shioda, Massachusetts General Hospital, United States of America

Funding: The Spanish Ministry of Science and Innovation with the funding of European Regional Development Fund-European Social Fund supported this research by grants-in-aid to LF-P (SAF2003-02117; SAF2006-07824) and MD (SAF2010- 22114-C02-01). LF-P was also supported by grants-in-aid from ACIISI (PI2007/033; PI2010/0110) and Alfredo Martin-Reyes Foundation (Arehucas)-FICIC. D.I-G and A.F-M are supported by the Danish Council for Independent Research and the Novo Nordisk Foundation. R.S-F was recipient of a predoctoral fellowship from the MCYT (SAF2003-02117), ULPGC-ACIISI (ULPAPD-08/01-4), and ACIISI-FSE (SE --10/13). M.M-G is a recipient of predoctoral fellowship from ULPGC-MEC (AP2001-3499). The funders had no role in study design, data collection and analysis, decision to publish, or preparation of the manuscript.

Competing Interests: The authors have declared that no competing interests exist.

* E-mail: lfernandez@dcc.ulpgc.es

Introduction

17β-estradiol (E2), a major natural estrogen in mammals, has physiological actions not limited to reproductive organs in males [1,2,3,4,5,6]. Studies in patients with natural mutations in the human estrogen receptor alpha (ERα) [7,8] and aromatase [9,10] genes, and in the ERα (ERKO) and aromatase (ArKO) null mice models have shown that E2 can play a critical physiological role in males [1]. In particular, an insufficient E2 signaling in the ERKO and ArKO null mice models results in a metabolic syndrome-like phenotype with fatty liver due to a disruption in β-oxidation and increased lipogenesis, a phenotype that is reversed by physiological doses of E2. Moreover, both of these models exhibit a sexually dimorphic fatty liver that, notably, is male specific [1].

The effects of E2 in the liver can be explained through the direct actions of ER [2,11,12,13] or, indirectly, by modulating growth hormone (GH) physiology [14,15]. E2 can influence pituitary GH secretion but also GH direct actions in the liver. In particular, E2 induces the expression of Suppressor of Cytokine Signaling (SOCS)-2, which is a negative regulator of the GHR-JAK2-STAT5 signaling pathway [16]. Recently, we have identified SOCS2 as an important regulator of hepatic homeostasis (i.e., lipid and glucose metabolism and inflammation) under conditions of high-fat dietary stress [17]. The ability of GHR-JAK2-STAT5 signaling pathway to regulate hepatic lipid metabolism has also been highlighted in recent mouse genetic studies showing that hepatic inactivation of the GHR [18], its associated kinase, JAK2 [19] or its downstream signaling intermediary, STAT5b [20], leads to fatty liver. The metabolic influence of GH deficiency has also been well documented in humans by the development of a metabolic syndrome (i.e, increased visceral obesity, reduced lean body mass and fatty liver), a phenotype that is ameliorated by GH replacement therapy [21]. Notably, oral administration of pharmacological doses of E2 in humans inhibits GH-regulated endocrine (e.g., IGF-I) and metabolic (e.g., lipid oxidation, protein synthesis) effects [22,23] but these effects are attenuated when E2 is administered transdermally, suggesting that liver is the major target of regulatory cross-talk between estrogens and GH. However, the molecular characterization of the hepatic changes induced by long-term E2 treatment, when it is administered subcutaneously, and how they influence the liver response to male pattern of GH administration are not well understood.

Animal studies of hepatic effects of E2 or its interplay with GH actions have been focused on females [24,25]. Nonetheless, it is unclear if males exhibit equivalent responses, and there are reasons why such equivalence should not be presumed. In particular, gender dimorphism in GH secretion patterns develops soon after birth and the pituitary GH release maintains a sexually dimorphic liver function in adulthood [15], which may influence the nature of E2 effects in the livers of males and females. Several GH deficient models can be used to study the interplay between E2 and GH in males. Notably, the hypothyroid-orchidectomized (TXOX) rat model reaches very low or undetectable blood levels of GH and E2, which can be readily restored by hormone replacement treatment (HRT) [25,26,27], and shows systemic and hepatic metabolic disturbances with features that mimic deficiencies in E2 [2] and GH [28] (e.g., hypercholesterolemia, adiposity, fatty liver). In this study, we hypothesized that functional interplay between E2 and GH influences liver physiology in male. To test this hypothesis, we investigated the mechanisms of E2 and GH to regulate liver function at the molecular level. We studied gene expression profiles in liver tissue and correlated them with the changes in hepatic lipid content in TXOX rats before and after E2 and/or GH replacement. The results show that the interactions with GH contribute to multiple effects of E2 in male rat liver. Indeed, we found that E2 significantly influenced the GH-regulated endocrine, metabolic, immune, and gender specific responses in the liver. E2- and GH-induced changes in hepatic gene expression profiles were associated with changes in hepatic lipid composition. Finally, hepatic lipid profiles from E2- and/or GH-treated TXOX rats substantially differ from those observed in intact animals, indicating that the normal functions of thyroid glands and testes are an absolute requirement for physiological hepatic lipid homeostasis.

Material and Methods

Materials

Recombinant human GH was kindly donated by Pfizer laboratories (Spain). E2 benzoate, Tri-Reagent and, unless otherwise indicated, the rest of the products cited in this work were purchased from the Sigma Chemical Co. (St. Louis, MO).

Animal treatment

This study was carried out in strict accordance with the recommendations in the Guide for the Care and Use of Laboratory Animals of the University of Las Palmas de Gran Canaria and conducted in accordance with European and Spanish

laws and regulations. The protocol was approved by the Committee on the Ethics of Animal Experiments of the University of Las Palmas de G.C. (permit number: 2006-07824). All efforts were made to minimize suffering. Adult (2–3 months old) male Sprague-Dawley rats (n = 6 per group) were used throughout these experiments. Animals were kept under a constant dark/light cycle, and in a controlled temperature (21–23°C) environment, and had free access to autoclaved standard chow (A04 SAFE Panlab, Barcelona, Spain) and tap water throughout the experiment. The generation of hypothyroid animals was performed as previously described [27,29]. The goitrogenic drug methimazole (MMI; 0.05%) was added to the drinking water for 5 weeks starting on postnatal day (PND) 59 until sacrifice on PND94. Calcium chloride (1%) was included with MMI in the water to ensure adequate dietary calcium intake because hypothyroidism decreases food intake by up to 40% [25]. The MMI-containing water was changed twice per week. Two weeks after starting MMI administration, rats were orchidectomized (OX) or sham–operated to make TXOX or testis-intact hypothyroid (TX) groups, respectively. Six rats were not treated with MMI and were subjected to sham-surgery to provide euthyroid testis-intact controls (INTACT). Four days after OX, we began HRT with E2 benzoate (50 µg/kg; sc; 5 days per week, from Monday to Friday) (TXOXE2) or vehicle (0.2 ml corn oil; sc; 5 days per week, from Monday to Friday) (TXOX) to TXOX rats for 20 days [30,31] before hormonal replacement for 7 days with either E2 plus GH (TXOXE2GH) or vehicle plus GH (TXOXGH). GH (0.3 mg/kg/day) was administered as two daily sc injections at 12-h intervals (08:00h and 20:00h) to mimic the male-specific GH secretion [32,33]. TX and TXOX control animals received equivalent amounts of the vehicle alone. Hypothyroidism status was corroborated by monitoring the body weight gain at 7 day intervals and the serum levels of T4 and T3. Twenty-four hours (in the case of E2) or twelve hours (in the case of GH) after the last injection, the animals were killed by exsanguinations. On PND94, blood samples were collected and serum stored at −80°C until analysis. Portions of the liver were snap frozen in liquid nitrogen and stored at −80°C until processed for mRNA analysis.

Serum analysis

The blood was analyzed for T3, T4, glucose, cholesterol (CHO), triacylglycerols (TG), leptin, IGF-I, E2, and testosterone (T). Serum free T3 and T4 concentrations were measured in duplicate by enzyme immunoassay (Access Systems, Beckman Coulter, Inc), with a detection limit of 0.60 ng/dl and 88 ng/dl, respectively. Serum levels of glucose, CHO, and TG were quantified by using an Olympus AU2700 chemistry analyzer (Beckman Coulter Inc.). The immunoassay method was also used to determine serum levels of E2 and T by using the UniCel DxI 800 immunoassay system (Beckman Coulter Inc). Serum levels of leptin and IGF-I were determined by using rat immunoassays (Quantikine, R&D systems) according to manufacturer recommendations. The IGF-I and leptin assays included quality controls provided by the manufacturer, and the standard curves of the assays were performed in accordance with the manufacturer's provided samples. All the samples were assayed together and each sample was assayed in duplicate.

Hepatic lipid analysis

Liver lipids were analyzed following the procedures detailed in Fabelo et al. [34]. Briefly, total lipids were extracted with chloroform/methanol (2:1 v/v) containing 0.01% butylated hydroxytoluene (BHT) as an antioxidant. Lipid classes were separated by one-dimensional double-development high-performance thin-layer chromatography (HPTLC) using methyl acetate/isopropanol/chloroform/methanol/0.25% (w/v) KCl (5:5:5:2:1.8 by vol.) as the developing solvent system for the polar lipid classes and hexane/diethyl ether/acetic acid (22.5:2.5:0.25 by vol.) for the neutral lipid classes. Lipid classes were quantified by densitometry using a Shimadzu CS-9001PC spot scanner. Total and neutral lipid fractions were subjected to acid-catalyzed transmethylation for 16 h at 50°C using 1 ml of toluene and 2 ml of 1% sulfuric acid (v/v) in methanol. The resultant fatty acid methyl esters (FAME) were purified by TLC, and visualized by spraying with 1% iodine in chloroform [35]. FAME were separated, and quantified by using a Thermo gas chromatograph equipped with a flame ionization detector (250°C) and a fused silica capillary column Supelcowax TM 10 (30 m×0.32 mm I.D.). Individual FAMEs were identified by referring to authentic standards. Equal amounts of total lipids were used in all analyses. Throughout the manuscript, lipid nomenclature adhered LIPID MAPS classification system (http://www.lipidmaps.org/data/structure/index.html).

RNA isolation, cDNA microarray, probe preparation, and hybridization

Total RNA was isolated by homogenization of frozen rat tissues using a polytrone PT-2000 (Kinematica AG) and TriReagent (Sigma, St. Louis, MO) according to the protocol supplied by the manufacturer. All samples were treated with RNAse-free DNase (Promega, Madison, WI). RNA yields were measured by UV absorbance and the quality of total RNA was analyzed by using a 2100 Bioanalyzer (Agilent Technologies, Palo Alto, CA). A microarray containing 27000 rat 70-mer oligo probe sets produced at the KTH Microarray Center (www.biotech.kth.se) was used to evaluate the effects of hypothyroidism and hormonal replacement in TXOX animals on liver gene expression. Five µg of high-quality total RNA from the liver were reversed-transcribed, labeled, and hybridized following the manufacturer's protocol (Pronto Plus System, Promega). After 16 h of hybridization, the slides were washed and scanned using the GenePix Microarray Scanner (Axon Instruments, CA). Four independent hybridizations were performed comparing individual animals from the different experimental groups for a total of 4 analyses.

Microarray data processing and analysis

Image analysis was performed using the GenePix Pro 6.0 software (Axon Instruments, Union City, CA) as previously described [36]. The LOWESS (Locally Weighted Scatter Plot Smoother) method was used to normalize the raw intensity data [37]. If the measured probe sets were not present in at least 3 of the 4 chips, they were assumed to contain no information and therefore were eliminated to reduce data complexity. Differentially expressed genes were identified by using the SAM (Significance Analysis for Microarrays) statistical technique [38]. A q value was assigned for each of the detectable genes in the array. This value is similar to a P-value, measuring the lowest false discovery rate (FDR) at which differential expression of a gene is considered significant. A minimal FDR of 0.05 was assigned for each gene. In this work, a completed list of regulated genes is available as supplementary (S) files. An additional selection requirement was added to FDR based on absolute changes in the gene expression ratios. A value of 1.5 (50%) ($\log_2$ ratio $\geq |0.58|$) was chosen to describe ratios as up- or down-regulated. The microarray data discussed in this publication have been deposited in NCBIs Gene Expression Omnibus [39] and are accessible through GEO Series accession number GSE50014 (www.ncbi.nlm.nih.gov/geo). Functional and system biological network analyses were performed on

the basis of the Gene Ontology (GO) enrichment of differentially expressed genes in liver using DAVID [40], and the results were depicted using Cytoscape [41]. For the graphical representation, the significance cut-off was set to a p value <0.05 and a corrected q value (Benjamini) <0.1. GO graphs interpretation: node (inner circle) size corresponds to the number of genes up-regulated by GH or E2; node border (outer circle) size corresponds to the number of genes down-regulated by GH or E2; color of the node and border corresponds to the significance of the gene set for up or down regulated genes, respectively (dark red = significantly enriched, light red = enriched no significantly; grey = absent); edge size corresponds to the number of genes that overlap between the two connected gene sets. Green edges correspond to shared up-regulated genes and blue edges correspond to shared down-regulated genes.

Analysis of gene expression by real-time quantitative-PCR (qPCR)

The mRNA expression levels of genes were measured using qPCR. Briefly, 2 µg of total RNA was treated with RNase-free DNase I (Promega) to remove genomic DNA and reverse transcribed using iScript (Bio-Rad) according to the manufacturer's instructions. Two µl of cDNA served as a template in a 20 µl qPCR reaction mix containing the primers and SYBR Green PCR Master Mix (Diagenode, Belgium). Quantification of gene expression was performed according to the manufacturer's protocol using ABI PRISM 7000 SD RT-PCR. A relative standard curve was constructed with serial dilutions (1:1, 1:10, 1:100) using a pool of the cDNA generated from all animals used in the study. The amplification program consisted of 1 cycle of 95°C for 10 min, followed by 45 cycles of 95°C for 15 s, annealing for 10 s, and 72°C for 30 s. The fluorescent intensity was measured at a specific acquisition temperature for each gene. A dissociation protocol was performed to assess the specificity of the primers and the uniformity of the PCR generated products. The amplified PCR products were subjected to agarose electrophoresis to confirm their predicted size. Data were extracted and amplification plots generated with ABI SDS software. All amplifications were performed in duplicate, and C_t scores were averaged for subsequent calculations of relative expression values. The level of individual mRNA measured by qPCR was normalized to the level of the housekeeping gene cyclophilin by using the Pfaffl method [42]. Exon-specific primers (Table S1) were designed by the Primer 3 program [43].

Statistical analysis

The significance of differences between groups was tested by one-way ANOVA, followed by post hoc comparisons of group means according to the GraphPad Prism 5 program (GraphPad Software, San Diego, CA). Statistical significance was reported if $P<0.05$ was achieved. For graphing purposes in the qPCR analysis, the relative expression levels were scaled so that the expression level of the INTACT group equaled one. Lipids classes and main fatty acids were additionally submitted to factor analysis by means of Principal Component Analysis (PCA) [44]. Variable extraction was carried out based on the proportion of total variance explained by two principal components. Factor scores for principal component 1 in multivariate analyses of lipid classes, and fatty acids from total and neutral lipid are depicted. Factor scores were further analyzed by one-way and to assess statistical differences between treatments and by two-way ANOVA to evaluate the combined effects of hormonal treatments and their interactions, as we have previously reported [45].

Results

Estradiol inhibits the effects of GH on somatotropic-liver axis and induces negative regulators of GH-STAT5 signaling and a female-pattern of gene expression in adult hypothyroid-orchidectomized rat liver

Upon sacrifice on PND94, biochemical hypothyroidism was shown, and significantly ($P = 0.001$) lower or undetectable serum levels of T3 (ng/dl) [36.17±5.43 (INTACT); 7.92±4.84 (TXOX); 1.48±2.14 (TXOXE2); 0 (TXOXGH); 0.14±0.33 (TXOX-E2GH)] and T4 (ng/dl) [(1.90±0.17 (INTACT) *vs.* 0)] were found in all TXOX groups in comparison with the age-matched euthyroid control group (INTACT). Serum E2 levels (pg/ml) [6.33±5.57 (INTACT); 438.80±122.12 (TXOXE2); 363.10±107.95 (TXOXE2GH)] were increased ($P = 0.001$ *vs* INTACT) in E2-treated TXOX rats up to 5-10 times those observed in rats during pregnancy [46] or the proestrus phase of the reproductive cycle [12]. A first approach to assess the effects of E2 on TXOX rat liver was made through analysis of its influence on the somatotropic-liver axis. TXOX caused impaired growth, which was evident as a reduction in daily body weight gain (Fig. 1A). At the last measurement, on PND94, a significant difference ($P<0.001$) in body weight remained between INTACT (447±33 g) and TXOX (338±28 g) groups. Accordingly, TXOX showed reduced levels of hepatic IGF-I mRNA (Fig. 1B) and circulating IGF-I (Fig. 1C). Treatment of TXOX rats with GH, partially or totally, restored body weight gain (Fig.1A), liver IGF-I mRNA (Fig.1B), and circulating IGF-I (Fig.1C), whereas these effects were prevented in the presence of E2. Notably, E2 administration to TXOX rats increased hepatic IGF-I mRNA to the level of non-orchidectomized-hypothyroid (TX) rats (Fig.1B). Next, we carried out mRNA quantitative analysis of SOCS2, CIS, and FGF21, which are negative regulators of GHR-STAT5 signaling [47,48]. E2 treatment of TXOX rats induced the mRNA expression of SOCS2 (Fig. 1D) and CIS (Fig.1E), but only to a fraction of the levels obtained after GH treatment. Indeed, when E2 was used in combination with GH, a 2-3-fold reduced mRNA expression levels of SOCS2 (Fig. 1D) and CIS (Fig.1E) were observed compared with GH treatment alone, again demonstrating the inhibitory actions of E2 on GH hepatic actions. In contrast, the positive effect of GH on FGF21 mRNA was not affected in the presence of E2 (Fig.1F). The level of SOCS3 mRNA was induced by hypothyroidism itself (Fig.1G), whereas neither hypothyroidism nor hormonal replacement altered SOCS5 (Fig.1H). Finally, we measured GH-regulated gene markers of liver sexual dimorphism [15]. Figure 1I shows that the mRNA expression level of CYP2C11, a male-specific gene, was completely abolished in TXOX liver, whereas it was recovered by intermittent GH replacement. In contrast, E2 prevented the GH-induced mRNA expression levels of CYP2C11 (Fig.1I) and CYP2C13 (Fig.1J). Unlike intermittent GH, E2 induced the female-specific CYP2C12 gene in TXOX liver (Fig.1K). Our microarray analysis (see below) also showed that male predominant genes (e.g., CYP2C11, CYP2C13, CYP2E1, alpha-2u-globulin) were down-regulated by E2, whereas female predominant genes (e.g., CYP2C12, CYP2A1, CYP2C7) were induced [15]. Overall, these findings demonstrate that E2 influences the transcription of GHR-STAT5 targeted genes and induces a female-pattern of gene expression in TXOX rat liver.

Influence of estradiol on serum and hepatic lipids in hypothyroid-orchidectomized rats

The effects of E2 and GH on hepatic lipid composition were explored in TXOX rats by carrying out a quantitative analysis of lipid classes (Table 1) and fatty acids from total (Table 2) and neutral (Table 3) lipids. The development of a hypothyroid state in male rats was accompanied by altered circulating lipids: a 2-fold increase of total CHO and a 3-fold decrease of TG (Table 1). However, these changes were not prevented by E2 or GH replacement. Hypothyroidism increased the hepatic levels of CHO and decreased those of TG and diacylglycerols (DG) in comparison with the INTACT group (Table 1). However, the levels of free fatty acids (FFA) were not significantly affected by hypothyroidism, though the average values were lower than in the INTACT animals. Among fatty acids from neutral lipids (Table 3), TXOX rats contained increased levels of total saturated fatty acids (SFA) (due to the significant increase in18:0 content). Among monounsaturated fatty acids (MUFAs), 18:1n-9 (oleic acid), 16:1n-7 (palmitoleic acid), and 18:1n-7 (vaccenic acid) were all significantly reduced in neutral lipids (where these MUFAs are abundant) compared with INTACT animals. Interestingly, these changes paralleled the increase in 18:0. These findings suggest an alteration of Δ9 desaturase in TXOX animals. The most representative 18 carbon polyunsaturated fatty acids (PUFAs), 18:2n-6 (linoleic acid) and 18:3n-3 (linolenic acid) from total (Table 2) or neutral (Table 3) lipids, were unaffected in TXOX. Noticeably, TXOX rats exhibited increased levels of 20:4n-6 [arachidonic acid (AA)] and reduced levels of its elongated precursor 22:4n-6. This later effect on AA was a direct consequence of hypothyroidism and orquidectomy and could be restored by the hormonal therapies used here, especially when used in combination. Another physiologically relevant very long chain polyunsaturated fatty acid (VLCPUFA), namely 22:6n-3 (DHA), exhibited a significant reduction in total lipids from TXOX rats (Table 2). As expected, levels of essential fatty acids 20:4n-6 and 22:6n-3 in INTACT and TXOX animals were three-to-four times higher in total lipids than in neutral lipids, confirming their preferential location within membrane phospholipids. Moreover, the results from total lipids indicate that hypothyroidism and orquidectomy strongly affect the acylation of DHA-containing phospholipids in liver cells. These findings are physiologically relevant since depletion of membrane 22:6n-3 is known to severely impact physicochemical properties of cell membrane [45,49].

E2 treatment in TXOX rats brought about an important reduction in total SFA compared with the TXOX group, this effect was due to the significant reduction in 14:0 (myristic acid), 16:0 (palmitic acid), and 18:0 (stearic acid) from both total (Table 2) and neutral (Table 3) lipids. In parallel to these changes, FFA as a lipid class and DG were dramatically reduced by more than 75% and 38%, respectively, whereas TG increased significantly by 72% in comparison to TXOX rats. It is worth mentioning that E2 did not alter VLCPUFA metabolism in terms of the levels of 20:4n-6, 20:5n-3, and 22:6n-3, which remained similar to the values for neutral and total lipids in the TXOX group. However, E2 increased the levels of the essential n-6 precursor 18:2n-6 (but not the essential n-3 precursor 18:3n-3) in neutral lipids. Noticeably, E2 treatment partly reversed the effects of hypothyroidism on MUFAs. Thus, E2 significantly increased levels of 18:1n-9 (and also those of 16:1n-7 acid though to a lower extent) in total and neutral lipids (Table 2 and 3, respectively), which, together with the reductions in 16:0 and 18:0, point to an upregulation of stearoyl-CoA desaturase (Scd) genes in response to E2. Overall, these changes indicate that E2 stimulates β-oxidation of SFA in the

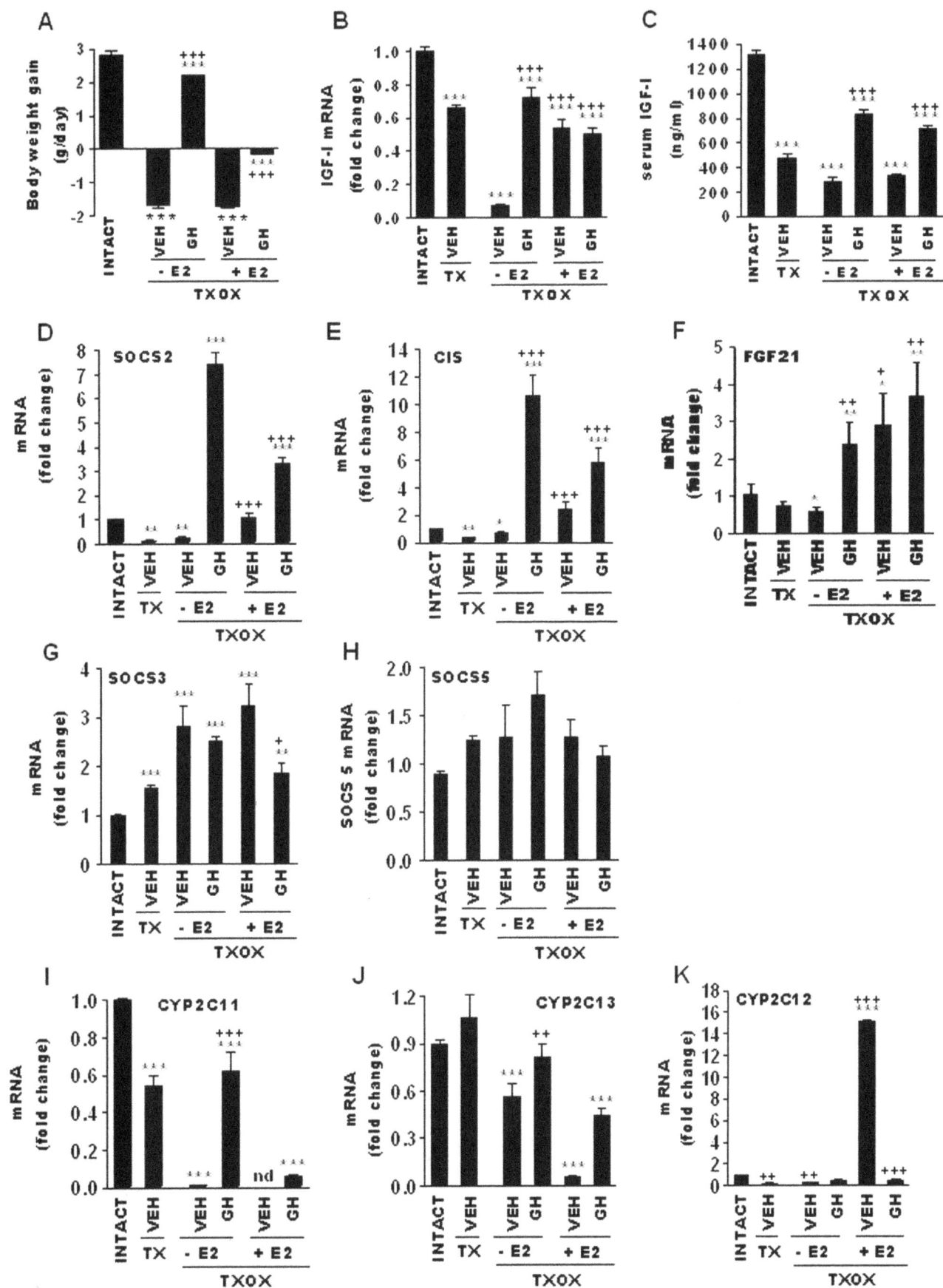

Figure 1. E2 inhibits the effects of GH on somatotropic-liver axis and induces negative regulators of GH-STAT5 signaling and a female-pattern of gene expression in hypothyroid-orchidectomized rat liver. Euthyroid testis-intact controls (INTACT) and hypothyroid-orchidectomized (TXOX) rats were described in Material and Methods. E2 or vehicle (VEH) administration to TXOX rats was performed for 20 days. Then, GH replacement during 7 days rats was carried out in TXOX in the absence (-E2) or in the presence (+E2) of E2. On PND90, body weight gain (A), hepatic IGF-I mRNA (B), circulating IGF-I (C), SOCS2 (D), CIS (E), FGF21 (F), SOCS3 (G), SOCS5 (H), CYP2C11 (I), CYP2C13 (J), and CYP2C12 (K) were measured by qPCR. Results are expressed as mean ± S.D. (n = 6). ***, P<0.001 for comparison with vehicle treated INTACT group; +, P<0.05, ++, P< 0.01, +++, P<0.001 for comparison with vehicle treated TXOX group.

liver while it increases the depots of n-6 polyunsaturated precursors in TG. Regarding other lipid classes, namely, CHO, CHO esters (CE), and phospholipids, we observed that E2 slightly reduces hepatic CHO levels (yet not statistically significant) compared with TXOX but induces a significant increase in the formation of CE (32% and 47% over TXOX and INTACT groups, respectively). This effect on CE formation is independent of hypothyroidism but totally determined by E2 and likely reflects an up-regulation of acyl:cholesterol acyltransferase (ACAT) activity (see discussion).

To some extent, GH administration to TXOX rats resembled the effects of E2 on hepatic lipid composition described above. Thus, GH increased TG and decreased FFA hepatic contents compared with TXOX rats. However, the levels of DG were notably increased by GH treatment in the TXOX group compared with the vehicle- or E2-treated TXOX group (Table 1). TG levels were also increased in serum from GH-treated rats (Table 1). Compared with the TXOX group, the hepatic levels of 18:0, 16:0 and total SFA in neutral lipids (Table 3) were reduced in GH-treated animals. Interestingly, unlike in E2-treated animals, the levels of 18:0 and 16:0 in neutral and total lipids in GH-treated rats approached those observed in INTACT animals. Taken together, these data suggest that GH induces a lipogenic effect in hypothyroid animals by mobilizing SFAs as FFA→DG→TG. Furthermore, when compared with vehicle- or E2-treated TXOX group, GH treatment induced a significant increase in 20:4n-6 and 22:6n-3 in total lipids (Table 2), which is reflected in the high levels of VLCPUFA observed in this group of animals. Because these later effects on LCPUFA were not observed in neutral lipids, the results point to a significant effect of GH on phospholipid remodeling, though the effect of E2 in modulating phospholipid acylation-reacylation has been established [34], similar effects induced by GH or by the combined treatment E2+GH represent a novel finding. Finally, CHO levels in the GH-treated TXOX group were returned to values observed in the INTACT group, but as in the E2-treated TXOX group, an increased CE level was present in response to GH (Table 1).

In the presence of E2, GH gave rise to a complex hepatic lipid phenotype. Thus, total SFA, 14:0, 16:0, and 18:0 in total lipids (Table 2), were reduced compared with TXOX and reached the levels observed in INTACT animals. In neutral lipids (Table 3), however, the contents of these fatty acids were significantly lower than in TXOX. Among VLCPUFA, AA was also decreased well below TXOX animals to values similar to those found in the INTACT group in neutral lipids (Table 3), while as in the TXOXGH group, DHA was significantly increased in the phospholipids of the TXOXE2GH group to achieve the levels in INTACT animals. A striking effect of the combined effects of E2 and GH is the complete restoration of MUFA levels from total and neutral lipids, an effect attributable to the increase in 18:1n9, likely through alteration of Δ9 desaturase expression. Interestingly, the combined E2+GH treatment gave raise to significantly higher MUFA levels compared to E2 and GH treatments individually. Given that MUFA levels of in GH-treated rats were identical to those in TXOX rats, and that E2 treatment increased MUFA (especially 18:1n-9) as compared for GH, the results suggest a

permissive action of GH on E2 effect. We have described above that E2 and GH increased hepatic CHO and CE, but the combined effect of the two hormones seemed to be additive with regard to CE, because its levels doubled those found in INTACT animals and were approximately 30% higher than in the E2 and GH groups (Table 1). Conversely, in the presence of E2, GH reduced the hepatic CHO content compared not only to the TXOX group but also in relation to the E2- or GH-treated TXOX groups, indicating an antagonistic hormonal interaction. Finally, the significant increase in hepatic TG by GH in the presence of E2, with the largest content among all groups (>96% compared with the INTACT and >133% compared with the TXOX group), points to a substantial stimulation of hepatic lipogenesis by the combination of E2 and GH.

Finally, we performed PCA to identify data that discriminate groups. Figure 2 shows the outcomes of PCA on lipid profiles for lipid classes and total and neutral lipids. For lipid classes (Fig. 2A), PC1 (principal component 1) was positively related to FFA, CHO, sphingomyelin (SM) and phosphatidylethanolamine (PE), and negatively correlated with CE and TG. On the other hand, for fatty acids from neutral (Fig. 2C) and total (Fig. 2B) lipids, PC1 was negatively related to C18 PUFA (18:2n-6, 18:3n-6 and 18:3n-3) in total lipids, and positively to saturated (14:0, 15:0 and 16:0) and n-7 MUFA. Therefore, PCA allowed a substantial simplification of lipid data for group discrimination. When we computed factor scores 1 and 2 from PC1 to obtain a group simplification for lipid classes (Fig. 2A) and fatty acids from total (Fig. 2B), and neutral lipids (Fig. 2C), these results allowed a neat discrimination between HRT conditions for all three analyses. INTACT rats are represented by a discrete cluster whereas the effects of E2 or GH on TXOX rats can be distinguished from the untreated TXOX group and display some degree of overlap with the combined E2 and GH treatments. Thus, in contrast with somatic growth in which E2 clearly prevented GH actions, the quantitative lipid analysis displays a more complex picture of the molecular changes induced by the different hormonal treatments. The E2 and GH effects on lipid contents display significant similarities but also treatment-dependent specific effects, such those leading to the regulation of DG content by GH.

Estradiol influences liver transcriptome in hypothyroid male rats

Next, we performed a genome wide gene expression analysis to better understand the influence of E2 on liver physiology. This experiment identified 634 genes that were differentially regulated in TXOX rats after E2 treatment (Table S2). Next, we identified the active biological processes from expression profiles by GO enrichment analysis [40] and system biological network [41] analysis. GO enrichment analysis of the 352 genes up-regulated by E2 revealed a significant over-representation of genes related to fatty acid metabolism whereas among the 282 down-regulated genes, over-representation of genes involved in steroid and xenobiotic metabolism was observed (Fig. 3). Accordingly, the genes up-regulated by E2 were clustered in cellular pathways (KEEG) related to the PPARα signaling (P = 1,4E-05; Bonferroni

Table 1. Effects of hormonal replacement on serum and hepatic lipid classes.

Lipid classes	INTACT		TXOX		TXOXE2		TXOXGH		TXOXE2GH	
	mean ± SEM		mean ± SEM		mean ± SEM		mean ± SEM		mean ± SEM	
Serum										
Glucose (g/dl)	141,2±4,7	a	126,8±6,2	ab	112,0±3,9	bc	107,7±7,1	bc	95,0±3,6	d
CHO (mg/dl)	67,2±3	a	126,5±10,1	b	141,2±12,9	b	129,3±12,9	b	143,5±6,4	b
TG (mg/dl)	133±16,3	a	44±5,1	b	52,6±7,5	b	43,8±6,5	b	77,3±7,3	b
Leptin (ng/ml)	2,8±0,5		2,6±1,1		1,1±0,4		2,1±0,3		1,2±0,6	
Hepatic (%)										
LPC	nd		nd		nd		nd		nd	
SM	0,9±0,0	a	1,0±0,0	a	0,5±0,0	b	0,7±0,2	ab	0,4±0,1	ab
PC	25,3±0,6	a	26,9±1,2	ab	30,6±0,6	b	28,5±0,7	b	26,9±1,1	ab
PS	2,3±0,0	a	1,8±0,2	a	3,3±0,1	b	2,7±0,3	b	2,9±0,2	ab
PG	5,5±0,1		5,3±0,3		5,3±0,3		4,5±0,5		5,6±0,4	
PE	17,4±0,3	a	18,1±0,6	a	16,1±0,5	a	17,6±1,6	ab	13,5±0,5	b
PLE	0,8±0,2		0,4±0,0		0,3±0,1		0,3±0,0		0,3±0,1	
DG	1,7±0,3	a	0,9±0,2	a	0,6±0,1	b	1,8±0,1	b	1,7±0,2	a
CHO	14,7±0,7	bc	17,1±0,3	bc	15,6±0,4	d	13,2±0,2	b	11,3±0,7	b
FFA	7,9±0,5	b	7,1±0,9	b	1,8±0,3	b	2,5±0,5	a	2,0±0,3	a
TG	8,3±0,4	a	7,0±0,4	a	12,1±0,6	bc	11,5±1,0	bc	16,4±3,0	c
CE	4,1±0,3	a	4,6±0,1	a	6,0±0,1	b	6,5±0,1	b	9,5±1,1	c
TPL	63,2±0,6		62,9±1,7		63,7±0,9		64,6±0,7		59±2,8	
TNL	36,8±0,6		37,1±1,7		36,3±0,9		35,4±0,7		41±2,8	

Euthyroid testis-intact controls (INTACT) and hypothyroid-orquidectomized (TXOX) rats were performed as described in Material and Methods. Long-term administration of E2 (TXOXE2) or vehicle (VEH) to TXOX rats was performed for 20 days. Then, VEH (TXOX), E2 (TXOXE2), GH (TXOXGH) or E2 plus GH (TXOXE2GH) replacements were performed for additional 7 days. On PND90, the animals were sacrificed and serum and hepatic lipid classes were measured. Data are expressed as mean ± SEM for serum (n = 5) or hepatic (n = 5) independent samples (different animals) and each independent sample was tested twice. Values represent weight percent of total lipid. Values were submitted to ANOVA followed by post hoc Tukey's test. Values in the same row with different lowercase letters are significantly different with $P<0.05$.

Table 2. Effects of hormonal replacement on fatty acid composition from liver total lipids.

TOTAL LIPIDS	INTACT mean ± SEM		TXOX mean ± SEM		TXOXE2 mean ± SEM		TXOXGH mean ± SEM		TXOXE2GH mean ± SEM	
Fatty acids										
C 14: 0	0,5±0,1	ab	0,5±0,1	b	0,3±0,0	a	0,5±0,0	ab	0,3±0,0	ab
C 16: 0	19,8±0,5	ab	20,2±0,6	b	18,5±0,1	a	19,1±0,3	ab	18,5±0,6	a
C 16:1 n-9	0,3±0,0		0,4±0,0		0,3±0,0		0,4±0,0		0,3±0,0	
C 16: 1 n-7	2,2±0,4		1,4±0,1		1,6±0,1		1,5±0,3		1,7±0,2	
C 18: 0	13,2±0,2	a	15,8±0,4	c	14,5±0,2	b	14,7±0,2	b	12,7±0,3	a
C 18: 1 n-9	10,6±0,5	a	10,3±0,4	a	12±0,5	ab	10,4±0,5	a	13,9±0,6	b
C 18: 1 n-7	5,6±0,4	b	2,6±0,3	a	2,8±0,1	a	3,0±0,2	a	2,9±0,1	a
C 18: 2 n-6	15,8±0,9	a	18,2±0,7		19,1±0,3		17,1±0,9	b	18,7±0,9	
C 18: 3 n-6	0,2±0,0	a	0,8±0,1	b	1,4±0,1	c	0,6±0,0	b	1,3±0,3	abc
C 18: 3 n-3	0,3±0,0		0,3±0,0		0,4±0,0		0,3±0,0		0,3±0,1	
C 18: 4 n-3	0,0±0,0	a	0,0±0,0	ab	0,1±0,0	b	0,0±0,0	ab	0,1±0,0	b
C 20: 3 n-6	0,7±0,0	b	1,2±0,2	c	0,4±0,0	a	0,8±0,0	b	0,5±0,0	a
C 20: 4 n-6	19±0,9	ab	18,1±0,2	a	17,5±0,2	b	20,2±0,4	b	18,8±1,4	ab
C 20: 5 n-3	0,2±0,0	a	0,4±0,1	b	0,4±0,0	b	0,3±0,1	ab	0,2±0,1	ab
C 22: 4 n-6	0,8±0,1		0,5±0,0		0,7±0,1		0,8±0,1		0,6±0,1	
C 22: 5 n-6	1,0±0,2		0,5±0,0		1,0±0,1		0,8±0,1		0,8±0,1	
C 24: 0	0,4±0,0	a	0,6±0,0	b	0,6±0,0	b	0,6±0,1	b	0,5±0,1	ab
C 22: 6 n-3	4,4±0,2	b	3,4±0,2	a	4,0±0,0	ab	4,4±0,3	b	4,2±0,1	b
Totals										
SFA	35,3±0,3		38,7±0,8		35,2±0,2		36,1±0,4		33,2±0,6	
MUFA	19,6±1,1	b	17,3±1,9	a	17,3±0,4	a	16,1±0,9	a	19,5±0,7	b
PUFA	43,7±1,5		43,2±2,0		46,9±0,3		47,0±10,9		46,9±0,8	
VLCPUFA	27±0,7		24,5±1,2		25,6±0,4		28,5±0,7		26,1±1,6	

Euthyroid testis-intact controls (INTACT) and hypothyroid-orquidectomized (TXOX) rats were performed as described in Material and Methods. Administration of E2 (TXOXE2) or vehicle (TXOX) rats was performed for 20 days. Then, VEH (TXOX), E2 (TXOXE2), GH (TXOXGH) or E2 plus GH (TXOXE2GH) replacements were performed for additional 7 days. On PND90, animals were sacrificed and the liver extracted for fatty acid analyses as described in Material and Methods. Data are expressed as mean ± SEM for 5 independent samples (different animals) and each independent sample was tested twice. Values represent weight percent of total lipid. Values were submitted to ANOVA followed by post hoc Tukey's test. Values in the same row with different lowercase letters are significantly different with P<0.05.

Table 3. Effects of hormonal replacement on fatty acid composition from hepatic neutral lipids.

NEUTRAL LIPIDS	INTACT		TXOX		TXOXE2		TXOXGH		TXOXE2GH	
	mean ± SEM		mean ± SEM		mean ± SEM		mean ± SEM		mean ± SEM	
Fatty acids (%)										
C 14:0	1,0±0,1	b	1,0±0,1	b	0,5±0,0	a	0,9±0,1	ab	0,6±0,1	ab
C 16:0	25,2±0,8	b	25,4±1,0	b	22,2±0,4	ab	22,9±0,6	ab	21,6±1,1	b
C 16:1 n-9	0,6±0,1	ab	0,7±0,0	b	0,5±0,0	a	0,7±0,0	ab	0,5±0,0	ab
C 16:1 n-7	4,1±0,8		2,5±0,1		2,8±0,2		2,8±0,5		3,3±0,6	
C 18:0	4,3±0,3	ab	7,7±0,7	c	4,2±0,2	ab	5,2±0,5	b	3,3±0,3	a
C 18:1 n-9	22,1±1,0	b	18,3±0,8	a	22,6±0,8	b	20,6±1,0	ab	26,1±0,3	c
C 18:1 n-7	6,3±0,7	b	3,3±0,3	a	3,1±0,1	a	3,5±0,3	a	3,7±0,2	a
C 18:2 n-6	20,9±1,7	b	22,3±1,3	ab	25,2±0,8	a	23,3±1,7	ab	25,1±0,8	a
C 18:3 n-6	0,2±0,0	a	1,1±0,1	b	2,3±0,1	c	1,1±0,1	b	1,8±0,5	abc
C 18:3 n-3	0,7±0,1		0,7±0,1		0,8±0,0		0,7±0,1		0,8±0,1	
C 18:4 n-3	0,0±0,0	a	0,1±0,0	abc	0,3±0,0	b	0,1±0,0	abc	0,2±0,0	c
C 20:3 n-6	0,4±0,0	b	0,6±0,1	c	0,3±0,0	a	0,6±0,0	a	0,5±0,1	bc
C 20:4 n-6	5,6±0,8	a	8,6±0,3	b	7,3±0,3	ab	7,8±0,5	ab	6,0±1,2	ab
C 20:5 n-3	0,3±0,0		0,4±0,1		0,5±0,1		0,5±0,2		0,3±0,1	
C 22:4 n-6	1,0±0,2		0,5±0,1		0,7±0,1		1,2±0,2		0,7±0,1	
C 22:5 n-6	0,6±0,1	a	0,3±0,1	a	0,4±0,1	ab	0,6±0,0	b	0,3±0,0	a
C 22:6 n-3	1,5±0,2	b	1,2±0,1	ab	1,1±0,0	ab	1,4±0,2	b	0,8±0,1	a
C 24:0	0,5±0,1	a	1,0±0,1	b	0,9±0,1	b	0,9±0,1	b	0,5±0,2	a
Totals										
SFA	32,1±0,8	b	36,5±1,2	b	28,9±0,3	ab	31,2±1,2	b	27,1±0,7	a
MUFA	34,4±2,3	b	25,8±1,1	b	29,8±0,8	a	28,5±1,7	ab	34,6±0,8	b
PUFA	32,5±3,0		37,1±1,6		40,5±0,9		39,2±2,1		37,8±1,2	
VLCPUFA	10,1±1,4		12,6±0,4		11,2±0,5		13,0±0,7		9,5±1,4	

Euthyroid testis-intact controls (INTACT) and hypothyroid-orquidectomized (TXOX) rats were performed as described in Material and Methods. Administration of E2 (TXOXE2) or vehicle (TXOX) to TXOX rats was performed for 20 days. Then, VEH (TXOX), E2 (TXOXE2), GH (TXOXGH) or E2 plus GH (TXOXE2GH) replacements were performed for additional 7 days. On PND90, animals were sacrificed and the liver extracted for fatty acid analyses as described in Material and Methods. Data are expressed as mean ± SEM for 5 independent samples (different animals) and each independent sample was tested twice. Values represent weight percent of total lipid. Values were submitted to ANOVA followed by post hoc Tukey's test. Values in the same row with different lowercase letters are significantly different with $P < 0.05$.

A) Lipid classes

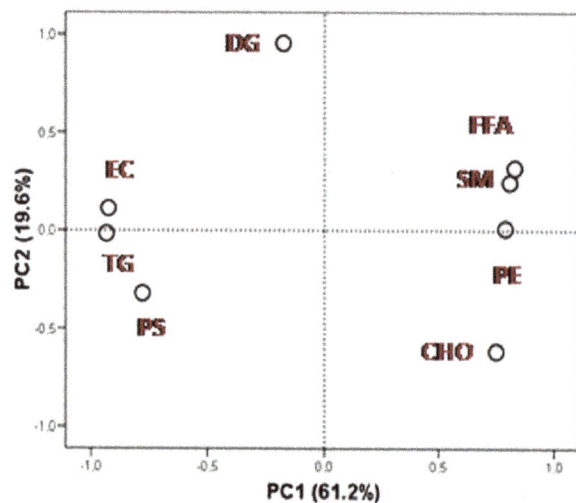

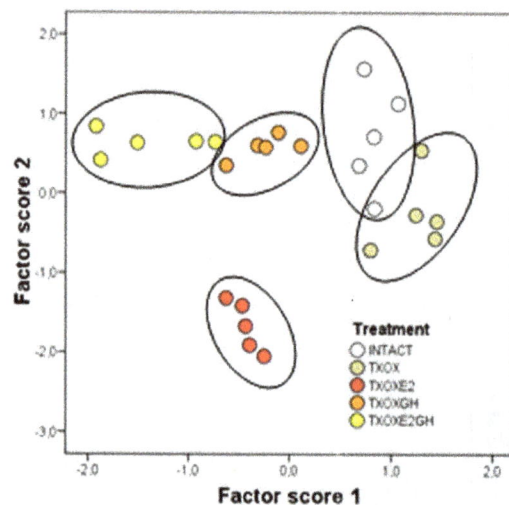

B) Fatty acids (total lipids)

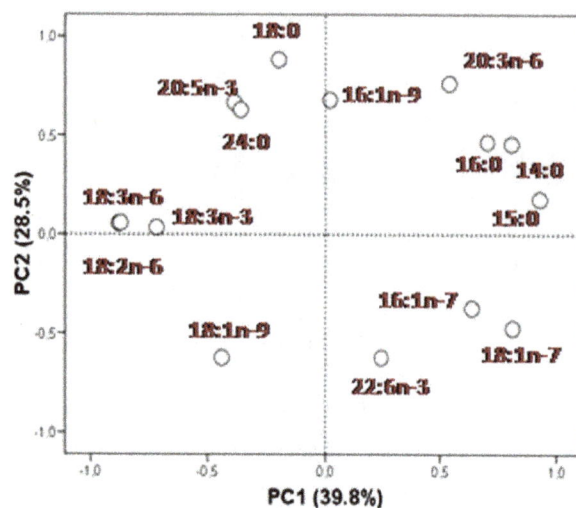

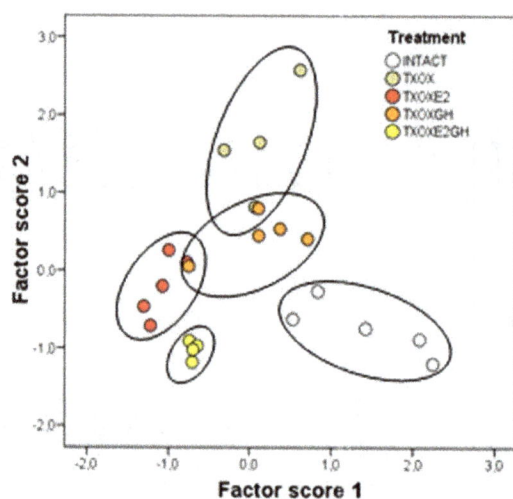

C) Fatty acids (neutral lipids)

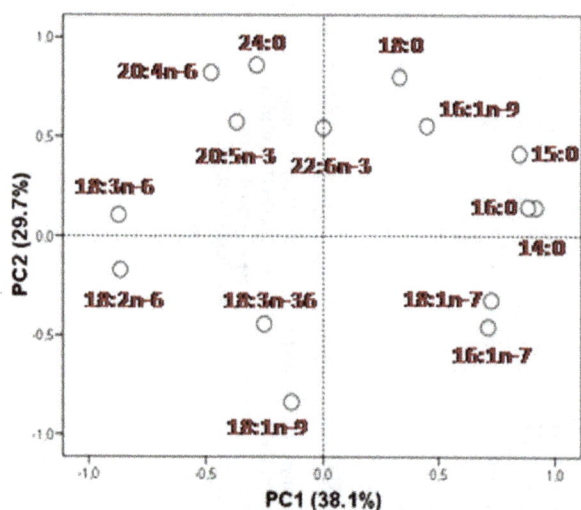

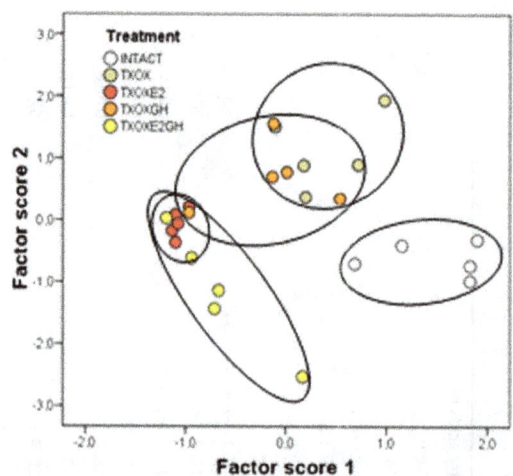

Figure 2. Principal component analysis for liver lipid composition. (A) Lipid classes, (B) Total lipids and (C) Neutral lipids. Left panels represent the factor loadings of principal components 1 (PC1) and 2 (PC2), and right panels the factor scores plots for PC1. Percent values in parentheses indicate the proportion of overall variance explained by each principal component. Each ellipse denotes a hormonal cluster. SM: sphingomyelin, PS: phosphatidylserine, PE: phosphatidylethanolamine, CHO: cholesterol, SE: sterol esters, FFA: free fatty acids, DG: diacylglycerols, TG: triacylglycerols.

= 0.002) and biosynthesis of unsaturated fatty acids (P = 8,2E-04; Bonferroni = 0.02). Among the genes up-regulated by E2 we observed PPARα and PPARα target genes [48,50] such as CYP4A1, CYP4A3, FGF21, carnitine palmitoyltransferase 2 (CPT-2), Scd1, LCFA-CoA ligase 4, acyl-CoA oxidase (ACOX1), acyl-CoA synthetase (ACS), fatty acid translocase (FAT/CD36),

angiopoietin-like 4 (ANGPTL4), ELOVL5, and BAAT. The fatty acid desaturases FAD6 and FAD1 were also upregulated by E2. In contrast, genes involved in the metabolism of C21-steroid hormones (P = 1,6E-05; Bonferroni = 0.001), metabolism of xenobiotic by cytochrome 450 (P = 1,2E-04; Bonferroni = 0.006), glutathione (P = 0.002; Bonferroni = 0.04), and androgen and estrogen

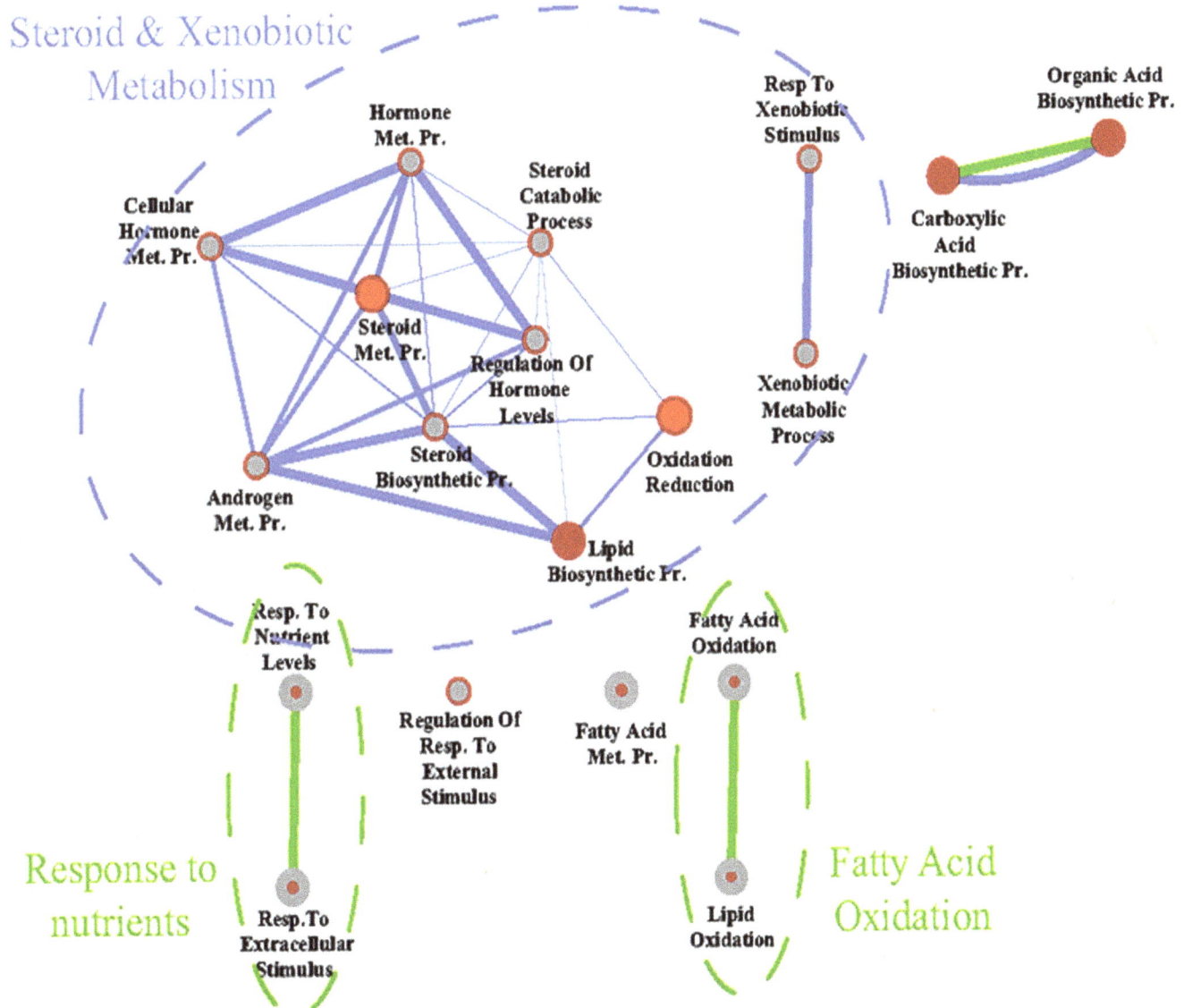

Figure 3. System biological network analyses of the effects of E2 on liver transcriptome in hypothyroid-orchidectomized rats. The genes differentially-expressed in the livers were identified by DNA microarrays as described under Material and Methods. Then, functional and system biological network analysis were performed on the basis of the GO enrichment of differentially-expressed genes in liver using DAVID, and the results depicted using Cytoscape. Node (inner circle) size corresponds to the number of genes up-regulated by E2; node border (outer circle) size corresponds to the number of genes down-regulated by E2; color of the node and border corresponds to the significance of the gene set for up or down regulated genes, respectively (dark red = significantly enriched, light red = enriched no significantly; grey = absent); edge size corresponds to the number of genes that overlap between the two connected gene sets. Green edges correspond to shared up-regulated genes and blue edges correspond to shared down-regulated genes.

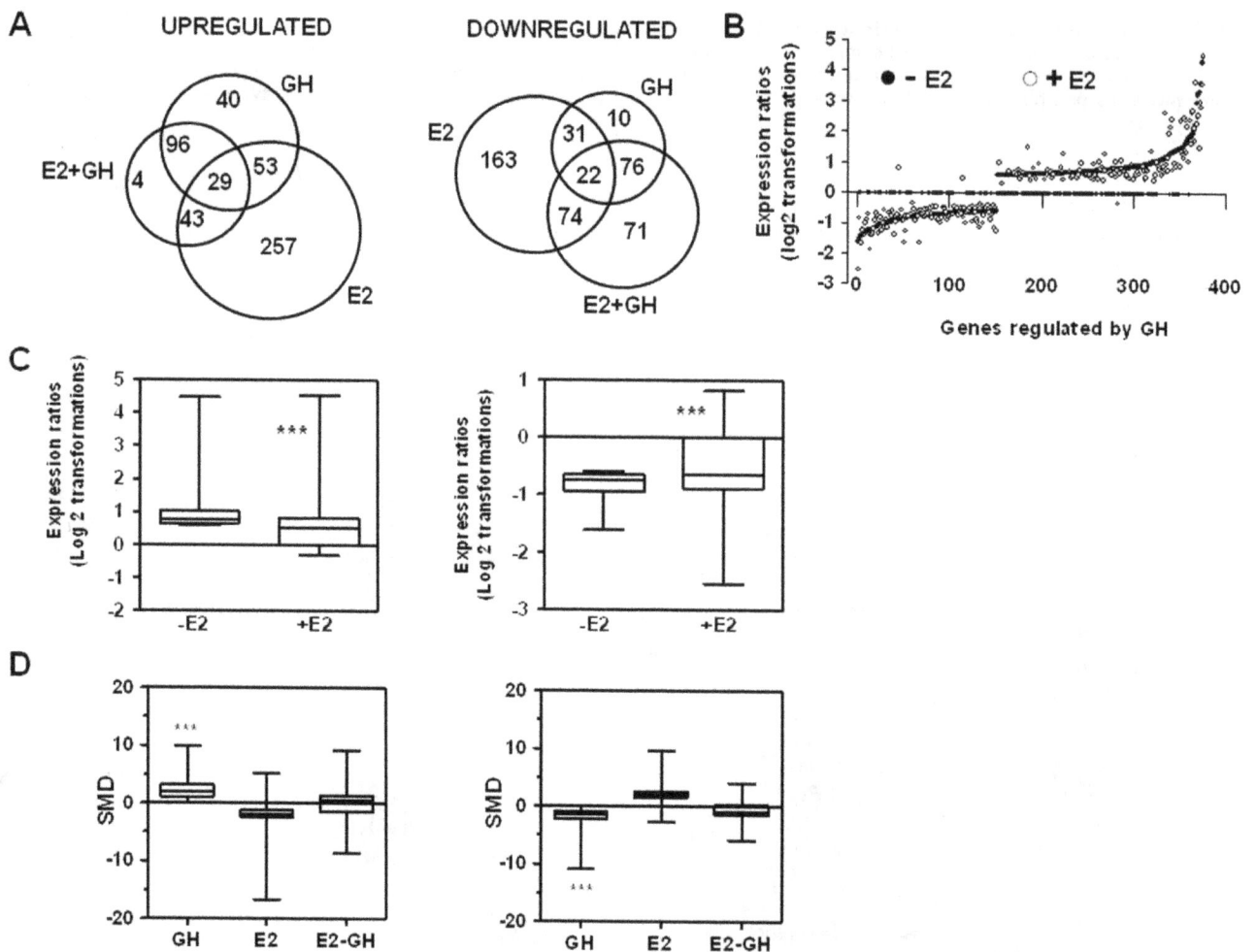

Figure 4. E2 influences the gene expression profiling regulated by GH in hypothyroid-orchidectomized rat liver. Hypothyroid-orchidectomized (TXOX) rats were injected with GH for 7 days in the absence (−E2) or in the presence (+E2) of E2. Differently expressed genes in the livers were identified by DNA microarrays as described under Material and Methods. (A) The number of genes regulated by GH in the absence of E2, by E2, or GH in the presence of E2. The overlapping areas show genes for which expression was altered by GH in the absence or presence of E2. (B) Individual genes are arranged along the X axis according to the value order of decreases and increases in gene expression measured in GH-treated TXOX rats in the absence of E2. The Y axis shows the log 2 ratio of the transcript signals in GH-treated TXOX rats in the absence (−E2) and in the presence (+E2) of E2. (C) Box plot shows a statistical evaluation of the differences in the mean expression changes induced by GH in the absence (−E2) or in the presence (+E2) of E2 for the set of genes induced and repressed by GH treatment in the absence of E2. (D) SAM multiclass analysis was performed to identify GH regulated genes whose mean expression (SMD) values were significantly different from E2 or E2 plus GH-treated TXOX rats. In Box plots, the lines connect the medians, the boxes cover the 25th to 75th percentiles, and the minimum and maximum values are shown by the ends of the bars. ***, $P<0.001$.

(P = 0.002; Bonferroni = 0.05) were significantly down-regulated by E2 [i.e., aldo-keto reductase 1D1 (Akr1d1), epoxide hydrolase 1; glutathione S-transferases (GST) (mu 2, mu 3, mu 4, mu 7, pi 1), CYP3A18, CYP2C23, CYP2E1, CYP17A1, 3β-HSD, 11β-HSD1, and estrogen sulfotransferase (SULT1E)]. Overall, these results reveal an extensive re-programing of liver's transcriptome by long-term E2 treatment of TXOX rats, particularly genes involved in the metabolism of fatty acids and endo-xenobiotics.

Estradiol influences GH-regulated liver transcriptome in hypothyroid male rats

Finally, we performed a genome wide gene expression analysis to better understand the interplay between E2 and GH in liver. We first defined the gene expression changes induced by GH replacement in TXOX rats (Table S3). Second, we analyzed the similarities in gene expression changes induced by treatment with

E2 or GH in TXOX rats and identified 869 significantly (FDR < 5%) regulated transcripts in both cases. A Spearman rank correlation test comparing the E2 and GH effects on these genes yields a positive correlation (r = 0.6287; $P<0.0001$) indicative of strong similarities between E2 and GH effects in the TXOX liver. Accordingly, 94% of these genes were regulated by E2 and GH in the same direction. Third, we obtained transcription profiles from TXOX rats simultaneously treated with both E2 and GH (Table S4). A comparative analysis of genes with altered expression levels in TXOXGH or TXOXE2GH groups revealed considerable reduction in the presence of E2 of the effects induced by GH in TXOX rats (Fig. 4A–B). This was a general phenomenon that affected a large fraction of GH-regulated genes. Accordingly, in the absence of E2, the average expression changes (log$_2$) across four independent hybridizations were 0.98 ± 0.04 and -0.82 ± 0.02 for GH-induced (Fig. 4C) and GH-repressed (Fig. 4D)

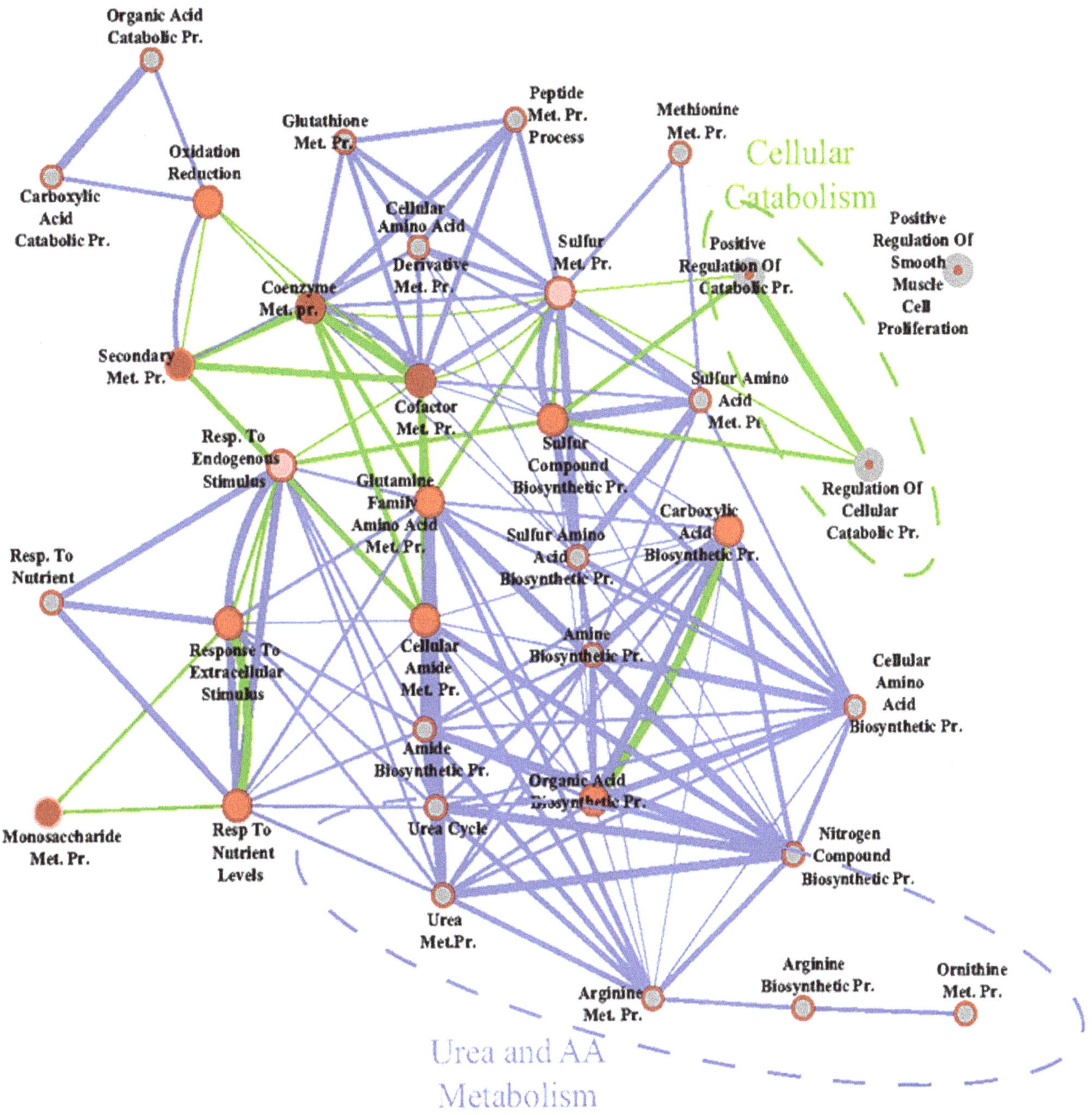

Figure 5. System biological network analyses on GH effects on liver transcriptome in hypothyroid-orchidectomized rats. The differentially-expressed genes in the livers were identified by DNA microarrays as described under Material and Methods. Then, functional and system biological network analysis were performed on the basis of the GO enrichment of differentially-expressed genes in liver using DAVID, and the results depicted using Cytoscape. Node (inner circle) size corresponds to the number of genes up-regulated by GH; node border (outer circle) size corresponds to the number of genes down-regulated by GH; color of the node and border corresponds to the significance of the gene set for up or down regulated genes, respectively (dark red = significantly enriched, light red = enriched no significantly; grey = absent); edge size corresponds to the number of genes that overlap between the two connected gene sets. Green edges correspond to shared up-regulated genes and blue edges correspond to shared down-regulated genes.

genes, respectively, whereas the average fold regulation for the same set of genes in the presence of E2 was 0.60±0.05 (Fig. 4C) and −0.61±0.04 (Fig. 4D). These differences were significant ($P<$ 0.001) and demonstrate an inhibition by E2 of the hepatic response to GH treatment. Finally, SAM multiclass analysis [38] identified genes regulated by GH whose mean expression values were significantly different from those in E2- or E2 plus GH-

treated TXOX rats (Fig. 4E and 4F and Table S5) (e.g., Snapc2, SLC13A2, Apo-H, TNFR, SULT1E1, CYP2C11, CYP2C12, EGFR, Hsd3b6, SPI, alpha 2u-globulin, FTO, Acacb, PPARα, ACOX1, SOCS5, Akr1c14).

Functional analysis [40] revealed that the biological processes over-represented in the list of genes up-regulated by GH in TXOX liver, were connected with the positive regulation of

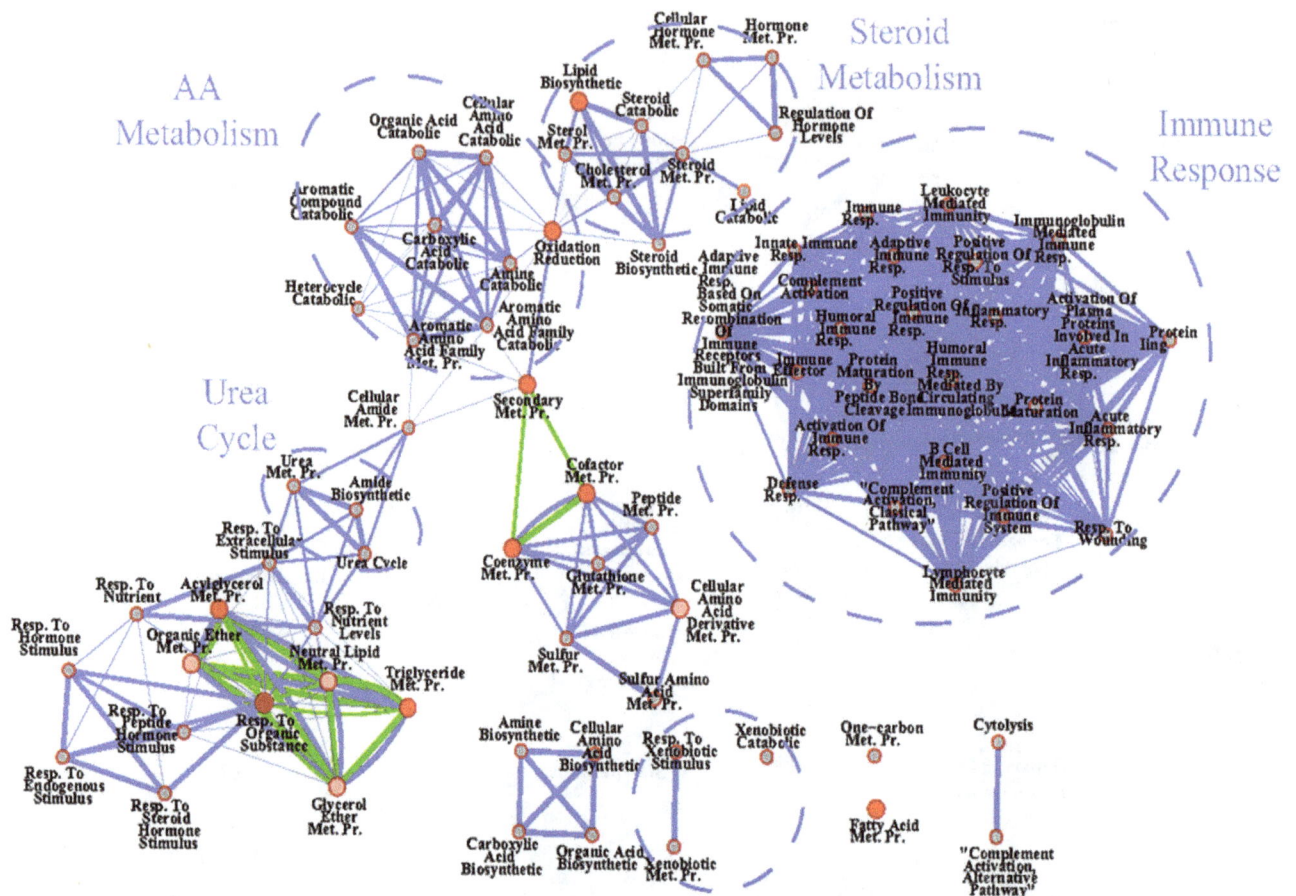

Figure 6. System biological network analyses on GH effects on liver transcriptome in E2-treated hypothyroid-orchidectomized rats. The differentially-expressed genes in the livers were identified by DNA microarrays as described under Material and Methods. Then, functional and system biological network analysis were performed on the basis of the GO enrichment of differentially-expressed genes in liver using DAVID, and the results depicted using Cytoscape. Node (inner circle) size corresponds to the number of genes up-regulated by E2-GH; node border (outer circle) size corresponds to the number of genes down-regulated by E2-GH; color of the node and border corresponds to the significance of the gene set for up or down regulated genes, respectively (dark red = significantly enriched, light red = enriched no significantly; grey = absent); edge size corresponds to the number of genes that overlap between the two connected gene sets. Green edges correspond to shared up-regulated genes and blue edges correspond to shared down-regulated genes.

cellular catabolism (e.g., GCLC, GCK, ABHD5, IGF-I, APOC2, HSP90AB1) whereas the metabolism of aminoacids and urea (e.g., CTH, ARG1, ASS1, OTC, ASL, CPS1, MAT1A, MTR, BHMT, PAH) was significantly down-regulated (Fig. 5). When these changes were mapped to cellular pathways (KEEG), we confirmed a significant connection with the metabolism of aminoacids and urea, metabolism of xenobiotics by P450, or PPARα signaling (Table 4). In contrast to E2, GH repressed the PPARα signaling pathway. Notably, in the presence of E2, the biological functions and network regulated by GH varied notably (Fig. 6). Thus, whereas GH still reduced the expression of genes involved in urea and amino acid metabolism in the presence of E2 [e.g., carbamoyl-phosphate synthetase-1 (CPS1), ornithine carbamoyl-transferase (OTC); argininosuccinate synthase (ASS1)], it was no longer able to induce the expression of genes related to cell catabolism [e.g., glutamate-cysteine ligase (GCLC) and APOC2] (Table 5 and Table S4). In contrast, genes involved in the metabolism of steroids and xenobiotics (e.g., Gst2, Gst-m3, Gst-m4, EPHX1, CYP2E1, Gst-p1, Gst-m7) downregulated by E2 but not by GH alone are even further inhibited by the combined treatment. Finally, when E2 was present, GH exerted a significant influence on immune system response (Fig. 6). In particular, the

mRNA expression levels of genes involved in complement activation, lymphocyte immune response, wounding or acute inflammatory response were significantly reduced (e.g., C9, C4a, AHCY, ASS1, MASP1, C5, CLU, CST3, AII3, C4BPB, SERPING1, C1S, C4BPA, C8B, SDC1, IL10RB, IldR1, SLC7A2, PIGR, sialomucin, IGFBP1, or CFI) (Table 5 and Table S4). Taken together, these data indicate that E2 and GH may be effective regulators of immune system contributing to the maintenance of immune homeostasis under conditions of immunological stress to reduce the susceptibility to stress-induced disease by negative immunoregulators.

Discussion

In this study, we show that E2 and GH replacements in hypothyroid male rats have a significant impact on lipid content and transcriptome in the liver and that E2 exerts a marked influence on GH-regulated endocrine, metabolic, immune, and gender specific responses in the liver.

Hypothyroidism impaired body weight gain and decreased circulating levels of IGF-I and biological markers of GH-STAT5b signaling activity in the liver (i.e., mRNA levels of IGF-I, ASL,

Table 4. Cellular pathways regulated by GH in hypothyroid-orchidectomized rat liver.

Pathway	up-regulated genes	down-regulated genes	P	B
Arginine and proline metabolism	GLUD-1	CPS-1, mitochondrial; OCT; ASS-1; arginase; aminotransferase	2,6E-05	3,1E-03
Urea cycle and metabolism of amino groups		CPS-1, mitochondrial; OCT; ASS-1; arginase	1,3E-03	3,8E-02
Glycine, serine and threonine metabolism	choline kinase α; glycyl-tRNA synthetase	serine dehydratase; betaine-homocysteine methyltransferase; cystathionase; glycine N-methyltransferase	1,6E-03	3,7E-02
Drug metabolism - Metabolism of xenobiotics by P450	CYP2C11; CYP2D2	CYP2A1; CYP2C23; ADH 1; GST (k1 and A3)	1,0E-03	4,0E-02
Methionine metabolism	methionine-tRNA synthetase	methionine adenosyltransferase I, alpha; betaine-homocysteine methyltransferase; cystathionase; 5-methyltetrahydrofolate-homocysteine methyltransferase	9,4E-04	5,4E-02
Retinol metabolism	CYP2C11	CYP2A1; CYP2C23; ADH 1; Dgat2	4,8E-03	9,1E-02
Caffeine metabolism	NAT-2	CYP2A1; urate oxidase	1,1E-02	1,7E-01
PPAR signaling	CPT-2; ANGPTL4	CYP27A1; CYP8B1; PEPK 1; ACOX-1	1,3E-02	1,8E-01
Cysteine metabolism	LDH-A	serine dehydratase; cystathionase	2,8E-02	2,9E-01
p53 signaling pathway	IGF-1; GADD-inducible, α and γ	cyclin B3; sestrin 1	4,4E-02	3,9E-01

The genes differentially-expressed in the livers were identified by DNA microarrays as described under Material and Methods. DAVID was used to identify the hepatic KEEG pathways that were affected by GH. The table shows pathway name, up- and down-regulated genes, P value, and corrected P value (Benjamini). Abbreviations: carbamoyl-phosphate synthetase-1 (CPS1); ornithine carbamoyltransferase (OTC); argininosuccinate synthase (ASS1); acyl-CoA oxidase 1, palmitoyl (ACOX-1); angiopoietin-like 4 (ANGPTL4); diacylglycerol O-acyltransferase 2 (Dgat2); glutamate dehydrogenase 1 (GLUD-1); Lactate dehydrogenase A (LDH-A); stearoyl-CoA desaturase (Scd); estrogen sulfotransferase (SULT).

SOCS2, CIS, and CYP2C11) [51]. These changes were totally or partially restored by intermittent GH administration to TXOX rats. However, the effects of GH were largely prevented by E2 which is in line with the negative effects of estrogens on continuously GH administration in hypophysectomized female rats [52]. The positive effects of E2 on hepatic SOCS2, CIS, and FGF21 transcripts (see Fig.1) suggest that E2 might prevent the activation of GH-STAT5b signaling in liver through induction of these negative regulators of GH signaling [14,48]. Similarly, estrogen administration in humans can prevent the GH-induced increase in IGF-I, IGFBP-3, lipid oxidation, and protein synthesis [22,23]. The effects of hypothyroidism on growth are associated, in part, with an increased hepatic amino acid catabolism and urea synthesis [53]. Biological network analysis shows that intermittent GH administration to TXOX rats causes a positive regulation of cellular catabolism, whereas the genes involved in the metabolism of amino acids and urea (i.e., OTC, ASS1, aminotransferases, and methyltransferases) are significantly down-regulated. This is in line with the positive effects of GH on nitrogen balance, which have been previously studied in hypophysectomized rats [54,55,56].

GH serves as an anabolic hormone that promotes lipolysis and prevents lipogenesis in adipose tissue, which increases the availability of FFA for energy expenditure [56]. E2 is also able to interfere with this process by preventing the induction of some genes related to fat utilization, such as ApoC2, which activates the enzyme LPL that hydrolyzes TG. Therefore, E2 actions in liver can impact the peripheral metabolic actions of GH.

Lipogenesis is often increased in situations of reduced energy expenditure such as hypothyroidism, GH deficiency, E2 deficiency, or aging [57]. Accordingly, our analysis of the hepatic lipid content revealed that TXOX rats contained significantly increased levels of total SFA compared to INTACT rats. E2 replacement did

not modify the mRNA expression levels of key regulators of hepatic lipogenesis [i.e., Sterol regulatory element binding protein (SREBP)1c, acetyl-Co A carboxylase alpha (ACC), fatty-acid synthase (FAS)] [58], whereas it activated a PPARα transcriptional program that promotes fatty acid catabolism in liver [50,59]. This was evidenced by the E2 increased expression of the PPARα gene itself and the PPARα target genes involved in the β/ω-oxidation of fatty acids (i.e., CTE-I, CPT-2, Fasd6, Fasd1, Fasd2, Scd1, ACOX1, ECH1, BAAT, FGF21, CYP4A1, CYP4A3) (Table S2). Accordingly, E2 replacement caused a significant reduction in SFAs. Overall these findings are indicative of a positive crosstalk between E2 and PPARα that is supported by multiple independent studies [2,60,61,62]. Interestingly, despite the increased expression of genes involved in β-oxidation, we detected a significant increase in hepatic TG content in E2 treated TXOX rats, which is likely explained by effects on lipid transport. The first step of long chain fatty acids uptake is its translocation across the plasma membrane. Notably, E2 increased the transcription of several known PPARα activated genes encoding proteins that have been implicated in fatty acids uptake and activation such as CD36, ACSL4 and SLC27A5 (FATP5) [63,64]. We have previously demonstrated that the fatty acid transporter CD36 is predominantly expressed in female rat livers and proposed that this sexual dimorphism depends on the GH secretion pattern, which can be influenced by E2 treatment. E2 also increased transcripts of the SLC27A5 gene which encodes FATP5, an fatty acid transporter that is an acyl-CoA synthetase (bile acid ligase) that catalyzes the conjugation of bile acids with amino acids before excretion into bile canaliculi [65]. Following fatty acids uptake, the first step for the intracellular use of long chain fatty acids is its esterification with CoA. This reaction is catalyzed by acyl-CoA synthetases such as ACSL4 which was also induced by E2 in TXOX rat liver. The produced

Table 5. E2 influences the cellular pathways regulated by GH in hypothyroid-orchidectomized rat liver.

Pathway name	up-regulated genes	down-regulated genes	P	B
Drug metabolism-Metabolism of xenobiotics by P450	FMO 3; CYP2C11; CYP2C24; NAT	CYP2A1; CYP2C23; CYP2E1; UGT (2B17; 2B36 and 1A); ADH 1; GST (mu 2; mu 3; mu 4; mu 7; pi 1; A3); uridine phosphorylase 2; epoxide hydrolase 1, microsomal	9,6E-10	1,2E-07
Complement and coagulation cascades	fibrinogen α chain	complement [C1; C4; C5; C8; C9]; serine (or cysteine) peptidase inhibitor; mannan-binding lectin serine peptidase 1	6,5E-07	2,8E-05
Retinol metabolism	CYP2C11; CYP2C24	CYP2A1; CYP2C23; CYP2A1; UGT (2B17; 2B36; and 1A); ADH A1; retinal pigment epithelium 65	1,0E-06	3,3E-05
Nitrogen metabolism	GLUD 1	CPS-1, mitochondrial; carbonic anhydrase (3 and 8); cystathionase; histidine ammonia lyase	1,7E-04	4,5E-03
PPAR signaling	CPT-1a; CPT-2; ANGPTL4	CYP7A1; CYP8B1; PEPK 1; CD36; LCA-CoA synthetase 1; Scd2	2,6E-04	5,7E-03
Fatty acid metabolism	CPT-1a; CPT-2	glutaryl-CoA dehydrogenase; ADH 1; enoyl CoA hydratase, short chain, 1, mitochondrial; ADH 2 (mitochondrial); LCA-CoA synthetase 1	4,4E-04	8,0E-03
Arginine and proline metabolism	GLUD1; spermidine synthase	CPS-1, mitochondrial; OCT; ASS-1; aminotransferase; ADH-2	1,5E-03	2,2E-02
Steroid hormone biosynthesis		CYP7A1; UGT (2B17; 2B36; 1A); SULT; HSD11B1	3,4E-03	4,3E-02
Tryptophan metabolism		tryptophan 2,3-dioxygenase; glutaryl-CoA dehydrogenase; kynurenine 3-monooxygenase; catalase; enoyl CoA hydratase, short chain, 1, mitochondrial; ADH-2	3,4E-03	4,3E-02
Glutathione metabolism	spermidine synthase	isocitrate dehydrogenase 1 (NADP+), soluble; GST (A3; mu2; mu 3; mu 4; mu 7; pi 1)	6,6E-03	7,4E-02

The genes differentially-expressed in the livers were identified by DNA microarrays as described under Material and Methods. DAVID was used to identify the hepatic KEEG pathways that were affected by GH in the presence of E2. The table shows pathway name, up- and down-regulated genes, P value, and corrected P value (Benjamini). Abbreviations: carbamoyl-phosphate synthetase-1 (CPS1); argininosuccinate synthase (ASS1); acyl-CoA oxidase 1, palmitoyl (ACOX-1); angiopoietin-like 4 (ANGPTL4); diacylglycerol O-acyltransferase 2 (Dgat2); glutamate dehydrogenase 1 (GLUD1); ornithine carbamoyltransferase (OTC); stearoyl-CoA desaturase (Scd); estrogen sulfotransferase (SULT).

acyl-CoAs are substrates for β-oxidation but also can prime the synthesis of TG, phospholipids, CE, and ceramides and therefore are also a primary source of signaling molecules [66]. The notion that E2 may regulate the formation of lipid signaling intermediaries is supported by the stimulation of fatty acids elongase-5 (Elovl5). Elovl functions with fatty acid desaturases to generate many of the long-chain PUFAs assimilated into cellular lipids (i.e., 20:4n-6 and 22:6n-3). However, it is worth mentioning that E2 administration did not alter VLCPUFA metabolism because the levels of 20:4n-6, 20:5n-3 and 22:6n-3 remained similar to values in the TXOX group. It has been reported that E2 might play a critical role in lipogenesis and Scd1 transcription [2], a gene that encodes a rate-limiting enzyme to generate MUFAs such as 18:1 n−9 and 16:1 n−7. Previous studies have reported that the absence of E2 or ERα in rats provoked a profound increase in lipogenesis and Scd1 transcription [67], which suggests that E2 inhibits Scd1 transcription. Interestingly, the antilipogenic effect of E2 therapy, while maintaining efficient TG export and reduced phospholipid transfer protein, has been reported to depend on hepatic ERα [13,61]. Our study, however, shows that E2 increased the Scd1 gene expression and that this effect was paralleled by reduced hepatic content of 18:0 and increased of 18:1 n−9 (the main product of SCD reaction) contents, in total and, especially, in neutral lipids compared with TXOX animals, which indicates that E2 modulates SCD1 activity in TXOX liver. Surprisingly, E2 also downregulated Scd2 gene expression in TXOX rat livers. The significance of this opposed transcriptional

regulation of Scd genes is unknown, but given that transcript levels of Scd1 are about 1800 times higher than that of Scd2 in the rat liver [68], changes in 18:1n-9 and 18:0 must be entirely attributed to variations in Scd1 gene expression. Overall, the changes in the lipid composition and gene expression profile seen in E2-treated TXOX rats support the finding that E2-PPARα functional interactions play a physiological role in the regulation of hepatic lipid metabolism.

E2 has the ability to reduce circulating CHO in women and in animal models fed on a high-fat diet [69]. However, E2 was unable to efficiently reverse hypercholesterolemia or hypotriglyceridemia in TXOX rats. This result may be due to the fact that E2 reduced expression levels of several transporters of CHO (and CE), including ApoB and ABCA1 in TXOX rats, which most likely contributed to maintaining an increased hepatic level CE. E2 may also induce intracellular CHO mobilization by modulating enzymes involved in CE and CHO synthesis and/or turnover [57,70]. Distinct enzymes can catalyze the CHO to CE conversion in liver: lecithin:cholesterol acyltransferase (LCAT), which uses phosphatidylcholine (PC) as a source of acyl changes and ACAT, which uses acyl-CoA. Because the levels of lysophosphatidylcholine (LPC) were undetectable in all groups, our initial conclusion was that E2 stimulated the ACAT2 reaction to increase CE. However, we did not detect changes in the expression level of the ACAT gene, which did not discard posttranslational modification of enzymes in the CE cycle in the liver from E2-treated TXOX rats.

An increased level of hepatic CE, together with the increased TG and decreased FFA hepatic contents in GH-treated TXOX rats, resemble the effects of E2 on hepatic lipid composition and suggest that some effects of E2 might be GH mediated. A striking consequence of the combined replacement with E2 and GH is the complete restoration of MUFA levels from total and neutral lipids, an effect attributable to the increase in 18:1n9, likely through alteration of Δ9 desaturase expression. Moreover, GH and E2 increased hepatic CE and the combined effect of the two hormones were additive with regard to CE because its levels doubled those found in INTACT animals and were approximately 30% higher than in the E2 or GH groups which indicates a more efficient hepatic CHO metabolism. Accordingly, in the presence of E2, GH reduced hepatic CHO content compared not only to the TXOX group but also in relation to the E2- or GH-treated TXOX groups. The hepatic content of TG was, however, significantly increased by GH in E2-pretreated TXOX rats, which suggests that combined treatment by E2 and GH dramatically enhances lipogenesis. It is known that in contrast with its lipolytic effects in adipose tissue, GH exerts lipogenic actions in liver through stimulation of SREBP1, which is usually accompanied by increased hepatic TG (VLDL) secretion [56]. Indeed, our lipid profiling analysis suggested that intermittent GH administration to TXOX rats increased lipogenesis in the liver. However, in contrast to the effects of a continuous infusion of GH in hypophisectomized rats [55], intermittent GH administration to TXOX rats did not increase SERBP1, whereas several genes involved in fatty acids transport (e.g., FABP) and the biosynthesis of unsaturated fatty acids from 18:2n-6 and 18:3n-3 (e.g., fatty acid desaturases 4, 5 and 6) were induced. Interestingly, intermittent GH administration to TXOX rats down-regulated the expression of the lipin gene, an SREBP1c target gene, which is critical in the regulation of cellular levels of DG and TG and a key regulator of fatty acid oxidation in adipose tissue, skeletal muscle, and liver tissue [71]. These findings support the hypothesis that the female pattern of GH administration is a more efficient stimulus to induce lipogenic effects in the liver than the male pattern [72,73]. Another mechanism whereby GH might promote lipogenesis in the liver is through the down-regulation of lipid oxidation. We have previously shown that continuous GH administration to hypophysectomized [55] and to old-intact [74] male rats inhibited PPARα. Accordingly, our lipidomic and genomic analysis showed that intermittent GH administration to TXOX rats also leads to down-regulation of the PPARα signaling pathway. In particular, GH represses the expression of PPARα itself, ACOX-1, CPT-1, FGF21, and several members of the CYP4A family, which are involved in fatty acid oxidation.

In summary, our study adds novel data that highlight the impact of subcutaneous E2 administration on liver physiology and its interplay with GH. These results highlight the role of E2 as a critical regulator of liver metabolism in mammals and add further weight to the hypothesis that E2 acts as an important regulator of GH actions in the liver. The E2-GH interplay in the liver is relevant because of the physiological roles that these hormones have in mammals and the widespread use of estrogen and estrogen-related compounds in human. Notably, this is the first study to demonstrate that hepatic lipid profiles are endowed with singular fingerprints that may be used to segregate different groups with altered hormone status. This includes different hormonal replacements (E2 or GH) that induced overlapping changes in gene expression. Therefore, liver lipid profiling can serve to identify cryptic hormone deficiencies or exposure to hormones or hormone-like substances.

Supporting Information

Table S1 Genes and primer sequences (5′- 3′) used for real-time PCR.

Table S2 Hepatic genes regulated by E2 in hypothyroid-orchidectomized rats. The administrations of vehicle (TXOX) or E2 (TXOXE2) to TXOX rats were described in Material and Methods. Then, differently expressed genes in the livers were identified by DNA microarrays. The analysis is based on the SAM statistical technique and differentially-expressed genes were discovered using a FDR less than 5% and a mean ratio of $\log 2 > |0.58|$. The table shows ENSEMBL gene ID, Unigene/ Refseq, gene symbol, gene description, R (TXOXE2/TXOX), SD, and q (%).

Table S3 Hepatic genes regulated by GH in hypothyroid-orchidectomized rats. The administrations of vehicle (TXOX) or GH (TXOXGH) to TXOX rats were described in Material and Methods. Then, differently expressed genes in the livers were identified by DNA microarrays. The analysis is based on the SAM statistical technique and differentially-expressed genes were discovered using a FDR less than 5% and a mean ratio of $\log 2 > |0.58|$. The table shows ENSEMBL gene ID, gene symbol, gene description, R (TXOXGH/TXOX), SD, and q (%).

Table S4 Hepatic genes regulated by GH in E2-treated hypothyroid-orchidectomized rats. The administrations of vehicle (TXOX) or E2 plus GH (TXOXE2GH) in TXOX rats were described in Material and Methods. Then, differently expressed genes in the livers were identified by DNA microarrays. The analysis is based on the SAM statistical technique and differentially-expressed genes were discovered using a FDR less than 5% and a mean ratio of $\log 2 > |0.58|$. The table shows ENSEMBL gene ID, gene symbol, gene description, R (TXOXE2GH/TXOX), SD, and q (%).

Table S5 Hepatic genes regulated by GH whose mean expression values are different from those in E2- and E2 plus GH-treated hypothyroid-orchidectomized rats. Hormonal replacements with GH (TXOXGH), E2 (TXOXE2) or E2 plus GH (TXOXE2GH) in TXOX rats were described in Material and Methods. Then, differently expressed genes in the livers were identified by DNA microarrays. SAM multiclass analysis was performed to identify GH regulated genes (TXOXGH) whose mean expression values were significantly different from E2 (TXOXE2)- or E2 plus GH (TXOXE2GH)-treated TXOX rats. The table shows ENSEMBL ID, Unigene/ Refseq, gene symbol, gene description, R (mean expression), and q (%).

Acknowledgments

The excellent technical assistance of Natalia Álvarez-Valtueña (Department of Clinical Biochemistry, Insular University Hospital of Gran Canaria) is greatly appreciated.

Author Contributions

Conceived and designed the experiments: LF-P MD AF-M DI JCD-C. Performed the experiments: LF-P RS-F MM-G CM-D IG BG MD. Analyzed the data: LF-P MM-G BG MD DI AF-M JCD-C. Wrote the paper: LF-P BG MD DI AF-M JCD-C.

References

1. Simpson ER, McInnes KJ (2005) Sex and fat—can one factor handle both? Cell Metab 2: 346–347.

2. Barros RP, Gustafsson JA (2011) Estrogen receptors and the metabolic network. Cell Metab 14: 289–299.

3. Maher AC, Akhtar M, Tarnopolsky MA (2010) Men supplemented with 17beta-estradiol have increased beta-oxidation capacity in skeletal muscle. Physiol Genomics 42: 342–347.

4. Devries MC, Hamadeh MJ, Graham TE, Tarnopolsky MA (2005) 17beta-estradiol supplementation decreases glucose rate of appearance and disappearance with no effect on glycogen utilization during moderate intensity exercise in men. J Clin Endocrinol Metab 90: 6218–6225.

5. Hamadeh MJ, Devries MC, Tarnopolsky MA (2005) Estrogen supplementation reduces whole body leucine and carbohydrate oxidation and increases lipid oxidation in men during endurance exercise. J Clin Endocrinol Metab 90: 3592–3599.

6. Diaz M, Ramirez CM, Marin R, Marrero-Alonso J, Gomez T, et al. (2004) Acute relaxation of mouse duodenum [correction of duodenun] by estrogens. Evidence for an estrogen receptor-independent modulation of muscle excitability. Eur J Pharmacol 501: 161–178.

7. Smith EP, Boyd J, Frank GR, Takahashi H, Cohen RM, et al. (1994) Estrogen resistance caused by a mutation in the estrogen-receptor gene in a man. N Engl J Med 331: 1056–1061.

8. Rochira V, Balestrieri A, Faustini-Fustini M, Carani C (2001) Role of estrogen on bone in the human male: insights from the natural models of congenital estrogen deficiency. Mol Cell Endocrinol 178: 215–220.

9. Rochira V, Balestrieri A, Faustini-Fustini M, Borgato S, Beck-Peccoz P, et al. (2002) Pituitary function in a man with congenital aromatase deficiency: effect of different doses of transdermal E2 on basal and stimulated pituitary hormones. J Clin Endocrinol Metab 87: 2857–2862.

10. Rochira V, Zirilli L, Maffei L, Premrou V, Aranda C, et al. (2010) Tall stature without growth hormone: four male patients with aromatase deficiency. J Clin Endocrinol Metab 95: 1626–1633.

11. Della Torre S, Rando G, Meda C, Stell A, Chambon P, et al. (2011) Amino acid-dependent activation of liver estrogen receptor alpha integrates metabolic and reproductive functions via IGF-1. Cell Metab 13: 205–214.

12. Villa A, Della Torre S, Stell A, Cook J, Brown M, et al. (2012) Tetradian oscillation of estrogen receptor alpha is necessary to prevent liver lipid deposition. Proc Natl Acad Sci U S A 109: 11806–11811.

13. Pedram A, Razandi M, O'Mahony F, Harvey H, Harvey BJ, et al. (2013) Estrogen reduces lipid content in the liver exclusively from membrane receptor signaling. Sci Signal 6: ra36.

14. Leung KC, Johannsson G, Leong GM, Ho KK (2004) Estrogen regulation of growth hormone action. Endocr Rev 25: 693–721.

15. Mode A, Gustafsson JA (2006) Sex and the liver - a journey through five decades. Drug Metab Rev 38: 197–207.

16. Rico-Bautista E, Flores-Morales A, Fernandez-Perez L (2006) Suppressor of cytokine signaling (SOCS) 2, a protein with multiple functions. Cytokine Growth Factor Rev 17: 431–439.

17. Zadjali F, Santana-Farre R, Vesterlund M, Carow B, Mirecki-Garrido M, et al. (2012) SOCS2 deletion protects against hepatic steatosis but worsens insulin resistance in high-fat-diet-fed mice. FASEB J 26: 3282–3291.

18. Fan Y, Menon RK, Cohen P, Hwang D, Clemens T, et al. (2009) Liver-specific deletion of the growth hormone receptor reveals essential role of growth hormone signaling in hepatic lipid metabolism. J Biol Chem 284: 19937–19944.

19. Sos BC, Harris C, Nordstrom SM, Tran JL, Balazs M, et al. (2011) Abrogation of growth hormone secretion rescues fatty liver in mice with hepatocyte-specific deletion of JAK2. J Clin Invest 121: 1412–1423.

20. Barclay JL, Nelson CN, Ishikawa M, Murray LA, Kerr LM, et al. (2011) GH-dependent STAT5 signaling plays an important role in hepatic lipid metabolism. Endocrinology 152: 181–192.

21. Cummings DE, Merriam GR (2003) Growth hormone therapy in adults. Annu Rev Med 54: 513–533.

22. Wolthers T, Hoffman DM, Nugent AG, Duncan MW, Umpleby M, et al. (2001) Oral estrogen antagonizes the metabolic actions of growth hormone in growth hormone-deficient women. Am J Physiol Endocrinol Metab 281: E1191–1196.

23. Munzer T, Rosen CJ, Harman SM, Pabst KM, St Clair C, et al. (2006) Effects of GH and/or sex steroids on circulating IGF-I and IGFBPs in healthy, aged women and men. Am J Physiol Endocrinol Metab 290: E1006–1013.

24. Fitts JM, Klein RM, Powers CA (1998) Comparison of tamoxifen effects on the actions of triiodothyronine or growth hormone in the ovariectomized-hypothyroid rat. J Pharmacol Exp Ther 286: 392–402.

25. Fitts JM, Klein RM, Powers CA (2001) Estrogen and tamoxifen interplay with T(3) in male rats: pharmacologically distinct classes of estrogen responses affecting growth, bone, and lipid metabolism, and their relation to serum GH and IGF-I. Endocrinology 142: 4223–4235.

26. Sap J, de Magistris L, Stunnenberg H, Vennstrom B (1990) A major thyroid hormone response element in the third intron of the rat growth hormone gene. EMBO J 9: 887–896.

27. Lopez-Guerra A, Chirino R, Navarro D, Fernandez L, Boada LD, et al. (1997) Estrogen antagonism on T3 and growth hormone control of the liver microsomal low-affinity glucocorticoid binding site (LAGS). J Steroid Biochem Mol Biol 63: 219–228.

28. Loria P, Carulli L, Bertolotti M, Lonardo A (2009) Endocrine and liver interaction: the role of endocrine pathways in NASH. Nat Rev Gastroenterol Hepatol 6: 236–247.

29. Nanto-Salonen K, Muller HL, Hoffman AR, Vu TH, Rosenfeld RG (1993) Mechanisms of thyroid hormone action on the insulin-like growth factor system: all thyroid hormone effects are not growth hormone mediated. Endocrinology 132: 781–788.

30. Takashima K, Mizukawa Y, Morishita K, Okuyama M, Kasahara T, et al. (2006) Effect of the difference in vehicles on gene expression in the rat liver—analysis of the control data in the Toxicogenomics Project Database. Life Sci 78: 2787–2796.

31. Geng XC, Li B, Zhang L, Song Y, Lin Z, et al. (2012) Corn oil as a vehicle in drug development exerts a dose-dependent effect on gene expression profiles in rat thymus. J Appl Toxicol 32: 850–857.

32. Oscarsson J, Olofsson SO, Bondjers G, Eden S (1989) Differential effects of continuous versus intermittent administration of growth hormone to hypophysectomized female rats on serum lipoproteins and their apoproteins. Endocrinology 125: 1638–1649.

33. Waxman DJ, Pampori NA, Ram PA, Agrawal AK, Shapiro BH (1991) Interpulse interval in circulating growth hormone patterns regulates sexually dimorphic expression of hepatic cytochrome P450. Proc Natl Acad Sci U S A 88: 6868–6872.

34. Fabelo N, Martin V, Gonzalez C, Alonso A, Diaz M (2012a) Effects of oestradiol on brain lipid class and Fatty Acid composition: comparison between pregnant and ovariectomised oestradiol-treated rats. J Neuroendocrinol 24: 292–309.

35. Almansa E, Sanchez JJ, Cozzi S, Rodriguez C, Diaz M (2003) Temperature-activity relationship for the intestinal Na+-K+-ATPase of Sparus aurata. A role for the phospholipid microenvironment? J Comp Physiol B 173: 231–237.

36. Henriquez-Hernandez LA, Flores-Morales A, Santana-Farre R, Axelson M, Nilsson P, et al. (2007) Role of pituitary hormones on 17alpha-ethinylestradiol-induced cholestasis in rat. J Pharmacol Exp Ther 320: 695–705.

37. Quackenbush J (2002) Microarray data normalization and transformation. Nat Genet 32 Suppl: 496–501.

38. Tusher VG, Tibshirani R, Chu G (2001) Significance analysis of microarrays applied to the ionizing radiation response. Proc Natl Acad Sci U S A 98: 5116–5121.

39. Edgar R, Domrachev M, Lash AE (2002) Gene Expression Omnibus: NCBI gene expression and hybridization array data repository. Nucleic Acids Res 30: 207–210.

40. Huang da W, Sherman BT, Lempicki RA (2009) Systematic and integrative analysis of large gene lists using DAVID bioinformatics resources. Nat Protoc 4: 44–57.

41. Cline MS, Smoot M, Cerami E, Kuchinsky A, Landys N, et al. (2007) Integration of biological networks and gene expression data using Cytoscape. Nat Protoc 2: 2366–2382.

42. Pfaffl MW (2001) A new mathematical model for relative quantification in real-time RT-PCR. Nucleic Acids Res 29: e45.

43. Rozen S, Skaletsky H (2000) Primer3 on the WWW for general users and for biologist programmers. Methods Mol Biol 132: 365–386.

44. Raykov T, Marcoulides GA (2008) Introduction to applied multivariate analysis. New York: Taylor & Francis.

45. Fabelo N, Martin V, Marin R, Santpere G, Aso E, et al. (2012b) Evidence for Premature Lipid Raft Aging in APP/PS1 Double-Transgenic Mice, a Model of Familial Alzheimer Disease. J Neuropathol Exp Neurol 71: 868–881.

46. Alonso A, Fernandez R, Ordonez P, Moreno M, Patterson AM, et al. (2006) Regulation of estrogen receptor alpha by estradiol in pregnant and estradiol treated rats. Steroids 71: 1052–1061.

47. Vidal OM, Merino R, Rico-Bautista E, Fernandez-Perez L, Chia DJ, et al. (2007) In vivo transcript profiling and phylogenetic analysis identifies suppressor of cytokine signaling 2 as a direct signal transducer and activator of transcription 5b target in liver. Mol Endocrinol 21: 293–311.

48. Inagaki T, Lin VY, Goetz R, Mohammadi M, Mangelsdorf DJ, et al. (2008) Inhibition of growth hormone signaling by the fasting-induced hormone FGF21. Cell Metab 8: 77–83.

49. Diaz ML, Fabelo N, Marin R (2012) Genotype-induced changes in biophysical properties of frontal cortex lipid raft from APP/PS1 transgenic mice. Front Physiol 3: 454.

50. Badman MK, Pissios P, Kennedy AR, Koukos G, Flier JS, et al. (2007) Hepatic fibroblast growth factor 21 is regulated by PPARalpha and is a key mediator of hepatic lipid metabolism in ketotic states. Cell Metab 5: 426–437.

51. Hosui A, Hennighausen L (2008) Genomic dissection of the cytokine-controlled STAT5 signaling network in liver. Physiol Genomics 34: 135–143.

52. Borski RJ, Tsai W, DeMott-Friberg R, Barkan AL (1996) Regulation of somatic growth and the somatotropic axis by gonadal steroids: primary effect on insulin-like growth factor I gene expression and secretion. Endocrinology 137: 3253–3259.

53. Hayase K, Yonekawa G, Yokogoshi H, Yoshida A (1991) Triiodothyronine administration affects urea synthesis in rats. J Nutr 121: 970–978.

54. Grofte T, Wolthers T, Jensen SA, Moller N, Jorgensen JO, et al. (1997) Effects of growth hormone and insulin-like growth factor-I singly and in combination on in vivo capacity of urea synthesis, gene expression of urea cycle enzymes, and organ nitrogen contents in rats. Hepatology 25: 964–969.

55. Flores-Morales A, Stahlberg N, Tollet-Egnell P, Lundeberg J, Malek RL, et al. (2001) Microarray analysis of the in vivo effects of hypophysectomy and growth hormone treatment on gene expression in the rat. Endocrinology 142: 3163–3176.

56. Vijayakumar A, Novosyadlyy R, Wu Y, Yakar S, LeRoith D (2010) Biological effects of growth hormone on carbohydrate and lipid metabolism. Growth Horm IGF Res 20: 1–7.

57. Debeer LJ, Mannaerts GP (1983) The mitochondrial and peroxisomal pathways of fatty acid oxidation in rat liver. Diabete Metab 9: 134–140.

58. Horton JD, Shah NA, Warrington JA, Anderson NN, Park SW, et al. (2003) Combined analysis of oligonucleotide microarray data from transgenic and knockout mice identifies direct SREBP target genes. Proc Natl Acad Sci U S A 100: 12027–12032.

59. Reddy JK, Hashimoto T (2001) Peroxisomal beta-oxidation and peroxisome proliferator-activated receptor alpha: an adaptive metabolic system. Annu Rev Nutr 21: 193–230.

60. Djouadi F, Weinheimer CJ, Saffitz JE, Pitchford C, Bastin J, et al. (1998) A gender-related defect in lipid metabolism and glucose homeostasis in peroxisome proliferator- activated receptor alpha- deficient mice. J Clin Invest 102: 1083–1091.

61. Zhu L, Brown WC, Cai Q, Krust A, Chambon P, et al. (2012) Estrogen treatment after ovariectomy protects against fatty liver and may improve pathway-selective insulin resistance. Diabetes 62: 424–434.

62. Gao H, Bryzgalova G, Hedman E, Khan A, Efendic S, et al. (2006) Long-term administration of estradiol decreases expression of hepatic lipogenic genes and improves insulin sensitivity in ob/ob mice: a possible mechanism is through direct regulation of signal transducer and activator of transcription 3. Mol Endocrinol 20: 1287–1299.

63. Anderson CM, Stahl A (2013) SLC27 fatty acid transport proteins. Mol Aspects Med 34: 516–528.

64. Su X, Abumrad NA (2009) Cellular fatty acid uptake: a pathway under construction. Trends Endocrinol Metab 20: 72–77.

65. Falany CN, Xie X, Wheeler JB, Wang J, Smith M, et al. (2002) Molecular cloning and expression of rat liver bile acid CoA ligase. J Lipid Res 43: 2062–2071.

66. Glatz JF, Luiken JJ, Bonen A (2010) Membrane fatty acid transporters as regulators of lipid metabolism: implications for metabolic disease. Physiol Rev 90: 367–417.

67. Faulds MH, Zhao C, Dahlman-Wright K, Gustafsson JA (2012) The diversity of sex steroid action: regulation of metabolism by estrogen signaling. J Endocrinol 212: 3–12.

68. Yamazaki T, Okada H, Sakamoto T, Sunaga K, Tsuda T, et al. (2012) Differential induction of stearoyl-CoA desaturase 1 and 2 genes by fibrates in the liver of rats. Biol Pharm Bull 35: 116–120.

69. Hodgin JB, Maeda N (2002) Minireview: estrogen and mouse models of atherosclerosis. Endocrinology 143: 4495–4501.

70. Jogl G, Hsiao YS, Tong L (2004) Structure and function of carnitine acyltransferases. Ann N Y Acad Sci 1033: 17–29.

71. Ishimoto K, Nakamura H, Tachibana K, Yamasaki D, Ota A, et al. (2009) Sterol-mediated regulation of human lipin 1 gene expression in hepatoblastoma cells. J Biol Chem 284: 22195–22205.

72. Sjoberg A, Oscarsson J, Boren J, Eden S, Olofsson SO (1996) Mode of growth hormone administration influences triacylglycerol synthesis and assembly of apolipoprotein B-containing lipoproteins in cultured rat hepatocytes. J Lipid Res 37: 275–289.

73. Tollet-Egnell P, Flores-Morales A, Stahlberg N, Malek RL, Lee N, et al. (2001) Gene expression profile of the aging process in rat liver: normalizing effects of growth hormone replacement. Mol Endocrinol 15: 308–318.

74. Tollet-Egnell P, Parini P, Stahlberg N, Lonnstedt I, Lee NH, et al. (2004) Growth hormone-mediated alteration of fuel metabolism in the aged rat as determined from transcript profiles. Physiol Genomics 16: 261–267.

10

Monocyte to Macrophage Differentiation Goes along with Modulation of the Plasmalogen Pattern through Transcriptional Regulation

Stefan Wallner, Margot Grandl, Tatiana Konovalova, Alexander Sigrüner, Thomas Kopf, Markus Peer, Evelyn Orsó, Gerhard Liebisch, Gerd Schmitz*

Institute of Clinical Chemistry and Laboratory Medicine, University Hospital Regensburg, University of Regensburg, Regensburg, Germany

Abstract

Background: Dysregulation of monocyte-macrophage differentiation is a hallmark of vascular and metabolic diseases and associated with persistent low grade inflammation. Plasmalogens represent ether lipids that play a role in diabesity and previous data show diminished plasmalogen levels in obese subjects. We therefore analyzed transcriptomic and lipidomic changes during monocyte-macrophage differentiation *in vitro* using a bioinformatic approach.

Methods: Elutriated monocytes from 13 healthy donors were differentiated *in vitro* to macrophages using rhM-CSF under serum-free conditions. Samples were taken on days 0, 1, 4 and 5 and analyzed for their lipidomic and transcriptomic profiles.

Results: Gene expression analysis showed strong regulation of lipidome-related transcripts. Enzymes involved in fatty acid desaturation and elongation were increasingly expressed, peroxisomal and ER stress related genes were induced. Total plasmalogen levels remained unchanged, while the PE plasmalogen species pattern became more similar to circulating granulocytes, showing decreases in PUFA and increases in MUFA. A partial least squares discriminant analysis (PLS/DA) revealed that PE plasmalogens discriminate the stage of monocyte-derived macrophage differentiation. Partial correlation analysis could predict novel potential key nodes including DOCK1, PDK4, GNPTAB and FAM126A that might be involved in regulating lipid and especially plasmalogen homeostasis during differentiation. An *in silico* transcription analysis of lipid related regulation revealed known motifs such as PPAR-gamma and KLF4 as well as novel candidates such as NFY, RNF96 and Zinc-finger proteins.

Conclusion: Monocyte to macrophage differentiation goes along with profound changes in the lipid-related transcriptome. This leads to an induction of fatty-acid desaturation and elongation. In their PE-plasmalogen profile macrophages become more similar to granulocytes than monocytes, indicating terminal phagocytic differentiation. Therefore PE plasmalogens may represent potential biomarkers for cell activation. For the underlying transcriptional network we were able to predict a range of novel central key nodes and underlying transcription factors using a bioinformatic approach.

Editor: Thomas Langmann, University of Cologne, Germany

Funding: The research leading to these results has received funding from the European Community's Seventh Framework Programme (FP7/2007–2013) under grant agreement n° 202272, IP-Project LipidomicNet. The project was also supported by the Deutsche Forschungsgemeinschaft (SFB-TR 13/A3), related to "The European Lipidomics Initiative; shaping the life sciences" a specific support action subsidized by the EC 2006-2007 (proposal number 013032). The results upon which this publication is based were partly funded by the Federal Ministry of Education and Research under the Project Number FKZ01KU1216J. The responsibility for the content of this publication lies with the author. The funders had no role in study design, data collection and analysis, decision to publish, or preparation of the manuscript.

Competing Interests: The authors have declared that no competing interests exist.

* E-mail: gerd.schmitz@ukr.de

Introduction

Macrophages are key players in innate immunity and play an important role in the development of atherosclerosis and insulin resistance in diabesity [1]. During atherogenesis, modified ApoB containing lipoproteins accumulate in atherosclerotic plaques and lead to chemotaxis and accumulation of monocytes in the subintima [1]. Under the pro-inflammatory influence of the local microenvironment these monocytes terminally differentiate to M1 or M2 macrophages or antigen presenting cells (APC) [2–4]. During early lesion growth macrophages develop resistance to apoptosis and oxidative stress, whereas in advanced lesions macrophage death contributes to the formation of a necrotic core [5]. Consequently metabolic syndrome correlates with persistent low grade inflammation as indicated by increased serum levels of IL-6, CRP and fibrinogen [6]. Moreover metabolic overload induces an ER-stress response and leads to the formation of reactive oxygen species (ROS) [7].

Lipids regulate biological processes either locally as membrane components or remotely as signaling molecules. The lipid composition of the plasma membrane determines membrane fluidity but direct lipid-protein interactions also play a role in cellular signaling [8–10]. Moreover the release of signaling lipids

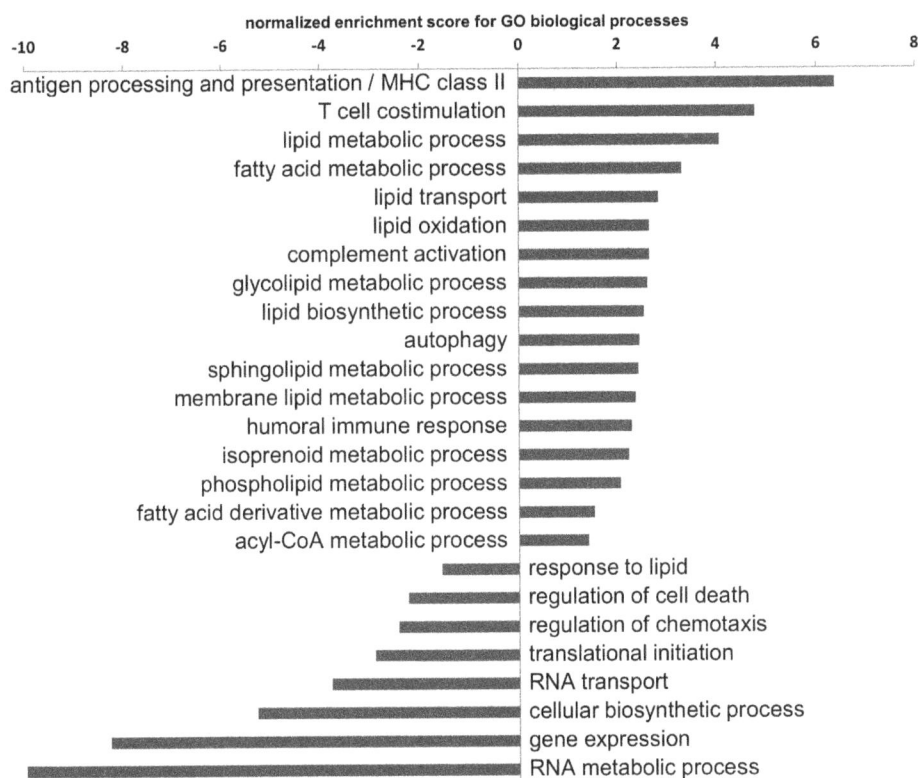

Figure 1. Gene ontology (GO) enrichment analysis of primary monocyte-macrophage differentiation. Normalized enrichment scores (NES) indicate the distribution of Gene Ontology categories across a list of genes ranked by hypergeometrical score (HGS). Higher enrichment scores indicate a shift of genes belonging to certain GO categories towards either end of the ranked list, representing up or down regulation (positive or negative values, respectively). Lipid related catergories are generally found to be shifted upwards.

from intracellular or membrane sources fulfills an important function in inflammatory signaling [11]. In this context especially eicosanoids, sphingosine-1-phosphate and lysophosphatidic acid are worth mentioning. Plasmalogens are a group of lipids that play a role in most of these tasks. In the plasma membrane they regulate membrane fluidity, via their vinyl-ether bond in sn-1 they act as anti-oxidants and in sn-2 position they carry the precursor residues for n-3 and n-6 prostanoid synthesis. Interestingly cleavage of this esther-bound alkyl chain is catalyzed by plasmalogen-selective phospholipase A2 (PLA2) [11]. In circulating monocytes plasmalogens represent around 15% of all cellular lipids [12]. They have also been shown to possess clinically significant correlations to vascular, metabolic and neurodegenerative diseases [13]. For example lower levels of plasmalogens were found in hypertensive patients and during aging in the aorta (even more pronounced in atherosclerotic aortas) [6,14]. Similarly plasmalogen depletion in red blood cell membranes has been proposed as a marker for oxidative stress and membrane rigidity and was suggested to be predictive for cardiovascular mortality [15]. Under these conditions plasmalogens may exert a scavenger function for reactive oxygen species in membranes, that could play a role during the ER stress response. On the other hand plasmalogen oxidation products such as alpha-hydroxyaldehydes and plasmalogen epoxides were found to accumulate in atherosclerotic lesions [16].

In a previous study we could demonstrate that a SREBP1 dependent induction of monocyte fatty acid synthesis is vital for monocyte-macrophage differentiation [17]. Detailed analysis of the changes in plasmalogen species and the underlying regulation

might therefore provide further insight into the role of this specific lipid classe in the differentiation process. Recent advances in mass spectrometry made high-throughput analysis of plasmalogen species feasible for use in research and in the clinical routine laboratory [18]. This technology allows to follow the regulation of individual species and analysis of regulatory patterns that might reflect pathophysiological adaptations in atherosclerotic lesions. The aim of the current study was therefore to analyze in detail the transcriptomic and lipidomic alterations during the course of monocyte-macrophage differentiation and to evaluate the potential of plasmalogens as marker lipids for differentiation of monocytes to inflammatory macrophages. A bioinformatic approach was used to examine the underlying transcriptional regulation and to reverse engineer gene regulatory networks to identify potential novel key nodes.

Materials and Methods

Materials

If not otherwise stated all materials including $1\alpha,25$-dihydroxyvitamin D_3 (VitD$_3$), all−trans−retinoic acid (ATRA) and ethanol were obtained from Sigma (Germany). Carrier-free macrophage colony-stimulating factor (M-CSF) and granulocyte macrophage colony-stimulating factor (GM-CSF) were both from R&D systems. HPLC grade methanol and chloroform were from Merck (Germany), plasmalogen standards from Avanti Polar Lipids (Alabaster, AL, USA).

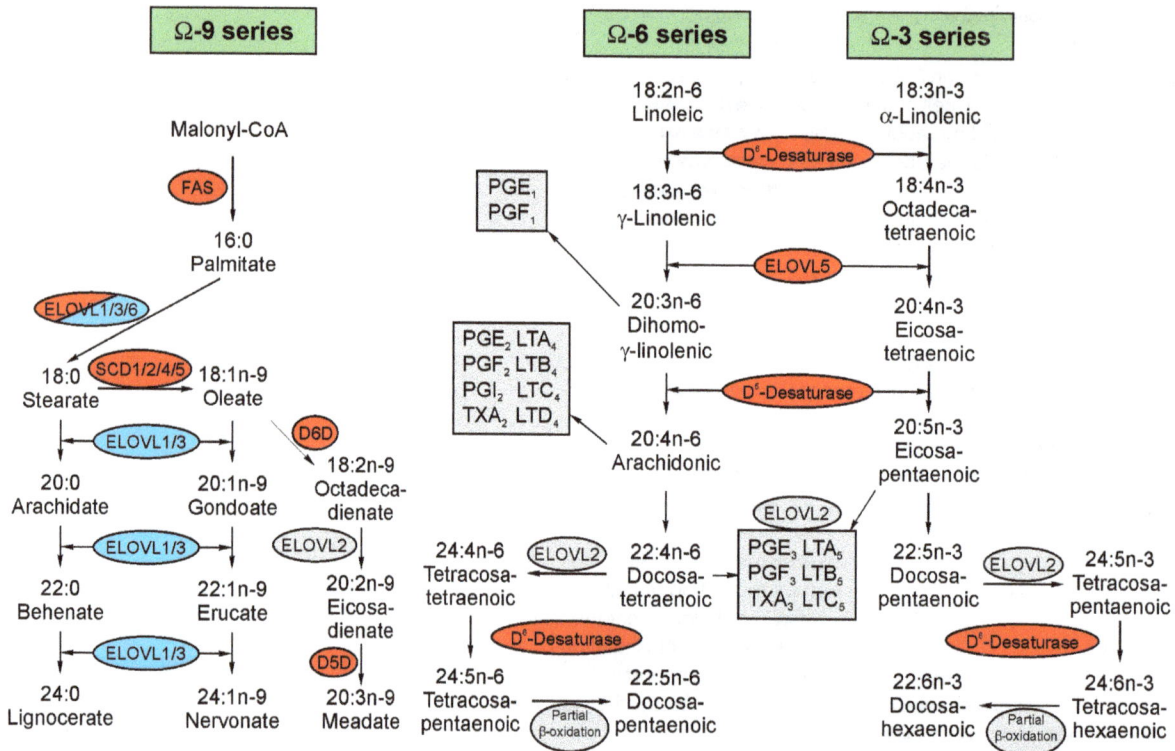

Figure 2. Schematic representation of fatty acid synthesis pathways. Fatty acid synthesis in the differentiation of primary human monocytes shows an induction of enzymes involved in the synthesis of palmitate (FAS), desaturation (SCD, D5D and D6D) and elongation (ELOVL5 & 6). Blue indicates significantly downregulated transcript levels, upregulated transcripts are shown in red and unchanged levels or unknown regulation in shown in grey.

Blood cell isolation and *in vitro* differentiation of human monocytes

Blood samples were obtained from thirteen healthy normolipidemic volunteers recruited from blood donors with apoE3/E3 genotype. Informed consent and approval of the Hospital Ethics Committee were obtained (Universitätsklinikum Regensburg, Ethikkommission der medizinischen Fakultät, proposal 08/119). Donors were fully informed of the possible complications and gave their written consent for the procedure. Blood cells were collected by leukapheresis in a Spectra cell separator (Gambro BCT, CO, USA) followed by subsequent counterflow elutriation as described elsewhere [19]. In brief, cells were elutriated in the following order: platelets, lymphocytes, monocytes and then granulocytes. Aliquots of the different cell fractions were analyzed for cell purity on a BD FACSCanto flow cytometer (Becton Dickinson) using BD FACSDiva Software. Cell numbers were determined on an ADVIA 120 automated cell counter (Siemens Healthcare Diagnostics GmbH, Germany). Elutriated monocytes were then seeded at a concentration of 1×10^6 cells/ml in serum-free macrophage medium (Invitrogen) that was supplemented with 50 ng/ml of recombinant human macrophage colony-stimulating factor (rhM-CSF, R&D Systems). 1×10^7 cells per dish were cultured in 10 cm Cell$^+$ plastic petri dishes (Sarstedt) for RNA isolation and 2×10^6 cells per well in Cell$^+$ 6-well plates (Sarstedt, USA) for triplicates in mass spectrometry. Incubation was performed at 37°C in a humidified atmosphere containing 5% CO_2 in an incubator. After washing with PBS, the cell pellets were stored at −80°C. Protein concentrations were measured according to Smith et al. [20] using the BCA Assay from Uptima-Interchim (France) with serial dilutions of bovine serum albumin as standard.

In vitro differentiation of HL-60 cells

HL-60 cells (ATCC; CCL-240) were grown in an incubator at 37°C (95% humidity, 5% CO_2) in Iscove's Modified Dulbecco's Medium (IMDM) (PAN-Biotech) containing 20% FBS (Biochrom) in polystyrene tissue culture flasks (Sarstedt). Cells were seeded at a density of 0.3×10^6 cells per ml and 50% of the medium was replaced once per week. Confluent cultures were split 1:3 to 1:4 once per week.

Differentiation experiments were performed in Cell$^+$ tissue culture flasks (Sarstedt). 1 µM working solutions of Vitamin D_3 (VitD$_3$)and all-trans retinoic acid (ATRA) were prepared in ethanol. Cells were seeded at 0.3×10^6 cells per ml. Monocytic differentiation was achieved in macrophage serum-free medium (SFM) supplemented with M-CSF (50 ng/ml) and 0.5 µM VitD$_3$. Granulocytic differentiation was carried out in macrophage SFM with GM-CSF (10 ng/ml) and 0.5 µM ATRA. Cells grown in the respective medium containing 0,05% ethanol were used as controls. Cells were harvested by scraping with a rubber policeman and centrifugation for 5 min at 500 g at room temperature. They were finally washed once with PBS. Cell number was determined by counting in a Casy automated cell counter (Roche). Differentiation phenotype and viability were monitored by flow cytometry and Pappenheim staining. Experiments were repeated in two biological replicates.

Gene expression studies

Cells from nine donors were harvested, washed in PBS, resuspended in buffer RLT and RNA was isolated using the RNeasy Mini Kit (Qiagen) according to the manufacturer's instructions. Purity and integrity of the RNA were determined on the Agilent 2100 Bioanalyzer with the RNA 6000 Nano LabChip reagent set (Agilent Technologies). RNA was quantified using a Nanodrop ND-1000 - UV/Vis spectrophotometer (PeqLab).

For gene array analysis we used a modified Agilent 4×44K microarray (014850) containing 205 free positions (Agilent Technologies). To these positions we added 201 probes, corresponding to 119 genes of the lipidome (LipidomicNet). 300 ng of total RNA were labeled with Cy3 using the Agilent Quick-Amp Labeling Kit - 1 color according to the manufacturer's instructions. cRNA was purified with the RNeasy Mini Kit (Qiagen). The amount and labeling efficiency of cRNA was measured with the NanoDrop ND-1000. For hybridization the Agilent Gene Expression Hybridisation Kit was used and arrays were incubated in Agilent SureHyb chambers for 17 hours in a hybridization oven at 65°C while rotating. Washing was performed according to the manufacturer's instructions. Scanning was done with the Agilent G2565CA Microarray Scanner System. The resulting TIFF files were processed with Agilent Feature Extraction software (10.7). Microarray data are available in the ArrayExpress database (www.ebi.ac.uk/arrayexpress) under accession number E-MTAB-2298.

Lipid mass spectrometry

Cell pellets were dissolved in 0.2% SDS solution. An aliquot corresponding to 100 μg was used for mass spectrometric lipid analysis. Lipid extraction was performed according to the method of Bligh and Dyer [21] in the presence of not naturally occurring lipid species as internal standards and the chloroform phase was dried in a vacuum centrifuge and dissolved in 10 mM ammonium acetate in methanol/chloroform (3:1 vol/vol). Samples were analyzed by ESI-MS/MS in positive ion mode by direct flow injection using the analytical setup and data analysis algorithms described previously by Liebisch et al. [22]. PE-based plasmalo-gens were quantified according to the principles described by Zemski Berry and Murphy [18]. For this purpose fragment ions of m/z 364, 380 and 382 were used for PE P-16:0, PE P-18:1 and PE P-18:0 species, respectively. After identification of relevant lipid species, selected ion monitoring analysis was performed. Material from all thirteen donors was analyzed for differentiation days 1, 4 and 5. 0d monocytes were additionally available for four donors. Lipid nomenclature is used as recommended in [23].

Statistical analysis
Partial least squares discrimination analysis (PLS/DA). Partial least square discrimination analysis was performed

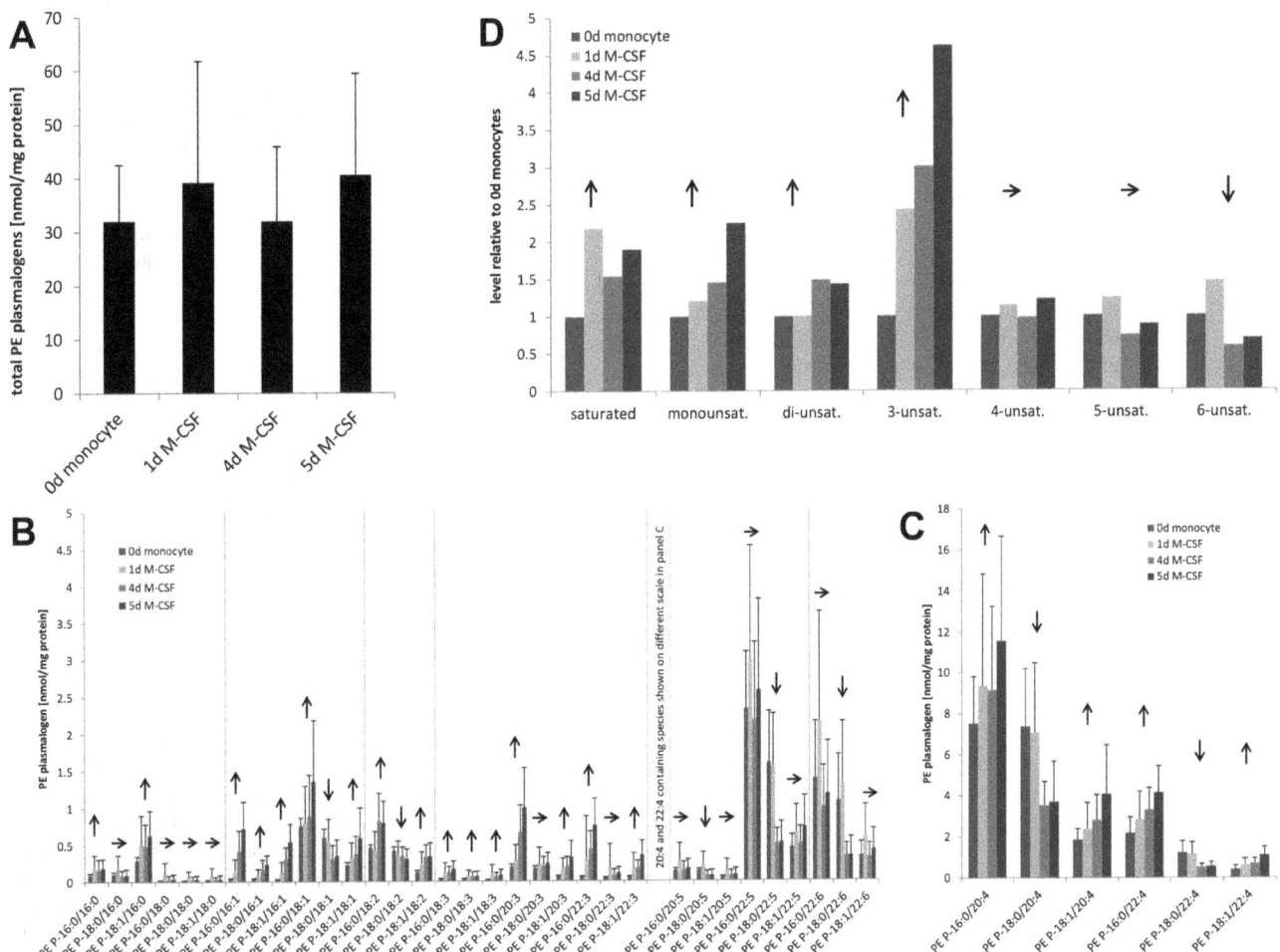

Figure 3. Regulation of PE plasmalogens during monocyte-macrophage differentiation. (A) Levels of total plasmalogens over thre course of 5 day primary monocyte macrophage differentiation. Individual PE plasmalogen species show specific changes during differentiation (B and C). Saturation specificity relative to day 0 is shown in panel (D).

and VIP values were calculated as described in [24–26]. The computation was performed using SPSS 20 (IBM) with the PLS extension module version 1.0.5 (IBM) under Python 2.7 with SciPy 0.11.0 (MLK) and NumPy 1.6.2 (MKL) on a 64 bit Windows 7 system (Microsoft).

Correlation analysis. Correlations between lipid species were calculated in SPSS 20 (IBM) using Pearson's product moment correlational analysis.

Evaluation of Agilent microarray data. Microarray experiments were analyzed according to the method described by Kondrakhin et al. [27]. With this method for each gene a so called hypergeometric score is calculated that considers the relative changes of an mRNA as well as the absolute signal intensity of the mRNA probe. Therefore relevant changes in gene expression can be identified more easily. Transcripts with calculated values <-4 or $>+4$ were considered significantly regulated. Gene Ontology enrichment analysis was performed on a ranked set of transcripts from microarray data with the BioUML software platform. For visualization only level 4 and level 5 categories were considered, primarily lipid related categories were chosen for visualization.

***In silico* transcription factor prediction.** Significantly expressed and regulated lipid related genes were subjected to transcription factor and virtual promoter analysis, predicting potential transcriptional regulators. *In silico* promoter analysis was carried out using the TRANSFAC database and the ExPlain software system (BioBase, Germany). For in silico analysis we focused on regions of 1000 bp upstream from the transcription start site, and used a threshold in order to minimize false positive errors.

Partial correlation and trancriptional network visualization. q-order partial correlation graphs (qpgraphs) were used to reverse engineer the molecular regulatory network from microarray and lipid data. All data was Z-transformed prior to analysis and average non-rejection rates (NRR) were calculated in R 2.15.3 [28] using the package qpgraph [29–31]. The NRR ranges from 0 to 1 and helps in deciding what edges are present or missing from a qpgraph. In our experiments a threshold value of 0.3 was used for the non-rejection rate. The resulting network was visualized in the editor yEd 3.12 (yWorks, Germany).

Results and Discussion

Adaptions in the lipid related transcriptome enable monocytes for their phagocytic and inflammatory function

Changes in the transcriptional profile during monocyte to macrophage differentiation were analyzed using microarrays for a comprehensive analysis of the cellular transcriptome at days 1, 4 and 5 of MCSF dependent *in vitro* differentiation. In total 4728 transcripts (34,1%) showed significant changes as indicated by a hypergeometrical score of <-4 or $>+4$ between days 1 and 4 of differentiation. Gene Ontology (biological process) analysis of the genearray data showed that lipid metabolism is among the most strongly positively enriched categories (Figure 1 and table S1). This is in line with our previously published results showing the importance of SREBP1/SP1 and SRF target genes in the differentiation process [17]. In the current study we therefore focused in depth on the regulation of genes involved in lipid metabolism, analyzed the lipidomic pattern by mass spectrometry and performed a bioinformatics analysis to study underlying regulatory mechanisms.

Fatty Acid Metabolism. Fatty acids are carboxylic acids with a long aliphatic tail. They are synthesized from acetyl-CoA and malonyl CoA precursors via six recurrent reactions that are catalyzed by the enzyme fatty acid synthase (FAS) until the 16-carbon product palmitic acid (C16:0) is produced. The main function of FAS is therefore to catalyze the synthesis of palmitate from acetyl-CoA and malonyl-CoA, in the presence of NADPH. The resulting saturated fatty acids are consecutively desaturated by desaturases or elongated by elongases at the endoplasmic-reticulum (ER). An overview over this process is given in Figure 2 with enzymes that were upregulated on the mRNA level shown in red and downregulated transcripts shown in blue. In our differentiation model transcripts for FAS were upregulated between days 1 and 4 and then remained stable at this level up to day 5, potentially leading to an increased synthesis of fatty acids. While the majority of cellular fatty acids that is synthesized via this pathway have a length of 16 to 18 carbon atoms, the cell also possesses the ability to elongate the carbon chain further using the ELOVL family of proteins. They catalyze the rate limiting condensation step in long chain fatty acid synthesis. ELOVL1 is ubiquitously expressed and catalyzes the condensation of very long chain ($>$C24) saturated or monounsaturated fatty acids [32]. During monocyte-macrophage differentiation, transcripts for ELOVL1 were downregulated. On the other hand transcripts for ELOVL5 and ELOVL6 were strongly induced. ELOVL5 processes a multitude of fatty acids including palmitoleic (C16:1, n-7), oleic (C18:1, n-9), γ-linolenic (C18:3, n-6), stearidonic (C18:4, n-3), arachidonic (C20:4, n-6) and eicosapentaenoic acid (EPA, C20:5, n-3) [33]. Adenoviral overexpression of ELOVL5 in primary rat hepatocytes has been reported to increase the elongation of arachidonic acid (20:4,n-6) and eicosapentaenoic acid (20:5,n-3) into adrenic acid (22:4,n-6) and docosapentaenoic acid (DPA, 22:5,n-3) respectively [34]. The monounsaturated fatty acids palmitic (16:0, preferentially) and palmitoleic acid (16:1) acid can be further elongated by ELOVL6 to stearic acid (18:0) and oleic acid (18:1) [35,36]. During phagocytic differentiation transcripts for ELOVL6 were only transiently upregulated on day 4 and dropped again on day 5. Additionally ELOVL6 is responsible for the elongation of C12 to C16 saturated and monounsaturated long chain fatty acids [36]. Furthermore peroxisomal trans-2-enoyl-CoA reductase was also strongly induced and is known to play a role in chain elongation of fatty acids, as well as estradiol 17-beta-dehydrogenase 12, an enzyme that also has 3-ketoacyl-CoA reductase activity suggesting a role in long fatty acid elongation. Transcripts for 3-hydroxyacyl-CoA dehydratases 1 and 3 were upregulated. These enzymes are responsible for the dehydration step in very long-chain fatty acid (VLCFA) synthesis [37].

Desaturases introduce double bonds between defined carbon atoms in fatty acyl chains. In humans four types of fatty acid desaturases could be identified, according to the position of the double bond introduced. In our experiments all analyzed fatty acid desaturases were upregulated with the notable exception of FADS3, which is interesting since the gene encoding FADS3 is clustered with FADS1 and FADS2 [38]. The conversion of palmitic acid (16:0) to palmitoleic acid (16:1), as well as the conversion of stearic acid (18:0) to oleic acid (18:1) is mediated by delta-9-desaturase (D9D), also called stearoyl-CoA desaturase-1 (SCD1) or FADS5. In this study the expression of this rate-limiting microsomal enzyme was found to be strongly upregulated during phagocytic differentiation of monocytes. The monounsaturated fatty acids produced by SCD are further used as substrates in the production of cholesteryl esters, triglycerides, phospholipids and wax esters (reviewed in [39]). FADS1 (D5D), as well as FADS2 (D6D) are enzymes that are required for the synthesis of omega-3 and omega-6 long-chain polyunsaturated fatty acids. They therefore play a role in the synthesis of arachidonic acid as a

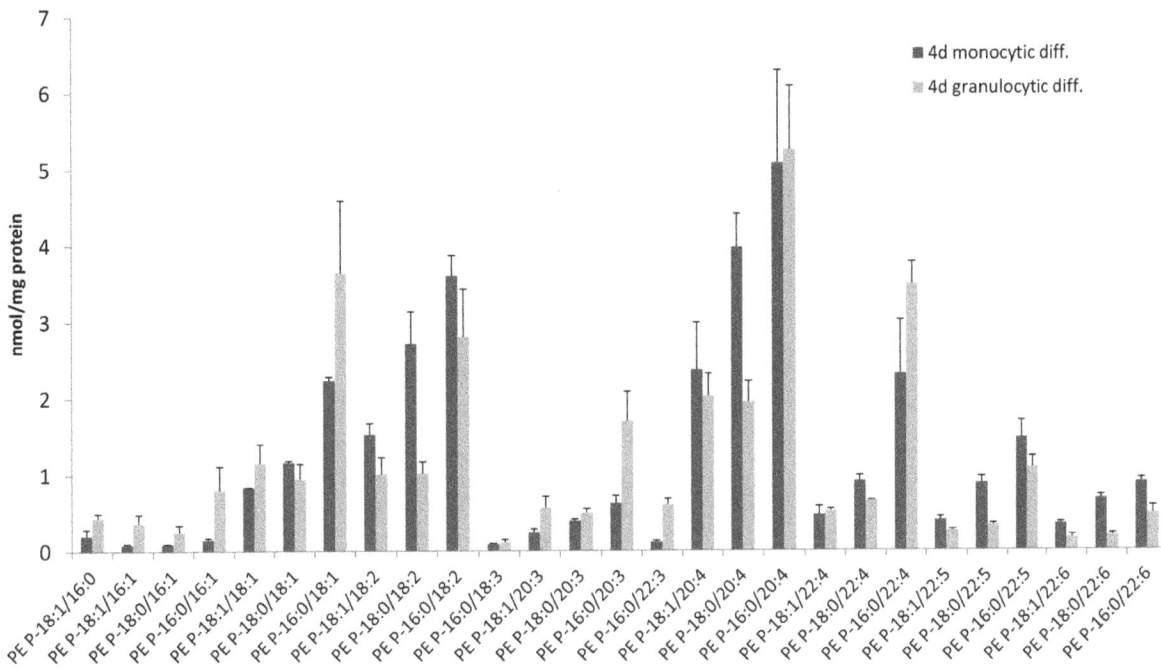

Figure 4. HL-60 model for monocyte and granulocyte differentiation. PE plasmalogen species in an *in vitro* model of monocytes and granulocytes using HL-60 cells show higher levels of species containing highly unsaturated acyl residues (>4 double bonds) in monocytic differentiation and higher levels of less unsaturated species (<4 double bonds), mono- or saturated species in granulocytic differentiation.

precursor molecule for eicosanoid biosynthesis in omega-6 fatty acid synthesis and in the omega-3 line of EPA and DHA production. FADS1 is a delta-5-desaturase and FADS2 a delta-6-desaturase. Delta-4 desaturases primarily play a role in sphingolipid metabolism and were not significantly regulated by differentiation.

The mitochondrial very long-chain specific acyl-CoA dehydrogenase (ACADVL) was strongly up-regulated. Its function is the degradation of very long fatty acids by catalyzing the first step of the mitochondrial fatty acid beta-oxidation pathway. It can accommodate substrates with chain lengths as long as 24 C-atoms [40]. At the transcript level it was greatly diminished during differentiation, indicating a decreased degradation of these long chain fatty acids. On the other hand the dehydrogenase ACADSB that is preferentially active on short and branched chain acyl-CoAs was induced on days 4 and 5.

Lipoprotein Metabolism. Lipoproteins are aggregates of apo-lipoprotein (apo) and lipid that allow the transport of lipids in the aqueous blood stream. Especially plasmalogens are known to correlate positively to serum HDL levels and to decrease with aging [41]. During monocyte-macrophage differentiation apoE was strongly induced. ApoE is important for metabolizing triglyceride rich lipoproteins, is involved in vascular remodeling [42] and has been described to be antiatherogenic and to mediate cholesterol efflux. It also plays a role in the immune function of macrophages. Apolipoprotein J/clusterin is involved in cholesterol export from foam cells [43]. Its transcript levels were strongly increased after differentiation. Apolipoprotein C-I and C-II were also strongly upregulated. ApoC-I in plasma is mainly bound to VLDL, chylomicrons, and HDL [44]. In macrophages it has recently also been described to be able to bind lipopolysaccharide (LPS) [45]. ApoC-I regulates cholesterol and phospholipid efflux and seems to be crucially involved in the development of atherosclerotic lesions [46]. ApoC-II plays a role in the catabolism

of triglyceride-rich lipoproteins and is also found on macrophages in atherosclerotic plaques [44]. It has the ability to activate lipoprotein-lipase (LPL). LPL is the primary enzyme for the hydrolysis of triacyl-glycerides in chylomicrons and VLDL. Its transcript was upregulated. Although LPL promotes foam cell formation and atherosclerosis [47], it is generally viewed as antiatherogenic [48]. Macrophage LPL together with its activator protein apoC-II is thought to modulate macrophage inflammatory capacity through release of n-3/n-6 fatty acids. Their increased expression during differentiation reflects the increased inflammatory capacity of the macrophage by being able to access fatty acids for the synthesis of inflammatory mediators. Another enzyme potentially complementing LPL in the hydrolysis of monoglycerides is monoglyceride lipase (MGLL). It was found upregulated, facilitating hydrolysis of intracellular triglyceride stores to fatty acids and glycerol. The expression level of the transcript for the LDL receptor (LDLR) was only slightly diminished on day 4 and significantly downregulated on day 5, for the VLDL-receptor (VLDLR) we found no changes in expression.

Accumulation of excess free cholesterol in macrophages contributes to the development of atherosclerosis by stimulating tumor necrosis factor-α (TNF-α) and interleukin-6 production [49]. This leads to ER-stress and apoptosis [50]. In general enzymes involved in SREBP2 dependent cholesterol biosynthesis were downregulated while proteins involed in cholesterol export were strongly induced. The levels of the transcripts for the rate-limiting enzyme in cholesterol synthesis HMG-CoA reductase [51] were unchanged on day 4 and downregulated on day 5. Also transcript levels for squalene oxidase, as well as lathosterol oxidase were downregulated on both days. Cholesterol export from macrophages is largely mediated by the transporter ABCA1, but also to a smaller extent by ABCG1 [52–54]. It was found to be protective against cardiovascular disease and fulfill an anti-inflammatory function [52,55,56]. Interestingly this role seems to

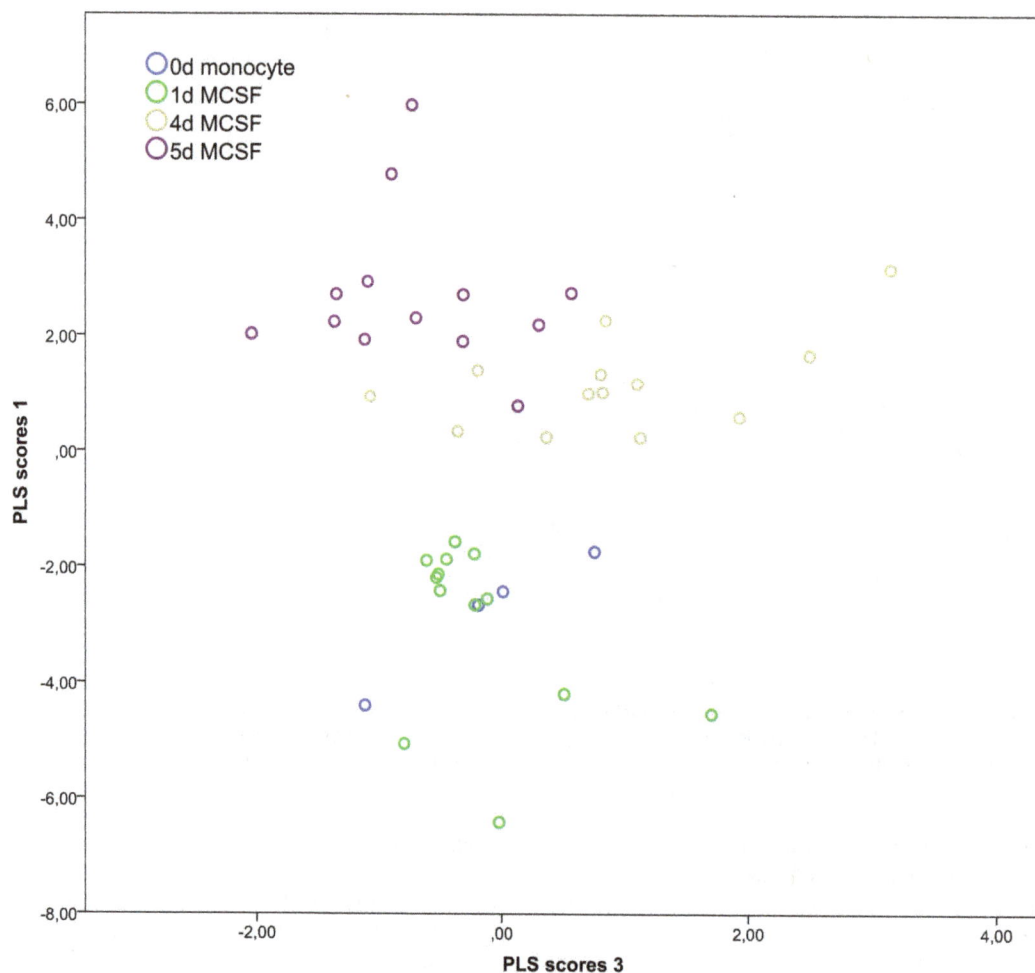

Figure 5. Partial least squares discriminant analysis (PLS/DA) during the course of primary monocyte-macrophage differentiation.
While latent variable LV1 is able to discriminate between days 0/1 and days 4/5 of differentiation, LV4 can differentiate between days 4 and 5.

be independent of cholesterol transport. Its expression was strongly induced by differentiation. The cytosolic acetoacetyl-CoA thiolase (ACAT2), the main enzyme mediating esterification of cholesterol in the liver [57] was downregulated on day 5.

Plasmalogens, Glycerophospholipids and Eicosanoid Biosynthesis. Plasmalogens are ether glycerol-phospholipids that are integral parts of the plasma membrane as well as storage lipids for acyl residues and cellular anti-oxidants. Their synthesis takes place in a non-redundant pathway in peroxisomes and at the ER-membrane. Interestingly, although the macrophage is prepared for an immune reaction, transcripts for enzymes in plasmalogen synthesis remained mostly unchanged. The rate limiting enzyme for this process, glyceronephosphate O-acyltransferase (GNPAT, also called DHAP-AT), was not regulated on the transcriptomic level. The acyl residues in phosphatidyl-ethanolamine (PE) plasmalogens are taken from the cellular free fatty acid pool. For this purpose the enzyme fatty acyl-CoA reductase 1 (FAR1) provides the long chain fatty alcohol by reduction of saturated fatty acyl-CoAs in the peroxisomal membrane. During monocyte-macrophage differentiation its mRNA was downregulated. The resulting alcohols are then further processed by alkyldihydroxyacetonephosphate synthase (AGPS/ADHAP-S) in the peroxisome, an enzyme whose transcript was downregulated on day 4 and increased on day 5. These findings are in agreement

with constant levels of total plasmalogens that were detected in lipid mass spectrometry (Figure 3A).

Eicosanoids are inflammatory mediators that are synthesized in macrophages from eicosapentaenoic acid (EPA, n-3), arachidonic acid (AA, n-6), and dihomo-gamma-linolenic acid (DGLA, n-6) [58,59]. Plasmalogens in the plasma membrane represent a major reservoir for the storage of these precursors in mammalian cells (reviewed in [11]). Hydrolysis of the ester-bond in the sn-2 position is the first step in eicosanoid synthesis from precursor acyl residues in plasmalogens. Phospholipase A_2 isoforms constitue a large family of enzymes specifically produced for this purpose. With the exception of PLA2G4A all PLA2 and PLD isoforms that were analyzed on the microarray showed either stable or increasing expression levels for their mRNA, while the endoperoxide synthases were downregulated. This points towards a tight regulation of eicosanoid production and processing. PLA2 is required for releasing precursor molecules for eicosanoid biosynthesis from either phosphatidylcholine (PC, esters) or plasmalogens (vinyl-ethers). Although plasmalogen specific PLA2 activity has been decribed in the literature, this enzymatic activity could not be associated with a specific isoform or transcript in the phospholipase family [60].

Eicosanoid production is subsequently mediated by the PG-endoperoxide synthases PTGS1 and PTGS2, also known as

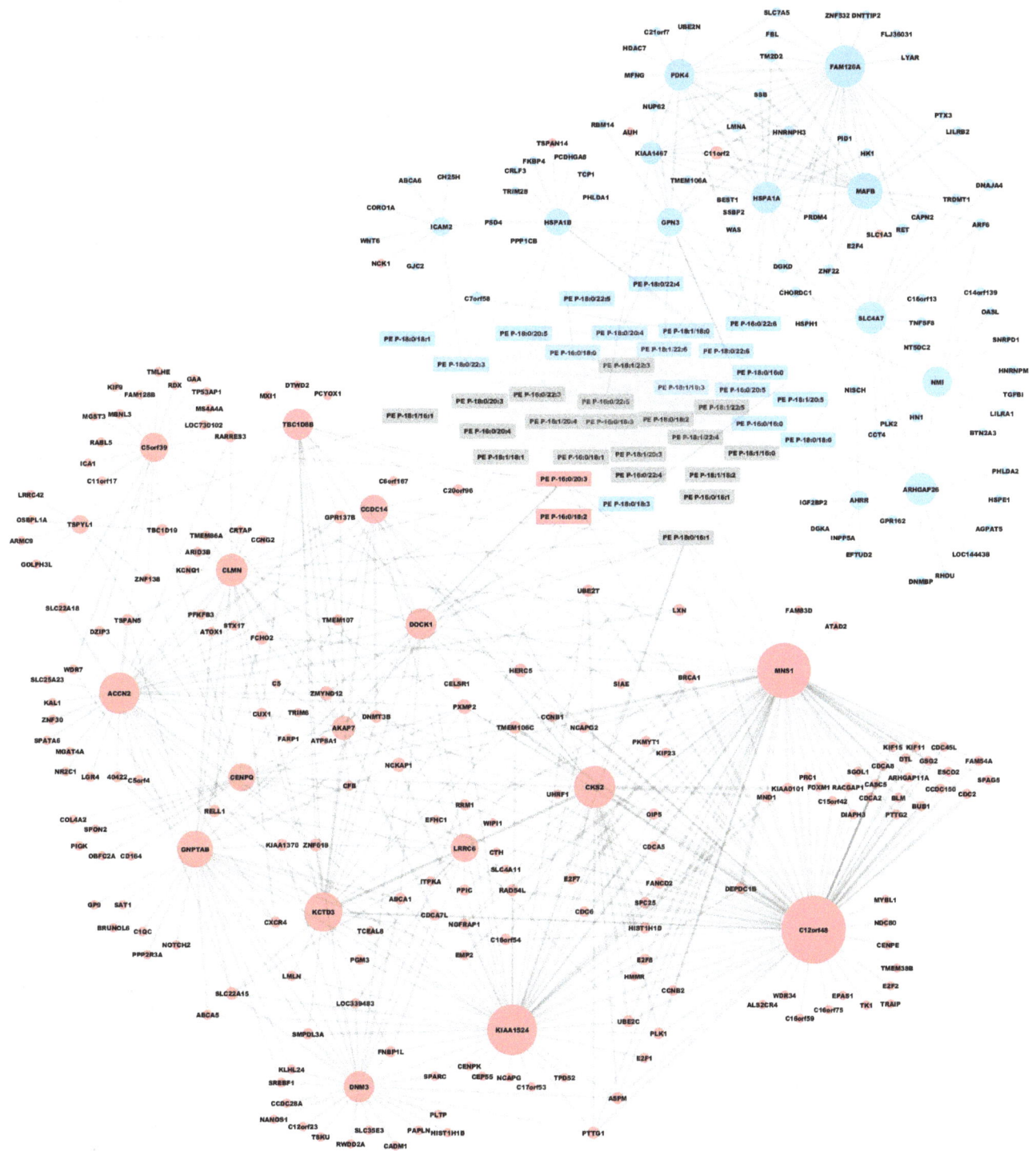

Figure 6. Molecular network graph of genes correlating to PE plasmalogens during primary monocyte-macropahge differentiation.
Plasmalogens are shown in the center (rectangles) and correlating transcripts are grouped around them (circles). Circle size represents the number of connected nodes, line width the strength of the correlation and color the tyoe of regulation (upregulated in red, downregulated in blue). Central key nodes can be identified that have a multitude of connections to other nodes.

cyclooxigenases COX1 and COX2 [61]. Transcripts for both enzymes were downregulated, reaching significance for PTGS1 and showing a trend for PTGS2 in the hypergeometrical score. On the other hand arachidonate 5-lipoxygenase (ALOX5), and leukotriene A4 hydrolase (LTA4H) were strongly induced.

In summary these changes in the gene expression profile reflect the major alterations in lipid metabolism: increases in fatty acid synthesis, elongation and desaturation as well as increased cholesterol production, utilization and export from phagocytic macrophages in comparison to blood monocytes. Since the modulated release of inflammatory mediators is a key contribution

Table 1. Results of an in silico transcription factor binding motif analysis.

Up-regulated genes	Down-regulated genes
GZF1	CHCH
LXR	PPARG
TATA	HIF2A
IRF	ZBTB
ZNF148	ELF ETS
ATF CREB	AP1
TFE	ZFX
GKLF	FKLF
USF	CNOT3
FOX	RNF96
	MYCMAX
	AHRARNTHIF
	E2F

Up- and down-regulated genes
AP2
NFY
P53
SP1SP4
EGR

Motives that were enriched in the 1000 bp promoter region upstream of up-regulated lipid related genes are shown on the left, binding motives that were enriched in promoters of down regulated genes on the right and motifs that were associated with up- and down-regulated genes at the bottom. Abbreviations (as used in the Transfac database): AHRARNTHIF: aryl hydrocarbon receptor/aryl hydrocarbon receptor nuclear translocator/hypoxia inducible factor, AP1: activator protein 1, AP2: activating protein 2, ATF CREB: activating transcription factor/cAMP response element binding, CHCH: Churchill, CNOT3: CCR4-NOT transcription complex subunit 3, E2F: E2F transcription factor, EGR: early growth response, ELF ETS: E74-like factor/E26 transformation-specific, FKLF: fetal β-like globin gene-activating Krüppel-like factor, FOX: forkhead box, GKLF: gut-enriched Krüppel-like factor, GZF1: GDNF-inducible zinc finger protein 1, HIF2A: hypoxia-inducible factor 2a, IRF: interferon regulatory factor, LXR: liver X receptor, MYCMAX: Myc/Max transcriptional complex, NFY: nuclear factor Y, p53: phosphoprotein p53, PPARG: peroxisome proliferator-activated receptor-gamma, RNF96: RING finger protein 96, SP1SP4: specificity protein 1 and 4, TATA: TATA box, TFE: Transcription Factor E (Family), USF: upstream transcription factor, ZBTB: zinc finger and BTB (broad complex, tramtrack, and bric-à-brac) domain-containing, ZFX: zinc finger X-chromosomal protein, ZNF148: zinc finger protein 148.

of macrophages in the development and sustenance of atherosclerotic plaques we next focused on an in depth examination of plasmalogen lipidomics.

The PE plasmalogen profile matures during monocyte macrophage differentiation

As already mentioned before plasmalogens regulate cell membrane fluidity and represent cellular stores for the precursor molecules of eicosanoid biosynthesis, mainly arachidonic acid [11]. Cellular PE-plasmalogen content was analyzed in monocytes and monocyte-derived macrophages at days 0, 1, 4 and 5 of an *in vitro* differentiation protocol. In general plasmalogen species containing AA in sn-2 position were most abundant. During the course of monocyte to macrophage differentiation, total plasmalogen levels did not change significantly (Figure 3A, p = 0.186, independent samples Kruskal-Wallis test due to varying number of experiments per time point). However, at the species level we found specific alterations in the plasmalogen species composition (Figure 3B and C). Especially minor plasmalogen species showed saturation dependent changes in concentration. As shown in Figure 3D, plasmalogen species containing poly-unsaturated (PUFA, five or six double bonds) or saturated acyl residues were typically downregulated while species with sn-2 bound residues with one, two or three double bonds were largely upregulated.

Species containing AA (20:4) did not show significant variations in their levels.

Clinically atherogenesis starts with increased deposition and modification of LDL in the subendothelium. Subsequently inflammatory cells such as monocytes are chemotactically attracted and migrate into the vessel wall. There they secrete reactive oxygen species (ROS) producing an environment with high levels of oxidative stress, that also affects the deposited LDL (reviewed in [62]). Interestingly LDL particles from obese metabolic syndrome patients and type 2 diabetics showed changes in their lipid composition that were similar to the changes we observed during monocyte-macrophage differentiation. Low density lipoprotein from obese subjects contained generally lower levels of PE plasmalogens and decreased proportions of PUFA in their cholesteryl esters, while saturated fatty acids (SFA) and mono-unsaturated fatty acids (MUFA) proportions were increased. Similarly LDL particles from type 2 diabetic patients contained decreased PUFA, had an unchanged MUFA content and showed increased SFA [63].

The PE plasmalogen pattern in human blood cells has been reported by our group before [12]. Interestingly the changes in the plasmalogen pattern during differentiation rendered a mature macrophage more similar to a blood granulocyte than the original monocyte. Since granulocytes represent fully mature cells from the same myeloid lineage as monocytes we suspected a general mode of regulation. In order to validate this pattern we used an *in vitro* model of HL-60 cells. Dependent on the stimulus, the pro-myelocytic cell line HL-60 can differentiate to a monocytic (Vit D_3) as well as granulocytic (all-trans retinoic acid, ATRA) phenotype. We therefore analyzed the plasmalogen species pattern of differentiating HL-60 cells (Figure 4) and performed correlation analysis.

The correlation of HL-60 monocytic and granulocytic differentiation was strongly positive (Pearson $r = 0.960$, $p < 0.001$). Furthermore we found a significant correlation between the changes in monocyte to macrophage differentiation and the regulation of plasmalogen species during monocytic differentiation of HL-60 cells (Pearson $r = -0.522$, $p = 0.005$). In contrast the changes found in HL-60 derived granulocytic cells were statisti-cally different from monocyte-macrophage differentiation (Pearson $r = -0.359$, $p = 0.066$).

Quantification of PE plasmalogen species allows discrimination of the differentiation state in a partial least squares discriminant analysis (PLS/DA)

Since the plasmalogen pattern undergoes very specific changes during phagocytic differentiation we hypothesized that it could serve as a biomarker for predicting the monocyte to macrophage differentiation state. To this end a PLS/DA analysis was performed on the data obtained from primary human monocytes and macrophages. PLS is a mathematical method that allows to analyze the influence of a matrix of predictors on an outcome variable.

In this approach 4 different latent variables were computed and plotted graphically to evaluate their power to differentiate the differentiation stage. As shown in Figure 5, latent variable 1 allowed the clear discrimination of cells on days 0 and 1 from more mature cells on days 4 and 5. On the other hand latent variable 3 permitted to partially discern between days 4 and 5. A distinction between day 0 and day 1 was not possible using this approach. Therefore while there was no clear difference in the PE-plasmalogen pattern between undifferentiated monocytes and monocytes that had been treated with rhMCSF for 1d, cells that had been cultured with M-CSF for 4 or 5 days were clearly different in their lipidomic profile.

We also determined the most influential lipid species by calculating so called Variable Importance for the Projection (VIP) values as shown in Figure S1. The PE-plasmalogen species that showed the strongest potential to discriminate between monocytes and macrophages (mean PLS VIP 1) were minor species containing docosapentaenoic acid (DPA, 22:5), docosahex-aenoic acid (DHA, 22:6) and palmitoleic acid (16:1) esterified in sn-2 position. On the other hand linoleic acid (18:2) containing species were best in discerning between days 4 and 5 of differentiation.

Chronic, systemic low-grade inflammatory diseases such as diabesity and the metabolic syndrome are characterized by a strong activation of tissue macrophages that chemotactically also attract additional monocytes from the circulating blood stream [64]. Therefore it is likely that such a chronic inflammatory reaction will also be reflected in the plasmalogen pattern of plasma samples or be detectable in isolated blood monocytes, rendering it a potential lipidomic biomarker.

Transcriptional network analysis for the identification of novel PE plasmalogen related key nodes

In order to identify novel players influencing the regulation of cellular plasmalogens we performed partial correlation analysis of

plasmalogen species with gene array data (Figure 6). This bioinformatic approach allows to reduce the complexity of the model by eliminating indirect interactions and therefore to better identify key nodes. Among the transcripts found to be central by this algorithm were well known factors proving the plausibility of the method. For example MAFB is a known inducer of monocytic differentiation [65] that showed up centrally. Also the 70 kDa heatshock proteins HSPA1B and HSPA1A were central nodes. Hsp70 proteins act as molecular chaperones that can maintain the survival of stressed cells [66–69]. They accomplish this by stabilizing lysosomal membranes through the binding of BMP [70]. For these proteins genotypic variations have been described that play a role in cytokine release [71].

Furthermore a range of proteins that have not yet been implicated in monocyte-macrophage differentiation were identi-fied. DOCK1 is a Rac activator that regulates myoblast fusion [72]. It is involved in regulating cell surface extensions and has the ability to bind phospholipids via a DHR1 domain [73]. Thereby it possesses the ability to interact with cell membranes, a function that is essential for differentiated macrophages. In our cells its transcript levels were significantly upregulated during the differ-ention process. Pyruvate dehydrogenase lipoamide kinase isozyme 4 (PDK4) is a mitochondrial enzyme regulating glucose metab-olism and therefore cellular energy homeostasis. Strongly decreased expression of PDK4 as found in our analysis increases metabolism and especially conversion of glucose to acetyl-CoA, thereby improving substrate availability for fatty acid synthesis. Moreover N-acetylglucosamine-1-phosphate transferase (GNPTAB, significantly upregulated during differentiation) is involved in synthesis of mannose-6-phosphate, that itself tags hydrolases for transfer to the lysosome. The expression of this enzyme could therefore have influence on the protein composition of the macrophage lysosomes. The protein encoded by FAM126A may play a part in the beta-catenin/Lef signaling pathway and has not been described to play a role in macrophages before. In our cells its mRNA levels were downregulated during differentiation. C12orf48 enhances PARP-1 activity, protecting from DNA damage [74], knockdown was anti-proliferative in HeLa cells [75]. It was upregulated. Meiosis-specific nuclear structural 1 (MNS1, significantly upregulated) has only been described to play a role in the regulation of meiosis before [76]. Its function in monocytes and macrophages has not been analyzed yet.

Further transcripts that were identified include ARHGAP26, a Rho GTPase involved in the formation of endocytotic vesicles, cell spreding and adhesion [77], ACCN2, a sodium channel [78] and KIAA1524, a phosphatase inhibitor that plays a role in oncogenesis [79].

Further studies are needed to clarify the role of these potential lipid related modulators of the differentiation process.

Identifying underlying transcription factors

Major regulatory changes in the cellular transcriptional profile are mediated by transcription factors. We used a bioinformatics approach therefore affecting all target genes that contain a specific binding motif in their promoter region at the same time. This search was based on transcripts listed in the Gene Ontology category lipid metabolism. Transcription factors that are predicted to bind to promoters of these genes are shown in Table 1. On the one hand this approach predicted well known factors essential in macrophage differentiation such as PPARG, KLF4 [80] and EGR1 [81,82] proving the feasibility of the algorithm. The binding motifs identified in this approach also allow binding of previously described factors that regulate lipid associated genes such as SREBP1, although it does not show up in the analysis directly. SREBP1 is known to interact with Sp1 at Sp1/Sp4 sites in regulating gene expression, as well as with Nuclear factor Y (NFY). Therefore combinations of these factors determine the regulation of distinct cellular pathways [83]. SREBP1 target genes include fatty acid synthase and the LDL-receptor. Transcript levels for both were strongly induced during the differentiation process. SREBP1 also posses the ability to cooperate with CREB-binding protein as a co-activator. This combinatorial regulation of multiple transcription factors is also likely to be the reason for the occurrence of the motifs for EGR and p53 in up- and downregulated genes at the same time. On the other hand an involvement of previously not published binding motives could be predicted for the first time. These include ZNF148, nuclear factor Y and RNF96. Since transcription factor binding motif analysis was based solely on lipid metabolism related transcripts, it is likely that some of the known effects of these transcription-factors on differentiation are mediated in turn by lipidomic changes in the cell.

Summary and Conclusion

During the differentiation process to mature macrophages monocytes undergo profound changes in their lipidomic and transcriptomic profile to prepare them for their phagocytic and inflammatory function.

Fatty acid synthesis pathways show a strong induction of transcripts involved in fatty acid desaturation and elongation. Lipid transport and lipoprotein related transcripts are differentially modulated partly in a pro-inflammatory way partly anti-inflammatory. On the one hand cholesterol biosynthesis is downregulated and export is strongly promoted. At the same time the cell prepares to facilitate the hydrolysis of triglycerides. In depth analysis was perfomed for plasmalogens. Plasmalogen biosynthesis pathways remained mostly unchanged on the mRNA level. This was reflected in lipidomic measurements showing constant total plasmalogen levels. Plasmalogens represent storage lipids for precursor acyl-residues in eicosanoid synthesis. Transcripts for the iso-enzymes involved in eicosanoid production were tightly regulated.

Lipidomic analysis of PE plasmalogens showed decreases in polyunsaturated species (PUFA) and increases in monounsaturated species (MUFA). These adaptions rendered monocyte derived macrophages more similar to mature granulocytes in their plasmalogen profile. This observation could be confirmed in an HL-60 model system for granulocytic and monocytic in vitro differentiation. The PE plasmalogen species pattern could be shown to be specific for monocytes and monocyte derived macrophages permitting a purely lipidomic discrimination of the differention stage. It might therefore also represent a potential biomarker indicating immune system activation, e.g. in patients with metabolic syndrome that suffer from chronic, low-grade inflammation.

A bioinformatic approach using partial correlation and molecular network reconstruction was able to identify novel potential hubs in the plasmalogen related transcriptional network, such as 70 kDa heatshock proteins, DOCK1, PDK4, GNPTAB, FAM126A, MNS1 and C12orf48. Further studies are currently underway to clarify the potential modulatory role of these proteins during differentiation. Transcription factor motif analysis revealed the involvement of several binding motifs with the remarkable involvement of motifs for NFY, P53, SP1 and EGR in up- and down- regulated genes. Therefore probably the interplay of multiple additional factors such as SREBP, determines the amount of transcription of lipid regulatory transcripts.

In summary we could for the first time show the specificity of the changes in the plasmalogen pattern during monocyte to macrophage differentiation. Furthermore using a bioinformatics approach we were able to predict novel regulatory hubs and transcription factors that might play central roles in the complex interplay of transcription-factors, transcripts and lipids during monocyte-macrophage differentiation.

Supporting Information

Figure S1 Variable importance in projection for PE plasmalogens during differentiation. The VIP value summarizes the contribution a variable makes to the model. Therefore higher values (especially values larger than 0.8) signify a more significant contribution of the respective lipid species to the estimation of the differentiation stage.

Table S1 GO classification of enriched transcripts during differentiation. Normalized enrichment scores (NES) were calculated to reflect the distribution of Gene Ontology categories across a list of genes ranked by hypergeometrical score (HGS). A higher enrichment score indicates a shift of genes belonging to certain GO categories towards either end of the ranked list, representing up or down regulation (positive or negative values, respectively).

Table S2 Hypergeometrical score of lipid related transcripts during differentiation. Agilent microarray data shows regulatory activity in fatty acid desaturation and elongation as well as changes in plasmalogen synthesis and degradation. Comparisons were made for day 4 versus day 1 (d4) and for day 5 versus day 1 (d5). Blue color indicates a significantly lower value following differentiation while red indicates significantly higher values.

Author Contributions

Conceived and designed the experiments: SW MG MP EO GS. Performed the experiments: SW MG AS MP GL. Analyzed the data: SW T. Konovalova AS MP GL GS. Contributed reagents/materials/analysis tools: T. Kopf. Wrote the paper: SW GS.

References

1. Moore KJ, Tabas I (2011) Macrophages in the pathogenesis of atherosclerosis. Cell 145: 341–355.
2. Mantovani A, Sica A, Sozzani S, Allavena P, Vecchi A, et al. (2004) The chemokine system in diverse forms of macrophage activation and polarization. Trends Immunol 25: 677–686.
3. Gordon S, Taylor PR (2005) Monocyte and macrophage heterogeneity. Nat Rev Immunol 5: 953–964.
4. Sica A, Mantovani A (2012) Macrophage plasticity and polarization: in vivo veritas. J Clin Invest 122: 787–795.
5. Gautier EL, Huby T, Witztum JL, Ouzilleau B, Miller ER, et al. (2009) Macrophage apoptosis exerts divergent effects on atherogenesis as a function of lesion stage. Circulation 119: 1795–1804.
6. Buddecke E, Andresen G (1959) [Studies on the chemistry of the arterial wall. IV. Quantitative determination of the acetalphosphatides (plasmalogens) in human aorta with consideration to arteriosclerosis]. Hoppe Seylers Z Physiol Chem 314: 38–45.
7. Hotamisligil GS (2006) Inflammation and metabolic disorders. Nature 444: 860–867.

8. Simons K, Toomre D (2000) Lipid rafts and signal transduction. Nat Rev Mol Cell Biol 1: 31–39.

9. Palsdottir H, Hunte C (2004) Lipids in membrane protein structures. Biochim Biophys Acta 1666: 2–18.

10. Engelman DM (2005) Membranes are more mosaic than fluid. Nature 438: 578–580.

11. Wallner S, Schmitz G (2011) Plasmalogens the neglected regulatory and scavenging lipid species. Chem Phys Lipids 164: 573–589.

12. Leidl K, Liebisch G, Richter D, Schmitz G (2008) Mass spectrometric analysis of lipid species of human circulating blood cells. Biochim Biophys Acta 1781: 655–664.

13. Yudkin JS, Stehouwer CD, Emeis JJ, Coppack SW (1999) C-reactive protein in healthy subjects: associations with obesity, insulin resistance, and endothelial dysfunction: a potential role for cytokines originating from adipose tissue? Arterioscler Thromb Vasc Biol 19: 972–978.

14. Graessler J, Schwudke D, Schwarz PE, Herzog R, Shevchenko A, et al. (2009) Top-down lipidomics reveals ether lipid deficiency in blood plasma of hypertensive patients. PLoS One 4: e6261.

15. Stenvinkel P, Diczfalusy U, Lindholm B, Heimburger O (2004) Phospholipid plasmalogen, a surrogate marker of oxidative stress, is associated with increased cardiovascular mortality in patients on renal replacement therapy. Nephrol Dial Transplant 19: 972–976.

16. Lessig J, Fuchs B (2009) Plasmalogens in biological systems: their role in oxidative processes in biological membranes, their contribution to pathological processes and aging and plasmalogen analysis. Curr Med Chem 16: 2021–2041.

17. Ecker J, Liebisch G, Englmaier M, Grandl M, Robenek H, et al. (2010) Induction of fatty acid synthesis is a key requirement for phagocytic differentiation of human monocytes. Proc Natl Acad Sci U S A 107: 7817–7822.

18. Zemski Berry KA, Murphy RC (2004) Electrospray ionization tandem mass spectrometry of glycerophosphoethanolamine plasmalogen phospholipids. J Am Soc Mass Spectrom 15: 1499–1508.

19. Muller G, Kerkhoff C, Hankowitz J, Pataki M, Kovacs E, et al. (1993) Effects of purinergic agents on human mononuclear phagocytes are differentiation dependent. Implications for atherogenesis. Arterioscler Thromb 13: 1317–1326.

20. Smith PK, Krohn RI, Hermanson GT, Mallia AK, Gartner FH, et al. (1985) Measurement of protein using bicinchoninic acid. Anal Biochem 150: 76–85.

21. Bligh EG, Dyer WJ (1959) A rapid method of total lipid extraction and purification. Can J Biochem Physiol 37: 911–917.

22. Liebisch G, Lieser B, Rathenberg J, Drobnik W, Schmitz G (2004) High-throughput quantification of phosphatidylcholine and sphingomyelin by electrospray ionization tandem mass spectrometry coupled with isotope correction algorithm. Biochim Biophys Acta 1686: 108–117.

23. Liebisch G, Vizcaino JA, Kofeler H, Trotzmuller M, Griffiths WJ, et al. (2013) Shorthand notation for lipid structures derived from mass spectrometry. J Lipid Res 54: 1523–1530.

24. Geladi P, Kowalski BR (1986) Partial least-squares regression: a tutorial. Analytica Chimica Acta 185: 1–17.

25. Barker M, Rayens W (2003) Partial least squares for discrimination. Journal of Chemometrics 17: 166–173.

26. Waterbeemd Hvd (1995) Chemometric methods in molecular design. 2.

27. Kondrakhin YV, Sharipov RN, Keld AE, Kolpakov FA (2008) Identification of differentially expressed genes by meta-analysis of microarray data on breast cancer. In Silico Biol 8: 383–411.

28. R Core Team (2013) R: A Language and Environment for Statistical Computing. [2.15.3]. R Foundation for Statistical Computing. Ref Type: Computer Program

29. Castelo R, Roverato A (2009) Reverse engineering molecular regulatory networks from microarray data with qp-graphs. J Comput Biol 16: 213–227.

30. Castelo R, Roverato A (2012) Inference of regulatory networks from microarray data with R and the bioconductor package qpgraph. Methods Mol Biol 802: 215–233.

31. Castelo R, Roverato A (2006) A robust procedure for Gaussian graphical model search from microarray data with p larger than n. JMLR 7: 2621–2650.

32. Jakobsson A, Westerberg R, Jacobsson A (2006) Fatty acid elongases in mammals: their regulation and roles in metabolism. Prog Lipid Res 45: 237–249.

33. Leonard AE, Bobik EG, Dorado J, Kroeger PE, Chuang LT, et al. (2000) Cloning of a human cDNA encoding a novel enzyme involved in the elongation of long-chain polyunsaturated fatty acids. Biochem J 350 Pt 3: 765–770.

34. Wang Y, Torres-Gonzalez M, Tripathy S, Botolin D, Christian B, et al. (2008) Elevated hepatic fatty acid elongase-5 activity affects multiple pathways controlling hepatic lipid and carbohydrate composition. J Lipid Res 49: 1538–1552.

35. Matsuzaka T, Shimano H, Yahagi N, Yoshikawa T, Amemiya-Kudo M, et al. (2002) Cloning and characterization of a mammalian fatty acyl-CoA elongase as a lipogenic enzyme regulated by SREBPs. J Lipid Res 43: 911–920.

36. Moon YA, Shah NA, Mohapatra S, Warrington JA, Horton JD (2001) Identification of a mammalian long chain fatty acyl elongase regulated by sterol regulatory element-binding proteins. J Biol Chem 276: 45358–45366.

37. Ikeda M, Kanao Y, Yamanaka M, Sakuraba H, Mizutani Y, et al. (2008) Characterization of four mammalian 3-hydroxyacyl-CoA dehydratases involved in very long-chain fatty acid synthesis. FEBS Lett 582: 2435–2440.

38. Marquardt A, Stohr H, White K, Weber BH (2000) cDNA cloning, genomic structure, and chromosomal localization of three members of the human fatty acid desaturase family. Genomics 66: 175–183.

39. Ntambi JM, Miyazaki M (2004) Regulation of stearoyl-CoA desaturases and role in metabolism. Prog Lipid Res 43: 91–104.

40. McAndrew RP, Wang Y, Mohsen AW, He M, Vockley J, et al. (2008) Structural basis for substrate fatty acyl chain specificity: crystal structure of human very-long-chain acyl-CoA dehydrogenase. J Biol Chem 283: 9435–9443.

41. Maeba R, Maeda T, Kinoshita M, Takao K, Takenaka H, et al. (2007) Plasmalogens in human serum positively correlate with high- density lipoprotein and decrease with aging. J Atheroscler Thromb 14: 12–18.

42. Seo HS, Lombardi DM, Polinsky P, Powell-Braxton L, Bunting S, et al. (1997) Peripheral vascular stenosis in apolipoprotein E-deficient mice. Potential roles of lipid deposition, medial atrophy, and adventitial inflammation. Arterioscler Thromb Vasc Biol 17: 3593–3601.

43. Gelissen IC, Hochgrebe T, Wilson MR, Easterbrook-Smith SB, Jessup W, et al. (1998) Apolipoprotein J (clusterin) induces cholesterol export from macrophage-foam cells: a potential anti-atherogenic function? Biochem J 331 (Pt 1): 231–237.

44. Jong MC, Hofker MH, Havekes LM (1999) Role of ApoCs in lipoprotein metabolism: functional differences between ApoC1, ApoC2, and ApoC3. Arterioscler Thromb Vasc Biol 19: 472–484.

45. Berbee JF, van der Hoogt CC, Kleemann R, Schippers EF, Kitchens RL, et al. (2006) Apolipoprotein CI stimulates the response to lipopolysaccharide and reduces mortality in gram-negative sepsis. FASEB J 20: 2162–2164.

46. Westerterp M, Van EM, de HW, Offerman EH, Van Berkel TJ, et al. (2007) Apolipoprotein CI aggravates atherosclerosis development in ApoE-knockout mice despite mediating cholesterol efflux from macrophages. Atherosclerosis 195: e9–16.

47. Babaev VR, Fazio S, Gleaves LA, Carter KJ, Semenkovich CF, et al. (1999) Macrophage lipoprotein lipase promotes foam cell formation and atherosclerosis in vivo. J Clin Invest 103: 1697–1705.

48. Tsutsumi K (2003) Lipoprotein lipase and atherosclerosis. Curr Vasc Pharmacol 1: 11–17.

49. Li Y, Schwabe RF, DeVries-Seimon T, Yao PM, Gerbod-Giannone MC, et al. (2005) Free cholesterol-loaded macrophages are an abundant source of tumor necrosis factor-alpha and interleukin-6: model of NF-kappaB- and map kinase-dependent inflammation in advanced atherosclerosis. J Biol Chem 280: 21763–21772.

50. Zhang K, Kaufman RJ (2008) From endoplasmic-reticulum stress to the inflammatory response. Nature 454: 455–462.

51. Tobert JA (2003) Lovastatin and beyond: the history of the HMG-CoA reductase inhibitors. Nat Rev Drug Discov 2: 517–526.

52. Yvan-Charvet L, Ranalletta M, Wang N, Han S, Terasaka N, et al. (2007) Combined deficiency of ABCA1 and ABCG1 promotes foam cell accumulation and accelerates atherosclerosis in mice. J Clin Invest 117: 3900–3908.

53. Oram JF, Lawn RM, Garvin MR, Wade DP (2000) ABCA1 is the cAMP-inducible apolipoprotein receptor that mediates cholesterol secretion from macrophages. J Biol Chem 275: 34508–34511.

54. Klucken J, Buchler C, Orso E, Kaminski WE, Porsch-Ozcurumez M, et al. (2000) ABCG1 (ABC8), the human homolog of the Drosophila white gene, is a regulator of macrophage cholesterol and phospholipid transport. Proc Natl Acad Sci U S A 97: 817–822.

55. Tang C, Liu Y, Kessler PS, Vaughan AM, Oram JF (2009) The macrophage cholesterol exporter ABCA1 functions as an anti-inflammatory receptor. J Biol Chem 284: 32336–32343.

56. Tang C, Oram JF (2009) The cell cholesterol exporter ABCA1 as a protector from cardiovascular disease and diabetes. Biochim Biophys Acta 1791: 563–572.

57. Parini P, Davis M, Lada AT, Erickson SK, Wright TL, et al. (2004) ACAT2 is localized to hepatocytes and is the major cholesterol-esterifying enzyme in human liver. Circulation 110: 2017–2023.

58. Calder PC (2012) Mechanisms of action of (n-3) fatty acids. J Nutr 142: 592S–599S.

59. Funk CD (2001) Prostaglandins and leukotrienes: advances in eicosanoid biology. Science 294: 1871–1875.

60. Farooqui AA, Yang HC, Hirashima Y, Horrocks LA (1999) Determination of plasmalogen-selective phospholipase A2 activity by radiochemical and fluorometric assay procedures. Methods Mol Biol 109: 39–47.

61. Bakhle YS (1999) Structure of COX-1 and COX-2 enzymes and their interaction with inhibitors. Drugs Today (Barc) 35: 237–250.

62. Hulsmans M, Holvoet P (2010) The vicious circle between oxidative stress and inflammation in atherosclerosis. J Cell Mol Med 14: 70–78.

63. Colas R, Sassolas A, Guichardant M, Cugnet-Anceau C, Moret M, et al. (2011) LDL from obese patients with the metabolic syndrome show increased lipid peroxidation and activate platelets. Diabetologia 54: 2931–2940.

64. Xu H, Barnes GT, Yang Q, Tan G, Yang D, et al. (2003) Chronic inflammation in fat plays a crucial role in the development of obesity-related insulin resistance. J Clin Invest 112: 1821–1830.

65. Kelly LM, Englmeier U, Lafon I, Sieweke MH, Graf T (2000) MafB is an inducer of monocytic differentiation. EMBO J 19: 1987–1997.

66. Meimaridou E, Gooljar SB, Chapple JP (2009) From hatching to dispatching: the multiple cellular roles of the Hsp70 molecular chaperone machinery. J Mol Endocrinol 42: 1–9.

67. Morano KA (2007) New tricks for an old dog: the evolving world of Hsp70. Ann N Y Acad Sci 1113: 1–14.

68. Tavaria M, Gabriele T, Kola I, Anderson RL (1996) A hitchhiker's guide to the human Hsp70 family. Cell Stress Chaperones 1: 23–28.

69. Wegele H, Muller L, Buchner J (2004) Hsp70 and Hsp90–a relay team for protein folding. Rev Physiol Biochem Pharmacol 151: 1–44.

70. Kirkegaard T, Roth AG, Petersen NH, Mahalka AK, Olsen OD, et al. (2010) Hsp70 stabilizes lysosomes and reverts Niemann-Pick disease-associated lysosomal pathology. Nature 463: 549–553.

71. Schroder O, Schulte KM, Ostermann P, Roher HD, Ekkernkamp A, et al. (2003) Heat shock protein 70 genotypes HSPA1B and HSPA1L influence cytokine concentrations and interfere with outcome after major injury. Crit Care Med 31: 73–79.

72. Laurin M, Fradet N, Blangy A, Hall A, Vuori K, et al. (2008) The atypical Rac activator Dock180 (Dock1) regulates myoblast fusion in vivo. Proc Natl Acad Sci U S A 105: 15446–15451.

73. Cote JF, Motoyama AB, Bush JA, Vuori K (2005) A novel and evolutionarily conserved PtdIns(3,4,5)P3-binding domain is necessary for DOCK180 signalling. Nat Cell Biol 7: 797–807.

74. Piao L, Nakagawa H, Ueda K, Chung S, Kashiwaya K, et al. (2011) C12orf48, termed PARP-1 binding protein, enhances poly(ADP-ribose) polymerase-1 (PARP-1) activity and protects pancreatic cancer cells from DNA damage. Genes Chromosomes Cancer 50: 13–24.

75. van DS, Cordeiro R, Craig T, van DJ, Wood SH, et al. (2012) GeneFriends: an online co-expression analysis tool to identify novel gene targets for aging and complex diseases. BMC Genomics 13: 535.

76. Hotta Y, Furukawa K, Tabata S (1995) Meiosis specific transcription and functional proteins. Adv Biophys 31: 101–115.

77. Doherty GJ, Ahlund MK, Howes MT, Moren B, Parton RG, et al. (2011) The endocytic protein GRAF1 is directed to cell-matrix adhesion sites and regulates cell spreading. Mol Biol Cell 22: 4380–4389.

78. Garcia-Anoveros J, Derfler B, Neville-Golden J, Hyman BT, Corey DP (1997) BNaC1 and BNaC2 constitute a new family of human neuronal sodium channels related to degenerins and epithelial sodium channels. Proc Natl Acad Sci U S A 94: 1459–1464.

79. Junttila MR, Puustinen P, Niemela M, Ahola R, Arnold H, et al. (2007) CIP2A inhibits PP2A in human malignancies. Cell 130: 51–62.

80. Liao X, Sharma N, Kapadia F, Zhou G, Lu Y, et al. (2011) Kruppel-like factor 4 regulates macrophage polarization. J Clin Invest 121: 2736–2749.

81. Krishnaraju K, Nguyen HQ, Liebermann DA, Hoffman B (1995) The zinc finger transcription factor Egr-1 potentiates macrophage differentiation of hematopoietic cells. Mol Cell Biol 15: 5499–5507.

82. Nguyen HQ, Hoffman-Liebermann B, Liebermann DA (1993) The zinc finger transcription factor Egr-1 is essential for and restricts differentiation along the macrophage lineage. Cell 72: 197–209.

83. Reed BD, Charos AE, Szekely AM, Weissman SM, Snyder M (2008) Genome-wide occupancy of SREBP1 and its partners NFY and SP1 reveals novel functional roles and combinatorial regulation of distinct classes of genes. PLoS Genet 4: e1000133.

11

Nitrogen-Deprivation Elevates Lipid Levels in *Symbiodinium* spp. by Lipid Droplet Accumulation: Morphological and Compositional Analyses

Pei-Luen Jiang[1,2]**, Buntora Pasaribu**[3]**, Chii-Shiarng Chen**[1,2,4]*

1 Graduate Institute of Marine Biotechnology, National Dong-Hwa University, Pingtung, Taiwan, **2** Taiwan Coral Research Center, National Museum of Marine Biology and Aquarium, Pingtung, Taiwan, **3** Graduate Institute of Biotechnology, National Chung-Hsing University, Taichung, Taiwan, **4** Department of Marine Biotechnology and Resources, National Sun Yat-Sen University, Kaohsiung, Taiwan

Abstract

Stable cnidarian-dinoflagellate (genus *Symbiodinium*) endosymbioses depend on the regulation of nutrient transport between *Symbiodinium* populations and their hosts. It has been previously shown that the host cytosol is a nitrogen-deficient environment for the intracellular *Symbiodinium* and may act to limit growth rates of symbionts during the symbiotic association. This study aimed to investigate the cell proliferation, as well as ultrastructural and lipid compositional changes, in free-living *Symbiodinium* spp. (clade B) upon nitrogen (N)-deprivation. The cell proliferation of the N-deprived cells decreased significantly. Furthermore, staining with a fluorescent probe, boron dipyrromethane 493/503 (BODIPY 493/503), indicated that lipid contents progressively accumulated in the N-deprived cells. Lipid analyses further showed that both triacylglycerol (TAG) and cholesterol ester (CE) were drastically enriched, with polyunsaturated fatty acids (PUFA; i.e., docosahexaenoic acid, heneicosapentaenoic acid, and oleic acid) became more abundant. Ultrastructural examinations showed that the increase in concentration of these lipid species was due to the accumulation of lipid droplets (LDs), a cellular feature that have previously shown to be pivotal in the maintenance of intact endosymbioses. Integrity of these stable LDs was maintained via electronegative repulsion and steric hindrance possibly provided by their surface proteins. Proteomic analyses of these LDs identified proteins putatively involved in lipid metabolism, signaling, stress response and energy metabolism. These results suggest that LDs production may be an adaptive response that enables *Symbiodinium* to maintain sufficient cellular energy stores for survival under the N-deprived conditions in the host cytoplasm.

Editor: Stephan Neil Witt, Louisiana State University Health Sciences Center, United States of America

Funding: This works was supported by National Science Council in Taiwan to C.S. Chen (NSC 101-2311-B-291-002-MY3) and PL Jiang (NSC 100-2313-B-291-003 and NSC 102-2313-B-291-001). The funders had no role in study design, data collection and analysis, decision to publish, or preparation of the manuscript.

Competing Interests: The authors have declared that no competing interests exist.

* E-mail: cchen@nmmba.gov.tw

Introduction

Symbiodinium spp., a unicellular dinoflagellate, is commonly found in mutualistic associations with invertebrates such as corals and anemones, and able to transfer more than 90% of its photosynthetically fixed carbon to the host cytoplasm in which it resides [1]. This intracellular symbiosis (i.e. the endosymbiosis) has been the focus of intensive research due to its pivotal role in maintaining the health of corals and homeostasis of the marine ecosystem.

Numerous studies have highlighted the crucial role of nutritional status and nutrient transfer in these endosymbioses [2]. The supply of organic carbon by *Symbiodinium* to the hosts and the recycling of essential nutrients during such associations have contributed to the success of coral reefs in nutrient-limited tropical seas [3]. On the other hand, the growth of *Symbiodinium* is dependent on nutrients from various sources including exogenous seawater, host catabolism, and host heterotrophy [4,5]. Nitrogen, which is one of the most important essential nutrients, can be excreted as ammonium by the host [6].

The host metabolism plays a significant role in regulating the nutritional status of its endosymbionts (i.e. the symbiotic *Symbiodinium*) [7]. For instance, it has been shown that the endosymbionts were present in nutrient-limited environments [8,9], particularly with respect to nitrogen sources. Using the infrared microspectroscopy, it has been shown that the concentration of nitrogenous compounds was significantly lower in endosymbionts, compared to that of free-living *Symbiodinium* [9]. It suggests that the cytoplasm of the host cell may be nitrogen-deficient and alter various physiologies of the endosymbiont. In examining the effect of nutrient supplement of host anemones on their endosymbionts, Zhu and colleagues have observed that there was an increase formation of lipid droplets (LDs) in endosymbionts after 45 days of nutrient starvation on host sea anemones [10]. However, the cellular and molecular mechanisms of this regulation remain to be elucidated.

Nitrogen deprivation represents an important source of stress for microalgae, which causes various changes in cellular metabolism and development. For example, treatments with nitrogen deprivation tend to elevate lipid production in algae [11,12], such as increases of cellular triacylglycerols (TAGs) stored in the cytoplasmic lipid droplets (LDs) [13]. TAGs, which are composed primarily of saturated and monounsaturated fatty acids, can be efficiently packed into the cell and generate more energy than carbohydrates upon oxidation, thus constituting the best reserve for rebuilding the cell upon returning to homeostatic conditions [14]. The proteins associated with these LDs have been the focus of investigations in order to elucidate the mechanism of LD formation [15]. Recent studies have shown that the LD-associated protein in *Chlorella* cells belong to the caleosin family [16,17]. Another lipid droplet protein, the major lipid droplet protein (MLDP), was also revealed in the LDs of various green microalga including *Chlamydomonas reinhardtii*, *Haematococcus pluvialis* [18,19] and *Dunaliella* [20].

In order to elucidate the mechanism that allows the endosymbionts to adapt to the nitrogen-limited environment of cnidarian hosts, the present study aims to first examine the cellular response of free-living *Symbiodinium* (clade B) to nitrogen deprivation treatment. To the best of our knowledge, the effect of nitrogen deprivation in free-living *Symbiodinium* spp. has never been described, mainly due to the relative difficulty of cultivating *Symbiodinium* spp. with synthetic culture media [21]. This is an unfortunate knowledge dearth, as nitrogen is an important nutrient required for the metabolism of *Symbiodinium* [22]. Under nitrogen deprivation, the specific goal here was to examine the changes in cellular biology, including the proliferation, lipid contents and ultrastructure in free-living *Symbiodinium* spp. Results show that there is increased formation of LDs, a phenomenon similar to that occur in symbiotic *Symbiodinium*. Subsequently, LDs of these N-deprived cells were purified for biochemical analyses for proteins and lipids. Results of the present study should provide important information in elucidating possible regulatory mechanisms underlying LDs formation in nitrogen-deficient environments such as those in the symbiotic association with Cnidaria.

Materials and Methods

Symbiodinium culture and the nitrogen-deprivation treatment

The free-living *Symbiodinium* spp. (clade B) used in this study were originally isolated from the sea anemone *Aiptasia pulchella* [9]. They were cultured in the f/2 medium [23] in filtered seawater (FSW) at room temperature under a photosynthetically active radiation (PAR) of 40 μmol m^{-2}s^{-1} in a 12-h light/12-h dark (12L/12D) cycle. Media were prepared with two different concentrations of nitrogen (as NaNO$_3$). Two new batch cultures were grown, one in nitrogen-deficient f/2 medium (no NaNO$_3$ was added to the medium) and the other in nitrogen-sufficient f/2 medium (0.882 mM NaNO$_3$).

Symbiodinium clade identification

The genetic identity (18S rDNA) of the cultured *Symbiodinium* was examined by PCR-RFLP (Polymerase chin reaction-Restriction fragment length polymorphism) analysis [24], and shown to be from clade B. *Symbiodinium* DNA was extracted using a plant genomic DNA extraction miniprep system (VIOGENE, Taipei). Basically, *Symbiodinium* nuclear small subunit (n18S-rDNA) was amplified by PCR) from 3 replicate extracts of each of the two cultures using the primers, ss5z (an equimolar mixture of the oligonucleotides 5'-GCAGTTATAATTTATTTGATGGTCAC-TGCTAC-3' and 5'-GCAGTTATAGTTTATTTGATGGTT-GCTGCTAC-3') and ss3z (5'-AGCACTGCGTCAGTCCGAA-TAATTCACCGG-3') and digested with the restriction enzyme, *Taq* I and *Sau*3A I (Promega, USA). Digestion products were separated by electrophoresis on 1.5% 0.5x TAE (Amresco, USA) agarose gels, to generate the RFLP pattern. RFLP pattern analysis

was compared to the literature [24] to assign each culture to one of the established *Symbiodinium* n18S-rDNA RFLP clades.

Cell density analysis

Symbiodinium proliferation was examined with hemocytometer-based cell counting. Cell densities were determined daily by placing an aliquot of well-mixed culture suspension on a Neubauer hemocytometer (Marienfel, Germany) under a Axioskop 2 Plus microscope (Zeiss, Germany) connected to a CCD (charge-coupled device) camera (Photometrics. USA)

Chlorophyll a and protein determinations

Symbiodinium were harvested by centrifugation at 12000 g for 10 min from the control and nitrogen-deprived culture. For each replicate during the analysis, a volume of culture containing 5×10^5 cells was used. Chlorophyll *a* was extracted in 90% acetone (v/v), and their amount was estimated spectrophotometrically as previously described [25]. The protein content was determined using the BCA protein assay kit (Invitrogen, USA) on the same samples used for chlorophyll quantification.

The isolation of lipid droplets (LDs) from nitrogen-deprived *Symbiodinium*

LDs were isolated from the nitrogen-deprived culture (1L) on days 5 and 7 by first homogenizing at 4°C in a "grinding buffer" (0.6 M sucrose in 10 mM sodium phosphate buffer, pH 7.5) according to a published procedure [26]. After filtration, each 8-ml portion of the homogenate was placed at the bottom of a 12-ml centrifuge tube, and 2-ml of "flotation buffer" (0.4 M sucrose in 10 mM sodium phosphate buffer, pH 7.5) was layered on top. The tube was centrifuged at 35,000×g for 60 min in a swinging-bucket rotor centrifuge (Beckman Coulter, USA). LDs on the top layer were collected and re-suspended in the "detergent washing solution" (0.4 M sucrose, 0.1% Triton X-100 in 10 mM sodium phosphate buffer, pH 7.5) to remove non-specifically associated proteins. After a further centrifugation, LDs on the top were collected and then re-suspended in the floating buffer to remove excess detergent. The washing step was repeated two more times. Finally, the purified LDs were re-suspended in the grinding buffer and stored at −20°C until further analysis.

Structural integrity of *Symbiodinium* LDs

The structural integrity of LDs isolated from *Symbiodinium* was assessed by examining the surface properties (steric hindrance and electrostatic repulsion) that accounted for the aggregation of LDs without fusion at pH 6.5 [26]. *Symbiodinium* LDs suspended in 5 mM sodium phosphate buffer, pH 7.5 or 6.5, were kept at 23°C for 6 hrs. To confirm that the steric hindrance was provided by surface proteins, a 2-ml preparation of *Symbiodinium* LDs was subjected to trypsin (2.5 μg; bovine pancreas type III, Sigma, USA) digestion at 37°C for 30 min.

Lipid analyses

All solvents for lipid analyses were analytic grade. Lipid contents of 3 replicates from the nitrogen-sufficient f/2 medium and nitrogen-deprived culture *Symbiodinium* cells were extracted by the Bligh and Dyer procedure [27]. Neutral lipids in the LDs isolated from *Symbiodinium* cells were extracted with 150 μl of chloroform/methanol (2:1, v/v). After centrifugation, the lower chloroform fraction was collected for the analysis by thin layer chromatography (TLC) (Analtech, USA) with the solvent system modified from previous reports [28,29]. Briefly, TLC was first developed to the R$_f$ = 1 position in hexane. The plate was air-dried and then developed to the top ($R_f = 1$) in benzene. The plate was air-dried and then developed to the top ($R_f = 0.5$) in hexane: diethyl ether: acetic acid (70:30:1 v/v/v). The lipid visualization on TLC plates was performed by staining with 0.03% Coomassie blue R 250 (Sigma, USA) dissolved in 20% methanol containing 0.5% acetic acid [30]. Concentrations of individual lipid species were then quantified using the Metamorph Image Processing system (Molecular Devices Inc., Toronto, Canada) based on calibration curves of individual lipid standards (Wax ester: Sigma-Aldrich, USA; TAGs [mixed triacylglycerides: tricaprin, tricaprylin, trilaurin, trimyristin, tripalmitin]: Sigma-Aldrich, USA; Cholesterol: Avanti Polar Lipids, USA; CE: Sigma-Aldrich, USA) co-run on the same TLC plate. To analyze the phospholipids in extracted lipids, chloroform: acetic acid: methanol: water (70:25:5:2, v/v/v/v) was used as the TLC solvent system.

Analyses of fatty acid composition in cellular TAGs by gas chromatography (GC)-mass spectrometry (MS)

Symbiodinium cells collected after five days of culture in either nitrogen-enriched (control) or nitrogen-deprivation f/2 medium were dried by lyophilization. Total lipids in dried cells were extracted and separated by TLC as previously described. TAGs on TLC plates was first identified by 0.001% primuline spraying (in 80% acetone), and then extracted by the Bligh and Dyer procedure [27]. Isolated TAGs were first saponified in 1N sodium hydroxide-methanol solution for 15 min at 80°C. The fatty acids were esterified in 14% boron trifluoride-methanol solution for 15 min at 100°C. After hexane extraction, the fatty acid methyl esters were analyzed on a gas chromatograph (GC, Varian CP-3800) and a mass spectrometer (Varian 320 MS) operated in full scan mode; scan range from 100 to 450 m/z. The column was a CP-Sil88 capillary column of length 20 m, 0.25 mm i.d., and the stationary phase had a film thickness of 0.2 μm. Helium was used as the carrier gas at a flow rate of 0.8 ml min^{-1}. The temperature program was as follows: held at 50°C for 1 min, 50–200°C with 8°C min^{-1}, held for 5 min, from 200–230°C with 20°C min^{-1}. Retention times and mass spectra were compared against the NIST02 library (National Institute of Standards and Technology, Gaithersburg, MD, USA) to identify fatty acids. Saturn GC/MS Workstation v6.9.3 software (Varian) was used to visualize spectra, integrate areas under peaks and search the library. Peaks of fatty acids were identified, and the relative amount of individual fatty acid was calculated by the integrated area percentage among total fatty acids.

Fluorescent microscopy

BODIPY 493/503 (4,4-difluoro-1,3,5,7-tetramethyl-4-bora-3a,4a- diaza-s-indacene) is a fluorescent lipophilic stain [31] widely used to label lipid droplets in plants. *Symbiodinium* and purified LDs were stained with 38.2 μM BODIPY 493/503(Invitrogen, USA) in the dark for 20 min at RT. The stained cells and LDs were visualized using the fluorescence microscope (Zeiss, Germany).

The transmission electron microscopy and imaging analysis

To investigate the intracellular accumulation of LDs, *Symbiodinium* cells under nitrogen-deprivation treatment were collected and fixed in 2.5% glutaraldehyde and 2% paraformaldehyde in 100 mM sodium phosphate containing and 5% sucrose (pH 7.3) for 2.5 h at 4°C. They were then rinsed with 100 mM sodium phosphate buffer at 4°C. Cells were then post-fixed in 1% OsO$_4$ in 50 mM sodium phosphate (pH 7.3) for 1 hr at 4°C. The cell

aliquots were then washed three times for 15 min each with the same buffer and dehydrated by a graded ethanol series (50, 70, 80, 90, 95 and 100%) before embedding in LR white Resin. Thin sections (70 nm) cut by a Leica Reichert Ultracut R were collected on nickel grids, post-stained with 2.5% uranyl acetate and 0.4% lead citrate, rinsed 3 times with water, and the samples were viewed on a JEM-1400 transmission electron microscope (JEOL, Japan). In order to determine the LDs area from the acquired images, the ratio of the actual length to pixel was first determined by distance calibration using the scale bar of the acquired transmission electron microscopy (TEM) image. Individual LDs were selected by threshold adjustment, and the area (μm^2) of individual LDs was calculated with Metamorph's region measurement function.

SDS-PAGE and Western blotting

Proteins from *Symbiodinium* cells and LDs were extracted with an equal volume of 2x sample buffer according to the suggestions in the Bio-Rad (Bio-Rad, USA) Trans-Blot instruction manual and resolved by SDS-PAGE using 15% (w/v) polyacrylamide in the separating gel and 4.75% polyacrylamide in the stacking gel [32]. After electrophoresis, the gel was stained with Coomassie Blue R-250 and then destained with methanol/acetic acid. For Western blotting, proteins were transferred from SDS-PAGE onto a nitrocellulose membrane in a Trans-Blot system (Bio-Rad, USA) according to the manufacturer's instructions. The membrane was subjected to immune-detection using a rabbit anti-ribulose-1,5-bisphosphate carboxylase/oxygenase (Rubisco) large subunit (1;2000 dilution; Cat. AS0037, Agrisera, Vannas, Sweden). After washing, the membrane was incubated with secondary antibodies conjugated with goat anti-rabbit horseradish peroxidase (HRP). The membrane was subsequently washed and resulting proteins visualized using SuperSignal West Pico Chemiluminescent substrate kits (Thermo Fisher Scientific, USA) according to the manufacturer's recommendations and visualized on a Brand Vilber Lourmat Model Fusion FX7 gel-doc under the chemiluminescent settings for 2 min.

In-gel digestion of the lipid droplet proteins in *Symbiodinium* spp.

Five protein bands of *Symbiodinium* lipid droplet resolved by SDS-PAGE were manually excised from the gel and ground into pieces. After washing with 50% acetonitrile and 50% acetonitrile/25 mM ammonium bicarbonate, the protein was reduced and alkylated at 56°C for 45 min in 10 mM dithiothreitol and 55 mM iodoacetamide in 25 mM ammonium bicarbonate, followed by overnight in-gel digestion with 0.1 μg in 15 μl of TPCK-treated modified porcine trypsin (Promega, USA) in the same buffer at 37°C. The supernatant containing tryptic peptides was combined with two more extracts of the gel by 50% acetonitrile/5% formic acid. The sample was analyzed by matrix-assisted laser desorption/ionization-mass spectrometry (MALDI-MS) and MALDI-MS/MS. All data were acquired by quadrupole-time-of-flight (Q-TOF) hybrid mass spectrometers (Micromass Q-Tof Ultima, Manchester, UK, and Applied Biosystems QSTAR, USA), in which α-cyano-4-hydroxycinnamic acid was used as the matrix. The low-energy collision-induced dissociation MS/MS product ion spectra acquired from Q-TOF Ultima and QSTAR were analyzed by Micromass ProteinLynx™ Global Server 2.0 and Applied Biosystems BioAnalyst™ data processing software, respectively. For protein identification, the acquired MS/MS spectra were automatically searched against the NCBInr database using the Mascot search program (www.matrixscience.com) restricted to all entries taxonomy. The mass tolerance parameter was 20 ppm, the MS/MS ion mass tolerance was 1 Da, and up to one missed cleavage was allowed. Variable modifications considered were methionine oxidation and cysteine carboxyamidomethylation. Positive identification of proteins was confirmed by observation of at least one of the following criteria: (i) the total number of matched peptides (mps) is more than 2, or (ii) the mps equals 2 with two different matched peptides, or (iii) the MOWSE score has to be higher than 70 which indicates identity or extensive homology (p<0.05).

Statistical analysis

All statistical analyses were performed using SigmaStat 3.5 (Systat software, Chicago, IL, USA). The results were expressed as mean±SD (standard deviation of the mean).

Results

Nitrogen-deprivation induces lipid content increase in *Symbiodinium*

To examine the effects of nitrogen deprivation on *Symbiodinium* growth and lipid accumulation, free-living *Symbiodinium* in f/2 medium at the early stationary phase were transferred to nitrogen-free f/2 medium. As shown, cells cultivated in the nitrogen-deprived medium proliferated more slowly than those cultivated in the normal medium, particularly after three days of experimentation (Fig. 1A).

The lipid accumulation in nitrogen-deprived *Symbiodinium* could be visualized by staining the cells with a neutral lipid specific dye BODIPY 493/503. In *Symbiodinium* grown in control medium, little lipid accumulation was observed as shown by very dim BODIPY staining (Fig. 1B). On the contrary, BODIPY fluorescence gradually increased in cells with the nitrogen-deprivation treatment (Fig. 1B; day 1 to day 9), indicating an increase of lipid accumulation.

The change in lipid content during nitrogen-deprivation treatment was then analyzed by TLC (Fig. 2). In the nitrogen-deprivation culture, *Symbiodinium* began to accumulate neutral lipids including TAGs and CEs (cholesterol esters); both of which were very low (<1 pg per cell) in control culture (Fig. 2A). Concentration of both TAGs and CEs increased to reach maximal levels (168.57 ± 4.93 and 13.5 ± 0.52 pg cell^{-1}, respectively) after seven days of nitrogen-deprivation treatment (Table 1).

As the amount of TAG was significantly increased by nitrogen-deprivation treatment, their fatty acid compositions in control and nitrogen-deprivation treatment for five days were further analyzed by GC-MS (Fig.3). In the control group of *Symbiodinium*, four major fatty acids were identified, such as C14:0, C16:0, C18:0 and C22:1. Among them, C16:0 and C18:0 were the most abundant, occupying approximately 38% and 32% of total fatty acids, respectively. After cultivation in nitrogen-deprived medium for 5 days, the percentages of C14:0 and C16:0 remained unchanged. Nevertheless, C18:0 drastically decreased from about 32% to 6.4%. On the other hand, apart from the C12:0 and C22:0, the

A

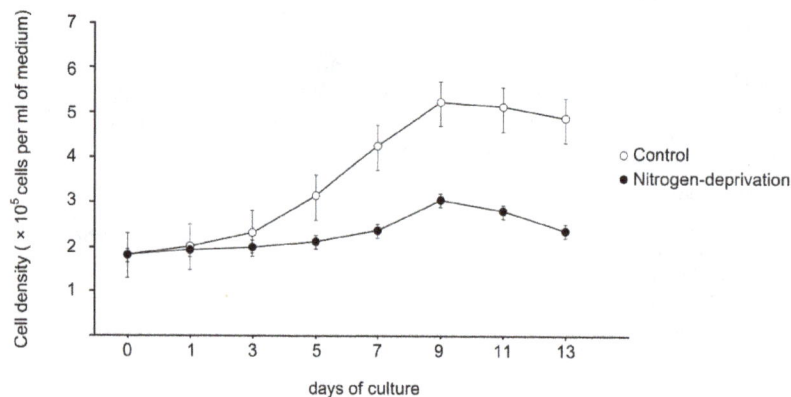

B

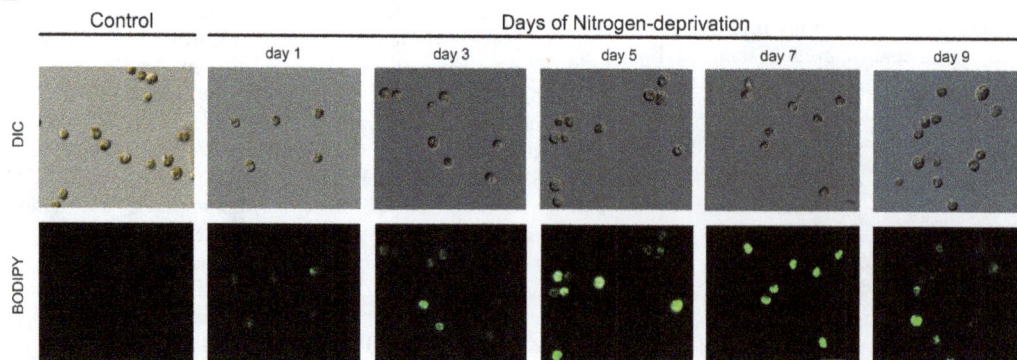

Figure 1. Effect of nitrogen-deprivation on the cell proliferation and lipid accumulation in *Symbiodinium*. (A) Growth of *Symbiodinium* cells cultivated in control versus nitrogen-deprivation media. The data represents mean ± SD (n=3). (B) The visualization of neutral lipid accumulation using BODIPY 493/503 in control vs. nitrogen-deprived cultures. Scale bar, 10 μm.

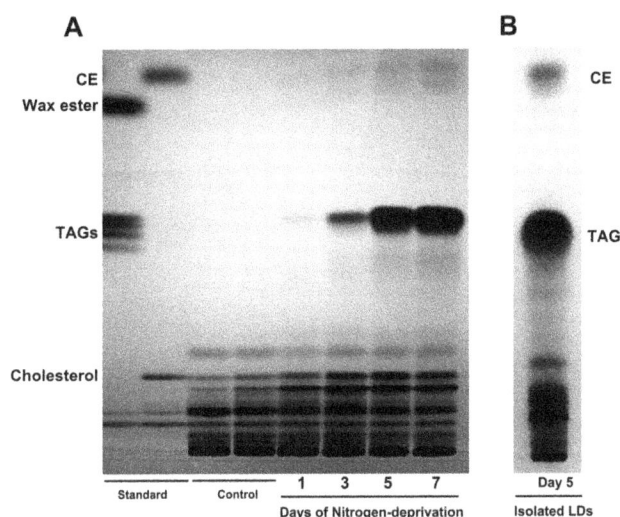

Figure 2. The TLC analysis of lipids extracted from _Symbiodinium_ spp. cells. The comparison between control and nitrogen-deprivation treated cells (A). The lipid content of the purified LDs from _Symbiodinium_ after five days of nitrogen deprivation is shown in (B).

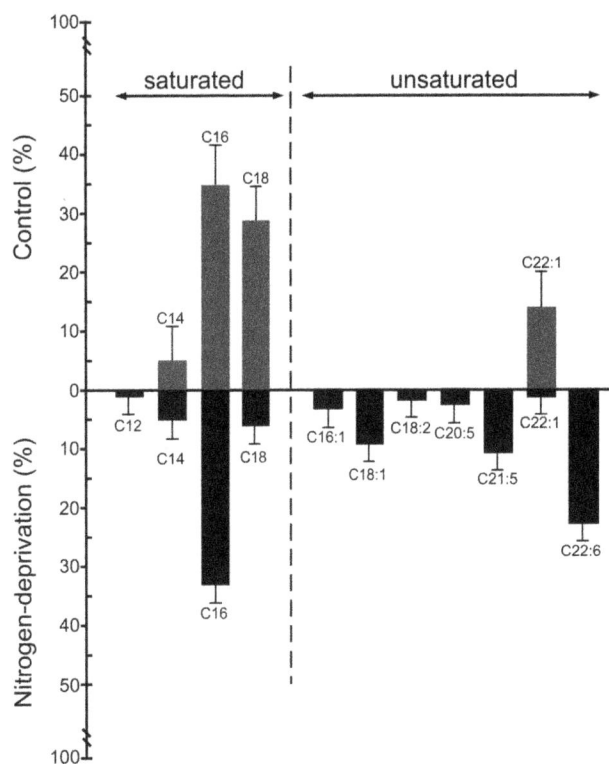

Figure 3. The change of fatty acid compositions in TAGs of _Symbiodinium_ after five days of nitrogen deprivation. Relative amounts (%) of fatty acid compositions in purified TAGs from total _Symbiodinium_ were determined (see the "Materials and methods" section). The data represents mean ± SD (n = 3).

amount of numerous polyunsaturated fatty acids (PUFAs) such as C16:1, C18:1, C18:2, C20:5, C21:5 and C22:6, increased significantly. In other words, the proportion of PUFAs over saturated fatty acids (SFA) in TAGs increased in _Symbiodinium_ cells under the nitrogen-deprived condition.

Ultrastructural changes in _Symbiodinium_ by nitrogen-deprivation

Morphological changes of _Symbiodinium_ after nitrogen-deprivation treatment were investigated by TEM (transmission electron microscopy). First, the average cell size (diameter) of _Symbiodinium_ cultivated in nitrogen-deprivation medium at day 5 and 7 increased significantly to 7.35 ± 0.86 µm and 6.96 ± 0.96 µm, respectively (_versus_ 6.54 ± 1.02 µm in control) (see Table 2). Secondly, the change in cell size was concurrent with cell wall thickness changes (Table 2 and Fig. 4). Specifically, the thickness of the cell wall increased three fold over the controls, from 0.08 ± 0.04 µm to 0.24 ± 0.06 µm at day 5, and then decreased to 0.17 ± 0.05 µm at the 7th day (see also the arrows in Figs. 4A, 4C, and 4E). Thirdly, there were 83.24% and 86.1% reductions in chlorophyll _a_ concentration relative to the controls in nitrogen-deprivation cells at day 5 and 7, respectively (Table 2). Finally, there were significant changes in the size and number of LDs after nitrogen-deprivation treatment (Table 2 and Fig. 4). A number of large LDs accumulated in the cytosol of _Symbiodinium_ (Figs. 4C and 4E) in nitrogen-deprivation samples. The LDs formation was

greatly induced by nitrogen-deprived at day 5, as shown in Figs. 4C-D. Furthermore, the sizes of these LDs significantly increased from 0.72 ± 1.02 µm^2 to 2.90 ± 1.88 µm^2 after 7 days of the nitrogen-deprivation treatment (Table 2). Moreover, numerous inclusion bodies, which are OsO_4 staining-negative, appeared inside the LDs at the 7th day of nitrogen-deprivation treatment (see the arrowhead in Fig. 4F).

Analysis of LD-associated proteins and lipids during the nitrogen-deprivation

LDs were isolated from the cultured _Symbiodinium_ incubated in nitrogen-deprivation medium for 5 days. It showed that the LDs purified from the _Symbiodinium_ cells maintained as individual particles in a medium of pH 7.5 at 23°C (Fig. 5A). An aggregation of these LDs was induced by lowering the pH of the medium to 6.5; most of these aggregates did not coalesce when left overnight

Table 1. TAGs and CEs accumulation in _Symbiodinium_ spp.

Lipid concentrations (pg/cell)	Control	Days of nitrogen-deprivation			
		Day 1	Day 3	Day 5	Day 7
TAGs	tr	1.90 ± 0.51^{ab}	96.55 ± 0.71^{c}	163.50 ± 5.54^{d}	168.57 ± 4.93^{de}
CEs	tr	tr	2.73 ± 0.93^{a}	8.95 ± 0.49^{b}	13.50 ± 0.52^{c}

Lipid contents of _Symbiodinium_ in the control and nitrogen-deprivation treatments were analyzed by TLC. Data are presented as mean±SD (N = 3). Superscript a-d denote statistical significance within control and nitrogen-starvation treatments (P<0.001). tr, trace (<1 pg).

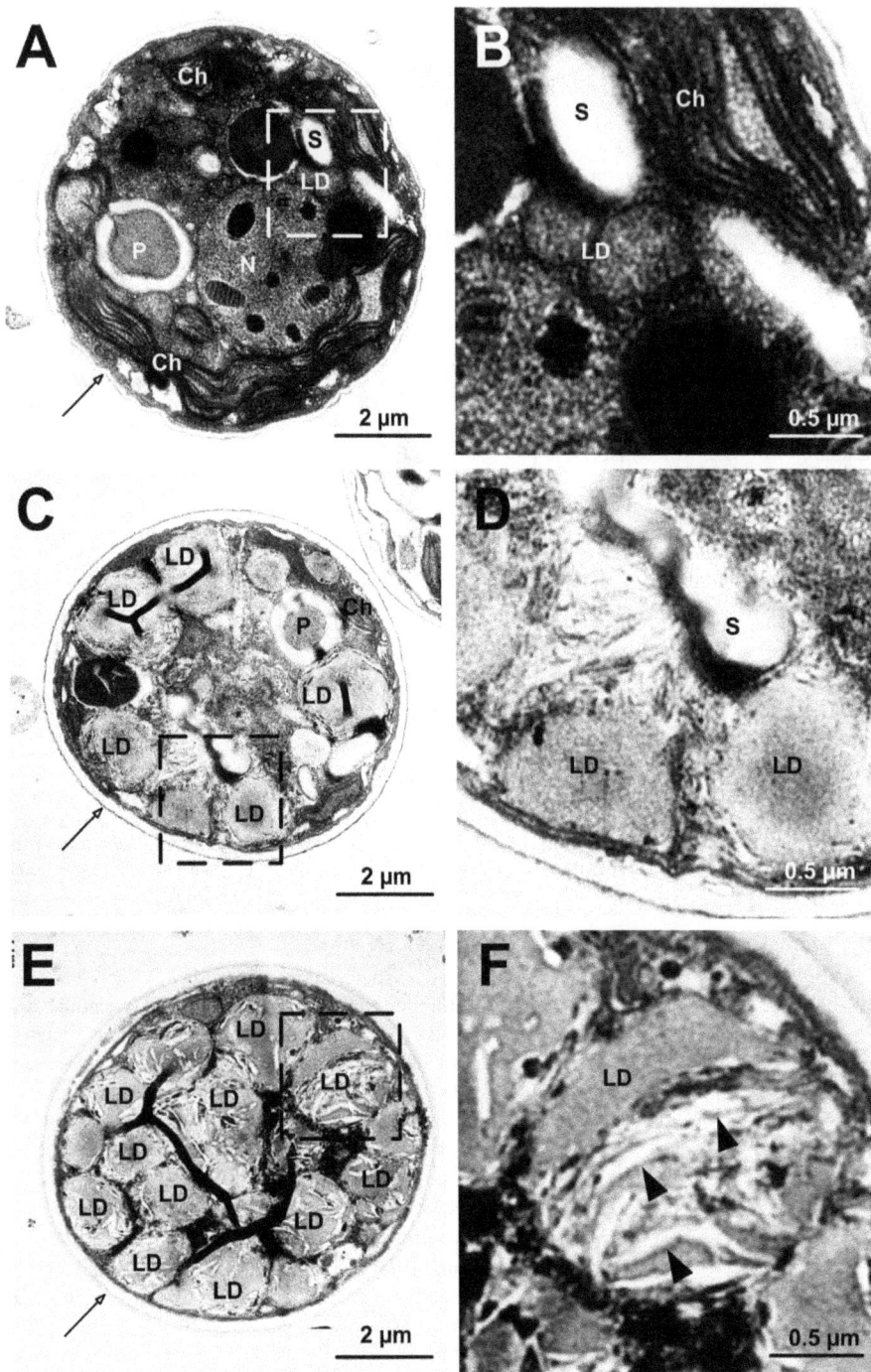

Figure 4. The ultrastructural examination of morphological changes and LD formation in *Symbiodinium* after nitrogen deprivation. Transmission electron micrographs of *Symbiodinium* in control (A, B) and nitrogen-deprivation media (five days: C–D; seven days: E–F). Insets in A, C and D were magnified as B, D and F, respectively. Arrows in A, C, and E indicated cell walls, while arrowheads in F indicated the OsO4-negative "inclusion bodies". Abbreviations: LD, lipid droplet; Ch, chloroplast; S, starch granule; P, pyrenoids; N, nucleolus.

at 23°C (Fig. 5B). After the trypsin treatment, LDs coalesced, floating rapidly and forming a transparent layer on the top of the reaction solution (Fig. 5C). Therefore, the structural integrity of the purified LDs was presumably maintained via the electrostatic repulsion and steric hindrance, in a manner similar to the stable LDs isolated from the *Chlorella* cells [16]. Furthermore, there was significant amount of membrane phospholipids (PLs) in purified LDs but not the lower layer of fractions after the detergent

washing (see the "Materials and methods" section) (Fig. 5D). This further indicates that the purified LDs have maintained their membrane integrity during the purification process.

The purity of the purified LDs was confirmed based on their absence of a chloroplast-specific protein, RuBisCO, by western blotting according to a published procedure [33] (Fig. 6A). The high lipid contents of the purified LDs were confirmed by strong BODIPY 493/503 staining (Fig. 6B). The TLC analysis showed

Table 2. Influence of duration after nitrogen depletion on *Symbiodinium* cells.

	Control	Starvation (Day 5)	Starvation (Day 7)
Cell size (diameter, μm)	6.54±1.02^a (n = 86)	7.35±0.86^b (n = 123)	6.96±0.96^c (n = 110)
Cell wall (thickness, μm)	0.08±0.04^a (n = 63)	0.24±0.06^b (n = 115)	0.17±0.05^c (n = 109)
Chl *a*/protein (μg 100cell^{-1}) Reduction of Chl *a* (%)	83.27±0.398 0%	3.98±0.011 95.22%	3.30±0.011 96.03%
LDs size (area, μm^2)	0.72±1.02^a (n = 117)	2.66±1.61^b (n = 436)	2.90±1.88^{b,c} (n = 497)

Impact of the nutrient regime in cell pattern between the normal growth and nitrogen starvation (i.e Cell size, LD size, Chl *a*, Cell wall) as analyzed by ANOVA. Upper level a–c denote statistical significance different between control growth, day 5(starvation) and day 7(starvation), respectively (P<0.001). Cell wall and cell size: n = number of cell analyzed, LD size: n= number of LDs analyzed.

that the purified LDs contained mainly neutral lipids such as TAGs and CEs (see Fig. 2B), which are both the major lipid component in whole cells as shown in Fig. 2A.

Five major LD-associated protein bands with an estimated molecular weight ranging from 17 to 84 kDa (Fig. 6A) were excised and analyzed by mass spectrometry. Seven proteins were identified, with a variety of cellular functions including lipid metabolism ("Sterol transfer protein" from the protein band 5, MW~17 kDa), signaling ("14-3-3 protein" from the protein band 3, MW~28 kDa and "ADP-ribosylation factor (ARF)" from the protein band 4, MW~19 kDa), stress responses ("Heat shock protein HSP90" from the protein band 1, MW~84 kDa and "Osmotically inducible protein Y" from the protein band 3, MW~28 kDa) and energy metabolism ("Mitochondria ATP synthase F1 alpha subunit-like protein 1" from the protein band 2, MW~59 kDa and "GTP-binding protein" from the protein band 4, MW~19 kDa) (Table 3).

Discussion

Effects of nitrogen-deprivation on *Symbiodinium* proliferation and morphology

The deprivation of nitrogen, one of the nutrient limitations that critically affect the cellular metabolism, has been shown to induce lipid accumulations in numerous strains of microalgae [13,34]. *Symbiodinium* spp. is an unique marine microalgae with two living status, either freely living in open ocean or symbiotically residing inside the gastrodermal cells of most marine cnidarian such as sea anemones and corals [1]. Here, we have shown for the first time that free-living *Symbiodinium* accumulate lipids in the form of LDs

when the nitrogen source of the culture medium was deprived. Positive growth of *Symbiodinium* spp. was observed after 3 days of the cultivation either under the normal condition or nitrogen starvation. However, it was observed that the slow growth pattern of *Symbiodinium* and accumulation of the LDs occurred under the nitrogen-deprivation (Fig. 1A). This has been reported in several algae that accumulated carbon metabolites under nitrogen-deficient condition, resulting in slow growth and lipid accumulation [35,36]. However, the nitrogen starvation may also induce an abrupt loss of electron transport capacity in the photosystem II, resulting in the change of the photochemical efficiency (*Fv/Fm*) and growth of *Symbiodinium* [37].

The ultrastructural observation indicated that, after the 7th day of nitrogen deprivation, numerous inclusion bodies started to appear in the LDs (Figs. 4E and 4F). The present study also demonstrated that there was a decreased lipid accumulation in cells on day 9 (Fig. 1B). It was presumed that during day 7 to day 9, *Symbiodinium* used the lipid in order to survive since the reserved nitrogen source had been used up. After 9 days of nitrogen starvation, the *Symbiodinium* cells entered the death phase (Fig. 1A).

Neutral lipids including TAG and CE, are the major lipid species in these LDs. Moreover, concurrently with drastic increases in size and number of LDs, detailed ultrastructural examinations indicated that the increases in cell size and cell wall thickness are typical features induced by the nitrogen deprivation (see Fig. 4 and Table 2). These morphological observations are similar to those reported on the morphological changes in green alga *Scenedesmus obtusiusculus* by nutrient starvation [38,39]. The increases in cell size and cell wall thickness were probably due to the nutrient deficiency, which led to a delay in cell cytokinesis

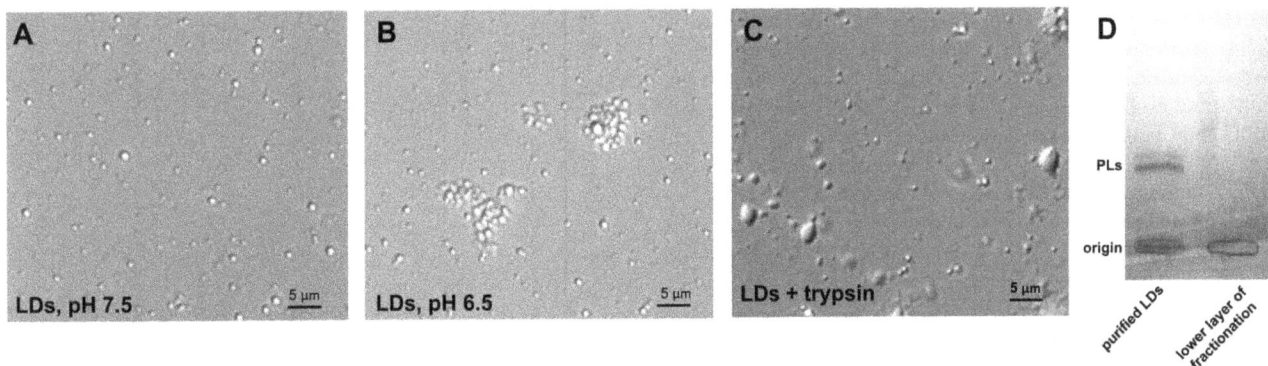

Figure 5. Light microscopy of the LDs purified from *Symbiodinium* cells after different treatments. The LDs were suspended in the (A) pH 7.5 grinding buffer, (B) pH 6.5 grinding buffer or (C) treated by the trypsin digestion. (D) Phospholipid analyses by TLC showing the presence of phospholipids (PLs) in purified LDs (the top layer during the centrifugation) but not lower layer fractions after detergent (0.1% Triton X-100) washing.

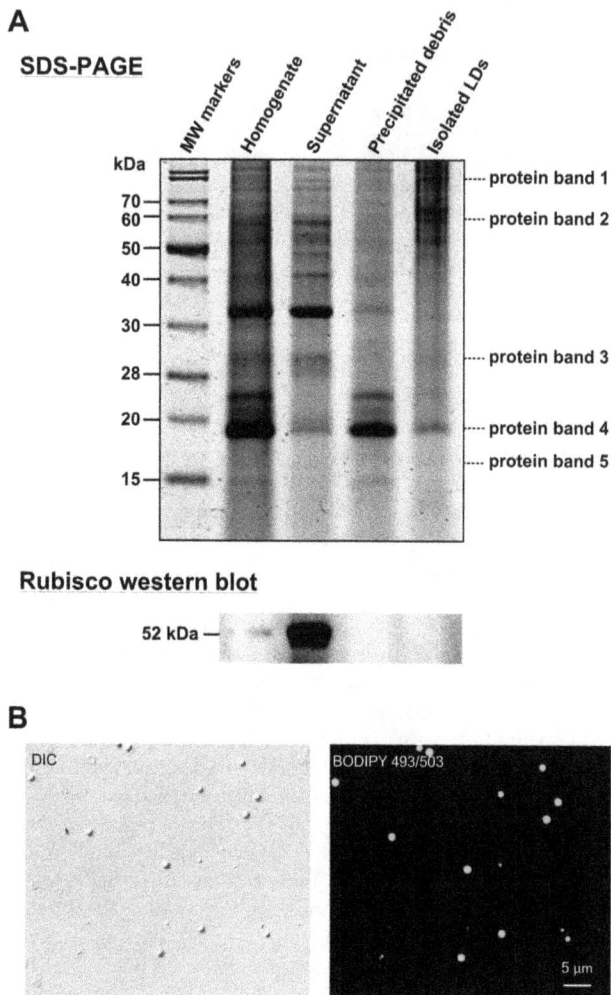

Figure 6. LDs purification and protein analyses. (A) SDS-PAGE analyses of isolated LDs fraction and the LD purity assessment by RuBisCO western blotting. *Symbiodinium* spp. cells harvested after five days of nitrogen deprivation were homogenized and fractionated to purify LDs as shown in the "Materials and methods" section. The purity of LDs was examined based on the absence of RuBisCO contamination by western blotting. Proteins bands 1 to 5 were excised for mass spectrometric analysis. (B) BODIPY 493/503 staining of isolated LDs displayed the abundance of neutral lipid.

[40,41]. Under severe nutrient limitation, cell division-arrested phytoplanktons were found to accumulate glycoproteins and other carbon contents in starch granules. As also shown in the present study, the proliferation of N-deprived *Symbiodinium* decreased significantly in comparison to the cells grown in normal medium (Fig. 1A). Similar inhibitory effects of nitrogen limitation on the proliferation and morphology of *Symbiodinium* in hospite have also been reported previously in two hermatypic corals (*Porites porites* and *Montastrea annularis*) and a sea anemone (*Aiptasia pallida*) [21,42].

The Carbon/Nitrogen (C/N) relocation during the nitrogen-deprivation and its implication for the endosymbiotic regulation

The decreased proliferation rates in most microalgae due to nitrogen deprivation have been reported to be concurrent with increases of total lipid levels [43,44]. This was also confirmed by

Table 3. Identification of lipid droplet proteins in *Symbiodinium* spp.

Protein name	Species/Taxonomy	Band No.	GI NO.	MS/mps(p)[a]	Sequence coverage (%)	Predicted MW(kDa)	Observed MW(kDa)	Found with LDs in other organisms	References
Lipid metabolism									
Sterol transfer family protein	*Pfiesteria piscicida*/dinoflagellate	5	112253295	72/1(1)	10	11.11	17	Yes	[64]
Signal-related proteins									
14-3-3 protein	*Perkinsus marinus*/protozoa	3	294885399	82/3(3)	9	27.39	28	Yes	[65]
ADP-ribosylation factor	Protozoan/eukaryotic	4	114131	79/1(1)	5	21.75	19	Yes	[66]
Stress response									
Chaperon heat shock protein 90 kDa	*Karlodinium micrum*/dinoflagellate	1	112253669	81/3(2)	5	81.51	84	Yes	[67]
Osmotically inducible protein Y	*Pseudomonas fluorescens*/bacteria	3	229592631	96/1(1)	8	21.02	28	No	[68]
Energy metabolism									
Mitochondrial ATP synthase F1 alpha subunit-like protein 1	*Karlodinium micrum*/dinoflagellate	2	319997184	85/3(2)	5	59.51	59	No	[69]
GTP-binding protein	*Helicobacter pylori*/bacteria	4	210134946	73/1(1)	5	19.00	19	Yes	[70]

a)MS/mps(p): Mowse score/number of total matched peptides (numbers of different matched peptides).

the present study, indicating that there is a reduced synthesis of new membrane constituents, in the nitrogen-deprived *Symbiodinium*. Here, the cell machinery shifted its metabolism to synthesize higher levels of TAGs and CEs for storage in LDs (Figs. 2–4).

The depletion of nitrogen could break the balance of carbon (C) and nitrogen (N) availability, switching the cellular metabolism toward the synthesis of carbon-containing compounds (i.e. lipids and/or carbohydrates); which then eventually leads to a higher cellular C:N ratio [45,46]. For example, the increase of lipid in microalgae upon nitrogen-deficiency could result from the partition of the excess carbon to intracellular lipid storage [47]. As shown in the present study, under nitrogen-deprived environment of free-living *Symbiodinium*, excess carbon allocates to the lipid pool and results in lipid accumulation as a form of LDs. Other environmental stresses, such as irradiance and various ecological changes, could also impact the C:N allocation in marine diatoms [48,49].

Besides amino acids, ammonium and nitrate have been shown to be sources of nitrogen in cultured and symbiotic *Symbiodinium* [6,50]. The investigation using the synchrotron radiation spectroscopy has further confirmed that symbiotic *Symbiodinium* (i.e. *in hospite*) contained less nitrogen compounds and more lipids than free-living *Symbiodinium* [9]. It has indicated the cytosol of the gastrodermal cell could be a nitrogen-limited environment, by which the symbiotic association with endosymbionts is regulated [51]. It is feasible that, by a mechanism of nitrogen limitation, cnidarian hosts regulate their symbiotic *Symbiodinium* to elevate lipid contents and store in a form of LDs [5,8,9].

The nitrogen-deprivation induced quantitative and qualitative changes of lipids in *Symbiodinium*

The most abundant fatty acid components in TAGs of *Symbiodinium* cultivated in the standard f/2 media (Fig. 3) were found to be myristic acid (C14), palmitic acid (C16), stearic acid (C18) and erucic (22:1), representing a fatty acid composition similar to that of higher plants [52]. These fatty acids are also common in other algae [19]. Palmitic acid (C16) and docosahexaenoic acid (C22:6) were the most abundant fatty acids in TAGs of nitrogen-deprived *Symbiodinium* cells. It has been proposed that TAG biosynthesis in microalgae may consist of three steps: (1) formation of acetyl coenzyme A in the cytoplasm, (2) elongation and desaturation of the carbon chain of fatty acids in the chloroplast, and (3) synthesis of TAG in the endoplasmic reticulum [53]. In the second step, C16 and C18 fatty acids are formed prior to the production of other long-chain fatty acids or PUFAs, which requires further elongation and desaturation. Therefore, the change of C16:0 and C18:0 levels in nitrogen-deprived *Symbiodinium* documented herein may imply the accelerated synthesis and

accumulation of PUFAs in the LDs, such as C16:1, C18:1, C18:2, C20:5, C21:5, C22:1, and C22:6. PUFAs are of the utmost importance for human metabolism [54]. For example, the long-chain docosahexaenoic acid (DHA) provides significant health benefits to the human population, particularly in preventing cardiac diseases such as arrhythmia, stroke and high blood pressure [55,56]. Accordingly, an increase in the PUFA production by microalgae has been shown to be important for many nutritional and pharmaceutical purposes [57].

The proteomics of LDs

Using the neutral lipid probe BODIPY 493/503, we were able to visualize the LDs isolated from cultured *Symbiodinium*, with 0.5–1 μm in diameter (Fig. 6B). Oil bodies of similar sizes (0.5–2 μm diameter) have previously been reported in plant seeds [15]. However, the protein compositions of LDs between the *Symbiodinium* and other plant seeds are obviously different, indicating distinct regulatory mechanisms [58]. For example, other LD-related proteins, such as oleosin, have been discovered to be abundant in microalgae and plant seeds [16,19,59]. Both N- and C- terminal domains of oleosins have been proposed to reside on the LD (or so called "Oil Bodies") surface to stabilize this organelle via steric hindrance and electronegative repulsion [60]. Nevertheless, oleosins were not identified in the LDs of *Symbiodinium* spp. as shown in this study, demonstrating a different formation mechanism during the endosymbiosis. As a consequence, the examination of protein compositions of LDs in nitrogen-deprived free-living *Symbiodinium* should provide new insights to elucidate the LD formation mechanism. Besides proteins responsible for lipid and energy metabolisms, other LD-associated proteins involving in signaling (14-3-3 protein and ARF) and stress response (HSP90 and the osmotically inducible protein Y) were also identified (see Table 3). The 14-3-3 proteins are a ubiquitous group of signaling proteins that involve in many regulatory functions [61] such as carbohydrate and lipid metabolisms in plants [62]. Furthermore, both ARF and GTP-binding proteins are key regulators in membrane trafficking, and thus play pivotal role in LDs biogenesis [63].

Acknowledgments

We would like to thank Dr. Anderson Mayfield for his careful proofreading of the manuscript, as well as insightful comments on its content.

Author Contributions

Conceived and designed the experiments: PLJ CSC. Performed the experiments: PLJ BP. Analyzed the data: PLJ CSC. Contributed reagents/materials/analysis tools: PLJ CSC. Wrote the paper: CSC PLJ.

References

1. Muscatine L (1980) Uptake, retention, and release of dissolved inorganic nutrients by marine alga-invertebrate associations. In: Cook CB, editor. In Cellular interactions in symbiosis and parasitism. Ohio state, pp. 229–244.
2. Belda CA, Lucas JS, Yellowlees D (1993) Nutrient limitation in the giant clam-zooxanthellae symbiosis: effect of nutrient supplements on growth of the symbiotic partners. Mar Biol 117:655–664.
3. Wang JT, Douglas AE (1998) Nitrogen recycling or nitrogen conservation in an alga-invertebrate symbiosis? J Exp Biol 201:2445–2453.
4. Szmant-Froelich A, Pilson MEQ (1984) Effects of feeding frequency and symbiosis with zooxanthellae on nitrogen metabolism and respiration of the coral *Astrangia danae*. Mar Biol 81:153–162.
5. Steen RG (1986) Evidence of heterotrophy by zooxanthellae in symbiosis with *Aiptasia pulchella*. Biol Bull 170:267–278.
6. Rahav O, Dubinsky Z, Achituv Y, Falkowski PG (1989) Ammonium metabolism in the zooxanthellate coral *Stylophora pistillata*. Proc R Soc Lond B Biol Sci 236:325–337.
7. Gordon BR, Leggat W (2010) Symbiodinium-invertebrate symbioses and the role of metabolomics. Mar Drugs 8: 2546–2568.
8. Cook CB, D'Elia CF (1987) Are natural populations of zooxanthellae ever nutrient-limited? Symbiosis 4:199–211.
9. Peng SE, Chen CS, Song YF, Huang HT, Jiang PL, et al. (2012) Assessment of metabolic modulation in free-living versus endosymbiotic *Symbiodinium* using synchrotron radiation-based infrared microspectroscopy. Bio Let. 23:434–437.
10. Zhu B, Pan K, Wang G (2010) Effects of host starvation on the symbiotic dinoflagellates from sea anemone *Stichodactyla mertensii*. Mar Eco 32:15–23.
11. Piorreck M, Baasch KH, Pohl P (1984) Biomass production, total protein, chlorophylls, lipids and fatty acids of freshwater green and blue-green algae under different nitrogen regimes. Phytochemist 23:207–216.
12. Li X, Hu HY, Yang J (2010) Lipid accumulation and nutrient removal properties of a newly isolated freshwater microalga, *Scenedesmus sp. LX1*, growing in secondary effluent. New Biotech 27:59–63.

13. Bigogno C, Khozin-Goldberg I, Adlerstein D,Cohen Z (2002) Biosynthesis of arachidonic acid in the oleaginous microalga *Parietochloris incise* (Chloropyccae): radiolabeling studies. Lipids 37:209–216.

14. Roessler PG (1990) Environmental control of glycerolipid metabolism in microalgae: commercial implications and future research directions. J Phyco. 26:393–399.

15. Frandsen GI, Mundy J, Tzen JTC (2001) Oil bodies and their associated proteins, oleosin and caleosin. Physiol Plant 112:301–307.

16. Lin IP, Jiang PL, Chen CS, Tzen JT (2012) A unique caleosin serving as the major integral protein in oil bodies isolated from *Chlorella* sp. cells cultured with limited nitrogen. Plant Physiol Biochem 61:80–87.

17. Pasaribu B, Lin IP, Chen CS, Lu CY, Jiang PL (2013) Nutrient limitation in *Auxenochlorella protothecoides* induces qualitative changes of fatty acid and expression of caleosin as a membrane protein associated with oil bodies. Biotechnol Lett DOI 10.1007/s10529-013-1332-1.

18. Nguyen HM, Baudet M, Cuine S, Adriano JM, Barthe D, et al. (2011) Proteomic profiling of oil bodies isolated from the unicellular green microalga *Chlamydomonas reinhardtii*: With focus on proteins involved in lipid metabolism. Proteomics 11:4266–4273.

19. Peled E, Leu S, Zarka A, Weiss M, Pick U, et al. (2011) Isolation of a novel oil globule protein from the green alga *Haematococcus pluvialis* (Chlorophyceae). Lipids 46:851–861.

20. Davidi L, Katz A, Pick U (2012) Characterization of major lipid droplet proteins from Dunaliella. Planta 236 :19–33.

21. Muller-Parker G, Lee KW, Cook CB (1996) Changes in the ultrastructure of symbiotic zooxanthellae (*Symbiodinium* sp., Dinophyceae) in fed and starved sea anemones maintained under high and low light. J Phyco 32:987–994.

22. Muller-Parker G, D'Elia CF (1998) Interactions between corals and their symbiotic algae. In: Birkeland C, editor. In Life and Death of Coral Reefs. New York: Chapman and Hall, pp.96–113.

23. Guillard RRL (1975) Culture of phytoplankton for feeding marine invertebrates. In: Smith WL, Chanley MH, editors. Culture of Marine Invertebrate Animals. New York: Plenum Press, pp.26–60.

24. Rowan R, Powers DA (1991) Molecular genetic identification of symbiotic dinoflagellates (zooxanthellae). Mar Ecol Prog Ser 71:65–73.

25. Jeffrey SW, Humphrey GF (1975) New spectrophotometric equations for determining chlorophylls a, b, c1 and c2in higher plants, algae and natural phytoplankton. Biochem Physiol Pflanzen 167: 191–194.

26. Tzen JTC, Peng CC, Cheng DJ, Chen ECF, Chiu JMH (1997) A new method for seed oil body purification and examination of oil body integrity following germination. J Biochem 121:762–768.

27. Bligh EG, Dyer WJ (1959) A rapid method for total lipid extraction and purification. Can J Biochem. Physiol. 37:911–917.

28. Oku H, Yamashiro H, Onaga K (2003) Lipid biosynthesis from [^{14}C]-glucose in the coral *Montipora digitata*. Fish Sci 69:625–631.

29. Fuchs B, Schiller J, Sub R, Schurenberg M (2007) A direct and simple method of coupling matrix-assisted laser desorption and ionization time-of-flight mass spectrometry (MALDI-TOF MS) to thin layer chromatography (TLC) for the analysis of phospholipids from egg yolk. Anal Bioanal Chem 389:827–834.

30. Abe A (1998) Modification of the Coomassie brilliant blue staining method for sphingolipid synthesis inhibitors on silica gel thin-layer plate. Anal Biochem 258:149–150.

31. Haugland R (1996) Handbook of Fluorescent Probes and Research Chemicals, Sixth Ed. Oregon: Molecular Probes Inc, pp. 13–19.

32. Laemmli UK (1970) Cleavage of structural proteins during the assembly of the head of bacteriophage T4. Nature 227:680–685.

33. Peng SE, Chen WNU, Chen HK, Lu CY, Mayfield AB, et al. (2011) Lipid bodies in coral-dinoflagellate endosymbiosis: Proteomic and ultrastructural studies. Proteomics 11:3540–3555.

34. Praveenkumar R, Shameera K, Mahalakshmi G, Akbarsha MA, Thajuddin N (2012) Influence of nutrient deprivations on lipid accumulation in a dominant indigenous microalga Chlorella sp. bum1 1008: evaluation for biodiesel production. Biomass Bioenerg 37:60–66.

35. Hu H, Gao K (2005) Response of Growth and Fatty Acid Compositions of *Nannochloropsis* sp. to Environmental Factors Under Elevated CO2 Concentration. Biotechnol Lett 28: 987–992.

36. Yeesang C, Cheirsilp B (2011) Effect of nitrogen, salt, and iron content in the growth medium and light intensity on lipid production by microalgae isolated from freshwater sources in Thailand. Bioresource Technol 102: 3034–3040.

37. Rodríguez-Román A, Iglesias-Prieto R (2005) Regulation of photochemical activity incultured symbiotic dinoflagellates under nitrate limitation and deprivation. Mar Biol 146: 1063–1073.

38. Tillberg JE, Rowley JR (1989) Physiological and structural effects of phosphorus starvation on the unicellular green algae *Scenedesmus*. Physio Plant 75:315–324.

39. Van Donk E, Hessen DO (1993) Grazing resistance in nutrient-stressed phytoplankton. Oecologia 93:508–511.

40. Mitchell SF, Trainor FR, Rich PH, Goulden CE (1992) Growth of *Daphnia magna* in the laboratory in relation to the nutritional status of its food species, *Chlamydomonas reinhardtii*. J Plankton Res 14:379–391.

41. Sterner RW, Hagemeier DD, Smith WL, Smith RF (1993) Phytoplankton nutrient limitation and food quality for *Daphnia*. Limnol Oceanogr 38:857–871.

42. Marubini F, Davies PS (1996) Nitrate increase zooxanthellae population density and reduces skeletogenesis in corals. Mar Bio 127:319–328.

43. Takagi M, Watanabe K, Yamaberi K, Yoshida T (2000) Limited feeding of potassium nitrate for intracellular lipid and triglyceride accumulation of *Nannochloris* sp. UTEX LB 1999. Appl Microbiol Biotechnol. 54:112–117.

44. Griffiths MJ, Harrison STL (2009) Lipid productivity as a key characteristic for choosing algal species for biodiesel production. J Appl Phycol 21:493–507.

45. Reitan KI, Rainuzzo JR, Olsen Y (1994) Effect of nutrient limitation on fatty acid and lipid content of marine microalgae. J Phycol 30:972–9.

46. Lynn SG, Kilham SS, Kreeger DA, Interlandi SJ (2000) Effect of nutrient availability on the biochemical and elemental stoichiometry in the freshwater diatom *Stephanodiscus minutulus* (Bacillariophyceae). J Phycol 36:510–22.

47. Martin T, Oswald O, Graham IA (2002) Arabidopsis seedling growth, storage lipid mobilization and photosynthetic gene expression are regulated by carbon:nitrogen availability. Plant Physiol 128:472–81.

48. Palmucci M, Ratti S, Giordano M (2011) Ecological and evolutionary implications of carbon allocation in marine phytoplankton as a function of nitrogen availability: a Fourier transform infrared spectroscopy approach. J Phycol 47: 313–323.

49. Norici A, Bazzoni AM, Pugnetti A, Raven JA, Giordano M (2011) Impact of irradiance on the C allocation in the coastal marine diatom Skeletonema marinoi Sarno and Zingone. Plant Cell Environ 34: 1666–1677.

50. Carroll S, Blanquet RS (1984) Alanine uptake by isolated zooxanthellae of the mangrove jellyfish, *Cassiopea xamachana*. I. Transport mechanisms and utilization. Biol Bull 166: 409–418.

51. Corzo A, Niell XF (1991) C/N ratio in response to nitrogen supply and light quality in *Ulva rigida* C. Agardh (Chlorophyta:Ulvophycea). Sci Mar 55:405–411.

52. Minzangi K, Kaaya AN, Kansiime F, Tabuti JRS, Samvura B, et al. (2011) Fatty acid composition of seed oils from selected wild plants of kahuzi-biega national park and surroundings, democratic republic of congo. African Journal of Food Science 5:219–226.

53. Deng X, Fei X, Li Y (2011) The effects of nutritional restriction on neutral lipid accumulation in Chlamydomonas and Chlorella. Afri J Microbiol Res 5:260–270.

54. Simopoulos AP (2002) Importance of the ratio of omega-6/omega-3 essential fatty acids. Biomed Pharmacother 56:365–379.

55. Romieu I, Tellez-Rojo MM, Lazo M, Manzano-Patino A, Cortez-Lugo M, et al. (2005) Omega-3 fatty acid prevents heart rate variability reductions associated with particulate matter. Am J Respir Crit Care Med 172:1534–1540.

56. Von Schacky C (2008) Omega-3 fatty acid: antiarrhythmic, proarrhythmic or both? Curr Opin Clin Nutr Metab Care 11:94–99.

57. Pereira H, Barreira L, Figueiredo F, Custódio L, Vizetto-Duarte C, et al. (2012) Polyunsaturated fatty acids of marine macroalgae: potential for nutritional and pharmaceutical applications. Mar Drugs 10:1920–1935.

58. Tzen JTC, Wang MMC, Chen JCF, Lin I,J, Chen MCM (2003) Seed oil body proteins: oleosin, caleosin, and steroleosin. Curr Topics Biochem Res 5:133–139.

59. Huang NL, Huang MD, Chen TL, Huang AH (2013) Oleosin of subcellular lipid droplets evolved in green algae. Plant Physiol 161:1862–1874.

60. Chen DH, Chyan CL, Jiang PL, Chen CS, Tzen JTC (2012) The same oleosin isoforms are present in oil bodies of rice embryo and aleurone layer while caleosin exists only in those of the embryo. Plant Physiol Biochem 60: 18024.

61. Ferl RJ, Manak MS, Reyes MF (2002) The 14-3-3s. Genome Biol 3:1–7.

62. Prescha A, Swiedrych A, Biernat J, Szopa J (2001) Increase in lipid content in potato tubers modified by 14-3-3 gene overexpression. J Agric Food Chem 49:3638–3643.

63. Tan R, Wang W, Wang S, Wang Z, Sun L, et al. (2013) Small GTPase Rab40c associates with lipid droplets and modulates the biogenesis of lipid droplets. PLoS One 8:e63213.

64. Prinz WA (2007) Non-vesicular sterol transport in cells. Prog Lipid Res 46: 297–314.

65. Yang L, Ding Y, Chen Y, Zhang S, Huo C, et al. (2012) The proteomics of lipid droplets: structure, dynamics, and functions of the organelle conserved from bacteria to humans. J Lipid Res 53: 1245–1253.

66. Bartz R, Seemann J, Zehmer K, Serrero G, Chapman DK, et al. (2007) Evidence that mono-ADP-ribosylation of CtBP1/BARS regulates lipid storage. Mol Biol Cell 18:3015–3025.

67. Zhang H, Hou Y, Miranda L, Campbell AD, Nancy R, et al. (2007) Spliced leader RNA trans-splicing in dinoflagellates. Proc Natl Acad Sci 11: 4618–4623.

68. Silby MW, Cerdeño-Tárraga AM, Vernikos GS, Giddens SR, Jackson RW, et al. (2009) Genomic and genetic analyses of diversity and plant interactions of *Pseudomonas* fluorescens. Genome Biol 10:R51

69. Danne JC, Waller RF (2011) Analysis of dinoflagellate mitochondrial protein sorting signals indicates a highly stable protein targeting system across eukaryotic diversity. J Mol Biol 408:643–653.

70. Fischer W, Windhager L, Rohrer S, Zeiller M, Karnholz A, et al. (2010) Strain-specific genes of Helicobacter pylori: genome evolution driven by a novel type IV secretion system and genomic island transfer. Nucleic Acids Res 38(18):6089–6101.

12

Influence of Fenofibrate Treatment on Triacylglycerides, Diacylglycerides and Fatty Acids in Fructose Fed Rats

Thomas Kopf[1], Hans-Ludwig Schaefer[2¤a], Martin Troetzmueller[3], Harald Koefeler[3], Mark Broenstrup[2¤b], Tatiana Konovalova[1], Gerd Schmitz[1]*

1 Institute for Clinical Chemistry and Laboratory Medicine, University Hospital Regensburg, Regensburg, Germany, **2** Sanofi-Aventis Germany, R&D DIAB Div./Biomarker & Diagnostics, Frankfurt, Germany, **3** Core Facility Mass Spectrometry, ZMF, Medical University Graz, Graz, Austria

Abstract

Fenofibrate (FF) lowers plasma triglycerides via PPARα activation. Here, we analyzed lipidomic changes upon FF treatment of fructose fed rats. Three groups with 6 animals each were defined as control, fructose-fed and fructose-fed/FF treated. Male Wistar Unilever Rats were subjected to 10% fructose-feeding for 20 days. On day 14, fenofibrate treatment (100 mg/kg p.o.) was initiated and maintained for 7 days. Lipid species in serum were analyzed using mass spectrometry (ESI-MS/MS; LC-FT-MS, GC-MS) on days 0, 14 and 20 in all three groups. In addition, lipid levels in liver and intestine were determined. Short-chain TAGs increased in serum and liver upon fructose-feeding, while almost all TAG-species decreased under FF treatment. Long-chain unsaturated DAG-levels (36:1, 36:2, 36:4, 38:3, 38:4, 38:5) increased upon FF treatment in rat liver and decreased in rat serum. FAs, especially short-chain FAs (12:0, 14:0, 16:0) increased during fructose-challenge. VLDL secretion increased upon fructose-feeding and together with FA-levels decreased to control levels during FF treatment. Fructose challenge of de novo fatty acid synthesis through fatty acid synthase (FAS) may enhance the release of FAs ≤16:0 chain length, a process reversed by FF-mediated PPARα-activation.

Editor: Matej Oresic, Steno Diabetes Center, Denmark

Funding: This work was supported by the European Community's Seventh Framework Programme (FP7/2007–2013) under grant agreement n° 202272, IP-Project LipidomicNet and BMBF ("SysMBo," sponsorship number 0315494C). The funders had no role in study design, data collection and analysis, decision to publish, or preparation of the manuscript.

Competing Interests: H.S. is currently employed by Infraserv GmbH & Co. Ho¨chst KG, Frankfurt; H.S. and M.B. were formerly employed by Sanofi-Aventis at the time of the study. There are no patents, products in development or marketed products to declare.

* Email: gerd.schmitz@klinik.uni-regensburg.de

¤a Current address: Infraserv GmbH & Co. Höchst KG, Frankfurt, Germany
¤b Current address: Helmholtz Centre for Infection Research, Braunschweig, Germany

Introduction

Fibrates such as fenofibrate (FF) are widely used in human medicine for their hypolipidemic effects [1,2]. They belong to a larger group of molecules called peroxisomal proliferators (PPs) [3]. PPs activate the peroxisome proliferator activated receptor α (PPARα) and modulate genes involved in lipid metabolism [4–7]. PPARα activation decreases several mediators of vascular damage such as lipotoxicity, inflammation, ROS, endothelial dysfunction, angiogenesis and thrombosis through modulation of cell signaling related to microvascular dysfunction [2]. PPARα modulators are linked to species-specific regulation of genes important in cell growth and differentiation [8]. In rodents, PPARα activation is associated with peroxisome proliferation and hepatocellular carcinoma [9]. However, due to species differences at the molecular level of PPARα regulation, humans might be resistant to liver cancer induced by PPARα agonists [9–11]. Oxidative stress due to excessive H_2O_2-generation upon FF treatment leads to lipid peroxidation and oxidative DNA damage and hepatocarcinogenesis in rodents [12–15]. PPARα function is coupled to several co-activators that interfere with NFkB signaling pathways

linked to carcinogenesis [16]. FF is found enriched in organs of absorption and elimination, the gut, liver and kidney [17]. The FIELD-Study (FF Intervention and Event Lowering in Diabetes) reported a robust and sustained decrease in plasma triglyceride-levels upon FF treatment [11]. In contrast, elevation of HDL-cholesterol and Apo A1 levels were less than expected and decreased progressively over the duration of the study[2]. A recent study suggests the possible application of HDL molecular composition for the stratification of patients that could potentially profit from FF treatment [18]. Recently it was shown that fish oil and FF treatment showed significant overlap in gene regulation, with fish oil down-regulating genes in cholesterol and fatty acid biosynthesis and FF down-regulating genes related to inflammation [19].

At the molecular level, PPARα forms a heterodimer with retinoid X receptor (RXR) upon agonist binding and stimulates the expression of various genes involved in FA β-oxidation and ω-oxidation, intracellular FA-transport and HDL-cholesterol metabolism [1,20,21]. Unsaturated FAs activate PPARα [22], and PPARα agonists inhibit the production of prostaglandin E2 (PGE_2) in vitro [23]. Through FF activation PPARα decreases triglyceride (TAG) and hepatic VLDL-production by enhancing FA-oxidation in the liver. In addition, it facilitates TAG-removal by stimulating LPL-production and enhances hepatic lipoprotein uptake due to the suppression of apo-C-III production [1]. FF also increases LDL particle size and reduces the prevalence of small dense LDL (sd LDL) and non-HDL-cholesterol containing lipoproteins related to apoB. PPARα activation increases HDL-C by stimulating the production of apo A-I and apo A-II [24,25].

Feeding previously fasted animals a low-fat/high-carbohydrate diet caused a marked induction of enzymes involved in catalyzing fatty acid desaturation [26], including ATP citrate lyase [27], FAS for lipogenesis [28], and SCD-1, which introduces a double bond in sn-9 position of a variety of fatty acyl CoA precursors [29]. SCD-1 is also a major player regulating the fatty acid composition of tissues [30], and SCD -/- mice are resistant to diet induced obesity [31]. Expression of SCD-1 in the liver of diabetic rats was found to be up-regulated, while Δ5-desaturase (FADS1) was not altered [32]. SCD-1 is regarded as a target for reversal of hepatic steatosis and insulin resistance [28]. FAS is a key enzyme of de novo lipogenesis, which catalyzes the terminal steps of FA-synthesis in the liver [28].

In rats, FF also induces carnitine-acetyltransferase (CAT), carnitine-palmitoyltransferase (CPT), fatty acyl oxidizing system (FAOS) and acetyl CoA oxidase 1 (Aco1) [12]. FF treatment of metabolic syndrome patients reduced plasma TAG-levels (30%) and also cholesterol (30%) in TAG-rich lipoproteins, together with a reduction of apo CIII and sdLDL and an increase in large LDL, but did not lower concentrations or turnover rates of NEFAs, nor did it change glucose or insulin responses to oral glucose challenge [33]. FF in this study modified FA-metabolism either in the liver or in TAG-rich lipoproteins but not in adipose tissue due to PPARα activation.

Dietary exposure to fructose may specifically provide lipid deposition in visceral adipose tissue, particularly in males, whereas glucose consumption appears to favor lipid deposition in subcutaneous adipose tissue [34]. Coingestion of fructose may elicit an unfavorable TAG-profile similar to fructose alone. Fructose feeding also induces ChREBP and increases the expression of lipogenic genes (FAS, ACC, SCD1).

The objective of this study was to investigate the effect of FF treatment on rats under metabolic overload conditions at the level of lipid species to get a more detailed insight into lipid metabolism than with the total level of lipid classes (like total TAG). Metabolic overload was induced in a rat model by fructose feeding, and serum, liver and jejunum samples of FF-treated animals and control groups were analyzed using ESI-MS/MS, ion trap LC-FT-MS and GC-MS techniques.

Materials and Methods

Study design

The animal studies were performed at Sanofi in Frankfurt-Höchst. The tissue and serum samples were taken from two different animal studies with the same basic design. The VLDL-data was derived from a third study. All experimental procedures were conducted according to the German Animal Protection Law. The animal studies were approved by the Sanofi-Aventis Deutschland GmbH institutional animal care and use committee and notified to the relevant authority. The institution is AAALAC accredited (AAALAC, 2012). Blood was drawn from the retro-orbital vein plexus under Isofluran CP (CP Pharma, Burgdorf, Germany) oxygen/nitric oxide anesthesia (3.5%, 2:1) Animals were anesthetized in a gas box. The animals were sacrificed by bleeding from the abdominal aorta under deep Isofluran CP (CP

Pharma, Burgdorf, Germany) oxygen/nitric oxide anesthesia (3.5%, 2:1). The anesthesia was initiated in gas box and sustained with mask. In all cases treatment was applied between 7:30 and 8:30 a. m. and blood samples were drawn 1 hour after treatment.

Tissue samples (study 1). Male Wistar rats (HsdCpb: WU) were obtained from Harlan Laboratories, NL and were used at the age of 10 weeks for the study. Animals were housed under controlled temperature (21–23°C), humidity (55%) in Macrolon type 4 cages (3 animals per cage) with a 12 h light-dark-circle with ad libitum access to drinking water. Animals were divided into 3 groups with group 1 receiving a standard diet (Ssniff), group 2 receiving a 10% fructose-challenge through drinking-water and group 3 receiving FF treatment through oral application and a 10%-fructose-challenge through drinking water. The study protocol ran for 14 days. In the FF treatment group, drug application (100 mg/kg p.o.) started at day 7 and lasted for 7 days. After the treatment period, the heart, the liver and the jejunum were extracted and stored at −20°C for lipid analysis.

Serum samples (study 2). Male SPRD rats were obtained from Charles River Laboratories (Wilmington MA) and were used at the age of 10 weeks for the study. Animals were housed under controlled temperature (21–23°C), humidity (55%) in Macrolon type 4 cages (3 animals per cage) with a 12 h light-dark-circle with ad libitum access to drinking water. Animals were divided into 3 groups with group 1 receiving a standard diet, group 2 receiving a 10% fructose-challenge through drinking-water and group 3 receiving FF treatment through oral application and a 10%-fructose-challenge through drinking water. The study protocol ran for 20 days. In the FF treatment group, drug application (100 mg/kg p.o.) started at day 14 and lasted for 7 days. Serum samples were taken before start of the fructose treatment (day 0), after 14 days of fructose treatment but before FF treatment (day 14) and at the end of the FF treatment (day 20). Samples were stored at −80°C until lipid analysis.

VLDL-secretion (study 3). Male Wistar rats (HsdCpb: WU) were obtained from Harlan Laboratories, NL and were used at the age of 10 weeks for the study. Animals were housed under controlled temperature (21–23°C), humidity (55%) in Macrolon type 4 cages (3 animals per cage) with a 12 h light-dark-circle with ad libitum access to drinking water. Animals were divided into 3 groups with group 1 receiving a standard diet, group 2 receiving a 10% fructose-challenge through drinking-water and group 3 receiving FF treatment through oral application and a 10%-fructose-challenge through drinking water. The study protocol ran for 14 days. In the FF treatment group, drug application (100 mg/kg p.o.) started at day 7 and lasted for 7 days. After 6 days the animals were fasted for approx. 23 h; then, Tyloxapol reagent (Sigma, 1:10 diluted with NaCl 0.9%) was applied and samples were taken after 1, 2, 4 and 6 hours.

Clinical chemistry

Serum levels of cholesterol, triglycerides, and phospholipids, as well as the safety variables: aspartate transaminase (ASAT), alanine transaminase (ALAT), alkaline phosphatase (AP) were determined on a Roche Cobas 6000 at Sanofi in Frankfurt Höchst using the respective Roche clinical chemistry kits for human diagnostics. Assays were performed according to the instructions from the suppliers.

Statistical analysis

Data are presented as mean±s.d.. Significant differences were calculated by an unpaired Mann-Whitney-U test. Two-point comparisons were performed. The two groups consisted of the fructose-free control group which was compared to the fructose-fed-group to establish the influence of the feeding itself. The second significance was determined between the fructose-fed-group and the Fenofibrate-treatment-group. For all statistical calculations, SPSS Statistics 19 software (IBM) was used. A $P <$ 0.05 was considered to be statistically significant ($P < 0.05$ is denoted by *; $P < 0.01$ is denoted by **).

Lipid measurements

Analysis of TAG/DAG-species was performed according to published methods [35]. In brief, 30 µl serum were extracted with a mixture of methyl tert-butyl ether, methanol and water (MTBE/ MeOH/H_2O; 10:3:2.5, v/v/v; 9.77 ml total volume) according to Matyash et al. [36]. Lipid extracts were spiked with a 1 µM mix of 18 LIPID MAPS internal standards (LM 6000, LM 6001) and 5 µl of spiked samples were injected onto a Thermo 1.9 µm Hypersil GOLD C18, 100×1 mm HPLC column mounted in an Accela HPLC instrument (Thermo Scientific). The final concentrations of internal TAG- and DAG-standard were 1 and 3 µM respectively. Solvent A was water with 1% ammonium acetate and 0.1% formic acid, solvent B was acetonitrile/2-propanol 5:2 (v/v) with 1% ammonia acetate and 0.1% formic acid. The gradient ran from 35 to 70% B in 4 min, then to 100% B in another 16 min and held there for further 10 min. The flow rate was 250 µl/min. Data acquisition was performed by FT-ICR-MS (LTQ-FT, Thermo Scientific) full scans at a resolution of 200k and <2 ppm mass accuracy with external calibration. From the FT-ICR-MS preview scan, the four most abundant m/z values were picked in data dependent acquisition (DDA) mode, fragmented in the linear ion trap and ejected at nominal mass resolution. Normalized collision energy was set to 35%, repeat count was two and exclusion duration was 60 s. Quantitative analysis of data acquired by the platform described above was carried out by Lipid Data Analyzer [37].

Total hydrolysate FAs were analyzed as FAMEs by GC/MS according to a recently published method [38]. Briefly, lipids were extracted with chloroform/methanol according to Bligh and Dyer [39]. Derivatization was performed as follows. 10 µl serum or cell homogenate corresponding to 50 µg protein were methylated in PTFE screw capped Pyrex tubes. 1 µg each of C13:0 and C21:0 iso were added as internal standards in 50 µl methanol. 200 µl of acetyl-chloride were added and the sample was shaken vigorously at 20°C for 14 h. Afterwards 5 ml 6% potassium carbonate solution was added. FAMEs were extracted with 500 µl n-hexane. 100 µl of the n-hexane top layer was transferred into a 500 µl auto-sampler vial and crimped. Analysis of the NEFA fraction was performed after Dole-extraction [40]. The NEFA-fraction was separated by SPE-fractionation by a newly developed method [41]. Derivatization and GC/MS-analysis of this fraction was performed with the method described above.

Results

Male Wistar rats were subjected to fructose feeding and subsequent treatment with fenofibrate (FF). Lipidomic analysis was performed using MS-techniques (for details see materials and methods).

General values

The clinical chemistry values for total triglycerides, total cholesterol, total phospholipids and body/liver weight as well as food/water consumption are shown in table 1. Also added are the values for the study safety parameters aspartate aminotransferase (ASAT), alanine aminotransferase (ALAT) and alkaline phosphatase (AP). Values are given as means with s.d. where appropriate.

The initial point of time was taken before the start of treatment, meaning the fructose-fed and the FF groups were on a fructose diet for one week at that time. These values were determined in the first rat study, in which the liver and jejunum lipid levels were measured. The final point of time was after one week of treatment with FF. Cholesterol, triglycerides and phospholipids were increased by fructose-feeding, while FF treatment reduced these levels below control levels. Body weight of the animals was approximately equal for the three groups at both times, while liver weight was slightly increased through fructose-feeding and highly increased after FF treatment. Water consumption was lower for the fructose-fed animals, while food consumption was almost doubled upon fructose-feeding and nearly back to control level upon FF treatment.

Triacylglycerols and diacylglycerols

The weight of liver and heart of the three groups are shown in Fig 1A. The weight of the heart did not change upon fructose feeding or FF treatment, while the weight of the liver increased under fructose feeding and nearly doubled upon FF treatment. Absolute TAG-levels were determined in rat serum at the beginning (day 0) and at the end of the study (day 16, Fig 1B). A significant increase of TAGs under fructose-feeding, and a significant decrease by ~25% under FF treatment was observed. The absolute levels of TAGs and DAGs (nmol/mg cell protein; Fig 1B) were analyzed in rat liver and jejunum at the end of the study. Without fructose-feeding rat jejunum contained about 10-fold higher TAG-levels than rat liver. Upon fructose-feeding TAG-levels decreased in the jejunum and increased in the liver, while FF slightly increased TAGs in the jejunum (~20%) and decreased TAGs in the liver (~50%) back to control group levels. DAG-levels in the liver increased upon fructose-feeding and more than doubled upon FF treatment. In contrast, baseline DAG-levels (Fig 1B) in the jejunum were 2-fold lower than in the liver, but did not change significantly under FF treatment. Fig 1C shows the sums of DAG- and TAG-species in the rat serum samples as an indicator for the total amount of DAGs/TAGs in rat serum. DAG-levels increased significantly by day 20 under fructose feeding and were reduced to control levels by FF treatment. The same was true for total TAGs, which increased upon fructose-feeding and decreased back to control levels by FF treatment.

DAG-species analysis

As a next step, individual DAG-species were determined in rat liver and rat serum at the three points of time. Fig 2A shows the DAG-species in rat liver. Given are the mean values of the sums of chain length of DAG-species (C32, C34, C36, C38) and the sums of degree of desaturation (saturated to hexa-desaturated). The complete data for the single DAG-species are shown in Fig. S1. During fructose-feeding, DAG-species containing 3 or more double bonds decreased, accompanied by an increase of species containing 2 or less double bonds. This increase is only present in the FA-species with 14, 16 or 18 carbons (C32, C34), while the C38-containing DAG-species decreased. FF-treatment caused variable changes of DAG-species. The short-chain DAG-species (32:0, 32:1 and 32:2; Fig. S1) decreased, as did DAG-species 34:0, 34:2 and 34:3, while DAG 34:1 increased. The longer-chain mono- and poly-unsaturated DAG-species 36:1, 36:2, 36:4, 38:3, 38:4 and 38:5 increased, while DAG 36:3, 38:2 and 38:6 decreased. The overall increase of DAG-levels was attributable to two species, DAG 34:1 and DAG 36:2, both may possibly contain FA 18:1.

It is interesting to note that only a few DAG-species (34:1, 34:2, 34:3, 36:2, 36:3, 36:4 and 38:4) were detectable in rat serum

Table 1. Clinical chemistry data, organ weight, food and water consumption.

	initial			final		
	fructose-free	fructose-fed	FF	fructose-free	fructose-fed	FF
Cholesterol [mmol/l]	2.18±0.39	2.63±0.40	2.45±0.49	2.25±0.50	2.40±0.41	1.38±0.32*
Triglycerides [mmol/l]	2.17±0.53	2.73±0.93	2.66±0.40	1.98±0.58	3.11±0.95**	1.41±0.36**
Phospholipids [mmol/l]	2.09±0.23	2.64±0.31	2.62±0.27	2.19±0.39	2.68±0.30**	1.71±0.18**
Body weight [g]	350±24.41	360±13.98	353±15.82	370±25.84	382±19.75	376±18.11
Liver weight [g liver/animal]	-	-	-	14.95±1.35	17.49±1.41*	24.03±2.88**
rel. water consumption/100 g BW [g/d/animal]	26.93	22.64	22.31	22.59	17.63	19.45
abs food consumption [g/d/animal]	37.82	61.39	77.35	36.03	69.80	46.88
ASAT [u/l]	145.17±36.59	120.17±26.88	115.17±22.89	111.00±32.60	78.50±6.72*	81.50±39.52
ALAT [u/l]	76.67±7.84	71.83±15.85	68.50±10.63	70.33±9.81	61.83±10.38	54.50±12.93
AP [u/l]	228.33±31.43	225.67±18.44	216.50±38.21	207.17±29.19	200.33±21.82	248.33±36.38*

Values determined in study 1. Values given are mean values of the 6 animals/group with standard deviation included. Significant changes are indicated using *: P<0.05; **: P<0.01; ***: P<0,001. ASAT: aspartate aminotransferase, ALAT: alanine aminotransferase, AP: alkaline phosphatase (AP).

(Fig 2B). There was no difference between the groups on day 0. After 14 days of fructose feeding DAG 34:1 and 36:2 increased, which may also contain FA 18:1, while DAG 36:4 decreased. FF treatment decreased DAG-species 34:2, 34:3, 36:3 and 36:4 to below control levels. It is also notable that DAG 36:2 (18:1/18:1 or 18:0/18:2) is not influenced by FF treatment in rat serum.

TAG-species analysis

TAG-species in the liver (Fig 3) are again displayed as means of sums of TAG-chain lengths (C46, C48, C50, C52, C54, C56, C58) and of sums of degree of desaturation (sat – deca). The levels of individual TAG-species are shown in Fig S2 (liver) and S3 (serum). Upon fructose-feeding TAG-species containing 4 or more double bonds decreased, while the species containing 3 or less double bonds increased. FF treatment decreased the concentration of

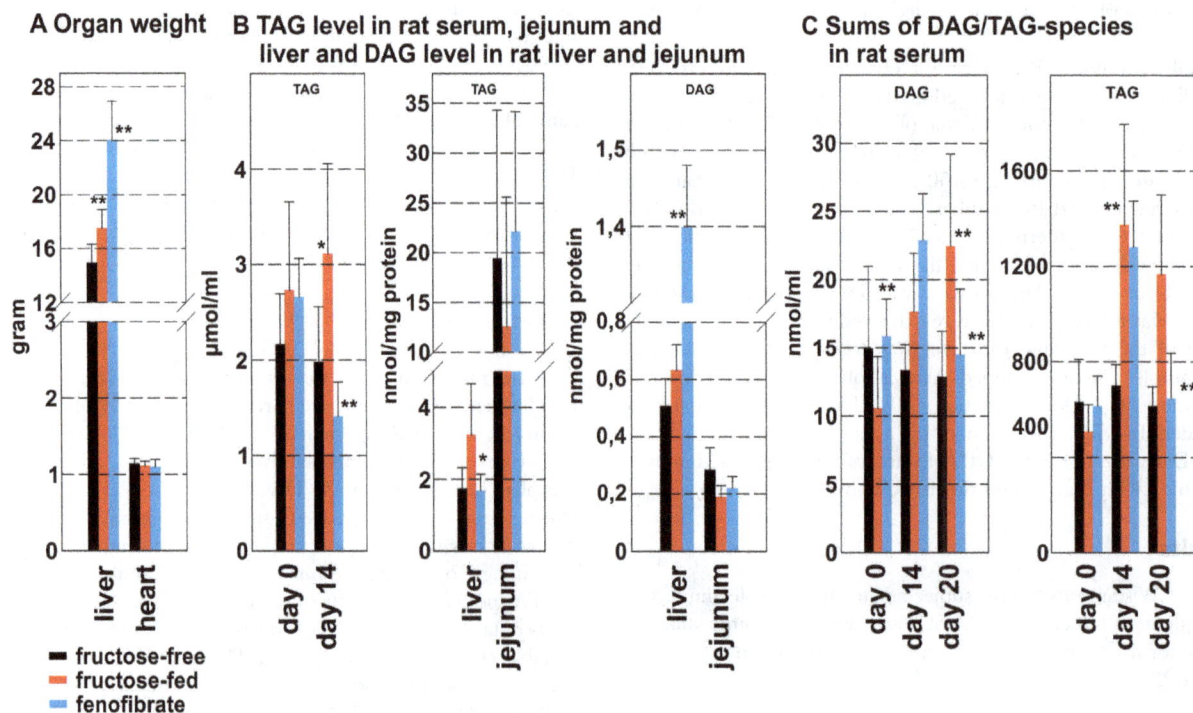

Figure 1. Tissue weights and overall TAG- and DAG-levels. A Organ weight of the animals after the end of treatment; **B** TAG-levels in rat liver, jejunum and serum and DAG-levels in rat liver and rat jejunum; **C** Sums of TAG- and DAG-species in rat serum. Tissue samples were taken during study 1, serum samples were taken during study 2; the control group is shown as black bar, the fructose-fed group is shown as red bar and the FF treated group is shown as a blue bar; values given are means±s.d.; significant changes are indicated using *: P<0.05; **: P<0.01; ***: P<0,001.

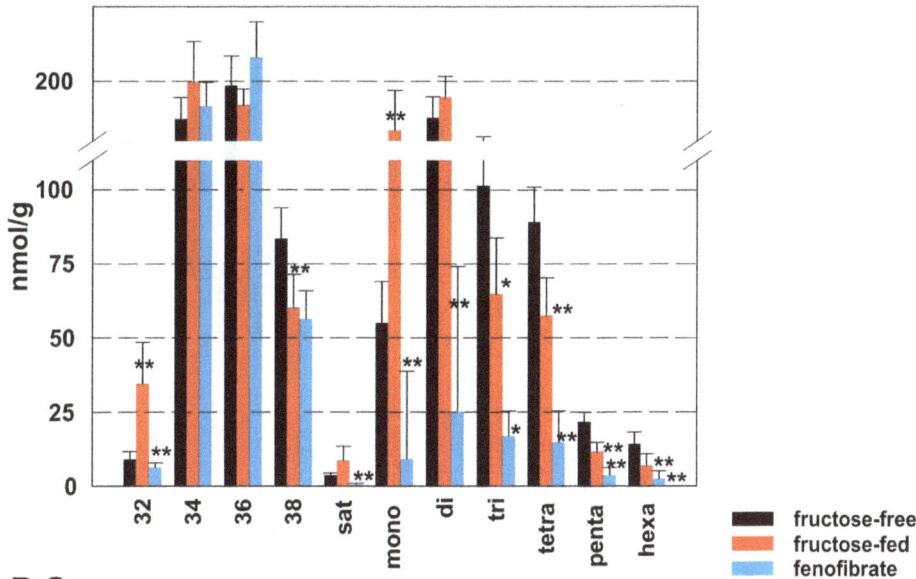

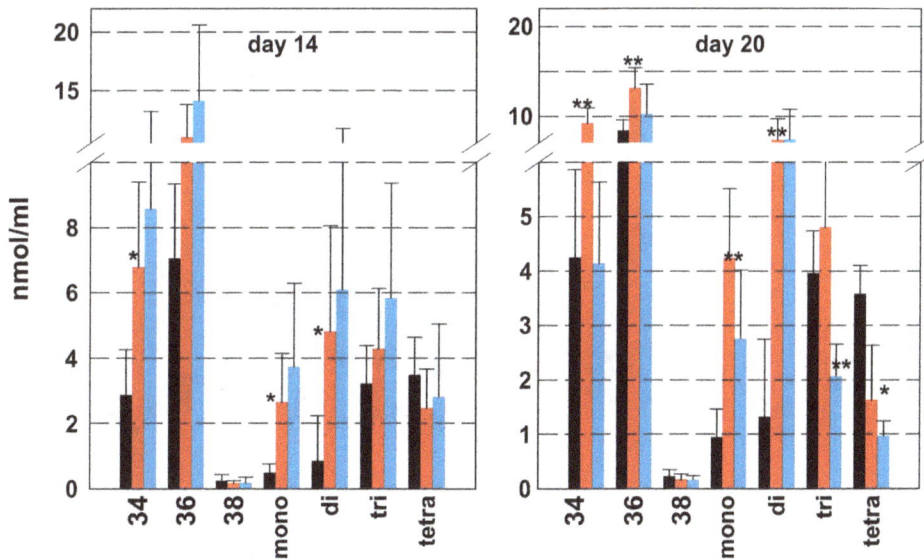

Figure 2. DAG-species in rat liver and serum. DAG-species in **A** rat liver and **B** rat serum. Values given are means±s.d. of the sums of the different chain lengths (C32 – C38) on the left and degree of desaturation (sat – hexa) on the right. Serum data is shown for day 14 and day 20; the control group is shown as black bar, the fructose-fed group is shown as red bar and the FF treated group is shown as a blue bar; significant changes are indicated using *: P<0.05; **: P<0.01; ***: P<0,001.

almost all TAG-species. However, it is interesting to note that the absolute levels of TAGs in the FF-group were lower than at baseline, except TAG 50:2, TAG 52:1, TAG 52:2, TAG 54:2 and TAG 54:3.

In contrast to DAG-species, there were more TAG-species detectable in rat serum than in the liver (Fig S3). The groups showed only marginal differences in TAG-species levels before treatment. After 14 days of fructose most of the TAG-species increased in rat serum with the exception of the highly unsaturated long-chain TAG-species (56 and more carbons combined with 6 or more double bonds, Fig S3). FF treatment completely reversed this species shift. With the exception of TAG 52:2, all TAG-species were reduced to levels well below the levels in the rats of the

untreated and unfed group. In total a clear trend is visible: The abundance of TAG-species C50, C52 and C54 is highest upon normal diet; fructose-feeding dramatically increased TAG-species C46, C48, C50 and C52 containing short-chain FAs (C12, C14, C16 and C18), while the TAG-species C56 and C58 containing long-chain FAs (C20 and C22) decreased.

Total hydrolysate fatty acids species analysis

In the next set of experiments, total hydrolysate fatty acid species levels (FA) in rat liver (Fig 4A) were determined. Concentrations of liver total hydrolysate FAs during fructose-feeding were lower for FAs 15:0, 17:0, 18:2 (n-6), 22:4 (n-6) and 22:5 (n-3). Interestingly, FA 14:0, which is generally not released

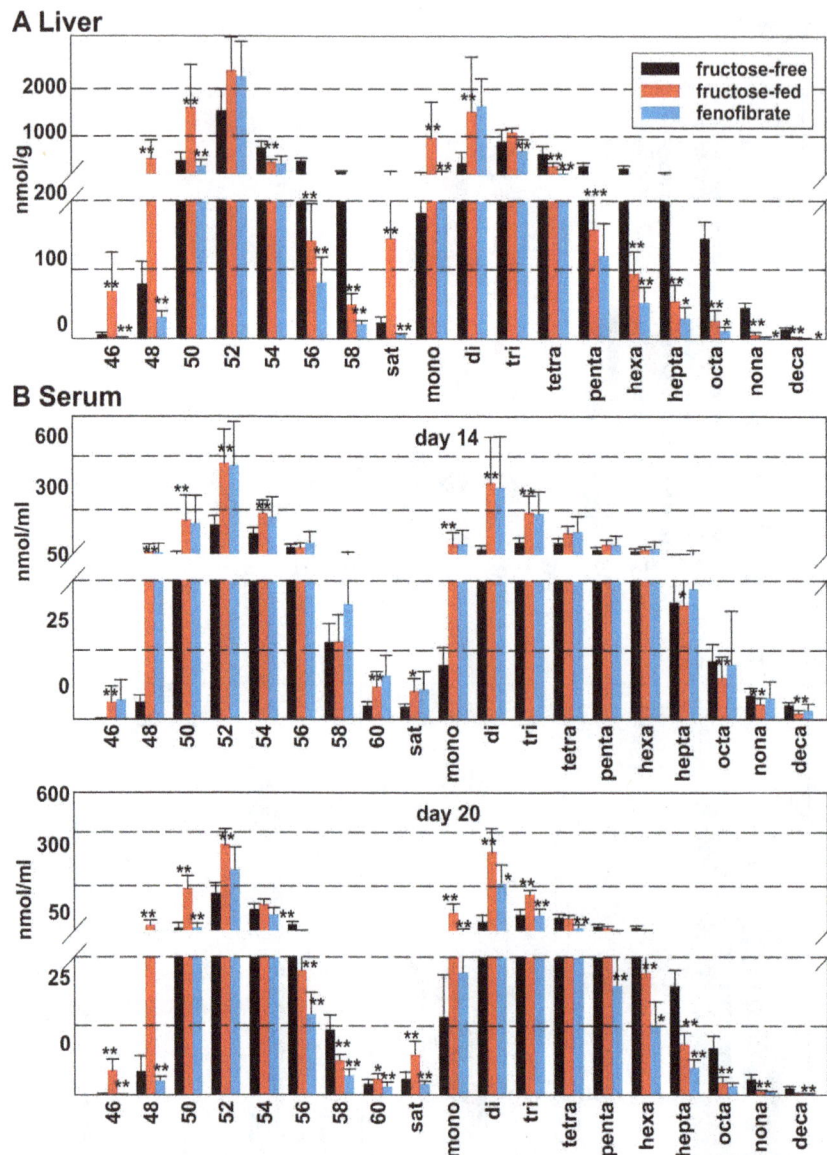

Figure 3. TAG-species in rat liver. Values given are means±s.d. of the sums of the different chain lengths (C46–C60) on the left and degree of desaturation (sat – deca) on the right. Serum data is shown for day 14 and day 20; the control group is shown as black bar, the fructose-fed group is shown as red bar and the FF treated group is shown as a blue bar; significant changes are indicated using *: P<0.05; **: P<0.01; ***: P<0,001.

by fatty acid synthase increased by about 30% during fructose-feeding. Also increased were FAs 16:1 (n-7), 18:1 (n-9) and 20:3 (n-6), with the other species displaying no change. During FF treatment the FA-species 14:0, 15.0, 17:0, 16:1 (n-7), 18:1 (n-7), 22:4 (n-6) and 22:5 (n-3) decreased, while FAs 16:0, 18:0, 18:1 (n-9), 20:3 (n-6), 20:4 (n-6) and 20:5 (n-3) increased.

In the jejunum, total hydrolysate FAs generally decreased under fructose-feeding, FF treatment tended to increase FA 14:0, FA 18:0, FA 20:0, and FA 22:0, while the other total hydrolysate FA-species remained unchanged. None of these changes reached significance.

In addition, more NEFA-species were present in the jejunum, but no significant changes were seen upon fructose-feeding or FF treatment (data not shown).

The concentrations of total FAs in serum were determined in all three groups at day 0, day 14 (before FF treatment), and at day 20 (Fig 4B). Total FA-levels at day 0 were higher in the fructose-

feeding group and the FF group. In the course of fructose-feeding, there was an increase in the short-chain FAs 14:0, 16:0, 16:1 (n-7) and 18:1 (n-7). In contrast, the abundance of long-chain FAs was reduced under fructose feeding, indicative of an impairment of elongase activity during metabolic (fructose) challenge. The increase of FA 18:1 (n-7) may be due to increased amounts of substrate FA 16:1 (n-7). Serum FA-levels of the fructose and the FF group were almost identical after 14 days of fructose-feeding and before FF treatment. The values at day 20 showed a further increase of short-chain FAs and a decrease of long-chain FAs upon fructose-feeding. FF treatment significantly reduced all FA levels, short-chain and long-chain, to or below baseline levels.

Non-esterified fatty acid species analysis

The non-esterified fatty acid (NEFA)-species in the rat liver that significantly changed upon treatment are shown in Fig 5A. NEFAs increased during fructose-feeding, with significantly higher

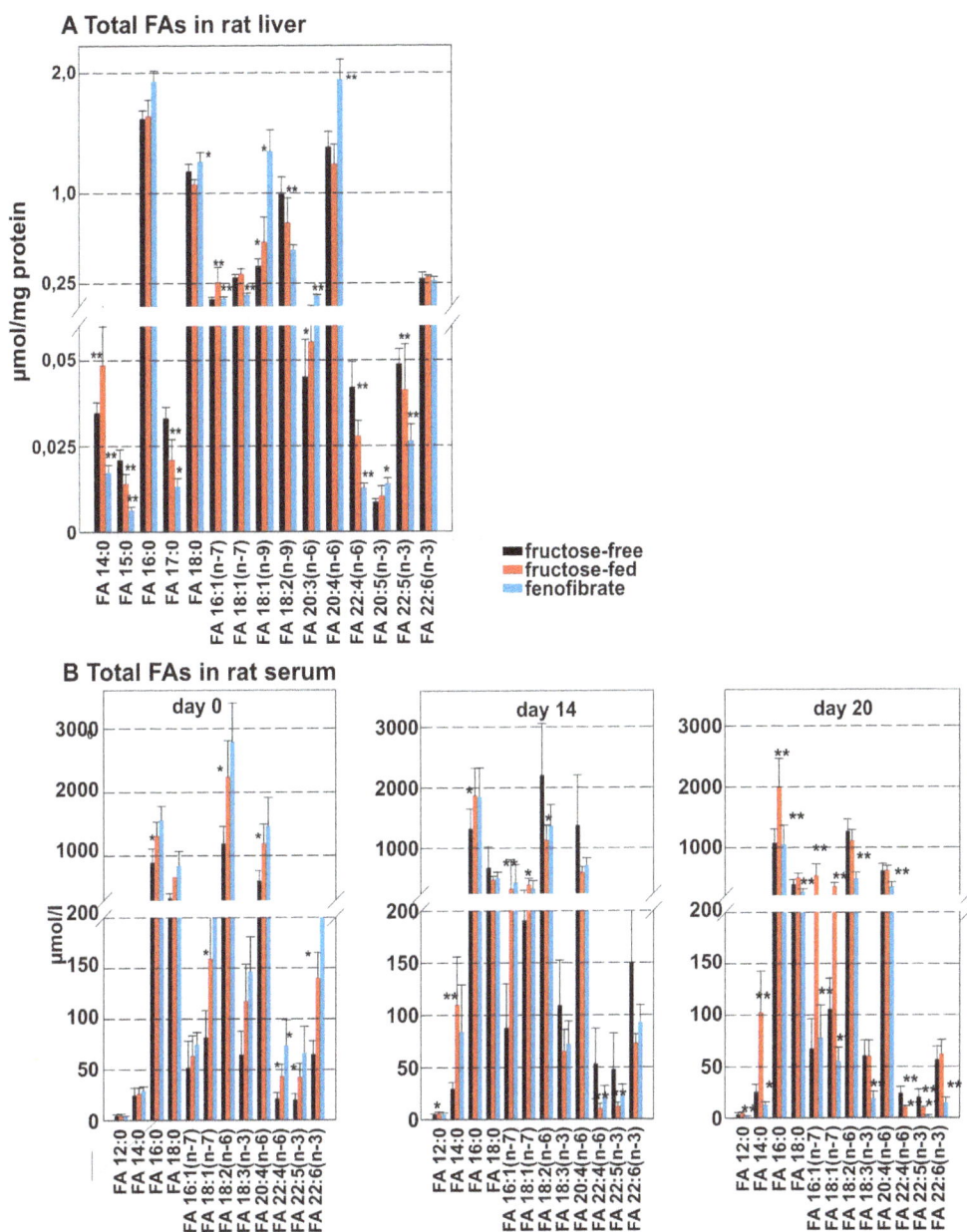

Figure 4. Total hydrolysate fatty acids in rat liver and serum. FA-species **A** total hydrolysate FA-species in rat liver **B** total hydrolysate FA-species in rat serum. The control group is shown as black bar the fructose-fed group is shown as red bar and the FF treated group is shown as a blue bar; values given are means±s.d.; significant changes are indicated using *: P<0.05; **: P<0.01; ***: P<0,001.

concentrations of FA 16:1 (n-7). Interestingly, during FF treatment FA 16:1 (n-7) decreased in rat liver, while FA 16:0, FA 18:0 FA, 18:1 (n-9) and FA 20:4 (n-6) increased significantly.

NEFAs at day 0 again showed higher levels for the fructose and the FF group (Fig 5B). During fructose feeding the NEFAs FA 14:0, FA 16:0, FA 16:1 (n-7), FA 18:0 and FA 18:1 (n-7) increased, while FA 18:2 (n-6) and FA 18:3 (n-7) decreased. NEFA-levels in the fructose-feeding and the FF group were identical before treatment. At day 20 the fructose-feeding group showed the same pattern compared to controls. Again, FF treatment reduced all NEFA levels to or below control group levels.

Interrelation analysis of TAG-, DAG- and total FA-species

The TAG/DAG- and FA-species were all quantified and the changes of species levels between the groups were plotted in an alternate display format (Fig 6). This figure contains just the section of TAGs with 54 C, DAGs with 36 C and FAs with 18 C to exemplify the principle. The complete displays are in Fig S4 and S5. With this information it is possible to determine graphically which FA and DAG-combinations might actually part of the corresponding TAG -species makeup. It is of course necessary to restrict the interpretation of this analysis to FA-species changes that reach significance.

We will give an example of this visualization. We have an increase of TAG 54:2 in the fructose-fed group compared to

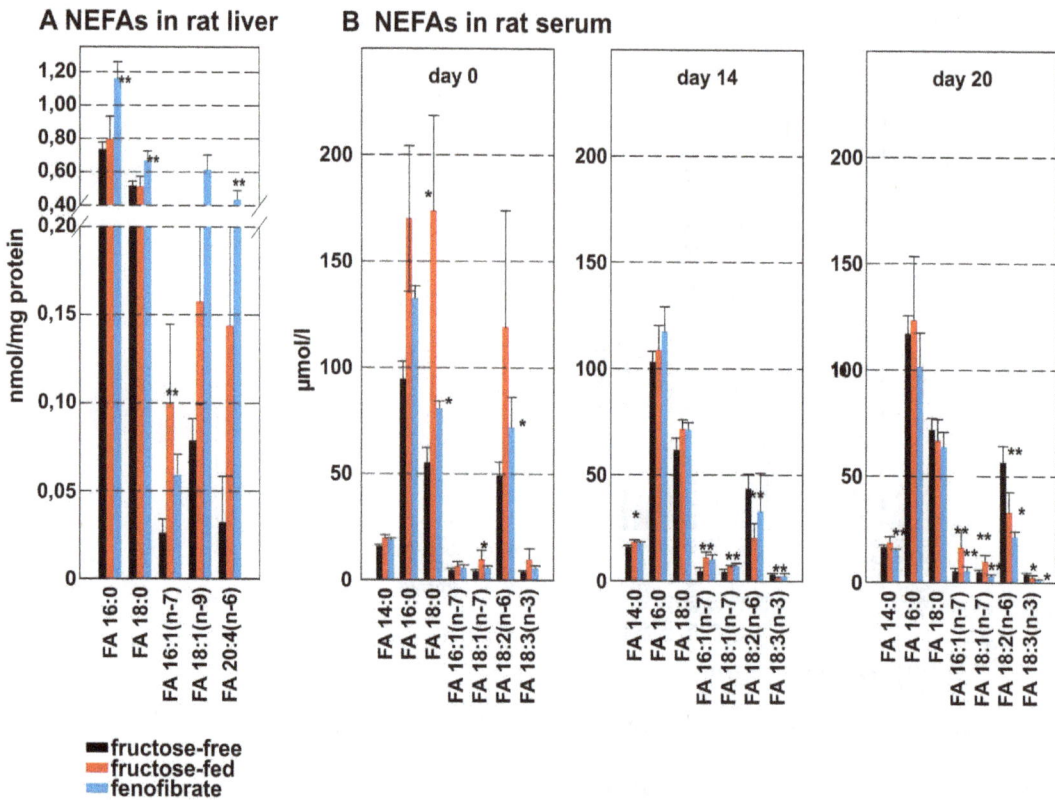

Figure 5. NEFA-species in rat liver and rat serum. A NEFA-species in rat liver homocysteine in rat serum; **B** NEFA-species in rat serum. The control group is shown as black bar, the fructose-fed group is shown as red bar and the FF treated group is shown as a blue bar; values given are means±s.d.; significant changes are indicated using *: P<0.05; **: P<0.01; ***: P<0,001.

A Rat liver fructose-fed - fructose-free

	36:1*	36:2**	36:3*	36:4*
54:7**				FA18:3
54:6**			FA18:3	FA18:2**
54:5**		FA18:3	FA18:2**	FA18:1*
54:4**	FA18:3	FA18:2**	FA18:1*	FA18:0
54:3	FA18:2**	FA18:1*	FA18:0	
54:2**	FA18:1*	FA18:0		

B Rat liver fenofibrate - fructose-fed

	36:1	36:2	36:3	36:4*
54:7				FA18:3
54:6			FA18:3	FA18:2
54:5		FA18:3	FA18:2	FA18:1*
54:4	FA18:3	FA18:2	FA18:1*	FA18:0*
54:3	FA18:2	FA18:1*	FA18:0*	
54:2	FA18:1*	FA18:0*		

C Rat serum fructose-fed - fructose-free

	36:2**	36:3	36:4*
54:7			FA18:3
54:6		FA18:3	FA18:2
54:5	FA18:3	FA18:2	FA18:1**
54:4	FA18:2	FA18:1**	FA18:0
54:3**	FA18:1**	FA18:0	
54:2**	FA18:0		

D Rat serum fenofibrate - fructose-fed

	36:2	36:3**	36:4
54:7**			FA18:3**
54:6*		FA18:3**	FA18:2**
54:5	FA18:3**	FA18:2**	FA18:1**
54:4	FA18:2**	FA18:1**	FA18:0**
54:3	FA18:1**	FA18:0**	
54:2*	FA18:0**		

Figure 6. Alternate display format of TAG/DAG-species and FA-species. A fructose-feeding vs control in rat liver; **B** FF treatment vs Fructose-feeding in rat liver; **C** fructose-feeding vs control in rat serum; **D** FF treatment vs Fructose-feeding in rat serum; An algorithm was developed to generate the alternate display form of DAG/TAG-species. In the diagrams the species shown in red have increased levels between the two respective groups, while a blue color indicates a decreased level of the respective species between the two groups. A black coloring indicates no change of the species between the two groups. All species have been colored if the difference of levels between the two groups were >5% of the total amount of the higher level, regardless of the significance of the change. For clarity, only the sections of the diagram containing the combinations of DAG C36 With FA C18 to yield TAG C54 are shown. If there is a fatty acid-DAG combination present with the same direction of change as the corresponding TAG-species, then this combination seems at least to contribute to the composition of the TAG-species. Significant changes are indicated using *: P<0.05; **: P<0.01; ***: P<0,001.

the controls in rat liver (Fig 6A). Considering the data of the DAG- and FA-species in this range (36 C for DAGs and 18 C for Fas), it seems most likely that TAG 54:2 consists of a combination of DAG 36:1 and FA 18:1 because both of those species are also increased, while FA 18:0, part of the potential combination DAG 36:2 and FA 18:0, is decreased. With these measurements of TAG/DAG-species it is possible to determine the most likely FA-combinations of the TAG/DAG-species (Fig S4).

It is also possible to determine the FA-species that converts a DAG-species into a corresponding TAG-species. For example, in the TAG-cluster containing 58 carbons (Fig S4), the species can be synthesized by esterification of DAG 38:5 with FA 20:0, 20:1, 20:2, 20:3, 20:4 or 20:5 respectively, to yield the resulting TAG-species (TAG 58:5, 58:6, 58:7, 58:8, 58:9 or 58:10). Of these six species TAG 58:9 and 58:10 decreased significantly under fructose feeding. The possible FA-DAG-combinations in which both components (FA and DAG) decreased are FA 20:4/DAG 38:5 for TAG 58:9; and FA 20:4/DAG 38:6 for TAG 58:10. These combinations are therefore likely to contribute to the compostion of those TAG-species. With this type of data analysis a relation of the up- or down-regulation of desaturation and elongation can be displayed graphically.

The same type of analysis has been performed for serum DAG/TAG/FA-species on day 20 (Fig S5). It is interesting to note that there were far less DAG-species present in rat serum but more TAG-species. Fructose-feeding induces the same pattern of increase/decrease of TAG-species in rat serum as in rat liver, as does FF treatment. Interestingly, all FA-species, except 20:3 decrease upon FF treatment.

VLDL secretion was also monitored (Fig 7). VLDL secretion after Triton treatment was doubled upon fructose feeding, while FF treatment decreased VLDL-secretion back to control levels.

Discussion

In the rat study presented here, most significant changes of lipid species occurred in rat liver, as a center of lipid synthesis and remodeling. TAGs are synthesized by esterification of DAGs through DGAT (consisting of two subtypes DGAT1 and DGAT2) with CoA-activated FAs being the substrates. The activity of DGAT is especially high in liver and intestine [42]. DGAT1 is most active at high substrate levels, which occur during high levels of lipolysis or during exposure to a high fat diet. DGAT2 on the other hand is most active at low substrate levels, which are connected to endogenous de novo FA-synthesis [43]. This endogenous synthesis is performed by fatty acid synthase, a multi-centered enzyme present in vivo as a dimer, which synthesizes fatty acids by incorporating C2-units until a chain-length of 16 carbons is reached and palmitate is released from the FAS-complex [44]. Fructose feeding increased FA 12:0 and 14:0, which are usually not liberated from FAS, in rat serum, while FF treatment reduced both species below control levels. Thus, fructose feeding seems to reduce the capacity of FAS for fatty acid binding of intermediates ≤C16:0 palmitate. This may be due to a metabolic overload of acetate caused by ingestion of fructose. The synthesis of other FA-species is then essentially achieved through elongation and desaturation of palmitate. Elongation involves elongases (ELOVL1/3/5/6) while desaturation is achieved by desaturases (SCD1-6, D5D, D6D) [45]. Under the influence of a diet high in carbohydrates (fructose-feeding) DGAT2 is the major enzyme active in the synthesis of TAGs as the storage form of FAs. In this study, TAG-levels in rat liver almost doubled upon fructose-feeding compared to standard diet and were decreased by FF-treatment to control levels. This indicates that DGAT2 inhibition is mediated by FF. It has been shown that SCD1 and DGAT2 are localized close to each other on the ER, indicating a participation of SCD1 in the synthesis of TAGs and the incorporation of MUFAs into TAGs [46]. This is in agreement with the presented data showing decreased FA16:1n7, the product of desaturation of FA16:0 by SCD1, in both NEFAs and total FAs. On the other hand, FA 18:1n9, which is also a product of SCD1, was increased in NEFAs and total FAs upon FF-treatment. This increase of FA 18:1n9 could be traced through most lipid species in liver and serum (e.g. CE 18:1) under FF-treatment. This might indicate that there is also SOAT1 activation by channeling oleate towards cholesteryl ester synthesis. Arachidonic acid (FA 20:4n6) is derived from linoleate through desaturation (D6D and D5D) and elongation (ELOVL5). In rat liver, FF-treatment significantly increased the level of the pro-inflammatory FA 20:4 (n-6) in total FAs and NEFAs, accompanied by a significant increase of the precursor FA20:3 (n-6) and a significant decrease of the corresponding precursor FA18:2 (n-6) in total FAs. In the NEFA-fraction, the FA18:2 (n-6) concentration remained unchanged compared to fructose-feeding, indicative of elongase (ELOVL5) and desaturase (D5D) activation towards the n-6-series induced by FF-treatment in rat liver. Most FA-species of the anti-inflammatory n-3-series decreased or remained below detection levels.

It is interesting to note that the weight of the liver increases significantly under FF treatment. The underlying mechanism is unclear. It might be due to the increased metabolic action induced by PPARα-activation resulting in an increased hepatic activity.

In order to verify the hypothesis described above, the linkage of transcriptional data to the lipid metabolite profiles is attempted as an extension of the present study.

Summary

The present study presents a detailed investigation of lipid species during fructose challenge and fenofibrate treatment in rats. To our knowledge, it is the most complete lipidomic analysis on FF effects conducted to date. In the first part, we determined the levels of TAG- and DAG-species as well as the amounts of total hydrolysis FA- and NEFA-species in rat liver and rat serum. Short-chain TAGs increased in serum and liver upon fructose-feeding and almost all TAG-species decreased under FF treatment, while longer-chain, more unsaturated DAG-levels (36:1, 36:2, 36:4,

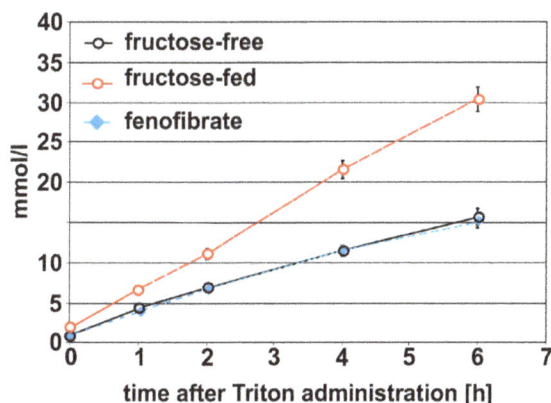

Figure 7. VLDL-secretion after FF treatment. The control group is shown in black, the fructose-fed group is shown in red and the FF treated group is shown in blue; values given are means±s.d.; significant changes are indicated using *: P<0.05; **: P<0.01; ***: P<0,001.

38:3, 38:4, 38:5) increased upon FF treatment in rat liver and decreased in rat serum. FAs, especially short-chain FAs (12:0, 14:0, 16:0) and VLDL secretion increased during fructose-challenge and decreased to control levels during FF treatment. Fructose challenge of de novo fatty acid synthesis through fatty acid synthase (FAS) may enhance the release of $\leq16:0$ chain length fatty acids, a process that is reversed by FF-mediated PPARα-activation.

In addition, we present a correlation diagram of the data that combines theoretical compositions of TAG and DAG with relative abundances of TAG, DAG and FA species, thereby providing a qualitative view on probable TAG and DAG compositions.

Supporting Information

Figure S1   DAG-species in rat liver and rat serum. The control group is shown as black bar, the fructose-fed group is shown as red bar and the FF treated group is shown as a blue bar; values given are means±s.d.; significant changes are indicated using *: $P<0.05$; **: $P<0.01$; ***: $P<0,001$.

Figure S2   TAG-species in rat liver. The control group is shown as black bar, the fructose-fed group is shown as red bar and the FF treated group is shown as a blue bar; values given are means±s.d.; significant changes are indicated using *: $P<0.05$; **: $P<0.01$; ***: $P<0,001$.

Figure S3   TAG-species in rat serum. The control group is shown as black bar, the fructose-fed group is shown as red bar and the FF treated group is shown as a blue bar; values given are means±s.d.; significant changes are indicated using *: $P<0.05$; **: $P<0.01$; ***: $P<0,001$.

Figure S4   Alternate display format of TAG/DAG-species and FA-species. A fructose-feeding vs control in rat liver; **B** FF treatment vs Fructose-feeding in rat liver; An algorithm was developed to generate the alternate display form of DAG/TAG-

species. In the diagrams the species shown in red were increased upon fructose-feeding compared to baseline, while a blue color indicates a decreased level of the respective species. For clarity, only the sections of the diagram containing the combinations of DAG C36 With FA C18 to yield TAG C54 are shown. If there is a fatty acid combination present with the same direction of change as the corresponding TAG-species, then this combination should be at least the biggest contributor to the TAG-species. Significant changes are indicated using *: $P<0.05$; **: $P<0.01$; ***: $P<0,001$.

Figure S5   Alternate display format of TAG/DAG-species and FA-species. A fructose-feeding vs control in rat serum; **B** FF treatment vs Fructose-feeding in rat serum; An algorithm was developed to generate the alternate display form of DAG/TAG-species. In the diagrams the species shown in red were increased upon fructose-feeding compared to baseline, while a blue color indicates a decreased level of the respective species. For clarity, only the sections of the diagram containing the combinations of DAG C36 With FA C18 to yield TAG C54 are shown. If there is a fatty acid combination present with the same direction of change as the corresponding TAG-species, then this combination should be at least the biggest contributor to the TAG-species. Significant changes are indicated using *: $P<0.05$; **: $P< 0.01$; ***: $P<0,001$.

Acknowledgments

We would like to thank our technicians Jolanthe Aiwanger, Simone Peschel, Doreen Müller and Bettina Hartl for their valuable contributions. This work was supported by the seventh framework program of the EU-funded "LipidomicNet" (proposal number 202272).

Author Contributions

Conceived and designed the experiments: T. Kopf MB HS. Performed the experiments: T. Kopf HS MT HK. Analyzed the data: T. Kopf MT HK. Contributed reagents/materials/analysis tools: T. Kopf T. Konovalova HK HS. Contributed to the writing of the manuscript: T. Kopf MB GS.

References

1. Staels B, Dallongeville J, Auwerx J, Schoonjans K, Leitersdorf E, et al. (1998) Mechanism of action of fibrates on lipid and lipoprotein metabolism. Circulation 98: 2088–2093.
2. Hiukka A, Maranghi M, Matikainen N, Taskinen MR (2010) PPARalpha: an emerging therapeutic target in diabetic microvascular damage. Nat Rev Endocrinol 6: 454–463. nrendo.2010.89 [pii];10.1038/nrendo.2010.89 [doi].
3. Klaunig JE, Babich MA, Baetcke KP, Cook JC, Corton JC, et al. (2003) PPARalpha agonist-induced rodent tumors: modes of action and human relevance. Crit Rev Toxicol 33: 655–780.
4. Ohta T, Masutomi N, Tsutsui N, Sakairi T, Mitchell M, et al. (2009) Untargeted metabolomic profiling as an evaluative tool of fenofibrate-induced toxicology in Fischer 344 male rats. Toxicol Pathol 37: 521–535.
5. Ueno T, Fukuda N, Nagase H, Tsunemi A, Tahira K, et al. (2009) Atherogenic dyslipidemia and altered hepatic gene expression in SHRSP.Z-Leprfa/IzmDmcr rats. Int J Mol Med 23: 313–320.
6. Galman C, Lundasen T, Kharitonenkov A, Bina HA, Eriksson M, et al. (2008) The circulating metabolic regulator FGF21 is induced by prolonged fasting and PPARalpha activation in man. Cell Metab 8: 169–174. S1550-4131(08)00207-6 [pii];10.1016/j.cmet.2008.06.014 [doi].
7. Cariello NF, Romach EH, Colton HM, Ni H, Yoon L, et al. (2005) Gene expression profiling of the PPAR-alpha agonist ciprofibrate in the cynomolgus monkey liver. Toxicol Sci 88: 250–264. kfi273 [pii];10.1093/toxsci/kfi273 [doi].
8. Shearer BG, Hoekstra WJ (2003) Recent advances in peroxisome proliferator-activated receptor science. Curr Med Chem 10: 267–280.
9. Peters JM, Cheung C, Gonzalez FJ (2005) Peroxisome proliferator-activated receptor-alpha and liver cancer: where do we stand? J Mol Med (Berl) 83: 774–785. 10.1007/s00109-005-0678-9 [doi].
10. Panigrahy D, Kaipainen A, Huang S, Butterfield CE, Barnes CM, et al. (2008) PPARalpha agonist fenofibrate suppresses tumor growth through direct and indirect angiogenesis inhibition. Proc Natl Acad Sci U S A 105: 985–990. 0711281105 [pii];10.1073/pnas.0711281105 [doi].
11. Keech A, Simes RJ, Barter P, Best J, Scott R, et al. (2005) Effects of long-term fenofibrate therapy on cardiovascular events in 9795 people with type 2 diabetes mellitus (the FIELD study): randomised controlled trial. Lancet 366: 1849–1861. S0140-6736(05)67667-2 [pii];10.1016/S0140-6736(05)67667-2 [doi].
12. Nishimura J, Dewa Y, Muguruma M, Kuroiwa Y, Yasuno H, et al. (2007) Effect of fenofibrate on oxidative DNA damage and on gene expression related to cell proliferation and apoptosis in rats. Toxicol Sci 97: 44–54.
13. Nishimura J, Dewa Y, Okamura T, Muguruma M, Jin M, et al. (2008) Possible involvement of oxidative stress in fenofibrate-induced hepatocarcinogenesis in rats. Arch Toxicol 82: 641–654.
14. O'Brien ML, Spear BT, Glauert HP (2005) Role of oxidative stress in peroxisome proliferator-mediated carcinogenesis. Crit Rev Toxicol 35: 61–88.
15. Seo KW, Kim KB, Kim YJ, Choi JY, Lee KT, et al. (2004) Comparison of oxidative stress and changes of xenobiotic metabolizing enzymes induced by phthalates in rats. Food Chem Toxicol 42: 107–114.
16. Michalik L, Desvergne B, Wahli W (2004) Peroxisome-proliferator-activated receptors and cancers: complex stories. Nat Rev Cancer 4: 61–70.
17. Caldwell KD (1988) Field-flow fractionation. Anal Chem 60: 959A–971A.
18. Yetukuri L, Huopaniemi I, Koivuniemi A, Maranghi M, Hiukka A, et al. (2011) High density lipoprotein structural changes and drug response in lipidomic profiles following the long-term fenofibrate therapy in the FIELD substudy. PLoS One 6: e23589. 10.1371/journal.pone.0023589 [doi];PONE-D-11-11221 [pii].
19. Lu Y, Boekschoten MV, Wopereis S, Muller M, Kersten S (2011) Comparative transcriptomic and metabolomic analysis of fenofibrate and fish oil treatments in mice. Physiol Genomics 43: 1307–1318. physiolgenomics.00100.2011 [pii];10.1152/physiolgenomics.00100.2011 [doi].
20. Lee CH, Olson P, Evans RM (2003) Minireview: lipid metabolism, metabolic diseases, and peroxisome proliferator-activated receptors. Endocrinology 144: 2201–2207. 10.1210/en.2003-0288 [doi].

21. Berger J, Moller DE (2002) The mechanisms of action of PPARs. Annu Rev Med 53: 409–435.
22. Hostetler HA, Petrescu AD, Kier AB, Schroeder F (2005) Peroxisome proliferator-activated receptor alpha interacts with high affinity and is conformationally responsive to endogenous ligands. J Biol Chem 280: 18667–18682.
23. Ledwith BJ, Pauley CJ, Wagner LK, Rokos CL, Alberts DW, et al. (1997) Induction of cyclooxygenase-2 expression by peroxisome proliferators and non-tetradecanoylphorbol 12,13-myristate-type tumor promoters in immortalized mouse liver cells. J Biol Chem 272: 3707–3714.
24. Duez H, Lefebvre B, Poulain P, Torra IP, Percevault F, et al. (2005) Regulation of human apoA-I by gemfibrozil and fenofibrate through selective peroxisome proliferator-activated receptor alpha modulation. Arterioscler Thromb Vasc Biol 25: 585–591.
25. Lefebvre P, Chinetti G, Fruchart JC, Staels B (2006) Sorting out the roles of PPAR alpha in energy metabolism and vascular homeostasis. J Clin Invest 116: 571–580.
26. Ntambi JM (1992) Dietary regulation of stearoyl-CoA desaturase 1 gene expression in mouse liver. J Biol Chem 267: 10925–10930.
27. Elshourbagy NA, Near JC, Kmetz PJ, Sathe GM, Southan C, et al. (1990) Rat ATP citrate-lyase. Molecular cloning and sequence analysis of a full-length cDNA and mRNA abundance as a function of diet, organ, and age. J Biol Chem 265: 1430–1435.
28. Flowers MT, Miyazaki M, Liu X, Ntambi JM (2006) Probing the role of stearoyl-CoA desaturase-1 in hepatic insulin resistance. J Clin Invest 116: 1478–1481.
29. Enoch HG, Catala A, Strittmatter P (1976) Mechanism of rat liver microsomal stearyl-CoA desaturase. Studies of the substrate specificity, enzyme-substrate interactions, and the function of lipid. J Biol Chem 251: 5095–5103.
30. Dobrzyn A, Ntambi JM (2004) The role of stearoyl-CoA desaturase in body weight regulation. Trends Cardiovasc Med 14: 77–81.
31. Ntambi JM, Miyazaki M, Stoehr JP, Lan H, Kendziorski CM, et al. (2002) Loss of stearoyl-CoA desaturase-1 function protects mice against adiposity. Proc Natl Acad Sci U S A 99: 11482–11486.
32. Montanaro MA, Rimoldi OJ, Igal RA, Montenegro S, Tarres MC, et al. (2003) Hepatic delta9, delta6, and delta5 desaturations in non-insulin-dependent diabetes mellitus eSS rats. Lipids 38: 827–832.
33. Vega GL, Cater NB, Hadizadeh DR, III, Meguro S, Grundy SM (2003) Free fatty acid metabolism during fenofibrate treatment of the metabolic syndrome. Clin Pharmacol Ther 74: 236–244.
34. Dekker MJ, Su Q, Baker C, Rutledge AC, Adeli K (2010) Fructose: a highly lipogenic nutrient implicated in insulin resistance, hepatic steatosis, and the metabolic syndrome. Am J Physiol Endocrinol Metab 299: E685–E694.
35. Fauland A, Kofeler H, Trotzmuller M, Knopf A, Hartler J, et al. (2011) A comprehensive method for lipid profiling by liquid chromatography-ion cyclotron resonance mass spectrometry. J Lipid Res 52: 2314–2322. jlr.D016550 [pii];10.1194/jlr.D016550 [doi].
36. Matyash V, Liebisch G, Kurzchalia TV, Shevchenko A, Schwudke D (2008) Lipid extraction by methyl-tert-butyl ether for high-throughput lipidomics. J Lipid Res 49: 1137–1146. D700041-JLR200 [pii];10.1194/jlr.D700041-JLR200 [doi].
37. Hartler J, Trotzmuller M, Chitraju C, Spener F, Kofeler HC, et al. (2011) Lipid Data Analyzer: unattended identification and quantitation of lipids in LC-MS data. Bioinformatics 27: 572–577. btq699 [pii];10.1093/bioinformatics/btq699 [doi].
38. Ecker J, Scherer M, Schmitz G, Liebisch G (2012) A rapid GC-MS method for quantification of positional and geometric isomers of fatty acid methyl esters. J Chromatogr B Analyt Technol Biomed Life Sci 897: 98–104. S1570-0232(12)00229-2 [pii];10.1016/j.jchromb.2012.04.015 [doi].
39. BLIGH EG, DYER WJ (1959) A rapid method of total lipid extraction and purification. Can J Biochem Physiol 37: 911–917.
40. DOLE VP, MEINERTZ H (1960) Microdetermination of long-chain fatty acids in plasma and tissues. J Biol Chem 235: 2595–2599.
41. Kopf T, Schmitz G (2013) Analysis of non-esterified fatty acids in human samples by solid-phase-extraction and gas chromatography/mass spectrometry. J Chromatogr B Analyt Technol Biomed Life Sci 938: 22–26. S1570-0232(13)00437-6 [pii];10.1016/j.jchromb.2013.08.016 [doi].
42. Yu YH, Ginsberg HN (2004) The role of acyl-CoA: diacylglycerol acyltransferase (DGAT) in energy metabolism. Ann Med 36: 252–261.
43. Yen CL, Stone SJ, Koliwad S, Harris C, Farese RV, Jr. (2008) Thematic review series: glycerolipids. DGAT enzymes and triacylglycerol biosynthesis. J Lipid Res 49: 2283–2301. R800018-JLR200 [pii];10.1194/jlr.R800018-JLR200 [doi].
44. Chirala SS, Wakil SJ (2004) Structure and function of animal fatty acid synthase. Lipids 39: 1045–1053.
45. Jump DB (2011) Fatty acid regulation of hepatic lipid metabolism. Curr Opin Clin Nutr Metab Care 14: 115–120. 10.1097/MCO.0b013e328342991c [doi].
46. Man WC, Miyazaki M, Chu K, Ntambi J (2006) Colocalization of SCD1 and DGAT2: implying preference for endogenous monounsaturated fatty acids in triglyceride synthesis. J Lipid Res 47: 1928–1939. M600172-JLR200 [pii];10.1194/jlr.M600172-JLR200 [doi].

13

A Systematic Evaluation of Short Tandem Repeats in Lipid Candidate Genes: Riding on the SNP-Wave

Claudia Lamina[1], Margot Haun[1], Stefan Coassin[1], Anita Kloss-Brandstätter[1], Christian Gieger[2], Annette Peters[3], Harald Grallert[4], Konstantin Strauch[2,5], Thomas Meitinger[6,7], Lyudmyla Kedenko[8], Bernhard Paulweber[8], Florian Kronenberg[1]*

1 Division of Genetic Epidemiology, Department of Medical Genetics, Molecular and Clinical Pharmacology, Innsbruck Medical University, Innsbruck, Austria, 2 Institute of Genetic Epidemiology, Helmholtz Zentrum München - German Research Center for Environmental Health (GmbH), Neuherberg, Germany, 3 Institute of Epidemiology II, Helmholtz Zentrum München – German Research Center for Environmental Health, Neuherberg, Germany, 4 Department of Molecular Epidemiology, Helmholtz Zentrum München, German Research Center for Environmental Health, Neuherberg, Germany, 5 Institute of Medical Informatics, Biometry and Epidemiology, Chair of Genetic Epidemiology, Ludwig-Maximilians-Universität, Munich, Germany, 6 Institute of Human Genetics, TechnischeUniversitätMünchen, Munich, Germany, 7 Institute of Human Genetics, Helmholtz Zentrum München – German Research Center for Environmental Health, Neuherberg, Germany, 8 First Department of Internal Medicine, Paracelsus Private Medical University Salzburg, Salzburg, Austria

Abstract

Structural genetic variants as short tandem repeats (STRs) are not targeted in SNP-based association studies and thus, their possible association signals are missed. We systematically searched for STRs in gene regions known to contribute to total cholesterol, HDL cholesterol, LDL cholesterol and triglyceride levels in two independent studies (KORA F4, n = 2553 and SAPHIR, n = 1648), resulting in 16 STRs that were finally evaluated. In a combined dataset of both studies, the sum of STR alleles was regressed on each phenotype, adjusted for age and sex. The association analyses were repeated for SNPs in a 200 kb region surrounding the respective STRs in the KORA F4 Study. Three STRs were significantly associated with total cholesterol (within *LDLR*, the *APOA1/C3/A4/A5/BUD13* gene region and *ABCG5/8*), five with HDL cholesterol (3 within *CETP*, one in *LPL* and one in*APOA1/C3/A4/A5/BUD13*), three with LDL cholesterol (*LDLR*, *ABCG5/8* and *CETP*) and two with triglycerides (*APOA1/C3/A4/A5/BUD13* and *LPL*). None of the investigated STRs, however, showed a significant association after adjusting for the lead or adjacent SNPs within that gene region. The evaluated STRs were found to be well tagged by the lead SNP within the respective gene regions. Therefore, the STRs reflect the association signals based on surrounding SNPs. In conclusion, none of the STRs contributed additionally to the SNP-based association signals identified in GWAS on lipid traits.

Editor: Reiner Albert Veitia, Institut Jacques Monod, France

Funding: This work was supported by the "Genomics of Lipid-associated Disorders – GOLD" of the "Austrian Genome Research Programme GEN-AU" to FK and by a grant from the Austrian Science Fund (FWF) P 26660-B13 to CL. The funders had no role in study design, data collection and analysis, decision to publish, or preparation of the manuscript.

* Email: Florian.Kronenberg@i-med.ac.at

Introduction

Plasma levels of total cholesterol, LDL cholesterol, HDL cholesterol and triglycerides are important risk factors for cardiovascular disease. They are also among the most intensely studied complex quantitative phenotypes in genetic association studies. A genome-wide meta-analysis by Teslovich et al. [1] identified 95 loci with significant influence on lipid levels. SNPs within these loci explain about 10–12% of phenotypic variance, corresponding to about 25–30% of the genetic variance. This is a rather high percentage compared to other complex quantitative phenotypes. Additional 62 variants have been identified recently [2], explaining further 1.6–2.6% depending on the phenotype. Nevertheless, a majority of the expected heritability of the lipid traits is still unexplained.

The impact of structural genetic variants has not been studied systematically, although it might contribute to some extent to the missing heritability. There are several types of structural variants that differ by the kind of variability and by their length [3]. Tandem repeats define a repetition of nucleotide motifs (2 to > 1000 bp), which are concatenated adjacent to each other. Copy-number variations (CNVs) consist of repeating motifs of more than 1000 base pairs (bp). Motifs consisting of less than 10 bp are called short tandem repeats (STRs), simple sequence repeats (SSR) or microsatellites. More specifically, di-, tri-, tetra-, penta- or hexa-nucleotide repeats are the most common STRs, referring to motifs of 2, 3, 4, 5 or 6 nucleotides. A comprehensive analysis within the 1000 Genomes project estimated that tandem repeat sites occupy about 1.25% of the human genome [4]. Mutations in short tandem repeats mostly originate from insertions or deletions of repeated motifs. In such tandem repeat regions, mutation rates are very high [5]. Therefore, a high percentage of short tandem repeats are highly polymorphic and multiallelic. In consequence,

STRs have been widely used in population genetic studies, genetic fingerprinting and as molecular markers for genetic association studies [6]. More specifically, STRs were used for hypothesis-free genome-wide linkage studies to derive new susceptibility loci in the era before genome-wide association studies (GWAS). They were neglected in favor of SNPs, as soon as genome-wide SNP chips became affordable and feasible. Although STRs were primarily regarded as non-functional markers in such studies, changes in length can have considerable impact on disease or disease-associated traits. For example, expansion of tandem repeats can lead to monogenetic disorders as Huntington disease or Fragile X syndrome [7]. Since STRs can also be found in promoter regions in a higher frequency than expected by chance, they might also have a high potential to modify gene regulation [8]. There is also accumulating evidence from candidate gene association studies that STRs are associated with susceptibility for complex diseases like schizophrenia, bipolar disorder, diabetes, cancer [9], asthma [10] and also cardiovascular disease [11,12] and related intermediate phenotypes [13]. There have also been some candidate gene studies on lipid phenotypes investigating the impact of STRs. Talmud et al. [14] identified one tetranucleotide repeat within the CETP promoter that was significantly associated with LDL size, triglycerides, and apolipoprotein B concentrations. This STR was further associated with HDL cholesterol levels [15].

Common GWAS as well as the vast majority of candidate gene studies, however, do not include structural variants such as short tandem repeats. Therefore, such associations are likely missed, thereby possibly contributing to the widely discussed missing heritability [9,16].

The intention of our study was to systematically evaluate short tandem repeats in known lipid gene regions on the lipid phenotypes total cholesterol, HDL cholesterol, LDL cholesterol and triglycerides. We hypothesized that short tandem repeats might explain an additional part of the genetic variation that is thought to be heritable. Therefore, we sorted all SNPs, which have been found in a big GWA meta-analysis to be associated with lipid phenotypes [1] by their strength of association, and selected 16 mostly unknown STRs within the genes lying near to the identified SNPs. We regressed these STRs on lipid phenotypes in two independent studies: one population-based study (KORA F4 Study) and one study in a healthy working population SAPHIR). Since a high fraction of tandem repeat polymorphisms have been shown to be well tagged by surrounding SNPs [4,17], we also evaluated whether our selected STRs can be tagged by surrounding SNPs or if they are rather independent. Therefore, the specific questions we address are as follows: 1) Are STRs selected from known lipid candidate genes associated with the respective lipid phenotypes? 2) Are these identified STR-lipid associations independent from genome-wide significant SNPs in the respective gene regions?

Materials and Methods

Study populations

The KORA F4 Study is a population-based sample from the general population living in the region of Augsburg, Southern Germany, which has evolved from the WHO MONICA study (Monitoring of Trends and Determinants of Cardiovascular Disease). Age-and sex-stratified samples were drawn in the years 1999–2001 (n = 4,261). A total of 3,080 subjects participated in a follow-up examination in 2006/08 (KORA F4) [18,19]. All participants are of European ancestry. Samples were available for 3063 participants.

The SAPHIR Study (Salzburg Atherosclerosis Prevention Program in subjects at High Individual Risk) is an observational study conducted in the years 1999–2002 involving 1770 healthy unrelated subjects (663 females and 1107 males). Study participants were recruited by health-screening programs in companies in and around the Austrian city Salzburg [20]. DNA samples and phenotypes were available for 1726 participants.

Ethics Statement

Study participants were examined according to the principles expressed in the Declaration of Helsinki. Participants from both studies provided written informed consent. The protocol of the KORA F4 Study was approved by the Ethical Committee of the "Bayrische Landesärztekammer", and the protocol of the SAPHIR Study was approved by the Ethical Committee of Land Salzburg. Study participant records were anonymized and de-identified prior to analysis.

Selection of Short Tandem Repeats

First, we sorted all SNPs from the 95 loci reported in Table 1 in Teslovich et al. [1] by their p-value, starting with the lowest. For each SNP, we picked the nearest gene and retrieved the corresponding sequence +/−10 kb up- and downstream of the gene. In case of gene clusters (*APOA1/C3/A4/A5/BUD13*) or overlapping genes (*ABCG5/8*) the sequences of all genes were included. We scanned these sequences for potential short tandem repeats (STRs) using the software SERV [21] (http://www.igs.cnrs-mrs.fr/SERV). SERV identifies potential STRs in nucleotide sequences. It uses the number of repeated units, the unit length and the purity to calculate a continuous score (named "Var-Score"), which provides an estimation of the repeat variability (with high scores correlating with a more pronounced variability).

In order to be considered for further experimental testing, a putative STR had to fulfill the following criteria: a SERV VarScore larger than 0.80, a repeat unit between 3 and 6 bp, a repeat purity larger than 97% (i.e. the percentage of sequence perfectly matching the identified repeat element) and a total repeat region length smaller than 200 bp. Dinucleotide repeats were not regarded because of their instability and the error-prone interpretation of the electrophoretic data when stutter peaks are present [22,23]. Repeat units greater than 5 nucleotides and repeats consisting only of C's or G's were excluded because of their difficult amplification in PCR analysis.

We decided a priori to genotype 16 STRs at a maximum since this number fits two multiplex PCRs. The 54 most strongly lipid-associated genes had to be screened consecutively to obtain the first 15 STRs which matched with the above described selection criteria for STR genotyping. All gene-regions that were considered in this search for potential STRs are given in Table S1 in File S1. The STR CETP_3 was additionally selected due to its known association with lipids [14,15], although the STR did not fulfill our bioinformatic criteria. Divergent from the prior selection criteria, we accepted a VarScore from 0.75 within *LIPG*, since the STR that actually fulfilled the criteria (VarScore = 1.76), caused unresolvable problems in the multiplex PCR and was therefore excluded. Only three out of the 15 loci were already known, two of them from forensics [24,25]. Only one penta-nucleotide-repeat in the *LPA* gene has already been subject of genetic association analyses [12]. Table 1 lists the selected 16 STRs and the corresponding gene regions together with the identified lead SNP based on Teslovich et al. [1] and their original association results.

Table 1. Gene regions for which STRs have been selected, reporting the magnitude of p-values taken from Teslovich et al. for all four investigated lipid phenotypes together with the respective lead SNP.

Gene/Gene region	STR in or near gene	Magnitude of p-value from the association with the following traits[a]				Lead-SNP[a]
		TC	LDL	HDL	TG	
ABCG5/8	ABCG5	10^{-45}	10^{-47}			rs4299376
APOA1/C3/A4/A5/BUD13	BUD13	10^{-57}	10^{-26}	10^{-47}	10^{-240}	rs964184
CETP	CETP_1	10^{-14}	10^{-13}	10^{-380}	10^{-12}	rs3764261
CETP	CETP_2	10^{-14}	10^{-13}	10^{-380}	10^{-12}	rs3764261
CETP	CETP_3	10^{-14}	10^{-13}	10^{-380}	10^{-12}	rs3764261
FRMD5	FRMD5				10^{-11}	rs2929282
HNF1A	HNF1A	10^{-14}	10^{-15}			rs1169288
HNF4A	HNF4A	10^{-13}		10^{-15}		rs1800961
JMJD1C	JMJD1C				10^{-12}	rs10761731
LDLR	LDLR	10^{-97}	10^{-117}			rs6511720
LIPG	LIPG	10^{-19}		10^{-49}		rs7241918
LPA	LPA	10^{-17}	10^{-17}	10^{-08}		rs1564348 (TC, LDL), rs1084651 (HDL)
LPL	LPL			10^{-98}	10^{-115}	rs12678919
SCARB1	SCARB1			10^{-14}		rs838880
TOP1	TOP1	10^{-17}	10^{-19}			rs6029526
TRIB1	TRIB1	10^{-36}	10^{-29}	10^{-19}	10^{-55}	rs2954029

[a] according to Teslovich et al. [1]; with the exception of LPA, the reported lead SNP in each gene region is the same for all associated lipid phenotypes.

Measurement of Short Tandem Repeats (STRs) - Multiplex PCR amplification

In both studies, the sixteen selected STRs were genotyped in two multiplex PCRs (see Table S2 & Figure S1 in File S1). Primers were designed with Visual OMP (DNA Software, Ann Arbor, MI) and HPLC-purified primers were purchased from Microsynth (Balgach, Switzerland). Forward primers were labeled with FAM, YY (Yakima Yellow), ATTO 550, and AT565.

The sequences were checked for existing null alleles, which could cause scoring errors and deviations from Hardy-Weinberg equilibrium by using the software Micro-Checker [26] (http://www.microchecker.hull.ac.uk/). The Hardy-Weinberg equilibrium was checked with ARLEQUIN [27] (http://cmpg.unibe.ch/software/arlequin3/).

The PCR was performed in 384 well plates in a total volume of 5 µl. The PCR mix contained 20 ng dried DNA, 2.5 µl Qiagen Multiplex PCR Plus Kit (Qiagen, Hilden, Germany), 1×Q solution (Qiagen, Hilden, Germany) and 1 µl primer-mix (final primer concentrations see Table 2). The amplification reaction for both PCRs was conducted on a DNA Engine Cycler (BioRad, Hercules, CA, USA) under following conditions: initial denaturation 95°C 15 min; 95°C 30 sec, 66°C 90 sec, 72°C 30 sec (30 cycles); final extension 68°C 30 min.

Measurement of Short Tandem Repeats (STRs) - Electrophoresis and data analysis

For the electrophoresis 1 µl PCR product was diluted 1:20, mixed with 8.8 µl HiDiformamide and 0.2 µl GeneScan-500 LIZ size standard (all Applied Biosystems, Foster City, CA) and denatured at 95°C for 5 minutes. The electrophoresis was run on an ABI 3730 s Genetic Analyzer using POP-7 polymer (both Applied Biosystems). Data were analyzed using GeneMapper-Software, version 4.1 (Applied Biosystems) (Figure S1 a and b in

File S1). For each STR at least two samples were subjected to Sanger sequencing to calibrate the allele calling algorithm in GeneMapper-Software.

SNP genotyping and imputation in KORA F4

For 2940 participants of the KORA F4 Study, a genome-wide SNP chip is available (Affymetrix Axiom). Quality control criteria for genotypes entering the genotype imputation were at least 97% call rate per person and 98% callrate per SNP, HWE (p-value ≥ $5 \times 10-6$) and a minor allele frequency of ≥0.01. Genotypes were imputed with the software IMPUTE (IMPUTE v2.3.0) based on the 1000 g phase 1 reference panel (all populations, 1000 G integrated phase 1 vers 3, March 2012) [28]. Therefore, a high density of SNPs was available for analysis in KORA F4 for each of the selected gene-regions. The specific gene regions were defined by the lead SNP according to Teslovich et al. (Table 1) ±100 kB. All SNPs lying within these regions with a MAF of >1% and imputation quality (info)>0.6 were extracted from the imputed genome-wide dataset. Genomic positions and LD refer to HG build 19 (1000 Genomes, phase 1 vers 3, March 2012).

Measurement of lipids

In KORA F4, total cholesterol (TC) was determined by cholesterol-esterase method (CHOL Flex, Dade-Behring, Germany), triglycerides (TG) and HDLC using the TGL Flex and AHDL Flex method (Dade-Behring), respectively, and LDLC was measured by a direct method (ALDL, Dade-Behring) [29].

All participants were fasting for at least 8 hours. For the present analysis, all participants taking lipid-lowering drugs were excluded. The analysis dataset in KORA F4 is thus based on 2553 participants with available STR measurements and genotypes derived from genome-wide SNP-chips and imputation.

Table 2. Characteristics and localization of selected STRs.

STR in gene	Repeat	Position of STR according to HG build 19			Position in gene	Minimum/Median/Maximum of number of repeats in	
		Chr	Start (bp)	End (bp)		KORA F4	SAPHIR
ABCG5	AAT	2	44060592	44060473	Intron1	6/12/18	8/12/17
BUD13	TAT	11	116638744	116638685	Intron 2	6/12/18	8/12/17
CETP_1	TAT	16	56989702	56989761	~5 kb upstream	3/9/17	3/9/16
CETP_2	TTTA	16	56997877	56997996	Intron 2	7/12/16	7/12/16
CETP_3	GAAA	16	56993722	56993961	promoter	37/48/56	38/48/56
FRMD5	GAT	15	44379990	44379931	Intron 1	11/15/20	11/15/20
HNF1A	ATCT	12	121425538	121425657	Intron 1	7/11/14	8/11/14
HNF4A	TTAT	20	42982780	42982839	~1.5 kb upstream	6/10/13	7/10/12
JMJD1C	TTG	10	65151334	65151215	Intron 1	8/12/15	8/12/14
LDLR	TTG	19	11202856	11202915	Intron 1	7/10/18	7/10/17
LIPG	AATA	18	47083781	47083900	~5 kb upstream	5/9/11	5/9/11
LPA	TTTA	6	161086738	161086619	Intron 1	5/8/12	4/8/11
LPL	TTAT	8	19815455	19815574	Intron 6	6/11/14	7/11/14
SCARB1	AAAGA	12	125255821	125255702	~6 kb downstream	5/10/28	5/10/28
TOP1	AAAT	20	39688036	39688095	Intron 2	7/12/17	8/12/16
TRIB1	AAAC	8	126440163	126440222	~2 kb upstream	7/11/14	7/11/14

In SAPHIR, blood samples were collected after an overnight fasting period. A complete lipoprotein profile including fasting TC, TG, HDLC and LDLC was determined using routine laboratory procedures (Roche Diagnostics GmbH, Mannheim, Germany). For statistical analysis, all participants taking lipid-lowering drugs were excluded. The analysis dataset in the SAPHIR Study is therefore based on 1648 participants.

Statistical methods

For each STR and individual, full allelic data was available. The association analysis was based on both, the allele-specific information as well as the sum of both alleles as the explaining variable. To account for the fact that there are two independent alleles from each individual in the allele-specific analysis, a robust standard error using a sandwich variance-covariance-matrix (function sandcov, R-package haplo.ccs) was calculated to derive the p-values for these analyses. Table 1 shows all loci together with the respective STRs that have been measured within these loci. Most of these loci have been shown to be associated with more than one lipid phenotype in Teslovich et al. and are evaluated on these same phenotypes in the present investigation: from the 16 measured STRs, 12 were regressed on total cholesterol, 10 on HDL cholesterol, 10 on LDL cholesterol and 8 on triglycerides. In KORA F4 and the SAPHIR Study, linear regression analyses were performed, regressing the sum of STR alleles on the respective lipid phenotypes. A linear mixed model assuming random intercepts was used to combine both datasets and derive a common effect estimate for each STR. Using Bonferroni correction, a p-value smaller than $0.05/40 = 0.00125$ was defined to be significant, accounting for the number of STRs studied in all phenotypes $(40 = 12+10+10+8)$.

To compare the STR results with the SNPs within each specific gene region, all SNPs in a region ± 100 kB around the lead SNP were extracted from the imputed genotype data in KORA F4 and regressed on the respective lipid phenotype of interest. An additive inheritance model was assumed for each SNP. Triglyceride levels were ln-transformed for all analyses due to the highly skewed distribution. All linear and linear mixed models were adjusted for age and sex.

To evaluate whether STRs are associated with the lipid phenotypes independently from the respective lead SNPs, the regression models from STRs were additionally adjusted for this same lead SNP. Pearson correlation coefficient r^2 is given as a measure of LD between STRs and SNPs.

Association analyses of SNPs were performed in SNPTest v.2.5 [30], all other analyses in R 3.0.1. The program LocusZoom [31] was used to create regional association plots for gene regions of interest.

Results

STR characteristics

The distribution of the STRs varies from 3 to 20 repeats, with the exception of CETP_3 (37–56 repeats). Figure S2 in File S1 shows a comparable distribution of the number of repeats based on alleles as well as on the sum of alleles for all STRs in the SAPHIR and KORA F4 studies. Minimum, median and maximum numbers of repeats for both studies are provided in Table 2.

STR association results with lipid phenotypes

All STR association results for KORA F4, SAPHIR and both studies combined are given in the Tables S3a)–d) in File S1. Since there was hardly any difference between allele-specific analysis and

analysis based on sum of alleles, the specific allelic information was discarded in favor of a denser representation taking the sum of both alleles in all subsequent analyses. For all four studied phenotypes, significant associations with STRs were observed for all four studied phenotypes (Table 3). In the combined analysis of KORA F4 and SAPHIR, three STRs were significantly associated with total cholesterol (LDLR, $p = 6.23E-07$; BUD13, $p = 1.42E-05$; ABCG5, $p = 6.51E-05$), five with HDL cholesterol (CETP_1, $p = 7.62E-12$; CETP_2, $p = 2.19E-14$; CETP_3, $p = 1.35E-28$; LPL, $p = 1.08e-05$, BUD13, $p = 1.43E-04$), three with LDL cholesterol (LDLR, $p = 9.96E-09$; ABCG5, $p = 4.70E-05$; CETP_1, $p = 2.32E-05$) and two with triglycerides (BUD13, $p = 1.04E-15$; LPL, $p = 5.34E-04$).

Comparing association results based on STRs with those based on SNPs

To put these findings into context, they were compared with results of SNP-association analyses in the respective gene regions. These analyses are only available in KORA F4, since no SNP microarray has been genotyped in SAPHIR. Table 3 shows the significantly associated STRs compared to the results for the respective lead SNP according to Teslovich et al [1]. For most loci given in Table 3, the lead SNPs were at least nominally significantly associated with their respective lipid phenotypes. Detailed results for all significantly and non-significantly associated STRs and the lead SNPs in the gene regions are given in Tables S4a)–d) in File S1. In general, none of the investigated STRs yielded a substantially lower p-value and therefore higher association peak than the lead SNP within their gene regions. For all significantly associated STRs, regional plots were created to further evaluate the association and LD structure of the SNPs within the gene regions of interest. The results of the STRs on KORA F4 were added to these plots to put them in context (Figures S3a)–k) in File S1. Association results of STRs adjusted for the lead SNP are given in Table 3 as well as Tables S4a)–d) in File S1.

For *LDLR*, for example, the STR falls in exactly the same genetic region as the conglomeration of SNPs constituting the highest association peak for total and LDL cholesterol (Figure 1). The assumption that the lead SNP rs6511720 tags the STR in *LDLR* can be verified by Figure 2: individuals with the common genotype GG generally have low repeat numbers. With increasing repeat numbers the number of rare alleles of that SNP increases. The correlation coefficient r^2 between this STR and rs6511720 is 0.866. Although the STR was highly associated with total cholesterol ($p = 6.27E-06$ in KORA F4) and LDL cholesterol ($p = 4.79E-08$ in KORA F4) there was no significant association anymore after adjusting for the lead SNP ($p = 0.611$, $p = 0.5831$).

For the *CETP* gene region, three STRs have been selected. All three of them are significantly associated with HDL cholesterol. However, the STRs pick up the association signal of their surrounding SNPs (Figure 3). CETP_1 is only marginally correlated with the lead SNP rs3764261 ($r^2 = 0.3008$, Figure 4A)). Still, the association signal with HDL cholesterol ($p = 1.15E-09$) vanishes after adjusting for that SNP ($p = 0.4610$). This same pattern can also be observed for CETP_3 which is highly correlated with the lead SNP ($r^2 = 0.8259$, Figure 4B)). All individuals with AA genotype are homozygote with repeat length 40 (sum of repeat length 80), for CA genotype sum of repeat length varies between 85 and 95, for CC genotype between 90 and 108. Thus, after adjustment for that SNP, the p-value of association vanishes from $p = 3.33E-19$ to $p = 0.9240$. CETP_2, however, seems to be independent from rs3764261 ($r^2 = 0.0463$). This STR is still significantly associated with HDL cholesterol after adjusting

Table 3. Results of linear mixed models and linear models on the investigated lipid phenotypes for all STRs which are significantly associated with lipids in KORA F4 and SAPHIR combined: 1) regression of the sum of STR alleles on lipids in KORA F4 and SAPHIR combined, 2) regression of the sum of STR alleles on lipids in SAPHIR, 3) regression of the sum of STR alleles on lipids in KORA F4, 4) regression of the minor allele using the lead SNP (LS) in the gene region on lipids in KORA F4, 5) regression of the minor allele using the best SNP in the gene region (LS+/−100 kB) on lipids in KORA F4.

STR/gene	SAPHIR & KORA F4 STR (sum of alleles)		SAPHIR study STR (sum of alleles)		KORA F4 study STR (sum of alleles)		KORA F4 study Lead SNP[a]		r² between lead SNP[a] and STR	KORA F4 study STR, adjusted for lead SNP	
	beta	p-value	beta	p-value	beta	p-value	beta	p-value		beta	p-value
Total Cholesterol											
LDLR	−1.3585	6.23E-07	−1.1168	0.0232	−1.4730	6.27E-06	−8.0470	2.01E-06	0.8660	−0.4522	0.6111
BUD13	1.2125	1.42E-05	1.8036	1.04e-04	0.8646	0.0132	5.4035	3.56E-04	0.5668	−0.0798	0.8800
ABCG5	1.2864	6.51E-05	1.9165	1.93e-04	0.8436	0.0417	2.5913	0.0234	0.1556	0.5433	0.2277
LDL Cholesterol											
LDLR	−1.4080	9.96E-09	−1.0408	0.0218	−1.5796	4.79E-08	−8.689	7.01E-09	0.8660	−0.4325	0.5831
ABCG5	1.1806	4.70E-05	1.6747	4.05e-04	0.8251	0.0246	1.8916	0.0622	0.1556	0.6331	0.1128
CETP_1	0.6998	2.32E-05	0.6395	0.0199	0.7400	3.36E-04	−2.2240	0.0265	0.3008	0.7265	0.0033
HDL Cholesterol											
CETP_1	−0.4513	7.62E-12	−0.3773	5.59e-04	−0.5014	1.15E-09	4.0230	3.67E-24	0.3008	−0.0715	0.4610
CETP_2	0.5595	2.19E-14	0.5267	1.14e-05	0.5798	3.37E-10	4.0230	3.67E-24	0.0463	0.4015	1.55e-05
CETP_3	−0.3393	1.35E-28	−0.3451	4.36e-11	−0.3359	3.33E-19	4.0230	3.67E-24	0.8259	−0.0085	0.9240
LPL	0.6805	1.08E-05	0.7685	0.0027	0.6140	0.0015	1.8011	0.0042	0.1609	0.4706	0.0259
BUD13	−0.3824	1.43E-04	−0.7041	3.60e-05	−0.1887	0.1278	−1.2328	0.0216	0.5668	0.0275	0.884
ln(Triglycerides)											
BUD13	0.0311	1.04E-15	0.0408	2.46e-10	0.0252	2.02E-07	0.1532	3.16e-13	0.5668	−0.0011	0.8810
LPL	−0.0208	5.34E-04	−0.0272	0.0052	−0.0167	0.0285	−0.0963	9.88E-05	0.1609	−0.0059	0.4772

All analyses are adjusted for age and sex.
[a]according to Teslovich et al. [1]; beta effect estimate for the lead SNP refers to the minor allele, assuming an additive model.

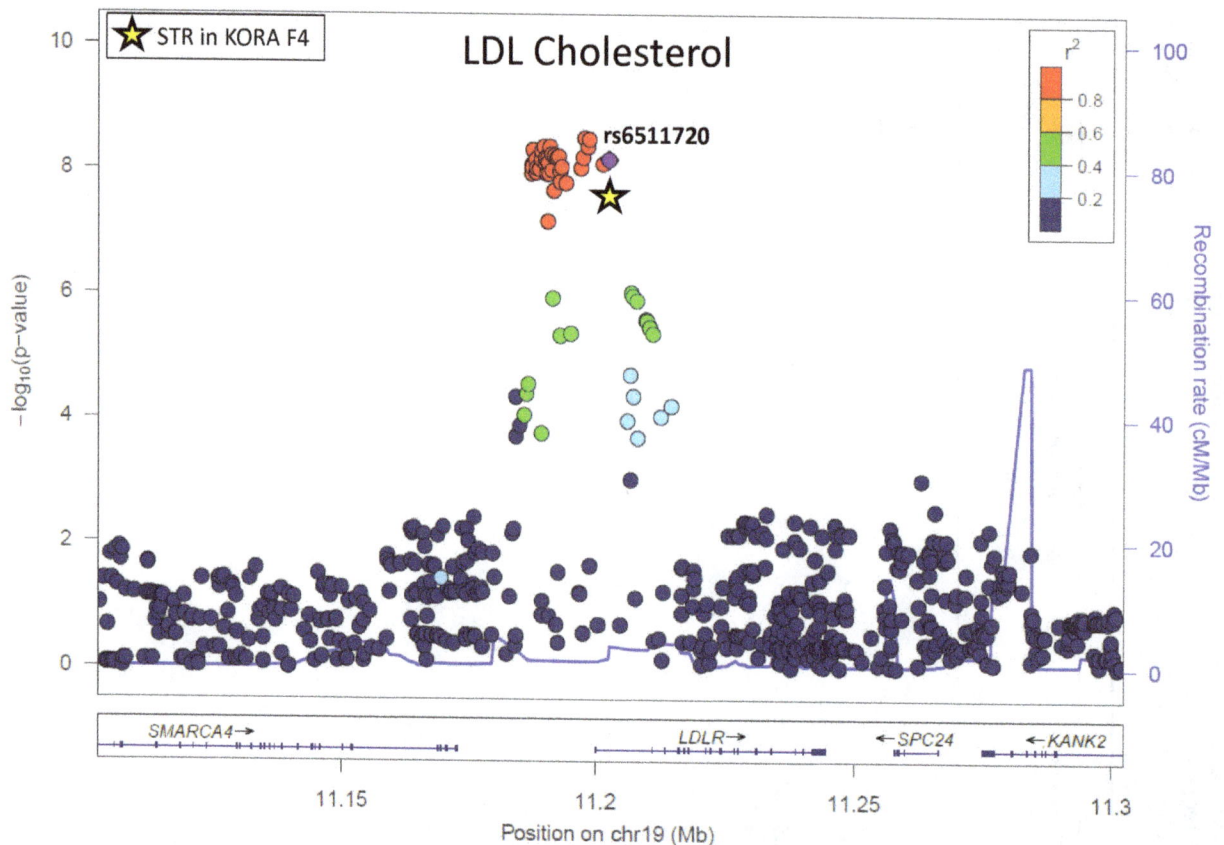

Figure 1. Regional plot showing the association of SNPs/STR in the *LDLR* region with LDL cholesterol. LD refers to the lead SNP according to Teslovich et al. [1] (rs6511720); p-values of the STR in KORA F4 is marked as a star.

for the lead SNP (p = 1.55E-05). From the regional plot (Figure 3) one might speculate that CETP_2 is tagged by the cluster of SNPs, which are independent from the lead SNP and are directly located downstream of CETP_2. To confirm this assumption, we set up a list of independent SNPs from the lead SNP ($r^2 < 0.1$) and selected the SNP with the lowest p-value with HDL cholesterol. This SNP (rs7203984, p = 2.48E-12) represents this cluster of SNPs directly next to CETP_2 and is highly correlated with this STR ($r^2 = 0.8471$, see Figure 4D). After adjusting for rs7203984, the

significant association of CETP_2 with HDL cholesterol (3.37E-10) is completely absent ($\beta = -0.0672$, p = 0.7747).

For all other STRs, which were shown to be significantly associated with the lipid phenotypes, similar observations can be made: The distribution of STRs depends on the genotypes of the respective lead SNPs (Figures S4 a)–h) in File S1). Therefore, after adjusting for the lead SNP, effect estimates are diminished and there are no significant associations of STRs with lipid phenotypes left (Table 3).

The published lead SNPs taken from Teslovich et al. [1] are not necessarily the SNPs with the lowest p-value in the KORA F4 study. All analyses were repeated based on the best SNP in KORA F4 within each extracted gene region. The best SNPs in KORA F4 are in high LD with the lead SNPs in the relevant gene regions (Figure S2 in File S1), and therefore, as expected, results did not change.

Discussion

Based on a published genome-wide search for lipid loci, we systematically selected STRs in or near genes at the most significant association peaks. We compared the association signals in these STRs with the association signals from SNPs genotyped by GWAS microarrays from the same gene regions. None of the investigated STRs, however, was significantly associated after adjusting for the lead SNPs within that gene region. Since these STRs have been found to be well tagged by SNPs, the STRs reflect the association signals based on surrounding SNPs.

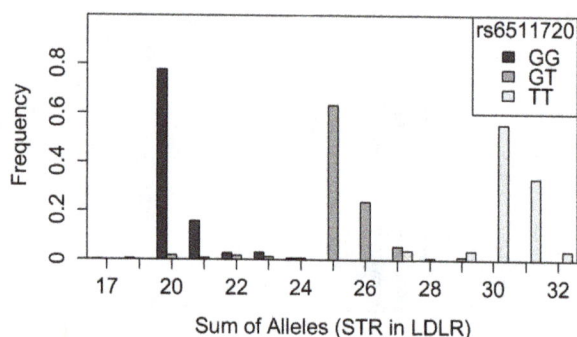

Figure 2. Distribution of sum of *LDLR*-STR-alleles, separated for rs6511720 genotypes.

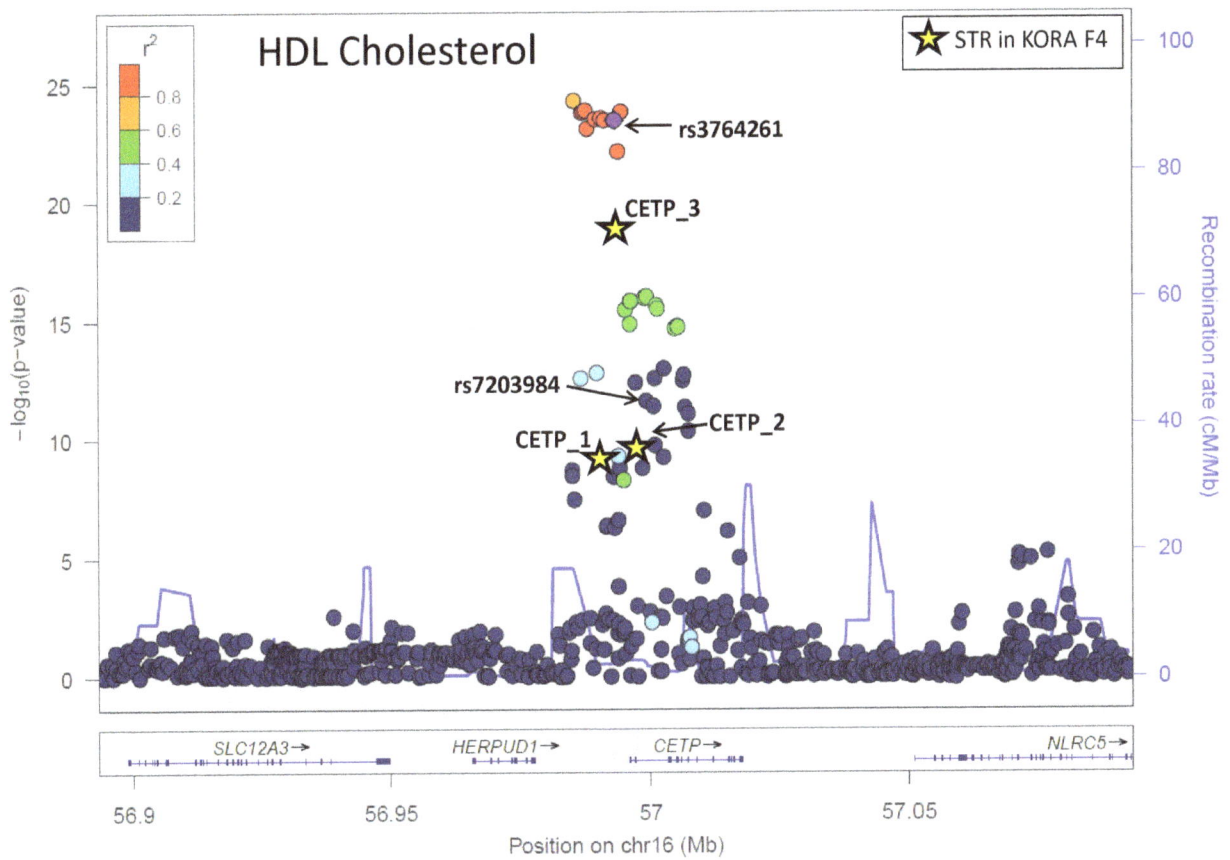

Figure 3. Regional plot showing the association of SNPs/STR in the *CETP* region with HDL cholesterol. LD refers to the lead SNP according to Teslovich et al. [1] (rs3764261); p-values of STRs in KORA F4 are marked as stars.

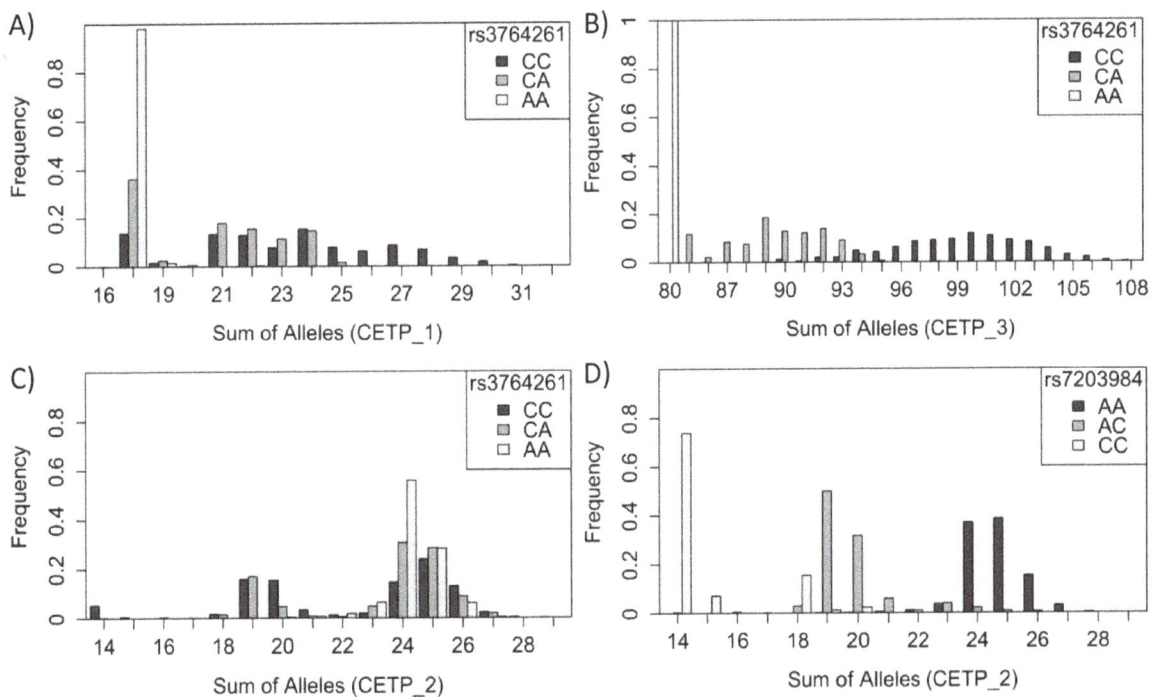

Figure 4. Distribution of sum of STR-alleles, separated for genotypes: A) Distribution of CETP_1, separated for rs3764261 genotypes, B) Distribution of CETP_3, separated for rs3764261 genotypes, C) Distribution of CETP_2, separated for rs3764261 genotypes, D) Distribution of CETP_2, separated for rs7203984 genotypes.

Therefore, none of the STRs contributed additionally to the SNP-based association signals.

In our analysis, three STRs were significantly associated with total cholesterol, five with HDL cholesterol, three with LDL cholesterol and two with triglycerides. We could replicate the association of the known STR in the CETP promoter region with HDL [15,32], but it was only nominally associated with LDL cholesterol and triglycerides. The other significant associations are all novel and have not been described elsewhere. The only other known STR from the literature within LPA did not yield any significant results with lipids in our investigation. This STR has been shown to be highly associated with lipoprotein(a) in the past [12,33], but not with any other phenotypes.

The finding of significant associations between STRs and lipids imposes the question, whether SNP-based analyses would have led to the same results. The significantly associated STRs are located within a high-LD-block with about the same p-value as these surrounding SNPs. This is especially the case for the STRs within LDLR (on total cholesterol and LDL cholesterol) and CETP (on HDL cholesterol). Additional adjustment for the lead SNP resulted in attenuation of STR association with the lipid phenotypes. Consequently, except CETP_2, none of the STRs was significantly associated anymore. Although CETP_2 was not correlated with the respective lead SNP in that region, it was correlated with an independent SNP in that gene region, representing the second highest association signal in CETP. Since this SNP and the adjacent SNP cluster was still genome-wide significant, these STR-tagging SNPs would have been identified in a genome-wide analysis based on SNPs anyway. Therefore, none of the STRs contributed additionally to the SNP-based association signals.

We cannot, however, clarify whether the STRs trigger the association of SNPs in the respective gene regions or the other way round. In a comprehensive and systematic investigation of 179 human genomes within the 1000 Genomes project, it was shown that 41% of tandem repeat polymorphisms are tagged by at least one SNP (with $r^2 > 0.8$) [4]. Therefore, it might be expected that a fair proportion of STRs drive the results of GWAS on common phenotypes.

The strength of our investigation is the systematic selection process of STRs in known lipid candidate genes. This approach took the location of the STRs in and around the respective candidate gene into account, their variability and a high repeat purity. To our knowledge, such an approach has not been reported before. So far, there was more interest in finding SNPs that are able to tag known associated STRs, since genotyping technologies for SNPs are easier and cheaper on a large scale.

It was surprising that STRs fulfilling our selection criteria could only be found in about 26% of the examined genes. Therefore, this approach cannot be imposed on any specific gene of interest. However, our selection strategy seems to follow to great extent what would have been expected from a random sample of polymorphic STRs. According to Payseur et al. [34], who examined two complete genome sequences, tetranucleotide repeats are the most variable tandem repeat polymorphisms, followed by di- and pentanucleotide repeats. Trinucleotide polymorphisms were shown to be the least variable polymorphisms. With our search strategy we selected tetra-nucleotide repeats most frequently (8×), followed by trinucleotide repeats (6×) and pentanucleotide repeats (2×). We excluded dinucleotide repeats due to the error-prone typing of this repeats class. Comparable to Payseur et al. [34], most repeat numbers varied

between 3 and 20 with only one exception, that is the knowledge-driven selection of the third STR in the CETP gene region. Trinucleotide repeats seem to be overrepresented, though, which is a consequence of the selection strategy: a high VarScore and high purity both favor a lower number of bases in the repeated regions.

Another strength of our investigation is that two independent studies, one population-based, one based on a healthy working population, were used to decrease false-positive findings. In one of these studies (KORA F4), 1000-Genomes-based imputed genotypes were available to compare and adjust the association results of STRs with surrounding SNPs. Unfortunately, we did not have SNP genotype data for the SAPHIR study available.

The necessity to restrict our evaluation to 16 STRs represents a major limitation of this work. A comprehensive analysis of all genome-wide significant regions originally identified in [1] was not possible due to the current technological limitations. Therefore, our results cannot be transferred directly to other STRs, other gene regions or other phenotypes. Nevertheless, in our study we covered already more than half of the most strongly lipid-associated genes. For many genes, however, no putative STRs according to our selection criteria were detected within the gene +/− 10 kB. This 10 kb cut-off, although necessarily arbitrary, was based on the assumption that this shall capture most promoter elements. For example, the ENCODE project [35] reported symmetrical presence of transcription factors +/−5 kb around the transcription factor binding site. In general, the potential impact of STRs on gene regulation is well-established. A well-known example illustrating the progressive disruption of regulatory elements by STR copies, is the so-called UGT1A1*28 polymorphism. This polymorphism consists of a TA repeat located exactly in the TATAA- element of the UGT1A1 promoter and increases the bilirubin levels in blood by reducing the UGT1A1 expression [13,36,37]. Our idea was to capture such STRs which have potential functional consequences affecting cis regulatory regions in proximity of the gene. However, our approach does not provide information for STRs affecting intergenic regulatory elements.

To conclude, from the 16 systematically selected STRs within known lipid susceptibility genes, several were highly associated with their respective phenotypes. However, all significantly associated STRs were well tagged by their surrounding SNPs. Thus, none of these STRs contributed additionally to the SNP-based association signal. Although STRs and other structural variants are neglected in SNP-based association studies and therefore, their associations are likely missed, this is not the case in our investigation based on known susceptibility genes on total cholesterol, HDL cholesterol, LDL cholesterol and triglycerides.

Author Contributions

Conceived and designed the experiments: CL MH SC AKB FK. Performed the experiments: CL MH SC AKB CG AP HG KS TM LK BP FK. Analyzed the data: CL MH SC AKB FK. Contributed reagents/materials/analysis tools: CL MH SC AKB CG AP HG KS TM LK BP FK. Contributed to the writing of the manuscript: CL MH SC AKB LK FK. Recruitment of cohorts: CL CG AP HG TM LK BP FK.

References

1. Teslovich TM, Musunuru K, Smith AV, Edmondson AC, Stylianou IM, et al. (2010) Biological, clinical and population relevance of 95 loci for blood lipids. Nature 466: 707–713. doi:10.1038/nature09270.

2. Willer CJ, Schmidt EM, Sengupta S, Peloso GM, Gustafsson S, et al. (2013) Discovery and refinement of loci associated with lipid levels. Nat Genet 45: 1274–1283. doi:10.1038/ng.2797.

3. Scherer SW, Lee C, Birney E, Altshuler DM, Eichler EE, et al. (2007) Challenges and standards in integrating surveys of structural variation. Nat Genet 39: S7–15. doi:10.1038/ng2093.

4. Montgomery SB, Goode DL, Kvikstad E, Albers CA, Zhang ZD, et al. (2013) The origin, evolution, and functional impact of short insertion-deletion variants identified in 179 human genomes. Genome Res 23: 749–761. doi:10.1101/gr.148718.112.

5. Ellegren H (2004) Microsatellites: simple sequences with complex evolution. Nat Rev Genet 5: 435–445. doi:10.1038/nrg1348.

6. Gulcher J (2012) Microsatellite markers for linkage and association studies. Cold Spring Harb Protoc 2012: 425–432. doi:10.1101/pdb.top068510.

7. Usdin K (2008) The biological effects of simple tandem repeats: lessons from the repeat expansion diseases. Genome Res 18: 1011–1019. doi:10.1101/gr.070409.107.

8. Sawaya S, Bagshaw A, Buschiazzo E, Kumar P, Chowdhury S, et al. (2013) Microsatellite tandem repeats are abundant in human promoters and are associated with regulatory elements. PLoS One 8: e54710. doi:10.1371/journal.pone.0054710.

9. Hannan AJ (2010) Tandem repeat polymorphisms: modulators of disease susceptibility and candidates for "missing heritability". Trends Genet 26: 59–65. doi:10.1016/j.tig.2009.11.008.

10. Hersberger M, Thun G-A, Imboden M, Brandstätter A, Waechter V, et al. (2010) Association of STR polymorphisms in CMA1 and IL-4 with asthma and atopy: the SAPALDIA cohort. Hum Immunol 71: 1154–1160. doi:10.1016/j.humimm.2010.08.008.

11. Kamstrup PR, Tybjaerg-Hansen A, Steffensen R, Nordestgaard BG (2008) Pentanucleotide repeat polymorphism, lipoprotein(a) levels, and risk of ischemic heart disease. J Clin Endocrinol Metab 93: 3769–3776. doi:10.1210/jc.2008-0830.

12. Trommsdorff M, Köchl S, Lingenhel A, Kronenberg F, Delport R, et al. (1995) A pentanucleotide repeat polymorphism in the 5′ control region of the apolipoprotein(a) gene is associated with lipoprotein(a) plasma concentrations in Caucasians. J Clin Invest 96: 150–157. doi:10.1172/JCI118015.

13. Lin J-P, O'Donnell CJ, Schwaiger JP, Cupples LA, Lingenhel A, et al. (2006) Association between the UGT1A1*28 allele, bilirubin levels, and coronary heart disease in the Framingham Heart Study. Circulation 114: 1476–1481. doi:10.1161/CIRCULATIONAHA.106.633206.

14. Talmud PJ, Edwards KL, Turner CM, Newman B, Palmen JM, et al. (2000) Linkage of the cholesteryl ester transfer protein (CETP) gene to LDL particle size: use of a novel tetranucleotide repeat within the CETP promoter. Circulation 101: 2461–2466.

15. Lira ME, Lloyd DB, Hallowell S, Milos PM, Thompson JF (2004) Highly polymorphic repeat region in the CETP promoter induces unusual DNA structure. Biochim Biophys Acta 1684: 38–45. doi:10.1016/j.bbalip.2004.06.002.

16. Manolio TA, Collins FS, Cox NJ, Goldstein DB, Hindorff LA, et al. (2009) Finding the missing heritability of complex diseases. Nature 461: 747–753. doi:10.1038/nature08494.

17. Mills RE, Pittard WS, Mullaney JM, Farooq U, Creasy TH, et al. (2011) Natural genetic variation caused by small insertions and deletions in the human genome. Genome Res 21: 830–839. doi:10.1101/gr.115907.110.

18. Wichmann H-E, Gieger C, Illig T (2005) KORA-gen–resource for population genetics, controls and a broad spectrum of disease phenotypes. Gesundheitswesen (Bundesverband der Ärzte des Öffentlichen Gesundheitsdienstes (Germany)) 67 Suppl 1: S26–30. doi:10.1055/s-2005-858226.

19. Lamina C, Forer L, Schönherr S, Kollerits B, Ried JS, et al. (2012) Evaluation of gene-obesity interaction effects on cholesterol levels: a genetic predisposition score on HDL-cholesterol is modified by obesity. Atherosclerosis 225: 363–369. doi:10.1016/j.atherosclerosis.2012.09.016.

20. Heid IM, Wagner SA, Gohlke H, Iglseder B, Mueller JC, et al. (2006) Genetic architecture of the APM1 gene and its influence on adiponectin plasma levels and parameters of the metabolic syndrome in 1,727 healthy Caucasians. Diabetes 55: 375–384.

21. Legendre M, Pochet N, Pak T, Verstrepen KJ (2007) Sequence-based estimation of minisatellite and microsatellite repeat variability. Genome Res 17: 1787–1796. doi:10.1101/gr.6554007.

22. Litt M, Hauge X, Sharma V (1993) Shadow bands seen when typing polymorphic dinucleotide repeats: some causes and cures. Biotechniques 15: 280–284.

23. Murray V, Monchawin C, England PR (1993) The determination of the sequences present in the shadow bands of a dinucleotide repeat PCR. Nucleic Acids Res 21: 2395–2398.

24. Budowle B, Lindsey JA, DeCou JA, Koons BW, Giusti AM, et al. (1995) Validation and population studies of the loci LDLR, GYPA, HBGG, D7S8, and Gc (PM loci), and HLA-DQ alpha using a multiplex amplification and typing procedure. J Forensic Sci 40: 45–54.

25. Pestoni C, Lareu MV, Rodríguez MS, Muñoz I, Barros F, et al. (1995) The use of the STRs HUMTH01, HUMVWA31/A, HUMF13A1, HUMFES/FPS, HUMLPL in forensic application: validation studies and population data for Galicia (NW Spain). Int J Legal Med 107: 283–290.

26. Van Oosterhout C, Hutchinson WF, Wills DPM, Shipley P (2004) micro-checker: software for identifying and correcting genotyping errors in microsatellite data. Molecular Ecology Notes 4: 535–538. doi:10.1111/j.1471-8286.2004.00684.x.

27. Excoffier L, Laval G, Schneider S (2005) Arlequin (version 3.0): an integrated software package for population genetics data analysis. Evol Bioinform Online 1: 47–50.

28. Marchini J, Howie B, Myers S, McVean G, Donnelly P (2007) A new multipoint method for genome-wide association studies by imputation of genotypes. Nat Genet 39: 906–913. doi:10.1038/ng2088.

29. Kollerits B, Coassin S, Beckmann ND, Teumer A, Döring A, et al. (2009) Genetic evidence for a role of adiponutrin in the metabolism of apolipoprotein B-containing lipoproteins. Hum Mol Genet 18: 4669–4676. doi:10.1093/hmg/ddp424.

30. Marchini J, Howie B (2010) Genotype imputation for genome-wide association studies. Nat Rev Genet 11: 499–511. doi:10.1038/nrg2796.

31. Pruim RJ, Welch RP, Sanna S, Teslovich TM, Chines PS, et al. (2010) LocusZoom: regional visualization of genome-wide association scan results. Bioinformatics 26: 2336–2337. doi:10.1093/bioinformatics/btq419.

32. Thompson JF, Lira ME, Durham LK, Clark RW, Bamberger MJ, et al. (2003) Polymorphisms in the CETP gene and association with CETP mass and HDL levels. Atherosclerosis 167: 195–204. doi:10.1016/S0021-9150(03)00005-4.

33. Røsby O, Berg K (2000) LPA gene: interaction between the apolipoprotein(a) size ('kringle IV' repeat) polymorphism and a pentanucleotide repeat polymorphism influences Lp(a) lipoprotein level. J Intern Med 247: 139–152.

34. Payseur BA, Jing P, Haasl RJ (2011) A genomic portrait of human microsatellite variation. Mol Biol Evol 28: 303–312. doi:10.1093/molbev/msq198.

35. Birney E, Stamatoyannopoulos JA, Dutta A, Guigó R, Gingeras TR, et al. (2007) Identification and analysis of functional elements in 1% of the human genome by the ENCODE pilot project. Nature 447: 799–816. doi:10.1038/nature05874.

36. Bosma PJ, Chowdhury JR, Bakker C, Gantla S, de Boer A, et al. (1995) The genetic basis of the reduced expression of bilirubin UDP-glucuronosyltransferase 1 in Gilbert's syndrome. N Engl J Med 333: 1171–1175. doi:10.1056/NEJM199511023331802.

37. Hsieh T-Y, Shiu T-Y, Huang S-M, Lin H-H, Lee T-C, et al. (2007) Molecular pathogenesis of Gilbert's syndrome: decreased TATA-binding protein binding affinity of UGT1A1 gene promoter. Pharmacogenet Genomics 17: 229–236. doi:10.1097/FPC.0b013e328012d0da.

14

Alternation in the Glycolipid Transfer Protein Expression Causes Changes in the Cellular Lipidome

Matti A. Kjellberg, Anders P. E. Backman, Henna Ohvo-Rekilä[¤], Peter Mattjus*

Biochemistry, Department of Biosciences, Åbo Akademi University, Turku, Finland

Abstract

The glycolipid transfer protein (GLTP) catalyzes the binding and transport of glycolipids, but not phospholipids or neutral lipids. With its all-alpha helical fold, it is the founding member for a new superfamily, however its biological role still remains unclear. We have analyzed changes in the HeLa cell lipidome in response to down- and up-regulation of GLTP expression. We used metabolic labeling and thin layer chromatography analysis, complemented with a lipidomics mass spectroscopic approach. HeLa cells were treated with GLTP siRNA or were transiently overexpressing the GLTP gene. We identified eight different lipid classes that changed as a result of the GLTP down- or up-regulation treatments; glucosylceramide, lactosylceramide, globotriaosylceramide, ceramide, sphingomyelin, cholesterol-esters, diacylglycerol and phosphatidylserine. We discovered that the amount of globotriaosylceramide (Gb$_3$) was extensively lowered after down-regulation of GLTP. Further, an up-regulation of GLTP caused a substantial increase in both the Gb$_3$ and glucosylceramide levels compared to the controls. Total galactosylceramide levels remained unchanged. Both lactosylceramide and ceramide showed small changes, an increase with increasing GLTP and a decrease in the HeLa cell GLTP knockdowns. The cholesterol-esters and diacylglycerol masses increased in cells that had upregulated GLTP protein levels, wheras down-regulation did not affect their amounts. For the glycerophospholipids, phosphatidylserine was the only species that was lower in GLTP overexpressing cells. Phosphatidylethanolamine, phosphatidylglyerol and phosphatidylinositol remained unaltered. A total of 142 lipid species were profiled and quantified using shotgun lipidomics analyses. This work provides for the first time insights into how alternations in the levels of a protein that binds and transfers glycolipids affects the cellular lipid metabolism. We discuss the observed changes in the lipidome and how these relate to GLTP. We suggest, that GLTP not only could be a significant player in cellular sphingolipid metabolism, but also could have a much broader role in the overall lipid metabolism.

Editor: Leah J. Siskind, University of Louisville, United States of America

Funding: This work was supported by the Academy of Finland, Sigrid Jusélius Foundation, Abo Akademi University, Magnus Ehrnrooths stiftelse and Medicinska Understödsföreningen Liv och Hälsa r.f. The funders had no role in study design, data collection and analysis, decision to publish, or preparation of the manuscript.

Competing Interests: The authors have declared that no competing interests exist.

* E-mail: pmattjus@abo.fi

¤ Current address: Perkin Elmer Life and Analytical Sciences, Wallac Oy, Turku, Finland

Introduction

Previous studies on the biological function of the glycolipid transfer protein (GLTP) have been focusing on events related to its lipid substrate class, the glycosphingolipids (GSLs) [1–7]. Little is known how other lipid classes are affected in cells that have altered GLTP expression. The glycolipid transfer protein is a cytosolic protein [1] that catalyzes the transport of both sphingoid and glycerol based glycolipids in vitro [8,9]. GLTP does not transfer phospholipids, sphingomyelin (SM) or neutral lipids [10]. GLTP with its all-alpha helical fold is the founding member for a new protein superfamily in eukaryotes [11]. GLTP is found in a broad range of species, fungi, red algae, plants and mammals [12]. Several homologues to mammalian GLTP have been found in various species [12–15], including the human four-phosphate adaptor protein, FAPP2 that contains a GLTP-motif, and has been shown to mediate the transfer of glucosylceramide (GlcCer) from early Golgi to distal Golgi compartments [5,15,16].

In a previous study we analyzed de novo GSL changes in cells overexpressing GLTP (transiently transfected) using ^{3}H-sphinga-nine metabolic labeling [1]. The results showed a significant increase in the de novo synthesis of GlcCer and a decrease in the SM synthesis. However, we did not detect any changes in the sphingolipid synthesis in GLTP-knockdown (RNA interference) cells compared to control cells [1]. In this work we have extended the analysis to also include other lipid classes. Moreover we have used FACS-sorted cells to be able to also detect even small changes that are often missed, if the transfection efficiency is not very effective. GLTP appears to respond to changes in the amount of newly synthesized GlcCer [17]. In another recent study we found that GLTP expression, both at the mRNA and protein levels, is elevated in cells that accumulate GlcCer, generated by brefeldin A treatments [17]. Also an 80% loss of GlcCer, caused by glucosylceramide synthase knockdown, resulted in a significant reduction in the expression of GLTP [17].

In the present study we used both metabolic labeling with TLC analysis and a lipidomic approach. The metabolic labeling of sphingolipids, phospholipids and sterols were initially used and complemented with a more detailed analysis of the individual lipid species by the MS approach. All in all, we investigated the changes in 15 different lipid classes and in 142 molecular lipid profiles of HeLa cells as a function of GLTP down- or up-regulation. The MS analysis was done by Zora Biosciences (Espoo, Finland) utilizing a shotgun lipidomics approach employing a hybrid triple quadrupole/linear ion trap mass spectrometer. We have discovered that the globotriaosylceramide level (Gb$_3$) showed a strong decrease after GLTP siRNA treatment, and that the up-regulation of GLTP caused an increase in both the Gb$_3$ and GlcCer levels compared to the control HeLa cells. The only glycerophospholipid that we detected changes in was phosphatidylserine (PS), whereas phosphatidylethanolamine (PE), phosphatidylglycerol (PG) and phosphatidylinositol (PI) levels remained unaltered.

Materials and Methods

Materials & Cell Culture

All chemical reagents were of analytical grade or higher. HeLa cervical carcinoma cells (ATCC CCL-2, LGC Standards) were cultured in 100 mm dishes (Greiner Bio-One GmbH, Germany) in Dulbecco's Modified Eagle's medium, Sigma (St. Louis, MO, USA) supplemented with penicillin (50 U/ml), streptomycin (50 U/ml), 4 mM L-glutamine and 10% fetal calf serum Sigma (St. Louis, MO, USA) prior to the experiments. The human GLTP (NM_016433.3) gene was cloned into a pcDNA3.1(+)-vector, Invitrogen (Carlsbad, CA, USA), between the cloning sites BamH1/EcoR1 (hGLTP) [1]. The empty green fluorescent protein pEGFPN1 vector, Clontech (Palo Alto, CA, USA) was used for fluorescence-activated cell sorting (FACS) transfection efficiency determination and sorting purposes of the GLTP overexpressing cells. Human GLTP Stealth RNAi and a Stealth RNAi negative control were obtained from Invitrogen (Carlsbad, CA, USA). Lipids standards for the TLC analysis where from Avanti Polar Lipids (Alabaster, USA) or Matreya LLC (Pleasant Gap, USA). The different lipid species of phosphatidylcholine (PC), phosphatidylethanolamine (PE), phosphatidylserine (PS), phosphatidylinositol (PI), phosphatidylglycerol (PG) and diacylglycerol (DAG) are listed with the two fatty acyl groups separated with a hyphen, e.g. PC 16:0–18:1. The ether-linked phospholipids are shown as PC O (alkyl), PE O (alkyl) and cholesteryl esters as CE. The acyl chains of ether-linked lipids and the N-linked acyl chains for sphingomyelin (SM), ceramide (Cer) and glycosphingolipids (GSLs) are shown after the slash. All samples, including the controls were cultured in the presence of the same amount of media and FCS and we assume that the small amount of FCS in the culture media is not affecting the cellular lipid metabolism [18].

Stealth RNAi Knockdown Experiments

The specific siRNA sequences (Invitrogen) were named #76; sense AUACAUUUCUUUCUCCACCUCCAGG and anti-sense CCUGGAGGUGGAGAAAGAAAUGUAU, #77; sense strand, ACUUAUAGGGUGCUGCGUACAGUGC and anti-sense strand, GCACUGUACGCAGCACCCUAUAAGU and #78; sense UUAAGCUCAGCGUUCAUCUGGGUGU and anti-sense ACACCCAGAUGAACGCUGAGCUUAA.

For the metabolic labeling and TLC analysis experiments a mixture of both expression constructs #76 and #78 of GLTP siRNA were used. The efficiency of the different siRNA constructs was analyzed both by qPCR and Western blot analysis. The qPCR analysis was performed as described previously [17]. For the MS analysis experiments HeLa cells, passage 8, were plated at 50% confluence in ten 60 mm dishes and transiently transfected with 75 pmol of GLTP siRNAs #77 with Lipofectamine 2000 (Invitrogen) according to the manufacturer's instructions for 24 h. Ten

different transfections were performed for each sample. After 24 h the cells were split (1/3) cultured for another 48 h in 100 mm dishes in DMEM medium supplemented with 2% fetal calf serum without penicillin/streptomycin and finally harvested for subsequent sorting. After FACS analysis the cells were counted, washed with PBS and aliquoted into two vials containing 5×10^6 cells each, dried and flash frozen in liquid nitrogen and stored at $-80°C$ awaiting lipidomics analysis. Zora Biosciences Ltd. (Espoo, Finland) analyzed the two samples once. Fluorescein isothiocyanate labeled (FITC) dsRNA oligomers, Block-iT Fluorescence Oligo (Invitrogen, Carlsbad, CA) were used to determine siRNA transfection efficiencies. The oligos were co-transfected with the GLTP siRNA for FACS, enabling us to harvest transfected HeLa cells for subsequent lipidomics analysis. Two Stealth RNAi negative universal control (UC) samples were also generated in parallel with the siRNA GLTP samples, and the two different samples were also analyzed once by Zora Biosciences.

GLTP Overexpression

5×10^6 HeLa cells (passage 10) were transiently transfected using a Gene Pulser II RF Module, Bio-Rad (Hercules, CA, USA) electroporator (30 ms), with 5 parallel samples with 10 µg pcDNA-GLTP or 10 µg control pcDNA 3.1(+)-vector, both co-transfected with 10 µg of the pEGFP vector. This was repeated twice. The pEGFP was used for FACS transfection efficiency determination and sorting purposes. After electroporation the cells were then incubated in 100 mm dishes in DMEM supplemented with 2% fetal calf serum. After 48 h the cells were trypsinated, washed and sorted. The pEGFP positive cells (between 30–40%) were sorted and pooled to generate sufficient number (5×10^6) of cells for one MS samples. This sample was analyzed once. One pcDNA 3.1(+)-vector mock control sample was also generated and analyzed once.

^{3}H-sphinganine and ^{3}H-acetate Labeling of Cells, Subsequent Treatments and Lipid Extraction

HeLa cells to be used for GLTP overexpression experiments were cultured to 95% confluency in 100 mm cell dishes (9–10×10^6. cells), and trypsinized. The cells were transfected using a Bio-Rad Gene Pulser II electroporation system, as described above, and transferred to 35 mm cell dishes in triplicates. An empty pcDNA 3.1(+)-vector was used as a control. The cells were left to overexpress the protein for 48 h after which they were harvested for further analysis. Only trace doses of precursor lipids were used. This does not alter the normal lipid metabolism of the cells. We added ^{3}H-acetate (5.3 µCi per milliliter) 6 h before harvesting the cells that were used to analyze the phospholipid expressions. Similarly, 1.0 µCi/ml of ^{3}H-sphinganine was used for the sphingolipid analysis experiments. HeLa cells to be treated with GLTP siRNA were grown to 50% confluency in 35 mm cell plates, after which a knockdown was performed, using Lipofectamine 2000 according to the manufacturer's instruction. 48 h after beginning the KD, we added 5.3 µCi/ml ^{3}H-acetate for 6 h, after which the cells were harvested for further analysis of phospholipids. Similarly, 1.0 µCi/ml of ^{3}H-sphinganine was used for the sphingolipid analysis experiments. The cells used for the labeling experiments were not sorted. Treated cells were washed three times in PBS. Total lipids were extracted directly from the culture dishes using hexane:2-propanol (3:2, v/v) and the extract was dried with nitrogen. The lipids were re-dissolved in hexane:2-propanol and analyzed on high performance thin-layer chromatography (HPTLC) silica plates (Whatman, UK) described below.

Identification of GSL Species with TLC

For GlcCer, GalCer and LacCer separation the solvent system chloroform:methanol:acetone:acetic acid:water (10:2:4:2:1) was used. For Gb$_3$ separation the solvent system 45:55:10, chloroform:methanol:0.2% CaCl$_2$ (in H$_2$O) was used. The GSL analysis was done using standards, run in parallel with the samples. Lipid migration was visualized using orcinol spray (0.2% orcinol in a 50% H$_2$SO$_4$ solution) and heating the plate on 120°C for 5 minutes. The lipid spots were scraped into Optiphase 'Hi phase' scintillation liquid (PerkinElmer-Wallac, Turku, Finland) and the radioactivity was measured using a liquid scintillation counter, 1216 Rackbeta (PerkinElmer-Wallac, Turku, Finland). After lipid extraction, the total cellular proteins were extracted with 0.1 M NaOH and the protein content was analyzed with the Lowry method [19]. Counts per minute obtained for the various experiments were normalized to total cellular protein. The results are displayed as the ^{3}H-sphinganine incorporation into various lipid normalized to the counts per minutes for the control cells.

Identification of Phospholipids with TLC

Around 35% of the total extracted lipids were applied to the HPTLC-plate, with 10 nmol of carrier lipids to help locate the correct bands [20]. The plates were developed chromatographically in an upright tank, using chloroform:methanol:acetic acid:water; 50:30:8:3 as the solvent until the elution front was approximately 1 cm from the edge of the plate. The plates were then stained using iodine vapor, spots marked with a pencil and scraped into scintillation vials. The results are displayed as the ^{3}H-acetate incorporation into the phospholipids, normalized to the counts per minutes for the control cells.

Flow Cytometry and Cell Sorting

The FITC (siRNA) and GFP (GLTP overexpression) positive cells were sorted from the untransfected cells to ensure a reliable lipidomics analysis. The flow cytometry was performed at the Cell Imaging Core facility at the Turku Centre for Biotechnology, Turku, Finland.

Western Blot

GLTP expression levels were analyzed by Western blotting. The cells were redissolved in a lysis buffer (50 mM NaH$_2$PO$_4$, 300 mM NaCl, 10 mM imidazole, 0.05% Tween-20, 0.5 mM PMSF, 1×Protease inhibitor cocktail (Sigma), 1 mM dithiotreithol, pH 8.0). The cells were sonicated using a Branson 250 probe sonifier (Emerson Industrial Automation, St. Louis, MO, USA), and the lysate protein concentration was determined using the method of Lowry [19]. The cell lysates were separated on SDS–PAGE and transferred onto a PVDF membrane. Immunoblots were carried out using antibodies against GLTP [1,21] and β-actin as a loading control. The rabbit anti-beta-actin antibody was from Rockland Immunochemicals (Gilbertsville, PA, USA) and the secondary peroxidase conjugated rabbit anti-goat antibody was from Thermo Scientific (Waltham, MA, USA). The detected proteins were visualized with the ECL chemiluminescence system (SuperSignal West Femto, Thermo Scientific) using X-ray film (Fujifilm, Tokyo, Japan).

Cell Sample Handling, Storage and Lipid Extraction

The extraction of the HeLa cell samples and the lipidomics analyses was conducted by Zora Biosciences Oy (Espoo, Finland) according to their standard operating procedures. The dry flash frozen HeLa cell samples with known cell count (5×10^6) were delivered and stored at −80°C prior to analysis. The samples were

thawed in a chilly environment upon starting the extraction. Briefly, the dry flash frozen HeLa cell samples were thawed in a chilly environment and re-suspended in ice-cold PBS prior to extraction. Cells were counted and all measurements were normalized to the cell number. A volume corresponding to 0.5 million cells was used for lipid extraction. Lipids were extracted using a modified Folch extraction procedure [22] performed on a Hamilton Microlab Star robot. Known amounts of deuterium-labeled or heptadecanoyl-based synthetic internal standards of SM, LPC, PC, PE, PS, PG, PA, DAG, CE, Cer, GlcCer, LacCer, and Gb$_3$, were added and used for quantification of the endogenous lipid species as described [23–25]. Following lipid extraction, samples were reconstituted in 1:2 (v/v) chloroform/methanol and stored at −20°C prior to MS analysis.

MS Analyses

The species of all phospholipids, SM, DAG and CE were analyzed by shotgun analysis on a hybrid triple quadrupole/linear ion trap mass spectrometer (QTRAP 5500, AB SCIEX, MA) equipped with a robotic nanoflow ion source (NanoMate HD, Advion Biosciences, NY) [26]. These analyses were performed using both positive and negative ion modes using multiple precursor ion scanning (MPIS) and neutral loss (NL) based methods [27,28], whereas CEs were analyzed in positive ion mode [29]. Sphingolipids were analyzed by reverse phase ultra-high pressure liquid chromatography (UHPLC) as previously described [30] using an Acquity BEH C18, 2.1×50 mm column with a particle size of 1.7 μm (Waters, Milford, MA) coupled to a hybrid triple quadrupole/linear ion trap mass spectrometer (QTRAP 5500, AB SCIEX, MA). A 25 min gradient using 10 mM ammonium acetate in water with 0.1% (v/v) formic acid (mobile phase A) and 10 mM ammonium acetate in 4:3 (v/v) acetonitrile:2-propanol containing 0.1% (v/v) formic acid (mobile phase B) was used. Quantification of sphingolipids was performed using multiple reaction monitoring. Lipidomic data is based on the analysis of each detected lipid class with one technical replicate for each cell sample. The samples, CTRL cells, GLTP siRNA and GLTP OE cells were in duplicates. The MS lipidomic analyses were performed in the Zora Biosciences laboratory that works according to Good Laboratory Practice, and the published validation data show less than 15% variation for most lipid species [31].

Lipidomics Data Processing

The MS data files were processed as previously described [24] using LipidView and MultiQuant software for producing a list of lipid names and peak areas. A stringent cutoff was applied for separating background noise from actual lipid peaks. Masses and counts of detected peaks were converted into a list of corresponding lipid names. Lipids were normalized to their respective internal standard [24] and the concentrations of molecular lipids are presented as pmol/500 000 cells.

Quality Control

Quality control samples were utilized to monitor the overall quality of the lipid extraction and mass spectrometry analyses [23] mainly to remove technical outliers and lipid species that were detected below the limit of quantification.

Statistical Significance

The statistical significance compared to the respective controls is indicated with asterisks. One asterisk (*) $p < 0.05$, two asterisks (**)

p<0.01 and three asterisks (***) p<0.005 indicate the statistical significance compared to the controls.

Results

GLTP Expression Levels after Down- and Up-regulation

qPCR analysis clearly showed a reduction of the GLTP gene expression as a function of time in the HeLa small interfering RNA (siRNA) transfected cells (Fig. S1A). All three siRNA sequences, termed #76, #77 and #78 used in this work were compared and normalized to the siRNA universal control (UC), Fig. S1A. Western blot analysis was also performed on the cell samples that were treated with the different siRNA constructs. The blots for normal control cells, UC, siRNA #76, #77, #78 and beta-actin are shown in Fig. S1B. Based on these control experiments, we found that the optimal GLTP siRNA concentrations to treat the HeLa cells were 75 nM, and gene sequence construct #77 was chosen for the MS lipidomics analysis. A mixture of both constructs #76 and #78 was used for the metabolic labeling experiments. Using #77 siRNA GLTP we achieved a yield of 94.0% FITC-positive cells for the knockdown transfection, and 90.5% of the UC transiently transfected cells had taken up the FITC-labeled dsRNA oligos. Cells for the metabolic labeling experiments were not sorted.

We also analyzed the GLTP levels in HeLa total cell lysates with GLTP up-regulated gene expression. Twenty-four hours after HeLa cells were transiently transfected with electroporation with the pcDNA-GLTP construct a very strong expression of GLTP was observed, with a slow decrease as a function of time (Fig. S2A, lower blot). Due to the low expression of endogenous GLTP the band in the control lane (C) is not visible in this blot (30 µg, loading, total cell lysate). Beta-actin was used as a loading control (Fig. S2A, upper blot). HeLa cells that were sorted and harvested for the MS lipidomics analysis were analyzed for expression of GLTP. In the left blot we show the endogenous expression of GLTP (lane 1) and the reduced GLTP expression in HeLa cells with GLTP knockdown, by siRNA (#77 siRNA GLTP gene construct), lane 2. A total of 80 µg total cell lysate was loaded, and beta-actin was used as the loading control, upper blot. The right blot shows the amount of GLTP in HeLa cells with GLTP overexpression (lane 4), and an invisible endogenous GLTP band in lane 3, due to the loading amount of just 10 µg total cell lysate, to avoid overloading of the OE sample. Beta-actin was used as a loading control (Fig. S2B, upper blots), lanes 1 and 2 were loaded with 80 µg of total cell lysate and lanes 3 and 4 with 10 µg of total cell lysate. No signs of abnormal morphological changes or cell death were observed after the down- and up-regulation treatments of the HeLa cells. The cells were routinely visualized during the whole culturing process.

Precursor Incorporation into GlcCer, GalCer, LacCer and Gb$_3$ and TLC Analysis

Previously we found using the same labeling technique that HeLa cells overexpressing GLTP showed an increase in the de novo GlcCer synthesis [1]. In the previous work the siRNA GLTP gene construct #77 was used, the same as we used in this work for the MS lipidomics approach. For the metabolic labeling experiments we used two gene sequence constructs together (#76 & #78). We analyzed, using ^{3}H-sphinganine incorporation, the levels of GlcCer and GalCer, because these two were not distinguishable using the MS protocol to be used in this study. Like previously, the levels of GalCer are at the same level as the control cells, regardless of the regulation of GLTP expression (Fig. 1A). The differences in the LacCer levels are not as

pronounced in the metabolic labeling experiments as in the MS analysis data. However it appears the LacCer increases with increased GLTP and vice versa (Fig. 1A). The analysis by TLC of the total de novo Gb$_3$ amount after ^{3}H-sphinganine incorporation in HeLa cells show a significant increase in the levels of Gb$_3$ compared to the levels in the controls. The GLTP siRNA cells also show a significant decrease in the incorporation of the radiolabeled precursor (Fig. 1A). Representative TLC plates showing the separation of the analyzed lipids, Fig. 1B. Right plate, shows the separation used for GlcCer, GalCer and LacCer using the solvent system 10:2:4:2:1 chloroform:methanol:acetone:acetic acid:H$_2$O (w/w). Left plate, used for Gb$_3$, were the 45:55:10 chloroform:-methanol:0.2% CaCl$_2$ (in H$_2$O) solvent system was used. During the chromatography Gb$_3$ separates into two bands. Heterogeneity in the composition of the fatty acid moieties and the Gb$_3$ head group structure (its isomer isoglobo-Gb$_3$) results in double bands [32,33]. Both bands were scraped and analyzed and termed Gb$_3$.

Precursor Incorporation into SM, PC, and PE and their TLC Analysis

With the use of ^{3}H-acetate incorporation we could detect a small decrease in all phospholipid levels for GLTP siRNA treated cells (Fig. 2, red bars). No significant deviations in the phospholipid profiles were observed in the GLTP overexpressing cells (Fig. 2, green bars). The solvent system for the separation of the phospholipids was chloroform:methanol:acetic acid:H$_2$O (50:30:8:3) (v/v) [34].

MS Analysis of the Changes in GSL Metabolism after Down- and Up-regulation of GLTP

To get a broader and more detailed picture of the effects the GLTP had on the HeLa cell lipid profile we performed a MS lipidomics analysis. The cells where grown under the same conditions as for the metabolic labeling experiments, with the exception that the GLTP siRNA gene sequence #77 constructs was used. The overexpressing HeLa cells were similar as to the labeling experiments. Cells for the MS analysis were sorted as described in the Material and Methods section.

GlcCer/GalCer

In the lipidomics analysis, on a mass level, there is a small increase in the total cerebrosides for the GLTP OE sample (Fig. 3A), and a larger increase in the total saturated sphinganine d18:1 base cerebrosides (Fig. 3A). This is probably due the higher levels of GlcCer seen in the metabolic labeling experiments (Fig. 1A). It is important to recall that this MS analysis is for both GlcCer/GalCer. See Table S1 for quantitative data with SD. For clarity, minor species and the SD have been left out. The molecular changes in the acyl chain compositions are shown in Fig. 3B.

LacCer

The total LacCer amounts did not change significantly in the cells that have up- or down-regulated GLTP (Fig. 3C), however the treated cells had deviating levels of LacCer compared to the control cells. For the LacCer d18:0 sphinganine base species 22:0, both siRNA and OE, and the 24:0 N-linked acyl chain d18:0 base for the siRNA sample is much lower that the control (Fig. 3D).

Cer

The total amount of the sphingolipid precursor Cer did not differ significantly from the control cells, however the d18:0 sphinganine-based Cers in the GLTP siRNA cells were 25% lower

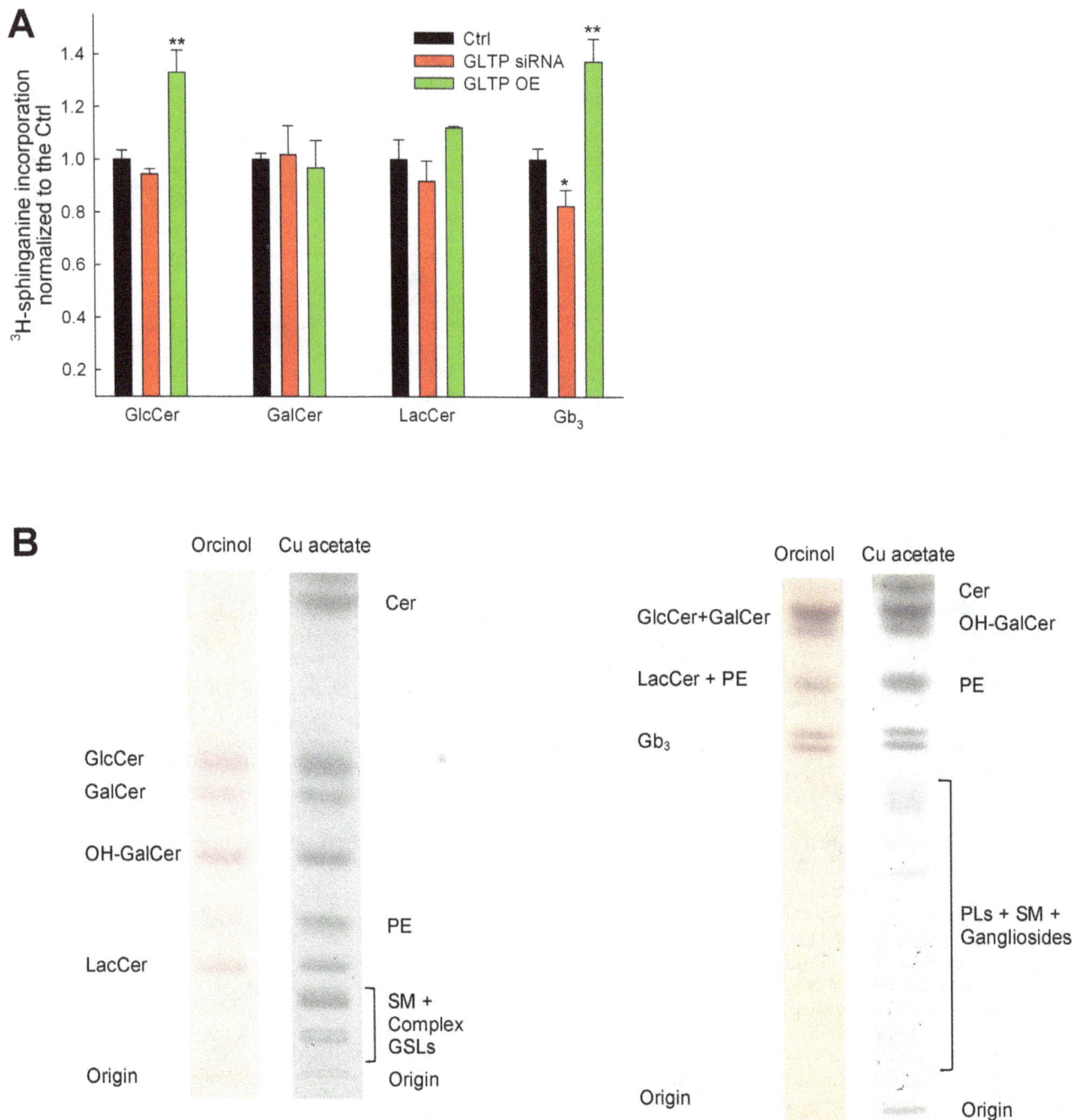

Figure 1. TLC analysis of metabolically labeled GSLs. (A) Metabolic labeling of HeLa cells, normal (black), GLTP siRNA (red) and overexpression of GLTP (green). ^{3}H-sphinganine incorporation into the sphingolipids was analyzed using TLC, and a mixture of both gene sequence #76 and #78 of the siRNA GLTP constructs were used. The TLC data for the incorporation of the radiolabeled ^{3}H-sphinganine are from at least three different experiments, and the results are normalized to the controls. **(B)** Representative TLC plates illustrating a typical separation of the analyzed lipids. The plates were first stained with the carbohydrate specific orcinol, followed by 3% Cu acetate charring to visualize all lipids. Left plate, shows the effective separation of GlcCer, GalCer and LacCer using the solvent system 10:2:4:2:1 chloroform:methanol:acetone:acetic acid:H$_2$O (v/v). Right plate, for effective Gb$_3$ separation from other GSLs we used the 45:55:10 chloroform:methanol:0.2% CaCl$_2$ (in H$_2$O) (v/v) solvent system. The statistical significance compared to the respective controls is indicated with asterisks. One asterisk (*) $p<0.05$ and two asterisks (**) $p<0.01$ indicate the statistical significance compared to the controls.

(Fig. 3E) and the GLTP OE cells had a reduction of more than 50% compared to the controls (Fig. 3E). It should be noted that the amount of d18:0 base Cers in the HeLa cells were only one-tenth to that of the d18:1 base, explaining the only minor differences in the total Cer levels (Table S1).

Gb$_3$

The completely new finding that the amount of Gb$_3$ decreased with in the GLTP siRNA treated cells can also be seen in the MS analysis (Fig. 4A, red bars). It should be noted that the amount of Gb$_3$ in HeLa cells was 4-fold higher than GlcCer and GalCer

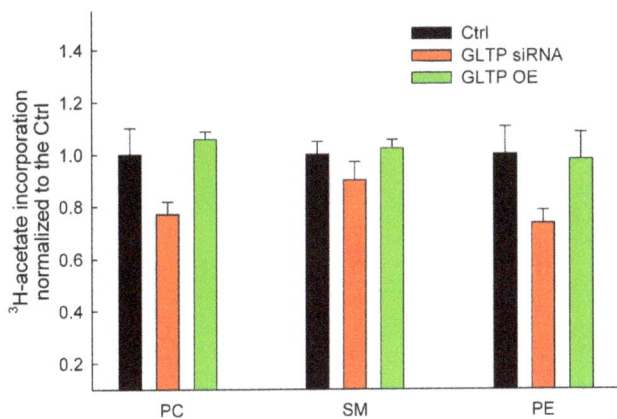

Figure 2. Analysis of metabolically labeled PC, SM and PE after alternation in the GLTP expression levels. Three phospholipid fractions were analyzed using ^{3}H-acetate incorporation of HeLa cells with different expression levels of GLTP. The TLC data for the incorporation of the radiolabeled ^{3}H-acetate are from at least three different experiments, and the results were normalized to the controls. The statistical significance compared to the respective controls is indicated with asterisks. Two asterisks (**) $p<0.01$ indicate the statistical significance compared to the controls.

together, and 6-fold higher compared to LacCer (Table S2). This is in well agreement with previous reports regarding the balance between Gb$_3$ and monohexosylceramides in HeLa cells. [6,35]. The decrease in the Gb$_3$ is in line with the data seen in the FAPP2 knockdown HeLa cells in the very resent study by D'Angelo and co-workers [5]. It appears that all molecular Gb$_3$ lipids that were analyzed were lower (Fig. 4A). The amounts of Gb$_3$ in the GLTP OE cells showed a small increase compared to the control cells (Fig. 4A, green bar).

Cholesterol Esters and Diacylglycerol Changes after Up-regulation of GLTP Expression

After knockdown of GLTP we did see a slight but not significant increase for the total CE's and a small decrease in the DAG amounts (Fig. 4B & 4C, red bars). The levels of CE and DAG on the other hand increased significantly in the HeLa cells overexpressing GLTP (Fig. 4B & 4C, green bars). The quantitative data for the molecular CE's and the DAGs are presented in Table S2.

Sphingomyelin and Phosphatidylcholine Levels after Regulation of GLTP Expression

Both in the down- and up-regulated GLTP HeLa cells we found in the MS analysis that the total SM mass decreased compared to the control cells (Fig. 5A). This is comparable to our previous finding that an up-regulation of GLTP causes a decrease in de novo SM synthesis using the metabolic labeling [1]. However we detected a small decrease in the de novo synthesis of SM in the GLTP knockdown cells using ^{3}H-acetate incorporation (Fig. 2). The data for the individual molecular lipids of SM is presented in the Table S3. The levels of total PC in the siRNA or GLTP overexpressing cells did not differ significantly from the control HeLa cells (Fig. 5B). The data for the individual molecular lipids of PC is presented in the Table S3. The ether-linked PC 18:0/16:0 species was also analyzed with MS but we could not detect any significant changes (Fig. 5B, yellow and blue bars).

Changes in the PE, PS, PI and PG Metabolism after Down- and Up-regulation of GLTP Expression in HeLa Cells

MS analysis of the total amounts of PE in the treated HeLa cells did not significantly change (Fig. 6A, Table S4). Quantitative analysis shows that PS was significantly reduced in the GLTP overexpressing cells, with a decrease of 50% (Fig. 6B, green bar). The GLTP siRNA cells did show a small increase in the PS mass, however not significant (Fig. 6B, red bar). The distribution of the hydrocarbon chains in PS of HeLa cells shows clearly that all acyl chain species decrease in the GLTP overexpressing HeLa cells (Fig. 6B, green bars). The amount of both PG and PI for the two treated HeLa cell types did not show any significant difference compared to the control cells (Fig. 6C & 6D). The molecular species level changes for PG and PI are shown in Table S4. The ether-linked PE (Table S4) species was also analyzed but we could not detect any significant changes (Fig. 6A, yellow and blue bars).

The Total Lipid Levels and the Distribution of the Amide-linked Acyl Chains in the Sphingolipids in HeLa Cells

The amounts for all 15 different lipid classes were summarized for each HeLa cell type and show a very similar total amount (Fig. 7A). The abundance of the different lipid classes and their changes are summarized in Fig. 7B. Analysis of the distribution of hydrocarbon chains in the sphingolipids of HeLa cells showed that the most abundant acyl chain was palmitic acid (16:0), followed by the unsaturated nervonic acid (24:1) and lignoceric acid (24:0) (Fig. 7C). Interestingly, the amounts of palmitic acid were reduced significantly in both the siRNA and GLTP overexpressing cells. Odd-carbon, especially 15:0 and 23:0 carbon chains were also found in the sphingolipids of HeLa cells, but with minor amounts (Fig. 7C). Changes in the saturated and the degree of unsaturation for the phospholipid species are presented in Fig. 5D. PS with fatty acids with 1 unsaturation are much lower and it appears that PS with 2 or more double bonds in their fatty acids are almost lost in HeLa cells overexpressing GLTP (Fig. 7D). There is also a loss of saturated fatty acids in PC for the GLTP overexpressing cells. The fatty acids with two or more double bonds in PE is also less in GLTP overexpressing cells, and elevated in the GLTP siRNA cells.

Discussion

We present for the first time a comprehensive lipid analysis of 15 different lipid classes, and how they are affected by the GLTP gene down- or up-regulation. We discuss the changes in the lipidome that we observe and how these might relate to GLTP. In particular, a completely new finding is that the globotriaosylceramide Gb$_3$ is significantly lower when GLTP is knocked down. Upon up-regulation of GLTP a significant increase in the Gb$_3$ levels is also observed. A similar observation was recently found by D'Angelo and coworkers for down-regulating the other member in the GLTP family, FAPP2 [16]. FAPP2 depletion selectively inhibited the synthesis of Gb$_3$. FAPP2 down-regulation did not decrease the synthesis of gangliosides GM$_3$. In this study, and with this approach, we have not analyzed the changes in the ganglioside families after alternation in the GLTP expressions.

The glycolipid transfer protein family members GLTP and FAPP2 have been connected to many cellular lipid events, and logically the sphingolipid classes have been in focus. Here we present a much broader analysis of the HeLa cell lipidome, covering also membrane phospholipids, cholesterol-esters and diacylglycerols. The compositional changes of 142 lipid molecular species of HeLa cells were analyzed commercially (Zora Biosciences) using shotgun lipidomics. The changes observed in the MS

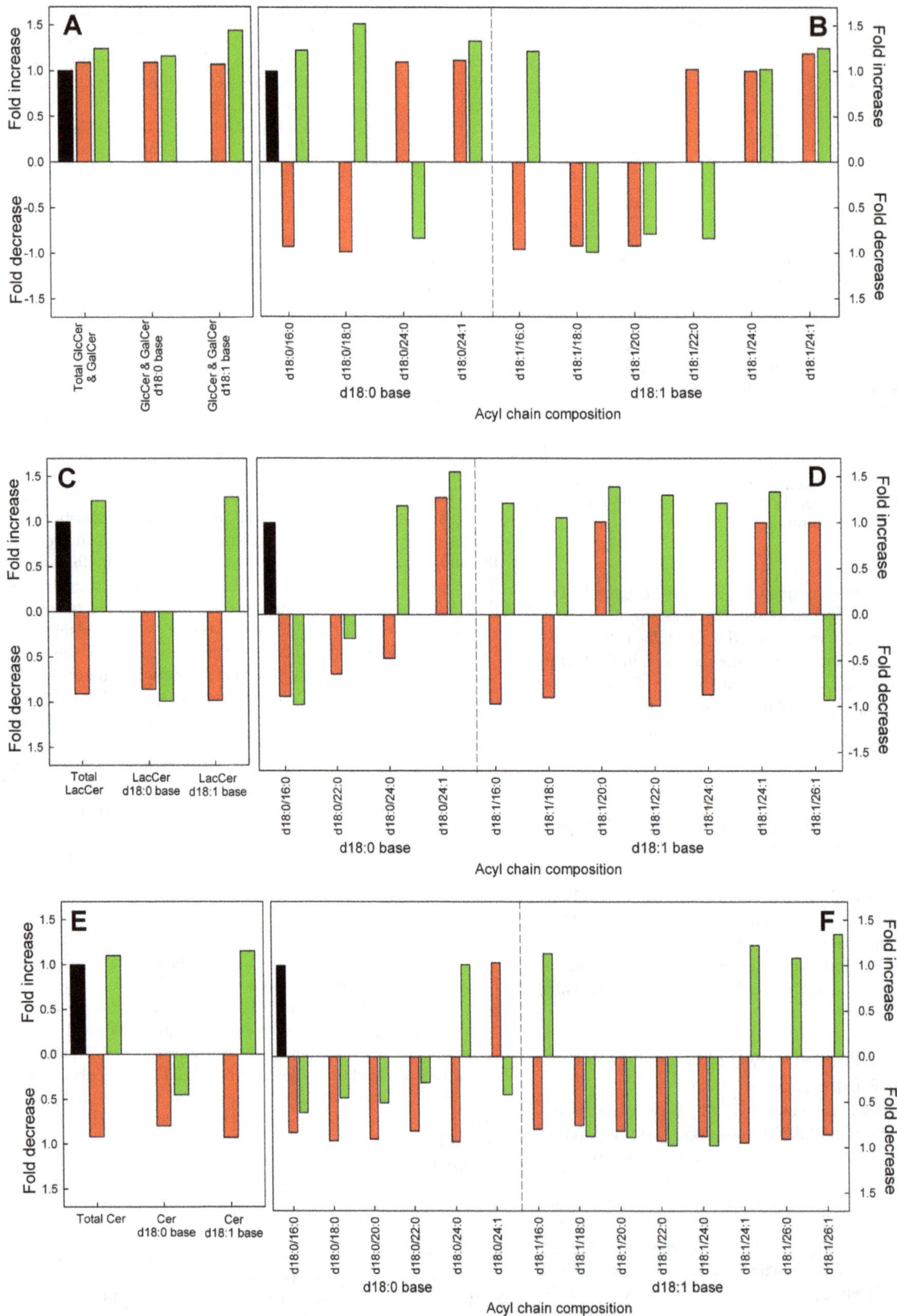

Figure 3. Relative change in GlcCer/Galcer, LacCer and Cer after knockdown or overexpression of GLTP in HeLa cells normalized to control cells. (**A**) Changes in the total masses of GlcCer/GalCer and GlcCer/GalCer with d18:0 and d18:1 base respectively. (**B**) Changes in the GlcCer/GalCer species after siRNA or OE treatments, compared to the control. (**C**) Changes in the total masses of LacCer and LacCer with d18:0 and d18:1 base respectively and (**D**) changes in the LacCer species. (**E**) Changes in the total masses of Cer and Cer with d18:0 and D18:1 base respectively, with (**F**) showing the changes in the Cer species. The data for siRNA GLTP (red bars) and GLTP OE (green bars) were normalized to the levels in the mock

control cells (black bars). The abbreviations for the lipid classes are given in "Materials and Methods". For clarity, minor species are not presented in the graph. For full list of species and quantitative data with SD, see Table S1).

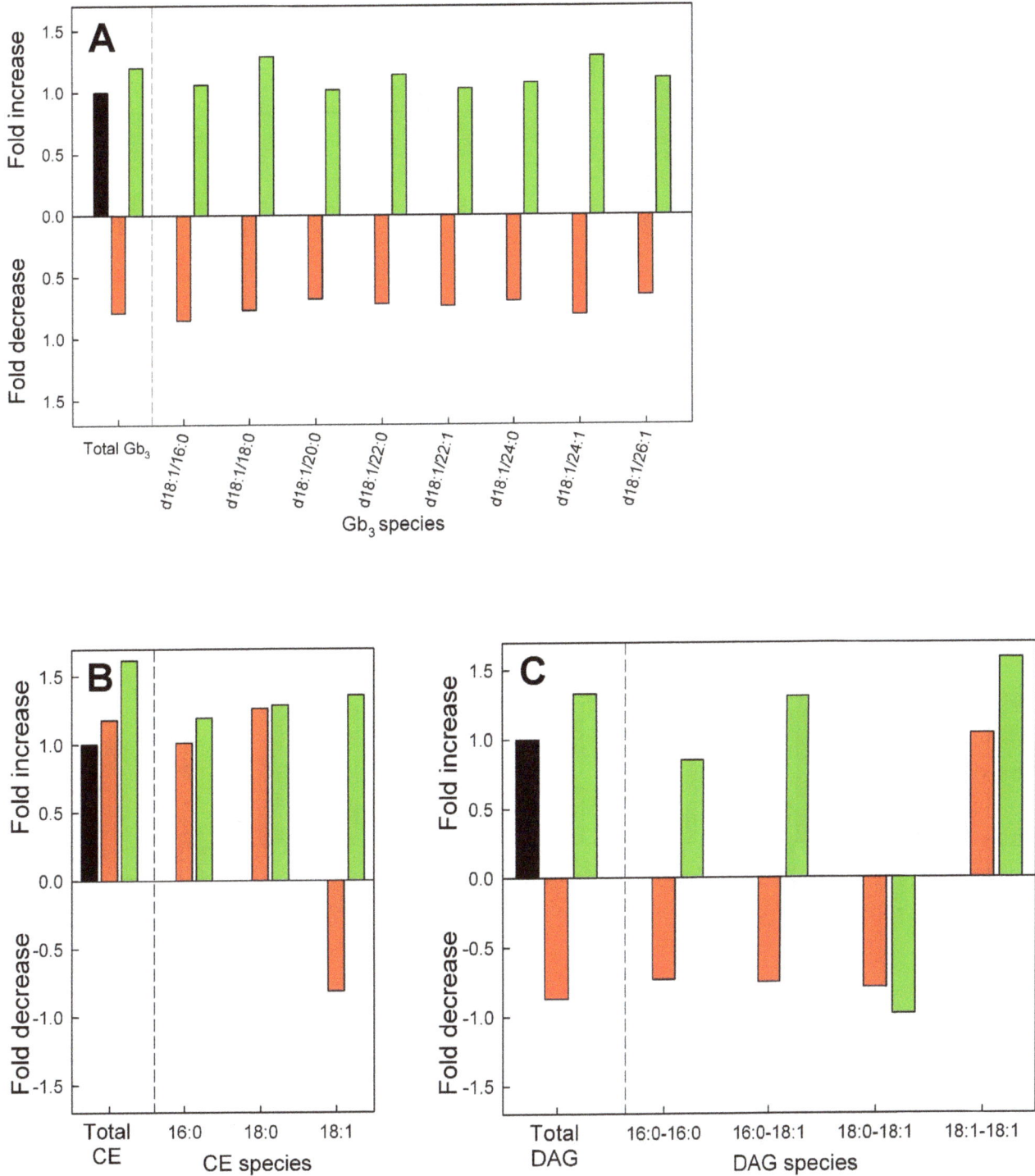

Figure 4. Relative change in Gb$_3$, CE and DAG after knockdown or overexpression of GLTP in HeLa cells compared to control cells.
(**A**) Changes in total (left side) and molecular species of Gb$_3$ in HeLa cells subjected to knockdown (red bars) or overexpression (green bars) of the GLTP gene normalized to the levels in the mock control cells (black bars). The Gb$_3$ level for the GLTP siRNA and OE sample is significantly different (p<0.01) than the control sample. (**B**) Relative changes in the total and molecular species of CE's and (**D**) DAG's as a function of GLTP expression levels. The abbreviations for the lipid classes are given in "Materials and Methods". For clarity, minor species are not presented in the graph. For full list of species and quantitative data with SD, see Table S2.

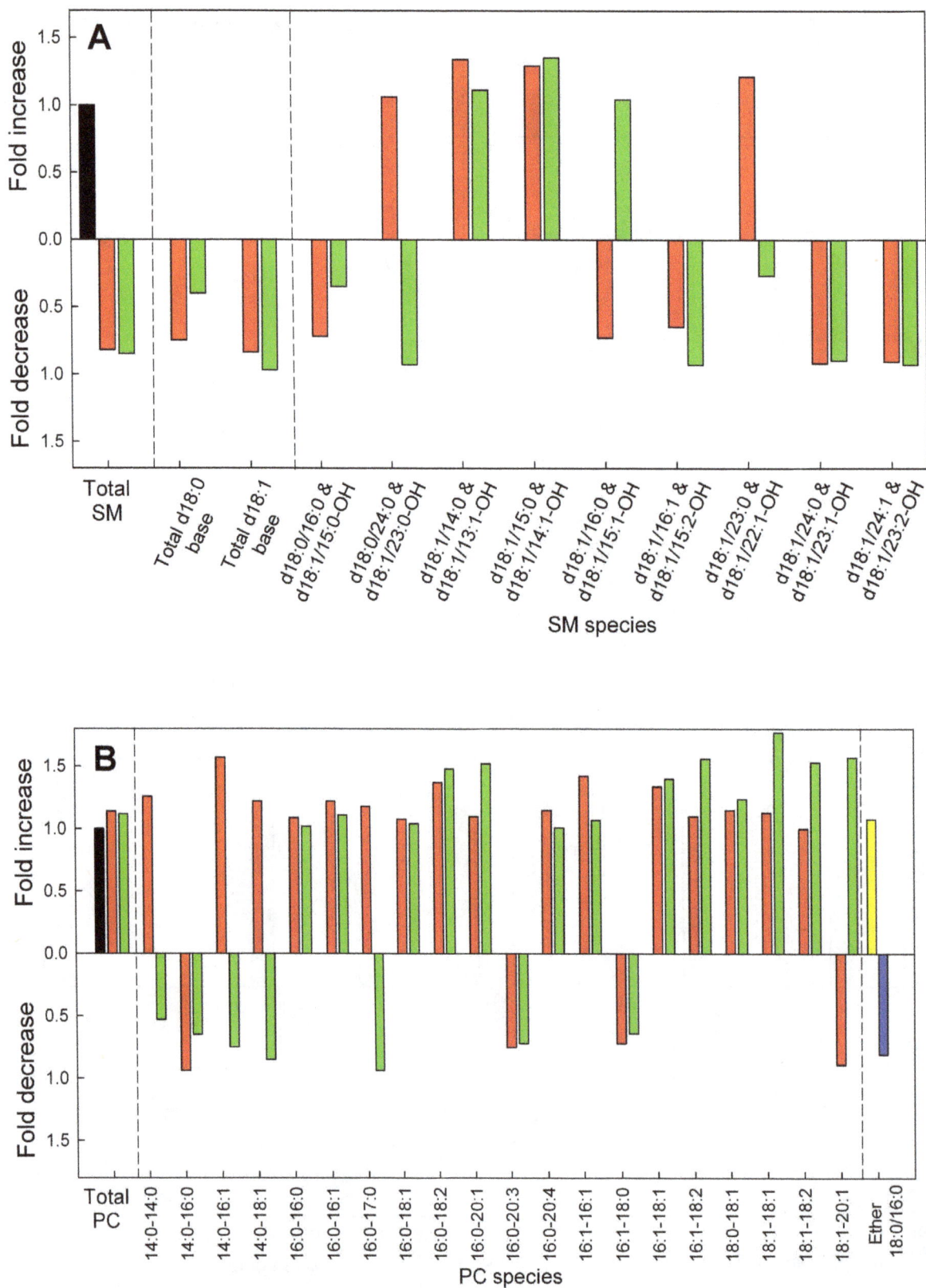

Figure 5. Relative changes in SM, PC and ether-PC after knockdown or overexpression of GLTP in HeLa cells compared to control cells. The relative changes in the masses of (**A**) SM and (**B**) PC in HeLa cells subjected to knockdown (red bars) or overexpression (green bars) of the GLTP gene normalized to the levels in the mock control cells (black bars). The changes for the ether linked PCs is on the far right, yellow represent the GLTP siRNA sample and the blue bar the GLTP OE sample. The abbreviations for the lipid classes are given in "Materials and Methods". For clarity, minor species are not presented in the graph. For full list of species and quantitative data with SD, see Table S3.

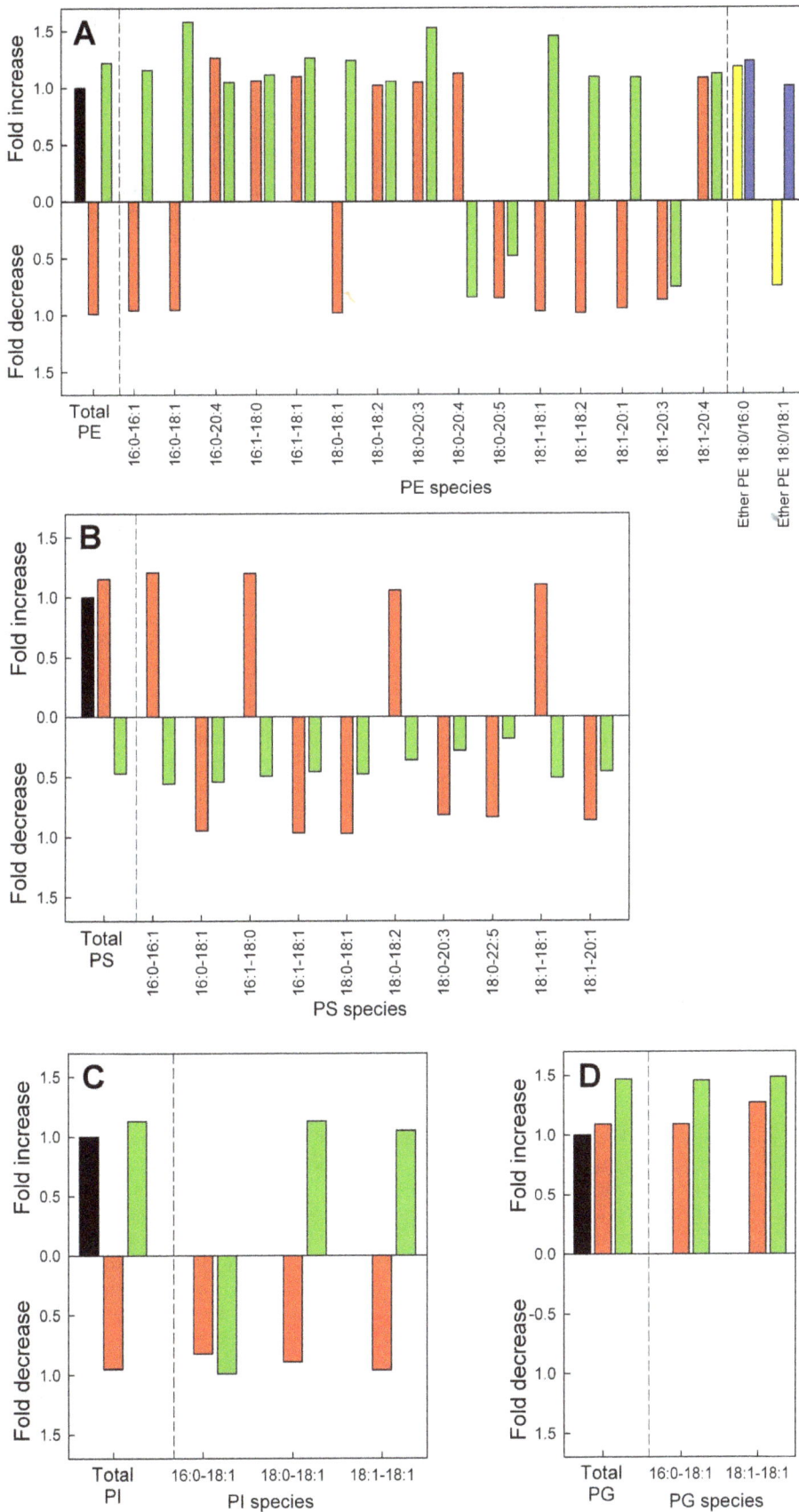

Figure 6. Relative changes in phosphoglycerides after knockdown or overexpression of GLTP in HeLa cells compared to control cells. Changes in the total masses and the molecular species of (**A**) PE, (**B**) PS, (**C**) PI and (**D**) PG in HeLa cells subjected to knockdown (red bars) or

analysis were well in agreement with the metabolic labeling and thin layer chromatography analysis that initiated this study. It should be mentioned that the metabolic labeling and TLC analysis determine the amount of all species in the lipid class, whereas the MS analysis has a cut-off limit and does not always determine the amounts of all acyl chain specific species within some of the lipid classes, especially those with more complex and minor acyl chain components.

When we analyzed the impact of down-regulation of GLTP on the synthesis of GSLs in HeLa cells we found that Gb_3 decreases in the MS analysis with 25% and almost 25% in the ^{3}H-sphinganine metabolic labeling experiments. The up-regulation of GLTP on the other hand, caused an increase in the synthesis of Gb_3 with about 20% in the MS analysis and almost 40% using ^{3}H-sphinganine metabolic labeling. Total GlcCer levels were higher in GLTP overexpressing cells, with GalCer remaining unchanged. A small increase seems to be present in the overexpressing HeLa cells, however not significant. A knockdown of GLTP also showed a somewhat lower quantity of LacCer. Clearly the situation with the expression of GLTP and the LacCer levels needs to be investigated further. D'Angelo and co-workers reported recently that the silencing of FAPP2 caused a dramatic drop in the LacCer and Gb_3 levels but not in the GM_3 synthesis in HeLa cells [16]. If the LacCer synthesis is silenced, the synthesis of all three GSLs is inhibited. These results indicate that there are at least two different pools of LacCer in the Golgi apparatus, one for gangliosides, and one for globosides, and that the lactosylceramide synthase is localize both to the Golgi cisternae and to the trans-Golgi network [16]. Previous reports from the Lingwood laboratory also suggested that neutral and acid glycosphingolipids are synthesized from distinct precursor glycosphingolipid pools [36]. These results further indicated that the precursor, GlcCer is transported to the LacCer synthase for globoside synthesis by FAPP2, and that GlcCer destined for LacCer synthase in the early Golgi is transported by other mechanisms, probably vesicular trafficking [5]. Globotriaosylceramide, ceramide trihexoside is also known as CD77 and is a cluster of differentiation. Lactosylceramide 4-alpha-galactosyltransferase (Gb_3 synthase) is the type II membrane protein that adds an additional galactose to lactosylceramide in the late Golgi to generate Gb_3 [37].

How can GLTP be linked to changes in Gb_3 levels? When GLTP is overexpressed more GlcCer for Gb_3 synthesis is generated. A close examination of our data indicated that the LacCer levels are also somewhat higher than the controls, though not significantly, this both for the MS data and the metabolic labeling experiments (Fig. 1B & Fig. 6A). Speculatively, increased levels of Gb_3 could be a consequence of a higher activity of GLTP in the ER/early Golgi in directing or sensing more GlcCer for FAPP2, followed by an increased movement of GlcCer to the trans-Golgi. A lower GLTP activity, as in the GLTP siRNA HeLa cells show less Gb_3, analogously less GlcCer would be presented to FAPP2, and less GlcCer would arrive at the trans-Golgi to be converted to LacCer and subsequently to Gb_3. A higher GlcCer in GLTP overexpressing cells could be a consequence of the elevated synthesis of Gb_3, to compensate for maintained levels of other higher GSLs, such as gangliosides, that do not depend on FAPP2 GlcCer transfer activity. For Cer, we only detected lower levels, in GLTP over-expressing cells, for the d18:0 bases, whereas the d18:1 bases and total Cer remained at the same level as the

controls. However, a strong decrease in total SM is seen in both down- and up-regulated HeLa cells. It appears that the pool of Cer for the altered GlcCer synthesis would come from the regulation of SM synthesis, probably also involving the Cer transporter CERT. There is experimental evidence for an indirect link between GLTP expression to sphingolipid homeostasis through Cer. Brown and colleagues have shown that Cer induced GLTP promoter activity and raised transcript levels (mRNA) in HeLa cells [3]. In HeLa cells, added C_6-Cer (dissolved in DMSO) altered the in vivo binding affinity of the transcription factors Sp1 and Sp3 for the GLTP promoter and decreased Sp3 acetylation. Whether the externally added short Cers also affect the expression of GLTP on a protein level, is not known.

The sphingolipid homeostasis and its close connection to the cholesterol homeostasis are regulated in a very complicated manner, involving lipid- synthases, sensors and transporters. One common feature for the involved sensors and transporters is that they all are capable of binding to the VAP proteins (VAMP-associated proteins) [38,39]. The binding to VAPs in the ER occurs through FFAT-like motifs (two phenylalanines in an acidic tract) [40]. The VAPs are highly conserved ER trans-membrane proteins involved in diverse functions. They regulate lipid transport and homeostasis, vesicular trafficking, and are involved in the unfolded protein response [39]. It is therefor likely that GLTP with its FFAT-like motif [2,40], and capacity to bind to VAP proteins, is also involved in the intricate regulation of sphingolipid homeostasis.

The synthesis of SM is mediated by at least two SM synthases, SMS1 and SMS2 [41]. They catalyze the transfer of a phosphocholine group from PC to the hydroxyl group of Cer at C1. This generates SM and DAG. SM synthesis is largely affected and regulated by the activity of CERT [42]. CERTs role is to transport Cer to the trans-Golgi from the ER [43,44]. CERT has also been shown to bind and transfer DAG to a minor extent [43,45]. It has also been suggested that DAG is transported back from the Golgi to ER by CERT [45]. The changes that we see in the DAG levels as a function of GLTP up- and down-regulation, could be a response caused by DAGs ability to be bound by CERT and its VAP association. Likewise, the response in SM levels could also be caused by CERTs activity with VAP. Changes in the Cer and SM as well cholesterol metabolism in the cell is sensed not only by CERT but also by the oxysterol binding protein, OSBP [46]. OSBP, CERT and yet Nir2, a PI/PC transfer protein, all act in concert, to maintain DAG levels in the Golgi [47–49]. In the Golgi, protein kinase D associates with the membrane via DAG, and directly phosphorylates CERT, inhibiting its Cer transport activity [50]. Protein kinase D phosphorylation of OSBP results in Golgi fragmentation, that inhibits CERT binding to Golgi membranes [51]. Dephosphorylation increases the Cer transfer mediated by CERT [50,52].

In cells overexpressing GLTP we detected a significant decrease in PS expression as well as an increase in the CE levels. The synthesis of PS in mammalian cells takes place in the ER, by a base-exchange reaction, where PS synthase-1 (PSS-1) primarily uses PC as a substrate and PS synthase-2 (PSS-2), PE as the base for exchange with serine [53]. Neither PC nor PE changed significantly, regardless of GLTP expression, despite the large decrease in PS levels. Both PSS-1 and PSS-2 are integral ER membrane proteins sub-compartmentalized to the mitochondria-

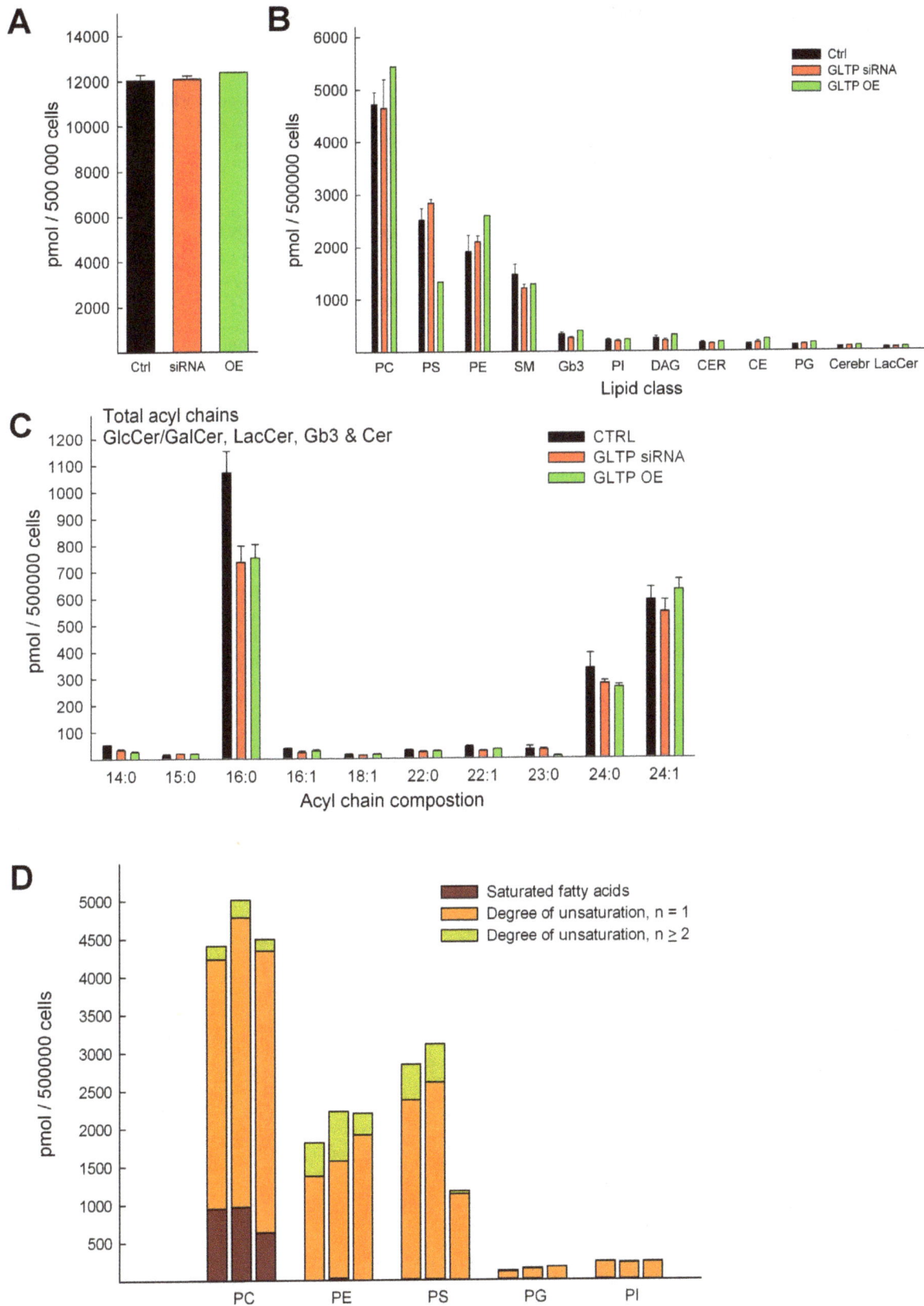

Figure 7. Summarized lipid amounts. (**A**) Comparison of the total lipid content summarized (from the MS data) between CTRL (black bars), GLTP siRNA cells (red bars) and GLTP overexpressing cells (green bars). Lipid quantities were calculated on the basis of the corresponding IS and the amounts (pmol/500000 cells) represented the sum of all individual molecular species of all lipid classes. (**B**) Comparison of the abundance of the lipid classes analyzed in this study. (**C**) Changes in the distribution of the amide-linked hydrocarbon chains in sphingolipids (GlcCer/GalCer, LacCer, Gb$_3$, Cer and SM) of normal, GLTP siRNA and GLTP overexpressing HeLa cells. (**D**) Changes in the degree of unsaturation of the acyl chains in the different

phospholipid classes relative to GLTP expression. Left control, middle GLTP siRNA and right GLTP OE HeLa cell samples, n is the sum of degree of unsaturation of fatty acid chains in the phospholipids.

associated membranes, MAM fractions. MAMs are a region of ER closely associated with the mitochondria. PSS-1 and PSS-2 do not seem to have an FFAT-like domain and have to our knowledge not been reported to bind to the VAP-proteins. The enzymatic activities required for synthesis of triacylgycerols, CE, and free cholesterol have also been located to the MAM fractions [54]. In addition, the synthesis of GlcCer has been reported to occur also in the MAMs [55] and not only in the cis-Golgi [56,57]. Furthermore, Meyer and de Groot suggested that PS might serve as the serine donor in the initial condensation of serine and palmitoyl coenzyme A catalyzed by serine palmitoyltransferase, the initial step in Cer biosynthesis [58]. This metabolic pathway together with the lipid metabolism in the MAMs would connect PS and CE to the Cer and GlcCer biosynthesis and could as discussed above be impacted by the expression of GLTP and its capacity to bind to the VAP-proteins. In a previous study Gao and coworkers speculate that overexpression of GLTP in HeLa cells would limit the availability of GlcCer for generation of higher GSLs, and the net effect would be reduced levels of complex GSLs [4]. GLTP would function as a sink protecting GlcCer from being processed. However, here we show that overexpression rather significantly elevates at least Gb_3 and to some extent also LacCer, and would rather function as an activator for the synthesis of Gb_3.

Summarizing, the changed in the lipidome that we observe in this work is likely to be a consequence of GLTPs involvement, as a glycolipid binder, sensor or transporter together with the other VAP protein binding players. In a broader perspective, since different species of Cer are destined for different sphingolipid end products [59–62], we speculate that GLTP might play a role in orchestrating the transfer of different GlcCer species to different destinations, with connections to the synthesis of Cer precursors or their transfer by vesicular means or by CERT. Perhaps the GlcCer destined to be transported via FAPP2 to the pool of LacCer specifically destined to Gb_3 synthesis is sensed/controlled by GLTP. With this work we put forward new data suggesting that GLTP would be a significant player in not only the sphingolipid metabolism but also could have a much broader role in the lipid metabolism in the cell.

Supporting Information

Figure S1 Efficiency of small interfering RNA on the GLTP protein expression in HeLa cells at different time intervals. (A) qPCR analysis of GLTP mRNA expression levels (in percent) after treatment with different GLTP siRNA sequences. All three siRNA sequences, termed #76 (red), #77 (green) and #78 (yellow) used in this work were compared and normalized to the scrambled siRNA universal control (UC, black). The three different sequences are described in the Materials and Methods section. **(B)** Western blotting analysis shows a reduced protein expression in HeLa siRNA transfected cells compared both the control and the UC samples. The GLTP expression was analyzed at different time intervals, after siRNA treatment. Immunoblot against human GLTP (upper blot) in normal HeLa cells, UC control cells and different GLTP siRNA sequences, 50 μg of total cell lysates were used and a rabbit anti-GLTP antibody.

Figure S2 Efficiency of overexpression on the GLTP protein levels in HeLa cells. (A) The upper blot shows the beta-actin expression and lower blot the expression of GLTP as a function of time. All lanes are loaded with 30 μg total protein whole cell lysate. Note the endogenous expression of GLTP is low and not visible in the lower blot, first lane. **(B)** Western blot analysis of the expression of GLTP in the HeLa cells used for the MS lipidomics analysis. Left blot shows the endogenous expression of GLTP (lane 1) and the reduced protein expression in HeLa cells with GLTP knockdown, by siRNA (#77 siRNA GLTP gene construct), lane 2. A total of 80 μg total cell lysate was loaded, and beta-actin was used as the loading control, upper blot. The right blot shows the amount of GLTP in HeLa cells with GLTP overexpression (lane 4), and an invisible endogenous GLTP band in lane 3, due to the loading amount of just 10 μg total cell lysate.

Table S1 The amounts for the molecular lipids in HeLa cells are presented as pmol/500000 cells for CTRL cells down- (GLTP siRNA) and up-regulated (GLTP OE) cell samples. The values for the CTRL are averages of thee mock samples, GLTP siRNA averages of two samples, and GLTP OE values from one sample.

Table S2 The amounts for the molecular lipids in HeLa cells are presented as pmol/500000 cells for CTRL cells down- (GLTP siRNA) and up-regulated (GLTP OE) cell samples. The values for the CTRL are averages of thee mock samples, GLTP siRNA averages of two samples, and GLTP OE values from one sample. Blank values means that the concentration is below the quantification limit or a value is not included in the final data set because of quality control (QC, see Materials and Methods) cutoff.

Table S3 The amounts for the molecular lipids in HeLa cells are presented as pmol/500000 cells for CTRL cells down- (GLTP siRNA) and up-regulated (GLTP OE) cell samples. The values for the CTRL are averages of thee mock samples, GLTP siRNA averages of two samples, and GLTP OE values from one sample. Blank values means that the concentration is below the quantification limit or a value is not included in the final data set because of quality control (QC, see Materials and Methods) cutoff.

Table S4 The amounts for the molecular lipids in HeLa cells are presented as pmol/500000 cells for CTRL cells down- (GLTP siRNA) and up-regulated (GLTP OE) cell samples. The values for the CTRL are averages of thee mock samples, GLTP siRNA averages of two samples, and GLTP OE values from one sample. Blank values means that the concentration is below the quantification limit or a value is not included in the final data set because of quality control (QC, see Materials and Methods) cutoff.

Acknowledgments

We thank Jessica Tuuf and Pia Roos-Mattjus for critical comments. Zora Biosciences Ltd did the lipidomics analysis.

Author Contributions

Conceived and designed the experiments: MAK HOR APEB PM. Performed the experiments: MAK HOR APEB. Analyzed the data:

MAK HOR APEB PM. Contributed reagents/materials/analysis tools: MAK HOR APEB PM. Wrote the paper: MAK APEB PM.

References

1. Tuuf J, Mattjus P (2007) Human glycolipid transfer protein-intracellular localization and effects on the sphingolipid synthesis. Biochim Biophys Acta 1771: 1353–1363.

2. Tuuf J, Wistbacka L, Mattjus P (2009) The glycolipid transfer protein interacts with the vesicle-associated membrane protein-associated protein VAP-A. Biochem Biophys Res Commun 388: 395–399.

3. Zou X, Gao Y, Ruvolo VR, Gardner TL, Ruvolo PP, et al. (2010) Human glycolipid transfer protein gene (GLTP) expression is regulated by Sp1 and Sp3: Involvement of the bioactive sphingolipid ceramide. J Biol Chem 286: 1301–1311.

4. Gao Y, Chung T, Zou X, Pike HM, Brown RE (2011) Human glycolipid transfer protein (GLTP) expression modulates cell shape. PLoS One 6: e19990-e19990.

5. D'Angelo G, Polishchuk E, Di Tullio G, Santoro M, Di Campli A, et al. (2007) Glycosphingolipid synthesis requires FAPP2 transfer of glucosylceramide. Nature 449: 62–67.

6. Halter D, Neumann S, van Dijk SM, Wolthoorn J, de Maziere AM, et al. (2007) Pre- and post-Golgi translocation of glucosylceramide in glycosphingolipid synthesis. J Cell Biol 179: 101–115.

7. Tuuf J, Mattjus P (2013) Membranes and mammalian glycolipid transferring proteins. Chem Phys Lipids.

8. Yamada K, Abe A, Sasaki T (1986) Glycolipid transfer protein from pig brain transfers glycolipids with beta-linked sugars but not with alpha-linked sugars at the sugar-lipid linkage. Biochim Biophys Acta 879: 345–349.

9. Mattjus P (2009) Glycolipid transfer proteins and membrane interaction. Biochim Biophys Acta 1788: 267–272.

10. Brown RE, Stephenson FA, Markello T, Barenholz Y, Thompson TE (1985) Properties of a specific glycolipid transfer protein from bovine brain. Chem Phys Lipids 38: 79–93.

11. Malinina L, Malakhova ML, Teplov A, Brown RE, Patel DJ (2004) Structural basis for glycosphingolipid transfer specificity. Nature 430: 1048–1053.

12. West G, Viitanen L, Alm C, Mattjus P, Salminen TA, et al. (2008) Identification of a glycosphingolipid transfer protein GLTP1 in Arabidopsis thaliana. FEBS J 275: 3421–3437.

13. Saupe SJ, Descamps C, Turcq B, Begueret J (1994) Inactivation of the Podospora anserina vegetative incompatibility locus het-c, whose product resembles a glycolipid transfer protein, drastically impairs ascospore production. Proc Natl Acad Sci U S A 91: 5927–5931.

14. Brodersen P, Petersen M, Pike HM, Olszak B, Skov S, et al. (2002) Knockout of Arabidopsis accelerated-cell-death 11 encoding a sphingosine transfer protein causes activation of programmed cell death and defense. Genes Dev 16: 490–502.

15. Godi A, Di Campli A, Konstantakopoulos A, Di Tullio G, Alessi DR, et al. (2004) FAPPs control Golgi-to-cell-surface membrane traffic by binding to ARF and PtdIns(4)P. Nat Cell Biol 6: 393–404.

16. D'Angelo G, Uemura T, Chuang CC, Polishchuk E, Santoro M, et al. (2013) Vesicular and non-vesicular transport feed distinct glycosylation pathways in the Golgi. Nature 501: 116–120.

17. Kjellberg MA, Mattjus P (2013) Glycolipid transfer protein expression is affected by glycosphingolipid synthesis. PLoS One 8: e70283.

18. Chigorno V, Giannotta C, Ottico E, Sciannamblo M, Mikulak J, et al. (2005) Sphingolipid uptake by cultured cells: complex aggregates of cell sphingolipids with serum proteins and lipoproteins are rapidly catabolized. J Biol Chem 280: 2668–2675.

19. Lowry OH, Rosebrough NJ, Farr AL, Randall RJ (1951) Protein measurement with the Folin phenol reagent. J Biol Chem 193: 265–275.

20. Allan D, Cockcroft S (1982) A modified procedure for thin-layer chromatography of phospholipids. J Lipid Res 23: 1373–1374.

21. Tuuf J, Kjellberg MA, Molotkovsky JG, Hanada K, Mattjus P (2010) The intermembrane ceramide transport catalyzed by CERT is sensitive to the lipid environment. Biochim Biophys Acta.

22. Ekroos K (2008) Unraveling Glycerophospholipidomes by Lipidomics. In: Wang F, editor. Biomarker Methods in Drug Discovery and Development: Humana Press. 369–384.

23. Jung HR, Sylvänne T, Koistinen KM, Tarasov K, Kauhanen D, et al. (2011) High throughput quantitative molecular lipidomics. Biochim Biophys Acta 1811: 925–934.

24. Ejsing CS, Duchoslav E, Sampaio J, Simons K, Bonner R, et al. (2006) Automated identification and quantification of glycerophospholipid molecular species by multiple precursor ion scanning. Anal Chem 78: 6202–6214.

25. Bergan J, Skotland T, Sylvänne T, Simolin H, Ekroos K, et al. (2013) The ether lipid precursor hexadecylglycerol causes major changes in the lipidome of HEp-2 cells. PLoS One 8: e75904.

26. Ståhlman M, Ejsing CS, Tarasov K, Perman J, Boren J, et al. (2009) High-throughput shotgun lipidomics by quadrupole time-of-flight mass spectrometry. J Chromatogr B Analyt Technol Biomed Life Sci 877: 2664–2672.

27. Ekroos K, Chernushevich IV, Simons K, Shevchenko A (2002) Quantitative profiling of phospholipids by multiple precursor ion scanning on a hybrid quadrupole time-of-flight mass spectrometer. Anal Chem 74: 941–949.

28. Ekroos K, Ejsing CS, Bahr U, Karas M, Simons K, et al. (2003) Charting molecular composition of phosphatidylcholines by fatty acid scanning and ion trap MS3 fragmentation. J Lipid Res 44: 2181–2192.

29. Liebisch G, Binder M, Schifferer R, Langmann T, Schulz B, et al. (2006) High throughput quantification of cholesterol and cholesteryl ester by electrospray ionization tandem mass spectrometry (ESI-MS/MS). Biochim Biophys Acta 1761: 121–128.

30. Merrill AH, Sullards MC, Allegood JC, Kelly S, Wang E (2005) Sphingolipi-domics: high-throughput, structure-specific, and quantitative analysis of sphingolipids by liquid chromatography tandem mass spectrometry. Methods 36: 207–224.

31. Heiskanen LA, Suoniemi M, Ta HX, Tarasov K, Ekroos K (2013) Long-term performance and stability of molecular shotgun lipidomic analysis of human plasma samples. Anal Chem 85: 8757–8763.

32. Pellizzari A, Pang H, Lingwood CA (1992) Binding of verocytotoxin 1 to its receptor is influenced by differences in receptor fatty acid content. Biochemistry 31: 1363–1370.

33. Adlercreutz D, Weadge JT, Petersen BO, Duus JO, Dovichi NJ, et al. (2010) Enzymatic synthesis of Gb3 and iGb3 ceramides. Carbohydr Res 345: 1384–1388.

34. Skipski VP, Peterson RF, Barclay M (1964) Quantitative Analysis of Phospholipids by Thin-Layer Chromatography. Biochem J 90: 374–378.

35. Yokoyama K, Suzuki M, Kawashima I, Karasawa K, Nojima S, et al. (1997) Changes in composition of newly synthesized sphingolipids of HeLa cells during the cell cycle - Suppression of sphingomyelin and higher-glycosphingolipid synthesis and accumulation of ceramide and glucosylceramide in mitotic cells. Eur J Biochem 249: 450–455.

36. De Rosa MF, Sillence D, Ackerley C, Lingwood C (2004) Role of multiple drug resistance protein 1 in neutral but not acidic glycosphingolipid biosynthesis. J Biol Chem 279: 7867–7876.

37. Kojima Y, Fukumoto S, Furukawa K, Okajima T, Wiels J, et al. (2000) Molecular cloning of globotriaosylceramide/CD77 synthase, a glycosyltransfer-ase that initiates the synthesis of globo series glycosphingolipids. J Biol Chem 275: 15152–15156.

38. Perry RJ, Ridgway ND (2005) Molecular mechanisms and regulation of ceramide transport. Biochim Biophys Acta 1734: 220–234.

39. Lev S, Ben Halevy D, Peretti D, Dahan N (2008) The VAP protein family: from cellular functions to motor neuron disease. Trends Cell Biol 18: 282–290.

40. Mikitova V, Levine TP (2012) Analysis of the key elements of FFAT-like motifs identifies new proteins that potentially bind VAP on the ER, including two AKAPs and FAPP2. PLoS One 7: e30455-e30455.

41. Huitema K, Van Den Dikkenberg J, Brouwers JF, Holthuis JCM (2004) Identification of a family of animal sphingomyelin synthases. EMBO J 23: 33–44.

42. Kudo N, Kumagai K, Matsubara R, Kobayashi S, Hanada K, et al. (2010) Crystal structures of the CERT START domain with inhibitors provide insights into the mechanism of ceramide transfer. J Mol Biol 396: 245–251.

43. Hanada K, Kumagai K, Yasuda S, Miura Y, Kawano M, et al. (2003) Molecular machinery for non-vesicular trafficking of ceramide. Nature 426: 803–809.

44. Hanada K (2013) Co-evolution of sphingomyelin and the ceramide transport protein CERT. Biochim Biophys Acta.

45. Kumagai K, Yasuda S, Okemoto K, Nishijima M, Kobayashi S, et al. (2005) CERT mediates intermembrane transfer of various molecular species of ceramides. J Biol Chem 280: 6488–6495.

46. Perry RJ, Ridgway ND (2006) Oxysterol-binding protein and vesicle-associated membrane protein-associated protein are required for sterol-dependent activation of the ceramide transport protein. Mol Biol Cell 17: 2604–2616.

47. Litvak V, Dahan N, Ramachandran S, Sabanay H, Lev S (2005) Maintenance of the diacylglycerol level in the Golgi apparatus by the Nir2 protein is critical for Golgi secretory function. Nat Cell Biol 7: 225–234.

48. Peretti D, Dahan N, Shimoni E, Hirschberg K, Lev S (2008) Coordinated lipid transfer between the endoplasmic reticulum and the Golgi complex requires the VAP proteins and is essential for Golgi-mediated transport. Mol Biol Cell 19: 3871–3884.

49. Lagace TA, Ridgway ND (2013) The role of phospholipids in the biological activity and structure of the endoplasmic reticulum. Biochim Biophys Acta.

50. Fugmann T, Hausser A, Schoffler P, Schmid S, Pfizenmaier K, et al. (2007) Regulation of secretory transport by protein kinase D-mediated phosphorylation of the ceramide transfer protein. J Cell Biol 178: 15–22.

51. Nhek S, Ngo M, Yang X, Ng MM, Field SJ, et al. (2010) Regulation of oxysterol-binding protein Golgi localization through protein kinase D-mediated phosphorylation. Mol Biol Cell 21: 2327–2337.

52. Tomishige N, Kumagai K, Kusuda J, Nishijima M, Hanada K (2009) Casein kinase I{gamma}2 down-regulates trafficking of ceramide in the synthesis of sphingomyelin. Mol Biol Cell 20: 348–357.

53. Vance JE, Tasseva G (2013) Formation and function of phosphatidylserine and phosphatidylethanolamine in mammalian cells. Biochim Biophys Acta 1831: 543–554.

54. Rusinol AE, Cui Z, Chen MH, Vance JE (1994) A unique mitochondria-associated membrane fraction from rat liver has a high capacity for lipid synthesis and contains pre-Golgi secretory proteins including nascent lipoproteins. J Biol Chem 269: 27494–27502.

55. Ardail D, Popa I, Bodennec J, Louisot P, Schmitt D, et al. (2003) The mitochondria-associated endoplasmic-reticulum subcompartment (MAM fraction) of rat liver contains highly active sphingolipid-specific glycosyltransferases. The Biochemical journal 371: 1013–1019.

56. Futerman AH, Pagano RE (1991) Determination of the intracellular sites and topology of glucosylceramide synthesis in rat liver. Biochem J 280: 295–302.

57. Jeckel D, Karrenbauer A, Burger KNJ, van Meer G, Wieland FT (1992) Glucosylceramide is synthesized at the cytosolic surface of various golgi subfractions. J Cell Biol 117: 259–267.

58. Meyer SG, de Groot H (2003) [14C]serine from phosphatidylserine labels ceramide and sphingomyelin in L929 cells: evidence for a new metabolic relationship between glycerophospholipids and sphingolipids. Arch Biochem Biophys 410: 107–111.

59. Lahiri S, Futerman AH (2005) LASS5 is a bona fide dihydroceramide synthase that selectively utilizes palmitoyl-CoA as acyl donor. J Biol Chem 280: 33735–33738.

60. Pewzner-Jung Y, Ben-Dor S, Futerman AH (2006) When do lasses (longevity assurance genes) become CerS (ceramide synthases)?: Insights into the regulation of ceramide synthesis. J Biol Chem 281: 25001–25005.

61. Stiban J, Tidhar R, Futerman AH (2010) Ceramide synthases: roles in cell physiology and signaling. Adv Exp Med Biol 688: 60–71.

62. Tidhar R, Futerman AH (2013) The complexity of sphingolipid biosynthesis in the endoplasmic reticulum. Biochim Biophys Acta 1833: 2511–8251.

15

Maintenance or Collapse: Responses of Extraplastidic Membrane Lipid Composition to Desiccation in the Resurrection Plant *Paraisometrum mileense*

Aihua Li[2,3]⑨**, Dandan Wang**[2]⑨**, Buzhu Yu**[2]**, Xiaomei Yu**[2]**, Weiqi Li**[1,2]*

1 Key Laboratory of Biodiversity and Biogeography, Kunming Institute of Botany, Chinese Academy of Science, Kunming, China, **2** Plant Germplasm and Genomics Center, Germplasm Bank of Wild Species, Kunming Institute of Botany, Chinese Academy of Sciences, Kunming, China, **3** University of Chinese Academy of Sciences, Beijing, China

Abstract

Resurrection plants usually grow in specific or extreme habitats and have the capacity to survive almost complete water loss. We characterized the physiological and biochemical responses of *Paraisometrum mileense* to extreme desiccation and found that it is a resurrection plant. We profiled the changes in lipid molecular species during dehydration and rehydration in *P. mileense*, and compared these with corresponding changes in the desiccation-sensitive plant *Arabidopsis thaliana*. One day of desiccation was lethal for *A. thaliana* but not for *P. mileense*. After desiccation and subsequent rewatering, *A. thaliana* showed dramatic lipid degradation accompanied by large increases in levels of phosphatidic acid (PA) and diacylglycerol (DAG). In contrast, desiccation and rewatering of *P. mileense* significantly decreased the level of monogalactosyldiacylglycerol and increased the unsaturation of membrane lipids, without changing the level of extraplastidic lipids. Lethal desiccation in *P. mileense* caused massive lipid degradation, whereas the PA content remained at a low level similar to that of fresh leaves. Neither damage nor repair processes, nor increases in PA, occurred during non-lethal desiccation in *P. mileense*. The activity of phospholipase D, the main source of PA, was much lower in *P. mileense* than in *A. thaliana* under control conditions, or after either dehydration or rehydration. It was demonstrated that low rates of phospholipase D-mediated PA formation in *P. mileense* might limit its ability to degrade lipids to PA, thereby maintaining membrane integrity following desiccation.

Editor: Jin-Song Zhang, Institute of Genetics and Developmental Biology, Chinese Academy of Sciences, China

Funding: This research was supported by grants from the National Natural Science Foundation of China (31070262), Kunming Institute of Botany (KSCX2-EW-J-24), the Germplasm Bank of Wild Species, the CAS Innovation Program of Kunming Institute (540806321211), as well as the 100-Talents Program of CAS. The funders had no role in study design, data collection and analysis, decision to publish, or preparation of the manuscript.

Competing Interests: The authors have declared that no competing interests exist.

* Email: weiqili@mail.kib.ac.cn

⑨ These authors contributed equally to this work.

Introduction

Drought is a major factor that limits plant growth and yield. In most parts of the world, drought continuously affects crop production and is of growing concern given the increasing demand for food production by the expanding global population [1,2]. Most crops are sensitive to drought, and except their seeds and pollen grains, their tissues cannot withstand water stress below 20% relative water content (RWC) [3]. However, a small group of so-called resurrection plants can tolerate extreme loss of water (desiccation) to 10% RWC or less [3]. Upon rewatering, the vegetative tissues of resurrection plants can quickly revive from the quiescent state that they enter upon loss of almost all of their free water [4]. Resurrection plants are excellent models to explore the physiological, biochemical, and molecular basis of desiccation tolerance [5]. A better understanding of the unique features of resurrection plants might benefit efforts to improve crop yields under conditions of water deficit.

Resurrection plants exhibit a series of distinct morphological, physiological, biochemical, and genetic protective mechanisms to resist or respond to extreme desiccation. The folding and re-expansion of leaves are the most obvious morphological changes that occur during desiccation and subsequent rewatering; folding might prevent the production of reactive oxygen species (ROS) induced by light during drying and rehydration [4,6–9]. Inward shrinking of the cell wall and dehydration-induced membrane shrinking are typical responses of resurrection plants to desiccation [5,6,10]. In these plants, photosynthetic activity is retained during mild drought, is lost during severe desiccation, and returns upon subsequent rehydration [7,11–14]. Resurrection plants do not necessarily share the same physiological strategies, and sometimes even employ completely opposite strategies to deal with extreme desiccation. For example, the osmoprotectant proline is widely used to resist cellular dehydration in plants. However, whereas some resurrection plants accumulate proline following desiccation, others do not [15–17]. In addition, some resurrection plants (poikilochlorophyllous species) lose their chlorophyll and degrade their thylakoid membranes to prevent the production of photosynthetically generated ROS during dehydration [9,18]. Other resurrection plants (homoiochlorophyllous species), such as *Craterostigma plantagineum* and *Haberlea rhodopensis*, retain their chlorophyll and thylakoid structures [19,20].

The ability of resurrection plants to maintain antioxidant activity even after severe cellular dehydration is thought to account in large part for their distinctive capacity to resist desiccation [9,21]. Osmoregulatory substances, such as sucrose, alleviate cellular dehydration and oxidative stress in resurrection plants [22–25]. Many genes that function in drought tolerance have been cloned from resurrection plants and characterized [26–29]. Transformation of certain plants with some of these genes improves drought resistance significantly [30]. The powerful approaches of transcriptomics [31,32], proteomics [33–36], and metabolomics [1] have enabled extensive investigation of the mechanisms that resurrection plants use to resist severe dehydration at the levels of global changes in gene expression and the abundances of proteins and metabolites. Lipid metabolism during and following desiccation was recently reported in *Craterostigma plantagineum* [37]. However, especially given that the tolerance strategies used by resurrection plants are often species-specific, little is known about how molecular species of membrane lipids respond to severe dehydration and subsequent rehydration, and how the changes in lipid profiles contribute to ability to survive extreme desiccation.

Maintenance of membrane integrity and fluidity is of critical importance to ensure that resurrection plants can survive cellular dehydration [5,38]. Several membrane components, such as phosphatidic acid (PA), sphingolipids, and sterols, have particular effects on membrane permeability [39–41]. A widely accepted speculation about the desiccation tolerance of resurrection plants is that although they experience membrane damage during dehydration, they can then repair this damage during their subsequent rehydration [42]. This suggests significant changes in their membrane lipids during both desiccation and rewatering. Three types of lipid change are often observed in resurrection plants during desiccation: 1) decreases in the plastidic lipid monogalac-tosyldiacylglycerol (MGDG), which is thought to suppress formation of the non-lamellar membrane phase [37,42–44]; 2) increases in fatty acid desaturation, which enhances membrane fluidity and thus favors dehydration resistance [44–46]; and 3) increases in PA, which is proposed to be an early signal of cellular dehydration and a structural feature of membrane injury [37,42,47]. Many metabolic enzymes regulate the lipid changes induced by cellular dehydration. Phospholipases hydrolyze phospholipids at different positions to produce lyso-phospholipids, diacylglycerol (DAG), or PA [48]. Phospholipases might be the most important enzymes in resurrection plants because they contribute to many aspects of dehydration-induced changes in membrane lipids. For example, the increases in the levels of PA and DAG that occur as levels of other phospholipids decrease during dehydration in *Craterostigma plantagineum* [37] suggest that phopholipases C and D act in response to dehydration. Dramatic increases in the abundances of lyso-phospholipids caused by freezing-induced cellular dehydration [49] suggest roles for phospholipase A and/or B in responses to dehydration. In particular, the role of phospholipase D-mediated PA formation has been extensively studied in processes related to cellular dehydration [49,50]. However, recent reports indicate that rather than increasing, levels of PA might even decline following desiccation in *A. thaliana* [51]. These findings suggest that there are still unknown responses of lipid changes to dehydration.

Paraisometrum mileense W. T. Wang is the only species in the monotypic genus *Paraisometrum* W. T. Wang of family Gesneriaceae. It had been thought to be extinct for one hundred years, but was rediscovered by botanists from the Chinese Academy of Sciences [52]. *P. mileense* is a stemless perennial herb that is 10–20 cm tall [53]. *P. mileense* grows over limestone on rocky outcrops, and in a seasonally arid subtropical region where half of the year is the dry season, and the other half is the wet season; about 88% of the annual rainfall occurs during the wet season, and only about 110 mm of rain falls during the dry season [54]. Generally, resurrection plants can survive extremely harsh environments and are usually found in habitats with sporadic rainfall. These include rocky outcrops and arid zones within tropical and subtropical areas [55]. In general, these plants are small [56,57]. Among the approximately 300 angiosperm resurrection species, more than two dozen belong to the family Gesneriaceae [58]. Although *P. mileense* is probably highly tolerant of water deficit, it has never been established if it is a resurrection plant.

The present study used physiological and biochemical analyses to demonstrate that *P. mileense* is a resurrection plant. We used lipidomic analyses based on electrospray ionization mass spectrometry (ESI-MS/MS) [49,59,60] to explore changes of membrane molecular species in response to dehydration and subsequent rehydration in *P. mileense*, and to compare these with similarly treated desiccation-sensitive *Arabidopsis thaliana* plants. We observed dramatic degradation of membrane lipids during both dehydration and rehydration, but the degradation was markedly different between two treatments in *A. thaliana*. Whereas plastidic lipids were sensitive to non-lethal dehydration in *P. mileense*, extraplastidic lipids were very stable to desiccation. Notwithstanding the dramatic degradation of lipids that occurred upon lethal dehydration of *P. mileense*, the changes differed markedly from those in *A. thaliana*. Whereas both of the major intermediates of lipid metabolism, PA and DAG, were involved in lipid changes in *A. thaliana*, only DAG increased in *P. mileense* during lethal dehydration and rehydration. The responses of fatty acid desaturation and acyl chain length to dehydration and rehydration were also examined in both plant species. We propose that desiccation tolerance might involve avoiding damage and the need for repair, as well as appropriate regulation of phopholipase activities.

Results

P. mileense tolerates extreme desiccation and is a resurrection plant

To test the desiccation tolerance of *P. mileense*, we reduced the RWC in mature plants by drying them for two, three, four, or five days, and then rehydrating them for 24 h (Figure S1A). The RWC of *P. mileense* seedlings decreased gradually during dehydration. After two days of desiccation, the RWC decreased to about 40%, and then decreased further with additional days of drying, reaching a RWC as low as 3.7%. After rewatering for one day, the seedlings with 3.7% RWC regained their initial RWC, 97.2% (Figure S1C). The leaves curled inward gradually during desiccation and became completely folded until only the abaxial surfaces of the leaves were in the outer whorl; finally, they lost nearly all of their water (Figure S1A, C). As this took place, F_v/F_m (maximal quantum efficiency of photosystem II in the dark adapted state) nearly reached zero (Figure S1A); this indicated that photosynthesis ceased almost completely. During the subsequent day of rehydration, the leaves spread and expanded until the leaves became fully unfolded; eventually, the RWC almost returned to its initial level, at which point the photosynthetic activity returned to the level prior to dehydration. The photosynthetic activity of *P. mileense* leaves recovered to its original level within 24 h after having decreased to a level of nearly zero. For comparison, we also tested the desiccation tolerance of *A. thaliana* under the same conditions. Its leaves quickly lost water and shrank in a random pattern during dehydration. These leaves exhibited a complete loss of chlorophyll fluorescence, which was not recovered during rehydration (Figure S1B). These results show that *P. mileense* could tolerate extreme loss of water below 10% RWC, but that the same conditions were lethal for *A. thaliana*. This demonstrates that *P. mileense* is a resurrection plant.

Generally, leaves of resurrection plants survive desiccation only when they are dried on the plant, although the detached leaves in some resurrection plants also tolerate dehydration [11,46,61,62]. To test whether the detached leaves of *P. mileense* could tolerate dehydration, we dried leaf discs of *P. mileense* and *A. thaliana* side-by-side for one day and then rehydrated them for one day (Figure 1). The RWC of *P. mileense* discs decreased to 52% and then returned to the initial levels; their photosynthetic activity decreased to 0.53 of F_v/F_m value and then recovered to the level associated with normal photosynthetic function. For *A. thaliana* discs, in contrast, RWC decreased to 9.8% and did not recover; their photosynthetic activity was lost completely (Figure 1). We also examined changes in the RWCs of *P. mileense* and *A. thaliana* discs during the processes of complete desiccation and subsequent rehydration (Figure S2); the patterns of change of RWC of leaf discs were the same as those in intact plants (Figure S1C). Given the demonstrated ability of detached *P. mileense*

leaves to be resurrected upon rehydration, our subsequent studies focused exclusively on leaf discs of *P. mileense*.

We further characterized the resurrection of *P. mileense* by examining the contents of proline, soluble sucrose, and chlorophyll, as well as the changes in membrane lipid peroxidation and ion leakage during dehydration and subsequent rehydration. These were compared with those of similarly treated *A. thaliana* leaf discs. Slight accumulation of sucrose occurred in *P. mileense*, whereas the proline content remained low and unchanged during both dehydration and rehydration (Figure 2A, B). These findings are the same as those for other resurrection plants [11,22,24]. The changes in the peroxidation of membrane lipids, which were indicated by the amount of malondialdehyde (MDA) and ion leakage, were significantly lower in *P. mileense* than in *A. thaliana* (Figure 2C, D). These findings indicate that *P. mileense* suffered less damage than *A. thaliana* under the same stress conditions. These lines of evidence affirm the resurrection of *P. mileense* upon rehydration after extreme desiccation. The amounts of chlorophyll (a+b) and carotenoid (c+x) of *P. mileense* leaves decreased by half in comparison with those of the control under dark desiccation, but then returned to the control level after rewatering (Table 1). This means that *P. mileense* is a homoiochlorophyllous resurrection plant.

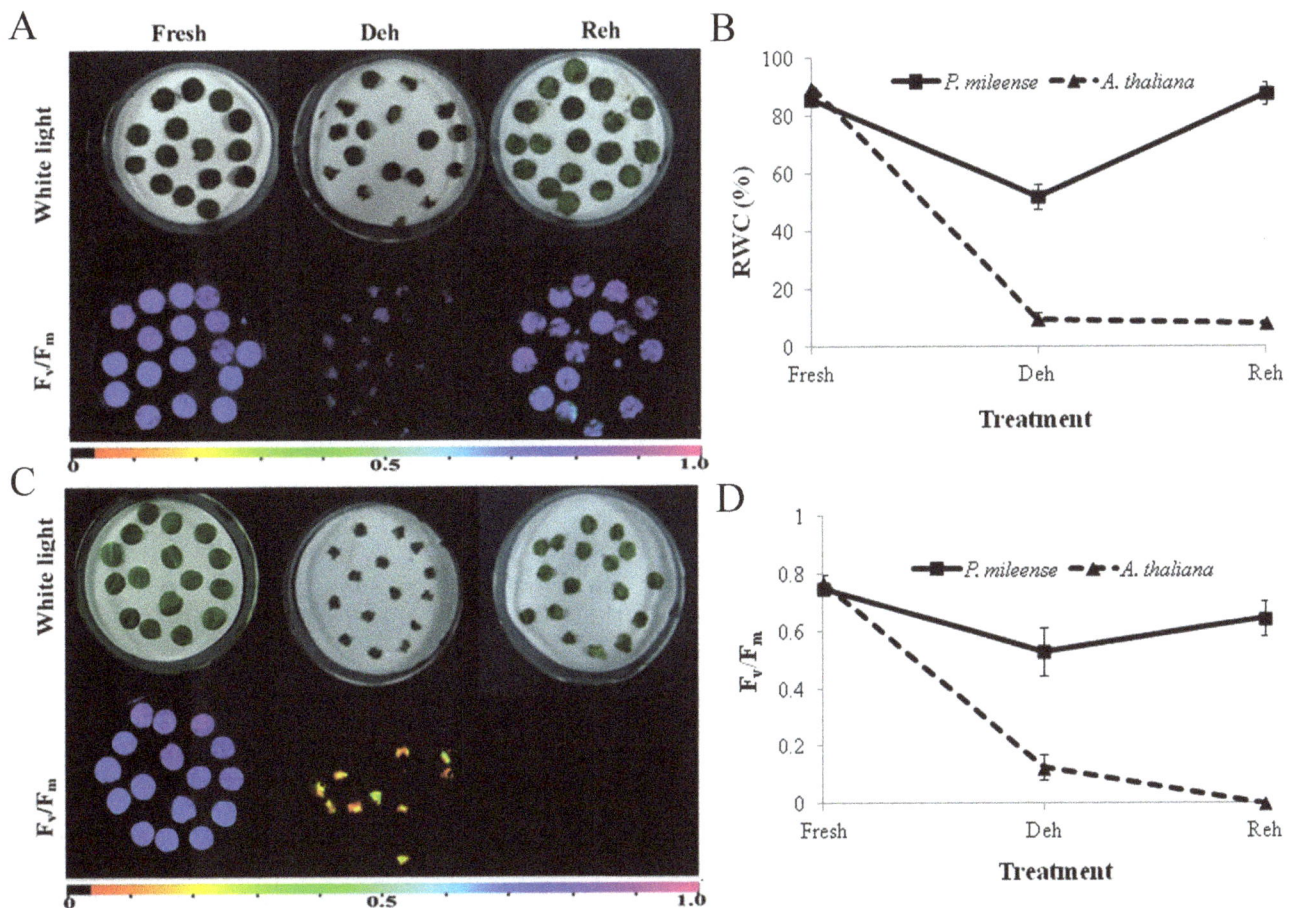

Figure 1. Dehydrated (Deh) and rehydrated (Reh) leaf discs of (A) *P. mileense* **and (B)** *A. thaliana.* White coloration (upper picture) or low F_v/F_m values for variable fluorescence (lower picture). The color bar at the bottom indicates F_v/F_m values. (C) Relative water content (RWC). (D) F_v/F_m. Values are means $\pm$ standard deviation ($n = 4$ or 5).

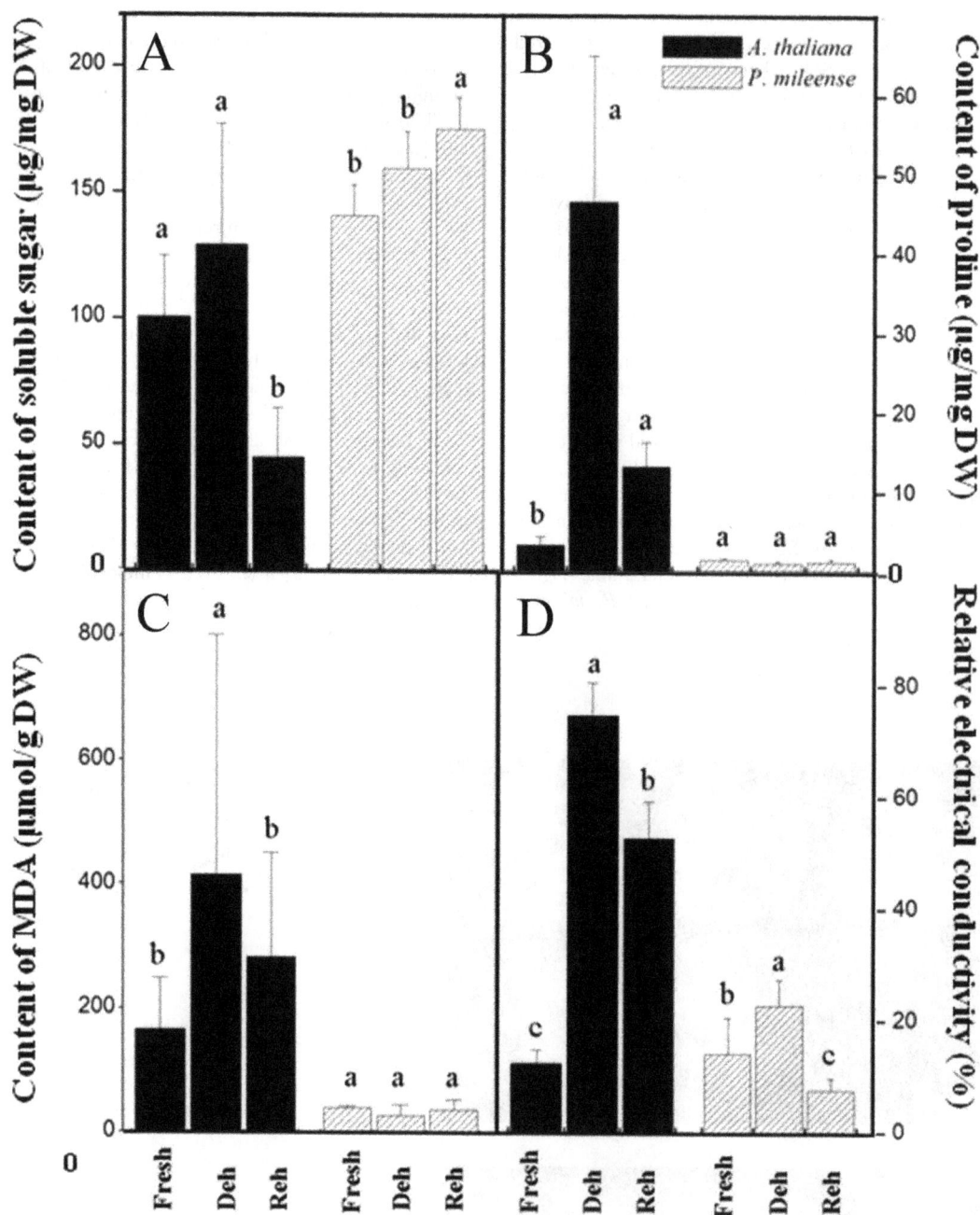

Figure 2. Changes in (A) soluble sugar, (B) proline, (C) malondialdehyde (MDA), and (D) relative electricity conductivity during dehydration (Deh) and rehydration (Reh) in *P. mileense* and *A. thaliana* leaves. Values are means ± standard deviation (n = 4 or 5). Values in the same bar type with different letters indicate that values are significantly different (P < 0.05).

Comparative profiling of molecular species of membrane lipids in leaves of *P. mileense* and *A. thaliana* under the same conditions of dehydration and rehydration

To analyze the changes in membrane lipids during desiccation, we first profiled the lipid molecular species of *P. mileense* and *A. thaliana* under control conditions (fresh leaves) and then after parallel treatment with dehydration for one day and subsequently rehydration for one day (Figure 1). More than 180 molecular species belonging to 12 classes of glycerolipids were quantified. The glycerolipids included six classes of phospholipids: phosphatidylglycerol (PG), phosphatidylcholine (PC), phosphatidylethanol-

amine (PE), phosphatidylinositol (PI), phosphatidylserine (PS), and PA; two classes of galactolipids: MGDG and digalactosyldiacylglycerol (DGDG); three classes of lyso-phospholipids: LysoPG, LysoPC, and LysoPE; and one class of neural glycerolipid: DAG. Every molecular species was presented as "acyl carbon atoms: double bonds" [63]. The changes in both the contents (nmol/mg dry weight, Figure 3A) and the composition (mol%, Figure 3B) of the lipids were subjected to clustering analysis. Total lipid amounts of each head-group class are shown in Table 2 and Table S1.

An overview of the lipid contents (Figure 3A) and compositions (Figure 3B) indicated four trends of lipid levels in the response to

Table 1. Changes of pigments during dehydration (Deh) and rehydration (Reh) in *P. mileense* and *A. thaliana* leaves.

Pigment Class	Species	Pigment content (µg/mg)		
		Fresh	Deh	Reh
Chl a	A. thaliana	9.26 ± 0.55^b	7.57 ± 0.42^c	10.76 ± 1.67^a
	P. mileense	4.98 ± 0.39^a	2.92 ± 0.99^b	4.74 ± 0.67^a
Chl b	A. thaliana	3.10 ± 0.19^b	2.49 ± 0.17^c	4.31 ± 0.62^a
	P. mileense	1.83 ± 0.23^a	0.88 ± 0.14^b	1.88 ± 0.27^a
Chl a/b	A. thaliana	2.99 ± 0.02^a	3.04 ± 0.05^a	2.50 ± 0.04^b
	P. mileense	2.73 ± 0.18^{ab}	2.85 ± 0.11^a	2.54 ± 0.01^b
Chl a+b	A. thaliana	12.36 ± 0.74^b	10.07 ± 0.58^c	15.07 ± 2.29^a
	P. mileense	6.81 ± 0.60^a	3.39 ± 0.58^b	6.62 ± 0.94^a
Carotenoid	A. thaliana	1.97 ± 0.10^b	1.79 ± 0.09^b	2.32 ± 0.23^a
	P. mileense	1.06 ± 0.10^a	0.62 ± 0.09^b	1.13 ± 0.07^a

Values in the same row with different letters are significantly different ($P<0.05$). Values are means $\pm$ SD ($n = 4$ or 5).

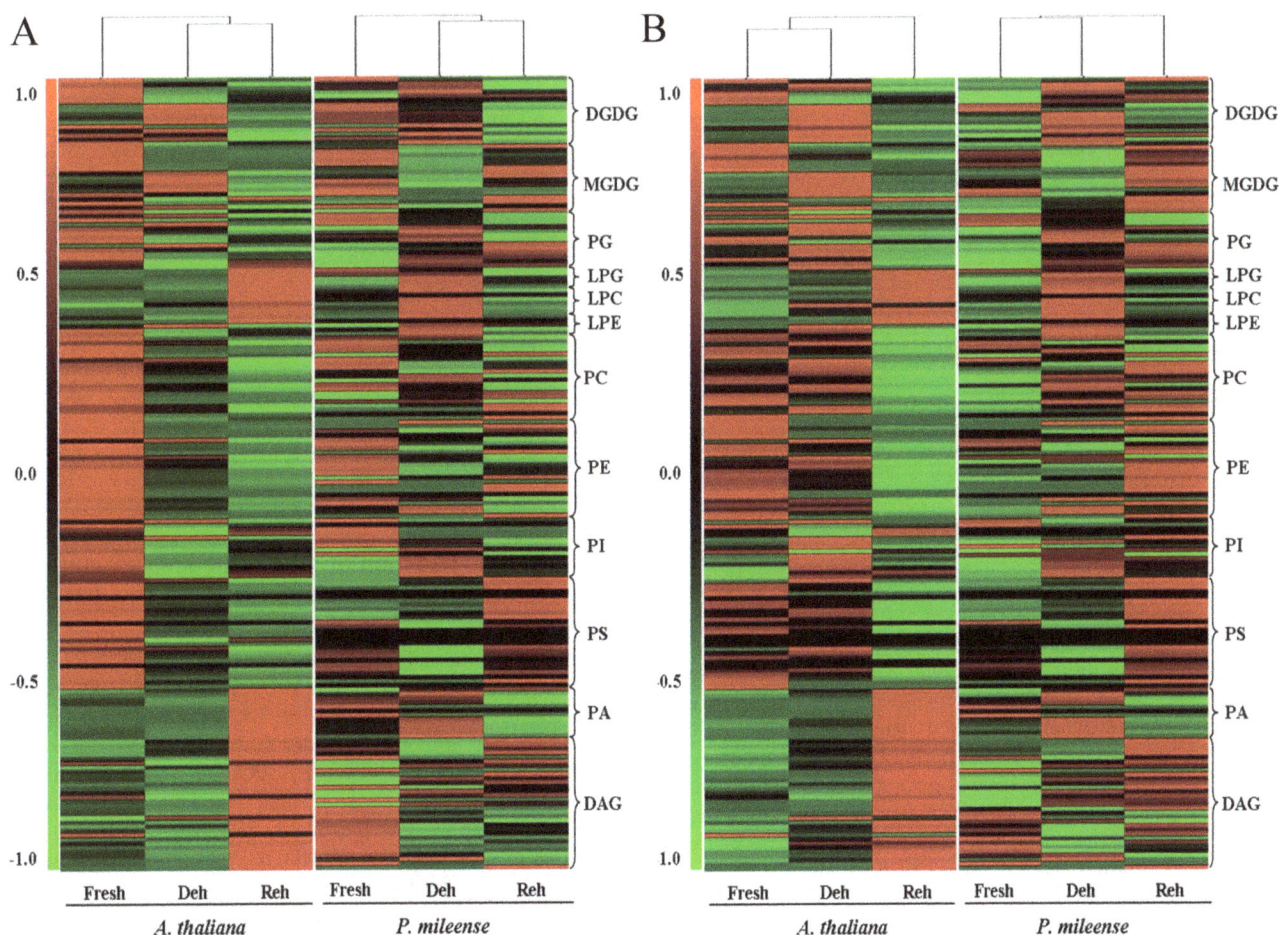

Figure 3. Hierarchical clustering analysis of lipid molecular species during dehydration (Deh) and rehydration (Reh) of *P. mileense* and *A. thaliana* leaves. (A) Contents (nmol/mg dry weight) of lipid molecular species. (B) Compositions (mol %) of lipid molecular species. The colored bar within a column represents the lipid molecular species in the corresponding plants and treatments. The color of each bar represents the abundance of the indicated lipid species, which is expressed relative to the change from the mean center of each lipid species within all treatments. Lipid species in the corresponding lipid classes were sorted using class (as indicated), total acyl carbons (within a class), and total double bonds (with class and total acyl carbons) in ascending order. Values are means ($n = 4$ or 5).

Table 2. Amount of lipid in each head-group class and total polar lipid during dehydration (Deh) and rehydration (Reh) of *P. mileense* and *A. thaliana* leaves.

Lipid class	Species	Lipid/dry weight (nmol/mg)			RC(F–D)(%)	RC(D–R)(%)
		Fresh	Deh	Reh		
DGDG	A. thaliana	54.35±2.94[a]	22.19±5.96[b]	4.61±1.15[c]	−59	−79
	P. mileense	23.37±0.63[a]	25.37±2.23[a]	19.62±2.42[b]	-	−23
MGDG	A. thaliana	261.4±15.8[a]	42.05±10.72[b]	5.62±0.56[c]	−84	−87
	P. mileense	58.19±2.32[a]	37.3±5.08[b]	37.34±6.98[b]	−36	-
PG	A. thaliana	23.95±2.44[a]	4.08±1.28[b]	3.02±0.18[b]	−83	-
	P. mileense	3.50±0.43[a]	3.90±0.12[a]	3.30±0.27[b]	-	−15
PC	A. thaliana	26.64±1.91[a]	5.87±1.70[b]	0.36±0.15[c]	−78	−94
	P. mileense	6.50±0.71[a]	5.82±1.20[b]	5.42±1.18[b]	−10	-
PE	A. thaliana	5.75±0.47[a]	0.95±0.32[b]	0.01±0.00[c]	−84	−99
	P. mileense	1.15±0.16[a]	0.93±0.15[b]	0.87±0.08[b]	−19	-
PI	A. thaliana	7.00±0.30[a]	2.93±0.15[b]	1.82±0.31[c]	−58	−38
	P. mileense	2.61±0.27[a]	2.62±0.23[a]	2.15±0.37[b]	-	−18
PS	A. thaliana	0.72±0.15[a]	0.14±0.06[b]	0.00±0.00[c]	−80	−97
	P. mileense	0.14±0.04[b]	0.09±0.02[c]	0.21±0.08[a]	−38	142
PA	A. thaliana	1.17±0.13[b]	0.43±0.10[b]	6.18±2.22[a]	-	1332
	P. mileense	0.50±0.05[a]	0.57±0.07[a]	0.31±0.08[b]	-	−45
LPG	A. thaliana	0.25±0.06[b]	0.33±0.12[b]	4.04±3.42[a]	-	1140
	P. mileense	0.05±0.01[a]	0.06±0.02[a]	0.03±0.02[b]	-	−47
LPC	A. thaliana	0.09±0.01[b]	0.10±0.01[b]	0.27±0.08[a]	-	179
	P. mileense	0.04±0.01[b]	0.09±0.03[a]	0.03±0.00[b]	127	−69
LPE	A. thaliana	0.10±0.01[a]	0.03±0.01[b]	0.03±0.01[b]	−70	-
	P. mileense	0.03±0.01[b]	0.06±0.02[a]	0.03±0.01[b]	119	−54
DAG	A. thaliana	14.71±7.22[b]	26.10±7.47[b]	102.8±26.4[a]	-	294
	P. mileense	6.80±1.31[a]	9.75±1.77[a]	8.44±2.70[a]	-	-
Total polar lipid	A. thaliana	381.4±20.4[a]	78.85±19.43[b]	26.73±6.80[c]	−79	−66
	P. mileense	92.80±9.10[a]	77.08±8.33[b]	69.41±10.86[b]	−17	-

The percentage relative change in lipids of dehydration RC (F–D) is the value for the difference between the values of Fresh and Deh discs, divided by the value of Fresh discs; that of rehydration RC (D–R) is the value for the significant difference between the values of Deh and Reh discs, divided by the value of Deh discs. Values in the same row with different letters are significantly different (P<0.05). Values are means ± standard deviation (n=4 or 5).

dehydration and rehydration. 1) There were clear differences between *P. mileense* and *A. thaliana*. These included differences between their fresh leaves and between their leaves subjected to dehydration and rehydration. 2) The changes in *P. mileense* were much smaller than those in *A. thaliana*. 3) Clustering of the lipid contents of dehydrated and rehydrated leaves (Figure 3A) suggested that the treatments caused significant changes in the relative abundances of lipids. 4) The lipid compositions of fresh and dehydrated leaves clustered together (Figure 3B); in other words, both species showed similar changes in lipid compositions after rehydration when these were compared with those of fresh leaves. This indicates the complexity of lipid metabolism in *P. mileense*, and that rewatering might not have simply restored the lipid composition back to that found in fresh leaves prior to dehydration. By contrast, in *A. thaliana*, major membrane damage occurred during rehydration but not during dehydration. The overall abundance of membrane lipids in *P. mileense* was significantly lower than that of *A. thaliana* (Table 2). This is consistent with a previous report that the lipid contents of resurrection plants are usually low [64]. The results of detailed data mining are presented in subsequent sections.

Differential degradation of membrane lipids occurred during lethal dehydration and subsequent rehydration of *A. thaliana*

After dehydration of *A. thaliana* leaves to 10% RWC, we found that the levels of seven of the twelve classes of lipid studied (including MGDG, PG, PC, PE, PI, PS, and LysoPE), decreased to less than half of their levels prior to drying; the abundances of four classes of lipid (PA, DAG, LysoPG, and LysoPC) remained unchanged; and one class of lipid (DGDG) became more abundant (Table 2). The increase in DGDG levels might have been due to galactosylation of MGDG; this would be consistent with the responses of other species to desiccation [37]. Upon subsequent rehydration, we found that the lipids that degraded during dehydration continued to decrease, and that the rates of degradation increased. Their relative changes during rehydration were greater than those during dehydration (Table 2). In contrast, lipid levels that were maintained during dehydration increased dramatically. Among these, PA and DAG increased 13.3-fold (from 0.44 to 6.18 nmol/mg) and 3.9-fold (from 26.1 to 102.76 nmol/mg), respectively (Table 2).

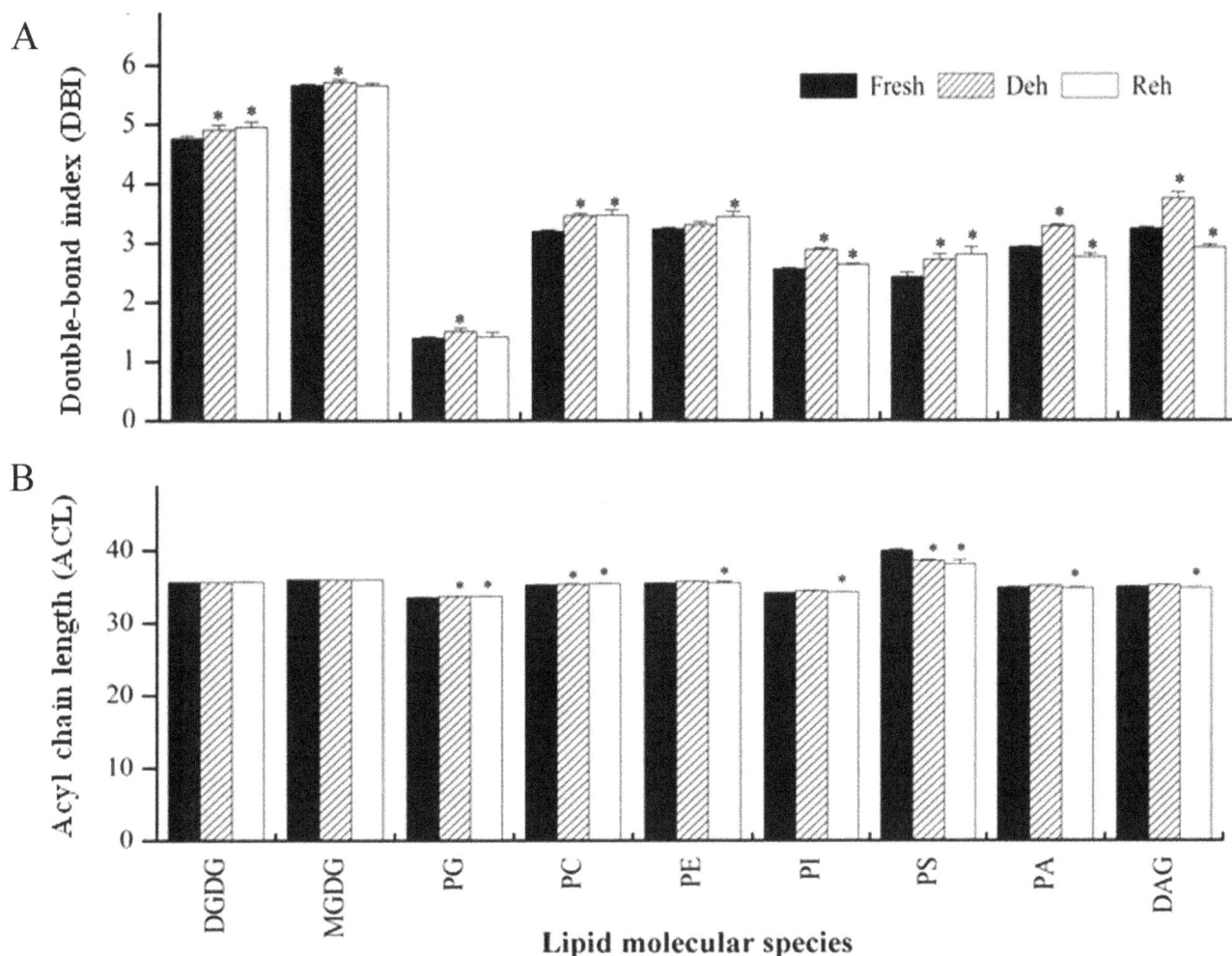

Figure 4. Changes in the (A) double-bond index (DBI) and (B) acyl chain length (ACL) of membrane lipids during dehydration (Deh) and rehydration (Reh) in *P. mileense* leaves. $\mathrm{DBI} = (\sum[\mathrm{N} \times \mathrm{mol\%lipid}])/100$, where N is the total number of double bonds in the two fatty acid chains of each glycerolipid molecule. $\mathrm{ACL} = (\sum[\mathrm{N} \times \mathrm{mol\%lipid}])/100$, where N is the total number of C atoms in the lipid molecular of each glycerolipid molecule. Bars with asterisks are significantly different ($P < 0.05$). Values are means ± standard deviation (n = 4 or 5).

To further examine which lipids were degraded, we compared acyl structures among the molecular species during rehydration to monitor potential turnover reactions [63]. We found that all degraded molecular species corresponded to the increases of PA molecular species with the same acyl structures, except for the individual PS molecular species that were present at very low levels (Figure S3B). For example, the decrease of 34:6 MGDG specifically corresponded to the increase of 34:6 PA; this suggests that 34:6 MGDG was converted to 34:6 PA. This is consistent with our previously described observations following freezing-induced dehydration of *A. thaliana* [49]. Interestingly, however, the acyl structures of increased DAG molecular species did not correspond to those of any other decreased lipid molecular species. This suggests that DAG was not directly derived from the major membrane lipids. These results indicate that membrane lipids of *A. thaliana* are degraded dramatically during both desiccation and rewatering, but that their degradation patterns varied and that most degradation occurred during rewatering. The results also suggest that the membrane lipids were degraded directly through PA and indirectly through DAG.

Extraplastidic membrane lipids essentially remained unchanged during dehydration and rehydration in *P. mileense*

Desiccation of *P. mileense* to 52.1% RWC (Figure 1C) caused only a small (17%) decrease in total membrane lipids (from 97.80 to 77.02 nmol/mg), which was much less than that observed in *A. thaliana* given the same treatment (Table 2). This change was mainly attributable to the decrease in the plastidic lipid MGDG (Table 2), whereas the abundances of other lipids remained essentially unchanged. In terms of the same acyl structures of molecular species, we found that changes in the levels of degraded MGDG corresponded to increases in the levels of DGDG and DAG (Figure S3A). This suggests the possible involvement of the lipid metabolism reaction mediated by SENSITIVE TO FREEZING 2 (SFR2), a galactolipid:galactolipid galactosyltransferase that reduces levels of the plastidic lipid MGDG. The decrease in the level of MGDG also increased the DGDG/MGDG ratio (Table S1), which favored plant survival following the imposition of water deficit [65]. The sensitivity of MGDG levels in *P. mileense* to drought is consistent with observations in other species [37,42–44,66]. However, the failure of levels of PA and DAG to increase

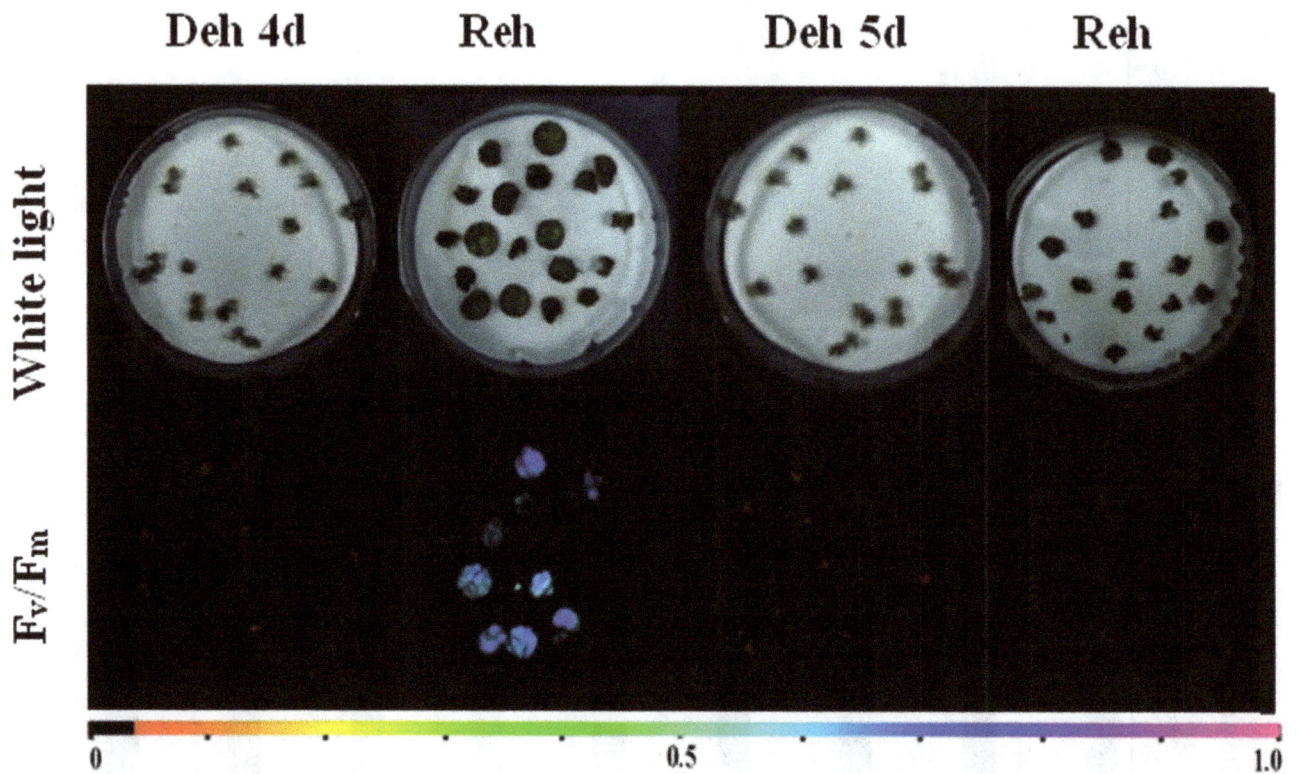

Figure 5. Half-lethal dehydration (Deh 4 d), lethal dehydration (Deh 5 d), and subsequent rehydration (Reh) of *P. mileense* **leaves.** White coloration (upper picture) or low F_v/F_m values for variable fluorescence (lower picture). The color bar at the bottom indicates F_v/F_m values.

substantially upon dehydration differs from observations made for other resurrection plants [37,42,47]. The absence of a change in the PC/PE ratio in *P. mileense* (Table S1) also distinguishes drought adaptation in *P. mileense* from the process in other plants [43,67,68]. During rewatering, the levels of both plastidic and extraplastidic lipids were maintained (Table 2), except for certain individual lipids, such as PA and PS. Levels of PA decreased very little, and basically remained at a very low level close to those found in fresh leaves (Table 2). The MGDG level remained low and exhibited no recovery following rehydration. These phenom-

Table 3. Lipid content in each head-group class and total polar lipid during half-lethal (Deh 4 d) or lethal dehydration (Deh 5 d) and rehydration for one day after dehydration for five days (Reh) in *P. mileense* leaves.

Lipid class	Lipid/dry weight (nmol/mg)		
	Deh 4 d	Deh 5 d	Reh
DGDG	25.27 ± 1.62^a	21.10 ± 5.79^a	12.28 ± 1.99^b
MGDG	15.51 ± 1.62^a	13.31 ± 2.98^a	6.48 ± 1.46^b
PG	4.15 ± 0.69^a	3.4 ± 0.93^a	2.46 ± 0.32^b
PC	4.86 ± 0.61^a	4.84 ± 1.38^a	0.51 ± 0.15^b
PE	0.69 ± 0.10^a	0.70 ± 0.37^a	0.04 ± 0.03^b
PI	3.83 ± 0.38^a	2.70 ± 0.78^b	0.88 ± 0.23^c
PS	0.04 ± 0.02^a	0.05 ± 0.03^a	0.01 ± 0.01^b
PA	0.73 ± 0.20^a	0.52 ± 0.31^a	0.52 ± 0.13^a
LPG	0.10 ± 0.02^c	0.29 ± 0.07^a	0.19 ± 0.07^b
LPC	0.35 ± 0.04^a	0.28 ± 0.08^b	0.04 ± 0.02^c
LPE	0.10 ± 0.02^a	0.07 ± 0.04^a	0.01 ± 0.00^b
DAG	31.30 ± 4.08^a	16.94 ± 5.41^b	28.15 ± 5.87^a
Total polar lipid	57.18 ± 6.81^a	47.36 ± 12.32^a	23.43 ± 4.21^b

Values in the same row with different letters are significantly different ($P<0.05$). Values are means $\pm$ standard deviation ($n = 4$ or 5).

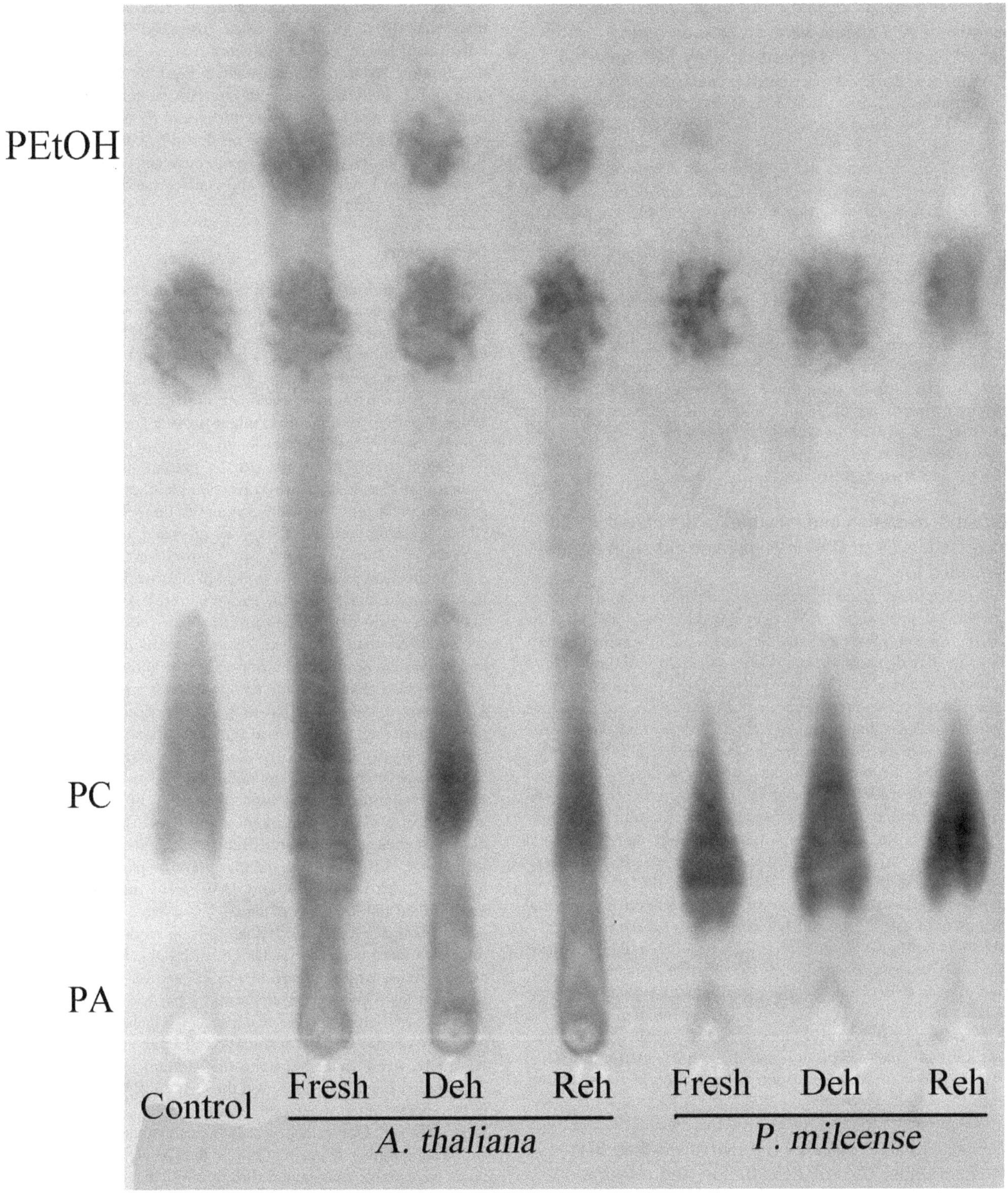

Figure 6. Transphosphatidylation activities of *A. thaliana* **and** *P. mileense.* Reaction products were separated by thin-layer chromatography and monitored by UV colorimetric analysis.

ena differ from those in other resurrection plants after rewatering [37]. Overall, these results indicate that levels of MGDG decreased in *P. mileense* in response to desiccation, but that levels of extraplastidic membrane lipids were maintained throughout the process of dehydration and rehydration; it also suggested that the immediate recovery of MGDG during rewatering was not required to restore viability after desiccation.

Membrane lipids became more unsaturated and the lengths of acyl chains were maintained during dehydration and subsequent recovery in *P. mileense*

Membrane fluidity is determined primarily by the level of lipid unsaturation [42] and the acyl chain lengths (ACL) [69] of the lipids. We calculated the double bond index (DBI) and ACL of membrane lipids to determine whether membrane fluidity responds to dehydration and rehydration in *P. mileense* and, if it does, to investigate how this occurs. During dehydration, the DBI of all lipid classes increased significantly (Figure 4A). For example, the DBI of DGDG increased 3.2%, from 4.76 to 4.91, and the DBI of PC increased 8.5%, from 3.19 to 3.46 (Table S2). During rehydration, the DBIs of DGDG, PC, PE, and PS remained high, and the DBIs of PI, PA, and DAG returned to the levels observed in the control (Figure 4A). The ACLs of lipids remained unchanged under most conditions, but some of them increased or decreased slightly during dehydration or rehydration (Figure 4B). These results show that *P. mileense* enhanced membrane fluidity by increasing the level of unsaturation of membrane lipids in order to deal with dehydration. These findings are consistent with observations in other plants, such as *Sporobolus stapfianus* [44,45] and *Boea hygroscopica* [46].

Lethal dehydration and rehydration dramatically increased levels of DAG in *P. mileense* although PA levels remained low

We then investigated the responses of membrane lipids to lethal dehydration in *P. mileense*. We dehydrated leaf discs of *P. mileense* for four days and five days, and then rehydrated them for one day. The four-day dehydration caused more than 70% of the leaf discs to lose their ability to revive, whereas all discs lost their capacity to revive after a five-day dehydration treatment (Figure 5). These findings indicate that persistent dehydration for four days causes lethal damage in the leaf discs of *P. mileense*. We found that, except for PA and DAG, all membrane lipids underwent significant degradation (Table 3). For example, the MGDG content decreased to 6.48 nmol/mg (Table 3), which was only about 10% of the level in fresh discs (Table 2), and PC content decreased to 0.51 nmol/mg (Table 3), which was also only about 10% of that of the fresh discs (Table 2). The PA level remained stable at a low level close to that of discs not subjected to a lethal treatment (Table 2); this differs from findings for the resurrection plant *Craterostigma plantagineum* [37]. In contrast, the DAG level increased dramatically following dehydration, reaching 2.5-fold that of fresh discs (Table 2). These results indicate that membrane deterioration occurred during lethal dehydration, and suggest that lipid degradation occurred through DAG pools and did not involve PA pools. This means that the patterns of lethal dehydration-induced degradation in *P. mileense* were distinct from those in *A. thaliana*.

PLD activity in *P. mileense* is much lower than that in *A. thaliana* under control, dehydration, and rehydration conditions

Given that PLD-mediated hydrolysis makes the greatest contribution to PA formation during water stress [37,42,47,48], we examined PLD activity to explore the reason why *P. mileense* maintains low levels of PA. We used thin-layer chromatography (TLC) to measure the levels of phosphatidylethanol (PEtOH), which is derived from a PLD-specific transphosphatidylation reaction in the presence of PC and ethanol [70]. Proteins isolated from fresh, dehydrated, and rehydrated leaves of *A. thaliana* produced significant amounts of PEtOH (Figure 6). Less PEtOH

was derived from the protein extracted from dehydrated leaves than that from protein extracts prepared from fresh and rehydrated leaves of *A. thaliana*. These data indicate that dehydration reduced the substantial level of PLD activity in hydrated *A. thaliana* tissues. In contrast, proteins isolated from fresh, dehydrated, and rehydrated leaves of *P. mileense* produced very much less PEtOH than was produced by comparably treated *A. thaliana* leaves. The results were consistent with the observed changes in PA levels (Table 2), and demonstrated that *P. mileense* was absent in PLD activity.

Discussion

The mechanisms responsible for the remarkable responses of resurrection plants to desiccation, particularly their changes of membrane lipids, have been reported extensively. However, the diversity of their habitats means that these plants likely adopt different strategies to adapt to desiccation. The novel and distinctive mechanisms of tolerance anticipated for plants indigenous to certain specific areas might provide considerable insight into the mechanisms of adaptation to environments where extreme desiccation occurs frequently. In the present study, we demonstrated that *P. mileense* is a resurrection plant. *P. mileense*, which grows on rocky outcrops with a six-month dry season in southwest China (a region famous for its plant diversity), could revive after desiccation below 10% RWC and showed several physiological and biochemical phenomena typical of resurrection plants. These include progressively inwardly curled leaves, less ion leakage than that observed in desiccation-sensitive plants, more soluble sugar accumulation than that observed in desiccation-sensitive plants, and no proline accumulation during dehydration.

We profiled changes in the composition of membrane lipids of *P. mileense* under non-lethal and lethal desiccation and subsequent rewatering. We also used the desiccation-sensitive plant *A. thaliana* for comparison in terms of both physiological and biochemical analyses. During non-lethal desiccation and subsequent rewatering, *P. mileense* responded by decreasing the abundance of MGDG and increasing the level of lipid unsaturation. Nonetheless, levels of its extraplastidic lipids remained largely unchanged; this response might prevent plasma membrane leakage. In particular, PA and DAG were maintained at low levels similar to those of fresh plants. Upon lethal desiccation, lipid composition decreased substantially owing to dramatic degradation, with large decreases in DGDG, MGDG, PE, PC, PS, PG, and PI and a large increase in DAG; however, the PA content remained low. The level of desiccation that was non-lethal to *P. mileense* was lethal for *A. thaliana*, in which the lipids were massively degraded. The degradation of lipids upon rehydration was more severe than that upon dehydration in *A. thaliana*; all degradation might have occurred through the PA and DAG pools in this species. Interestingly, there was no evidence of PLD activity in *P. mileense*. Our evidence thus indicates that *P. mileense* has two distinguishing features. One is that levels of extraplastidic lipids were stably maintained during non-lethal desiccation. The other is that PA was not involved in the process of lipid degradation, even following lethal membrane damage. These distinctive features might contribute to the capacity of *P. mileense* for resurrection upon rehydration after extreme desiccation.

Tolerance and avoidance are two basic strategies by which plants resist environmental stresses. The model of tolerance in resurrection plants was previously described as a two-step process [42]: destruction during desiccation and recovery from this destruction during rewatering. At cellular levels, the two-step process is like that the membrane lipid composition was damaged and then was reconstituted subsequently. A good example is provided in a recent report on *Craterostigma plantagineum* [37], in which membrane lipids changed during desiccation and returned to the levels observed in controls after rehydration. Changes in PA content were representative, increasing seven-fold and then quickly returning close to normal during dehydration and rehydration, respectively [37]. According to this model, resurrection depends mainly on the capacity of plants to repair the dehydration-induced damage rapidly during rehydration. In contrast, *P. mileense*, the species identified as a resurrection plant in the current study, uses tolerance to resist desiccation by maintaining the composition of the plasma membrane under severe dehydration (Table 2). Such desiccation is lethal to non-resurrection plants (Table 3). Thus, this study has revealed a completely novel model for tolerance in resurrection plants, in which the plasma membrane is maintained through severe dehydration. Damage is not repaired, and if dehydration becomes extreme, the plasma membrane collapses.

This "maintenance or collapse" model of desiccation tolerance is not only of interest to understand how *P. mileense* uses variations in the levels of extraplastidic lipids to deal with desiccation. The current study also raises questions regarding the underlying biochemical mechanisms and whether the effectiveness of the response is linked to the failure of *P. mileense* to increase the PA content, even following massive lipid degradation and membrane deterioration. Phospholipases are major regulators of membrane-lipid metabolism. The significant increases in lyso-phospholipids (Tables 2 and 3) suggest the involvement of phospholipases A and B in the response of *P. mileense* to desiccation. Given that the acyl structures of the accumulated DAG molecules did not match those of the phospholipids degraded in *P. mileense* and *A. thaliana*, DAG was not derived from PLC-mediated phospholipid hydrolysis. Therefore, PLC might not function in the desiccation-induced lipid metabolism, at least not directly. PA pools are a key feature in the lipid degradation pathway. PLD-mediated PA generation is the major pathway of PA formation following water-deficit [37,42,47,71]. As such, a stable low PA content, even after membrane deterioration (Table 2), suggested that *P. mileense* had limited PLD activity, as demonstrated by the transphosphatidylation assay (Figure 6). Another question raised here is that whether maintenance of extraplastidic lipids is the cause or the consequence of desiccation tolerance in *P. mileense*. Our evidence strongly implies that it is a protective mechanism against desiccation. Given that a limit in the size of PA pools can block lipid degradation and thus resist stress-induced damages on the membranes [72], the absence of PLD activity should inhibit lipid degradation during desiccation in *P. mileense*. This might enable *P. mileense* to maintain its extraplastidic lipid composition during desiccation.

In *A. thaliana*, desiccation caused significant lipid changes, but rewatering was associated with extensive lipid degradation (Table 2). In addition, the changes in lipid composition during rewatering were more complicated than those during desiccation (Figure 3B). Given that lethal cellular damage occurred during desiccation, these lines of evidence suggest that lethal damage and the degradation of the most abundant lipids are not synchronous. The reasons for this could be that the consequences of lethal damage developed during rewatering and that severe dehydration not only impacted on the cellular membrane but also reduced enzyme activity, including the activity of lipolytic enzymes. The latter attenuates lipid degradation. Recovery of the activity of lipolytic enzymes, such as PLD, upon rewatering caused substantial lipid hydrolysis in *A. thaliana*.

In summary, we have demonstrated that *P. mileense* is a resurrection plant and that it has a distinct manner of adjusting its membrane lipid composition to tolerate desiccation. *P. mileense* could maintain its extraplastidic lipids until lethal damage occurred. A stable low level of PA, which is associated with limited PLD activity regardless of whether *P. mileense* is subjected to non-lethal or lethal desiccation, accounts for its capacity to maintain the composition of its plasma membranes in order to tolerate severe dehydration. This paper thus presents a novel model of tolerance in resurrection plants in which repair does not occur during rewatering.

Materials and Methods

Ethics statement

The habitat of *P. mileense* used in this study is neither a site of conservation of the natural environment nor a private location. Access to this area did not require permission, and the species is not protected by Chinese law when we collected them in January 2008.

Plant materials

Specimens of the desiccation-tolerant plant *P. mileense* were collected from their natural habitat in Southern Yunnan (24°35′13.8″N, 103°31′59″E, 1960 m alt.), China. There, the plants grow on rocky slopes under trees in a forest. Plants were harvested together with the soil around them. After collection, the *P. mileense* plants were cultivated along with *A. thaliana* seedlings in a chamber at 20–23°C, with a light intensity of 120 μmol m^{-2}s^{-1}, a 12-h photoperiod, and a relative humidity of 60%. All plants were given adequate water until the beginning of the experiments. Mature and fully expanded leaves were selected for tests. Leaf discs with diameters of 1.5 cm were made from the leaves. Leaf discs of *P. mileense* and *A. thaliana* were each dehydrated for one day by incubation at 15°C in air with a relative humidity of 15%. The dehydrated leaf discs were incubated on wet filter papers in petri dishes for one day at room temperature in the dark to allow them to recover from dehydration.

Measurement of physiological indices

After dehydration and rehydration treatments, the RWC of leaf discs was determined by weighing them before and after oven drying (105°C for 17 h) and expressed as the percentage water content in water-saturated discs. The RWC of the leaf discs was calculated according to the formula [73]:

RWC(%) =

$100 \times [(\text{fresh weight} - \text{dry weight})/(\text{saturated weight} - \text{dry weight})]$.

Chlorophyll fluorescence was analyzed using an imaging chlorophyll fluorometer, the MAXI-Imaging Pulse-Amplitude (PAM) instrument (Walz, Germany). After the leaf discs had adapted to darkness for 20 minutes, the maximal quantum yield of photosystem II (PS II) photochemistry was determined on the basis of the initial fluorescence level (F_0) and the maximal fluorescence level (F_m), and was expressed as $F_v/F_m = (F_m - F_0)/F_m$.

To determine solute leakage, leaf discs were soaked in 5 ml of double-distilled water and shaken at room temperature for 4 h before aliquots were taken for leachate measurements. Samples were then boiled for 20 min and shaken for an additional hour prior to measuring the maximum conductivity using a DDBJ 350 conductivity meter. The injury index was calculated according to the formula: injury index $\% = (C/T) \times 100$, where T and C represent the conductivity of the leachate after and before boiling treatment, respectively. To estimate lipid peroxidation (MDA), leaf discs were homogenized in 3 ml of 10% trichloracetic acid. The homogenates were centrifuged at 14,000 g for 10 min at 4°C. The MDA content was calculated as described previously [74].

The proline content was determined according to the methods described previously [75] by measuring the quantity of the colored product of proline with ninhydric acid. The absorbance was read at 512 nm. The amount of proline was calculated using a standard curve and expressed as μg/mg dry weight. The solute sugar content was measured by the quantity of the colored reaction product of solute sugar with anthrone. The absorbance was read at 625 nm. The solute sugar content was calculated using a standard curve and expressed as μg/mg dry weight.

Lipid extraction, ESI-MS/MS, and data processing

The process of lipid extraction, sample analysis, and data processing were performed as described previously, with minor modifications [49,60]. Briefly, the leaf discs were dropped into 2 ml of isopropanol containing 0.01% butylated hydroxytoluene (BHT) at 75°C. The tissue was extracted three additional times with chloroform:methanol (2:1) containing 0.01% BHT, with 12 h agitation each time. The remaining plant tissue was dried overnight at 105°C and weighed. Lipid samples were analyzed on a triple quadrupole MS/MS equipped for ESI (Kansas Lipidomics Research Center, http://www.k-state.edu/lipid/lipidomics). The lipids in each class were quantified by comparison with two internal standards of the class. Data processing was performed as described previously [60]. Five replicates of each sampling time were analyzed. The Q test was performed on the total amount of each lipid class with different head group, and outliers were removed [63]. The data were subjected to one-way analysis of variance (ANOVA) with SPSS 16.0. Statistical significance was tested by Fisher's least significant difference (LSD) method. The double-bond index (DBI) was calculated according to the formula: DBI $= (\Sigma[N \times \text{mol}\% \text{ lipid}])/100$, where N is the number of double bonds in each lipid molecule [76]. Acyl chain length (ACL) was calculated using a formula derived from the DBI calculation above: ACL $= (\Sigma[NC \times \text{mol}\% \text{ lipid}])/100$, where NC is the number of acyl carbon atoms in each lipid molecule.

Protein extraction and PLD activity assay

Protein extraction and assays of PLD activity followed a previous procedure with minor adjustment [70,77]. Leaf discs were ground into a fine powder under liquid nitrogen. The powder was transferred to a centrifuge tube that contained 0.5 ml of homogenization buffer (50 mM Tris-HCl/pH 7.5, 10 mM KCl, 1 mM EDTA, 2 mM dithiothreitol, and 0.5 mM phenylmethylsulfonyl fluoride). The homogenate was centrifuged at 4800 g for 10 min at 4°C. The supernatant was referred to as the soluble total proteins. The total protein content was determined as instructions (Bio-Rad) [78]. The transphosphatidylation reaction mixture contained 100 mM MES (pH 6.5), 25 mM $CaCl_2$, 0.5 mM SDS, 1% (v/v) ethanol, enzyme solution containing 30 μg total protein, and 2 mM PC (egg yolk) in a total volume of 150 μl. After adding enzyme solution, the mixture was incubated at 30°C while shaking continuously (100 rpm) for 30 min, and the reaction was stopped by adding 1 ml chloroform: methanol (2:1) with 0.01% butylatedhydroxytoluene and vortexing the mixture. The chloroform and aqueous phases were separated by adding

100 μl of 2 M KCl. The aqueous phase was transferred into another tube and re-extracted by addition of 0.5 ml of chloroform. The chloroform phases from both tubes were combined and dried. The dried lipid samples were dissolved in 20 μl chloroform, and then spotted onto a TLC plate (silica gel G). The plate was developed with chloroform: methanol: NH_4OH (6.5:3.5:0.2). The spots of PEtOH, PC, and PA were monitored by UV colorimetric analysis after spraying with the color-developing agent primuline [77]. The measurements were repeated twice.

Supporting Information

Figure S1 Dehydrated (Deh) and rehydrated (Reh) seedlings of (A) *P. mileense* and (B) *A. thaliana*. White coloration (upper row) or low F_v/F_m values for variable fluorescence (lower row). The color bar at the bottom indicates F_v/F_m values. (C) Relative water content (RWC) values of *P. mileense* and *A. thaliana* following exposure to different periods of dehydration and rehydration.

Figure S2 Changes in the RWC values of leaf discs during different periods of dehydration of *P. mileense* and *A. thaliana*.

Figure S3 Changes in the molecular species of membrane lipids following the dehydration (Deh) and rehydration (Reh) of (A) *P. mileense* and (B) *A. thaliana* leaves. Values are means ± standard deviation ($n = 4$ or 5).

Table S1 Molar percentage of lipids in each head-group class during dehydration (Deh) and rehydration (Reh) in *P. mileense* and *A. thaliana* leaves. The ratios of lipid mol% content of PC/PE and DGDG/MGDG are also shown at the bottom of the table. Values in the same row with different letters are significantly different ($P < 0.05$). Values are means ± standard deviation ($n = 4$ or 5).

Table S2 Double-bond index (DBI) values of membrane lipids during dehydration (Deh) and rehydration (Reh) in *P. mileense* and *A. thaliana* leaves. DBI $= (\sum[N \times \text{mol}\% \text{ lipid}])/100$, where N is the total number of double bonds in the two fatty acid chains of each glycerolipid molecule. The percentage relative change in DBI of dehydration RC (F–D) is the value for the difference between the values of Fresh and Deh discs, divided by the value of Fresh discs; that of rehydration RC (D–R) is the value for the significant difference between the values of Deh and Reh discs, divided by the value of Deh discs. Values in the same row with different letters are significantly different ($P < 0.05$). Values are means ± standard deviation ($n = 4$ or 5).

Acknowledgments

The authors would like to thank Mary Roth for the acquisition and processing of the ESI-MS/MS data , Dr Ruth Welti and Dr Michael Möller for their critical reading of this paper.

Author Contributions

Conceived and designed the experiments: WQL AHL. Performed the experiments: AHL DDW BZY. Analyzed the data: AHL WQL. Contributed reagents/materials/analysis tools: DDW WQL. Wrote the paper: WQL AHL XMY.

References

1. Melvin JO, Guo LN, Alexander DC, Ryals JA, Bernard WM, et al. (2011) A sister group contrast using untargeted global metabolomic analysis delineates the biochemical regulation underlying desiccation tolerance in *Sporobolus stapfianus*. Plant Cell 23: 1231–1248.

2. Boyer JS (1982) Plant productivity and environment. Science 218: 443–448.

3. Proctor MC, Pence VC (2002) Vegetative tissues: bryophytes, vascular resurrection plants and vegetative propagules. In: Black M, Pritchard H, editors. Desiccation and Survival in Plants: Drying without Dying. pp. 207–237.

4. Gaff DF (1989) Responses of desiccation tolerant 'resurrection' plants to water stress. In: Kreeb KH, Richter H, Hinckley TM, editors. Structural and Functional Responses to Environmental Stress. The Hague: SPB Academic Publishing. pp. 308.

5. Vicré M, Farrant JM, Driouich A (2004) Insights into the cellular mechanisms of desiccation tolerance among angiosperm resurrection plant species. Plant Cell Environ 27: 1329–1340.

6. Dalla Vecchia F, El Asmar T, Calamassi R, Rascio N, Vazzana C (1998) Morphological and ultrastructural aspects of dehydration and rehydration in leaves of *Sporobolus stapfianus*. Plant Growth Regul 24: 219–228.

7. Farrant JM (2000) A comparison of mechanisms of desiccation tolerance among three angiosperm resurrection plant species. Plant Ecol 151: 29–39.

8. Farrant JM, Vander Willigen C, Loffell DA, Bartsch S, Whittaker A (2003) An investigation into the role of light during desiccation of three angiosperm resurrection plants. Plant Cell Environ 26: 1275–1286.

9. Sherwin HW, Farrant JM (1998) Protection mechanisms against excess light in the resurrection plants *Craterostigma wilmsii* and *Xerophyta viscosa*. Plant Growth Regul 24: 203–210.

10. Vander Willigen C, Pammenter NW, Jaffer MA, Mundree SG, Farrant JM (2003) An ultrastructural study using anhydrous fixation of *Eragrostis nindensis*, a resurrection grass with both desiccation-tolerant and -sensitive tissues. Funct Plant Biol 30: 281–290.

11. Georgieva K, Maslenkova L, Peeva V, Markovska Y, Stefanov D, et al. (2005) Comparative study on the changes in photosynthetic activity of the homoiochlorophyllous desiccation-tolerant *Haberlea rhodopensis* and desiccation-sensitive spinach leaves during desiccation and rehydration. Photosynth Res 85: 191–203.

12. Georgieva K, Szigeti Z, Sarvari E, Gaspar L, Maslenkova L, et al. (2007) Photosynthetic activity of homoiochlorophyllous desiccation tolerant plant *Haberlea rhodopensis* during dehydration and rehydration. Planta 225: 955–964.

13. Pandey V, Ranjan S, Deeba F, Pandey AK, Singh R, et al. (2010) Desiccation-induced physiological and biochemical changes in resurrection plant *Selaginella bryopteris*. J Plant Physiol 167: 1351–1359.

14. Evelin RP, Mihailova G, Petkova S, Tuba Z, Georgieva K (2012) Differences in physiological adaptation of *Haberlea rhodopensis* Friv. leaves and roots during dehydration–rehydration cycle. Acta Physiol Plant 34: 947–955.

15. Gaff DF, McGregor GR (1979) The effect of dehydration and rehydration on the nitrogen content of various fractions from resurrection plants. Physiol Plantarum 21: 92–99.

16. Tymms MJ, Gaff DF (1979) Proline accumulation during water stress in resurrection plants. J Exp Bot 30: 165–168.

17. Gechev TS, Dinakar C, Benina M, Toneva V, Bartels D (2012) Molecular mechanisms of desiccation tolerance in resurrection plants. Cell Mol Life Sci 69: 3175–3186.

18. Ingle RA, Collett H, Cooper K, Takahashi Y, Farrant JM, *et al.* (2008) Chloroplast biogenesis during rehydration of the resurrection plant *Xerophyta humilis*: parallels to the etioplast–chloroplast transition. Plant Cell Environ 31: 1813–1824.

19. Strasser RJ, Tsimilli-Michael M, Qiang S, Goltsev V (2010) Simultaneous *in vivo* recording of prompt and delayed fluorescence and 820-nm reflection changes during drying and after rehydration of the resurrection plant *Haberlea rhodopensis*. BBA-Biomembranes 1797: 1313–1326.

20. Tuba Z, Protor CF, Csintalan Z (1998) Ecophysiological responses of homoiochlorophyllous and poikilochlorophyllous desiccation tolerant plants: a comparison and an ecological perspective. Plant Growth Regul 24: 211–217.

21. Jovanovic Z, Rakic T, Stevanovic B, Radovic S (2011) Characterization of oxidative and antioxidative events during dehydration and rehydration of resurrection plant *Ramonda nathaliae*. Plant Growth Regul 64: 231–240.

22. Farrant JM, Brandt W, Lindsey GG (2007) An overview of mechanisms of desiccation tolerance in selected angiosperm resurrection plants. Plant Stress 1: 72–84.

23. Peters S, Mundree SG, Thomson JA, Farrant JM, Keller F (2007) Protection mechanisms in the resurrection plant *Xerophyta viscosa* (Baker): both sucrose and raffinose family oligosaccharides (RFOs) accumulate in leaves in response to water deficit. J Exp Bot 58: 1947–1956.

24. Bianchi G, Gamba A, Limiroli R, Pozzi N, Elster R, et al. (1993) The unusual sugar composition in leaves of the resurrection plant *Myrothamnus flabellifolia*. Physiol Plantarum 87: 223–226.

25. Muller J, Sprenger N, Bortlik K, Boller T, Wiemken A (1997) Desiccation increases sucrose levels in *Ramonda* and *Haberlea*, two genera of resurrection plants in the Gesneriaceae. Physiol Plantarum 100: 153–158.

26. Ingram J, Chandler JW, Gallagher L, Salamini F, Bartels D (1997) Analysis of cDNA clones encoding sucrose-phosphate synthase in relation to sugar interconversions associated with dehydration in the resurrection plant *Craterostigma plantagineum* Hochst. Plant Physiol 115: 113–121.

27. Apostolova E, Rashkova M, Anachkov N, Denev I, Toneva V, et al. (2012) Molecular cloning and characterization of cDNAs of the superoxide dismutase gene family in the resurrection plant *Haberlea rhodopensis*. Plant Physiol Bioch 55: 85–92.

28. Wang RH, Wang LL, Wang Z, Shang HH, Liu X, et al. (2008) Protein gene cloning and expression in resurrection plant *Boea hygrometrica*. Prog Nat Sci 18: 1111–1118.

29. Wang L, Shang H, Liu Y, Zheng M, Wu R, et al. (2009) A role for a cell wall localized glycine-rich protein in dehydration and rehydration of the resurrection plant *Boea hygrometrica*. Plant Biology 11: 837–848.

30. Wang Z, Zhu Y, Wang L, Liu X, Liu Y, et al. (2009) A WRKY transcription factor participates in dehydration tolerance in *Boea hygrometrica* by binding to the W-box elements of the galactinol synthase (BhGolS1) promoter. Planta 230: 1155–1166.

31. Gechev TS, Benina M, Obata T, Tohge T, Sujeeth N, et al. (2013) Molecular mechanisms of desiccation tolerance in the resurrection glacial relic *Haberlea rhodopensis*. Cell Mol Life Sci 70: 689–709.

32. Shen A, Denby K, Illing N (2007) Identification of different temporal classes of gene expression during a cycle of desiccation in the resurrection plant, *Xerophyta humilis*. S Afr J Bot 73: 494–494.

33. Jiang G, Wang Z, Shang H, Yang W, Hu Z, et al. (2007) Proteome analysis of leaves from the resurrection plant *Boea hygrometrica* in response to dehydration and rehydration. Planta 225: 1405–1420.

34. Ingle RA, Schmidt U, Farrant JM, Thomson JA, Mundree SG (2007) Proteomic analysis of leaf proteins during dehydration of the resurrection plant *Xerophyta viscosa*. Plant Cell Environ 30: 435–446.

35. Abdalla KO, Baker B, Rafudeen MS (2010) Proteomic analysis of nuclear proteins during dehydration of the resurrection plant *Xerophyta viscosa*. Plant Growth Regul 62: 279–292.

36. Oliver MJ, Jain R, Balbuena TS, Agrawal G, Gasulla F, et al. (2011) Proteome analysis of leaves of the desiccation-tolerant grass, *Sporobolus stapfianus*, in response to dehydration. Phytochemistry 72: 1273–1284.

37. Gasulla F, Dorp KV, Dombrink I, Zähringer U, Gisch N, et al. (2013) The role of lipid metabolism in the acquisition of desiccation tolerance in *Craterostigma plantagineum*: a comparative approach. Plant J 75: 726–741.

38. Moellering ER, Muthan B, Benning C (2010) Freezing tolerance in plants requires lipid remodeling at the outer chloroplast membrane. Science 330: 226–228.

39. Verkleij A, De Maagd R, Leunissen-Bijvelt J, De Kruijff B (1982) Divalent cations and chlorpromazine can induce non-bilayer structures in phosphatidic acid-containing model membranes. 684: 255–262.

40. Lynch DV, Dunn TM (2004) An introduction to plant sphingolipids and a review of recent advances in understanding their metabolism and function. 161: 677–702.

41. Beck JG, Mathieu D, Loudet C, Buchoux S, Dufourc EJ (2007) Plant sterols in "rafts": a better way to regulate membrane thermal shocks. 21: 1714–1723.

42. Quartacci MF, Glisic O, Stevanovic B, Navari-Izzo F (2002) Plasma membrane lipids in the resurrection plant *Ramonda serbica* following dehydration and rehydration. J Exp Bot 53: 2159–2166.

43. Kotlova ER, Sinyutina NF (2005) Change's in the content of individual lipid classes of a lichen *Peltigera aphthosa* during dehydration and subsequent rehydration. Russ J Plant Physl 52: 35–42.

44. Quartacci MF, Forli M, Rascio N, DallaVecchia F, Bochicchio A, et al. (1997) Desiccation-tolerant *Sporobolus stapfianus*: lipid composition and cellular ultrastructure during dehydration and rehydration. J Exp Bot 48: 1269–1279.

45. Neale AD, Blomstedt CK, Bronson P, Le TN, Guthridge K, et al. (2000) The isolation of genes from the resurrection grass *Sporobolus stapfianus* which are induced during severe drought stress. Plant Cell Environ 23: 265–277.

46. Navariizzo F, Ricci F, Vazzana C, Quartacci MF (1995) Unusual composition of thylakoid membranes of the resurrection plant *Boea hygroscopica*: changes in lipids upon dehydration and rehydration. Physiol Plantarum 94: 135–142.

47. Frank W, Munnik T, Kerkmann K, Salamini F, Bartels D (2000) Water deficit triggers phospholipase D activity in the resurrection plant *Craterostigma plantagineum*. Plant Cell 12: 111–124.

48. Wang X (2002) Phospholipase D in hormonal and stress signaling. Curr Opin Plant Biol 5: 408–414.

49. Li W, Wang R, Li M, Li L, Wang C, et al. (2008) Differential degradation of extraplastidic and plastidic lipids during freezing and post-freezing recovery in *Arabidopsis thaliana*. J Biol Chem 283: 461–468.

50. Hong Y, Devaiah SP, Bahn SC, Thamasandra BN, Li M, et al. (2009) Phospholipase Dε and phosphatidic acid enhance *Arabidopsis* nitrogen signaling and growth. Plant J 58: 376–387.

51. Hong Y, Pan X, Welti R, Wang X (2008) Phospholipase Dα3 is involved in the hyperosmotic response in *Arabidopsis*. Plant cell 20: 803–816.

52. Shui Y (2007) *Paraisometrum mileense* re-emerges in Yunnan. Science (Chinese) 1: 37–38.

53. Weitzman AL, Skog LE, Wen-tsai W, Kai-yu P, Zhen-yu L (1997) New taxa, new combinations, and notes on Chinese Gesneriaceae. Novon: 423–435.

54. Committee AoSYACC (2006) Annuals of shilin yizu autonomous county (in Chinese). Kunming: Yunnan National Press.

55. Rascio N, Rocca NL (2005) Resurrection plants: the puzzle of surviving extreme vegetative desiccation. Crit Rev Plant Sci 24: 209–225.

56. Bewley JD, Krochko JE (1982) Desiccation-tolerance. Physiological Plant Ecology II. Berlin Heidelberg: Springer. pp. 325–378.

57. Moore JP, Lindsey GG, Farrant JM, Brandt WF (2007) An overview of the biology of the desiccation-tolerant resurrection plant *Myrothamnus flabellifolia*. Ann Bot-London 99: 211–217.

58. Porembski S (2011) Evolution, diversity, and habitats of Poikilohydrous vascular plants. Plant Desiccation Tolerance. Berlin Heidelberg: Springer. pp. 139–156.

59. Welti R, Shah J, Li W, Li M, Chen J, et al. (2007) Plant lipidomics: discerning biological function by profiling plant complex lipids using mass spectrometry. Front Biosci 12: 2494.

60. Zhou Z, Marepally SR, Nune DS, Pallakollu P, Ragan G, et al. (2011) LipidomeDB data calculation environment: online processing of direct-infusion mass spectral data for lipid profiles. 46: 879–884.

61. Gaff DF, Churchill DM (1976) *Borya nitida* Labill.—an Australian species in the Liliaceae with desiccation-tolerant leaves. Aust J Bot 24: 209–224.

62. Oliver M (1996) Understanding disability: From theory to practice. St Martin's Press.

63. Welti R, Li WQ, Li MY, Sang YM, Biesiada H, et al. (2002) Profiling membrane lipids in plant stress responses - Role of phospholipase Dα in freezing-induced lipid changes in *Arabidopsis*. J Biol Chem 277: 31994–32002.

64. Stevanovic B, Thu PTA, Depaula FM, Dasilva JV (1992) Effects of dehydration and rehydration on the polar lipid and fatty acid composition of *Ramonda* species. Can J Bot 70: 107–113.

65. Zhang XD, Wang RP, Zhang FJ, Tao FQ, Li WQ (2012) Lipid profiling and tolerance to low-temperature stress in *Thellungiella salsuginea* in comparison with *Arabidopsis thaliana*. Biol Plantarum: 1–5.

66. Gigon A, Matos AR, Laffray D, Zuily-Fodil Y, Pham-Thi AT (2004) Effect of drought stress on lipid metabolism in the leaves of *Arabidopsis thaliana* (ecotype Columbia). Ann Bot-London 94: 345–351.

67. Liljenberg CS (1992) The effects of water deficit stress on plant membrane lipids. Prog Lipid Res 31: 335–343.

68. Toumi I, Gargouri M, Nouairi I, Moschou PN, Ben Salem-Fnayou A, et al. (2008) Water stress induced changes in the leaf lipid composition of four grapevine genotypes with different drought tolerance. Biol Plantarum 52: 161–164.

69. Denich TJ, Beaudette LA, Lee H, Trevors JT (2003) Effect of selected environmental and physico-chemical factors on bacterial cytoplasmic membranes. J Microbiol Meth 52: 149–182.

70. Pappan K, Zheng S, Wang X (1997) Identification and characterization of a novel plant phospholipase D that requires polyphosphoinositides and submicromolar calcium for activity in Arabidopsis. 272: 7048–7054.

71. Katagiri T, Takahashi S, Shinozaki K (2001) Involvement of a novel *Arabidopsis* phospholipase D, AtPLDδ, in dehydration-inducible accumulation of phosphatidic acid in stress signalling. Plant J 26: 595–605.

72. Jia Y, Tao F, Li W (2013) Lipid profiling demonstrates that suppressing *Arabidopsis* phospholipase Dδ retards ABA-promoted leaf senescence by attenuating lipid degradation. PLoS One 8: e65687.

73. Barrs HD, Weatherley PE (1962) A re-examination of the relative turgidity technique for estimating water deficits in leaves. Aust J Bio Sci 15: 413–428.

74. Dhindsa RS, Matowe W (1981) Drought tolerance in two mosses: correlated with enzymatic defence against lipid peroxidation. J Exp Bot 32: 79–91.

75. Bates LS, Waldren RP, Teare ID (1973) Rapid determination of free proline for water-stress studies. Plant Soil 39: 205–207.

76. Rawyler A, Pavelic D, Gianinazzi C, Oberson J, Braendle R (1999) Membrane lipid integrity relies on a threshold of ATP production rate in potato cell cultures submitted to anoxia. Plant Physiol 120: 293–300.

77. Wang X, Dyer JH, Zheng L (1993) Purification and immunological analysis of phospholipase D from castor bean endosperm. 306: 486–494.

78. Lee MH (1989) Phospholipase D of rice bran. II. The effects of the enzyme inhibitors and activators on the germination and growth of root and seedling of rice. 59: 35–43.

16

Identification of Unusual Phospholipid Fatty Acyl Compositions of *Acanthamoeba castellanii*

Marta Palusinska-Szysz[1], Magdalena Kania[2], Anna Turska-Szewczuk[1], Witold Danikiewicz[2], Ryszard Russa[1], Beate Fuchs[3]*

1 Department of Genetics and Microbiology, Maria Curie-Sklodowska University, Lublin, Poland, **2** Mass Spectrometry Group, Institute of Organic Chemistry, Polish Academy of Sciences, Warsaw, Poland, **3** Institute of Medical Physics and Biophysics, Medical Faculty, University of Leipzig, Leipzig, Germany

Abstract

Acanthamoeba are opportunistic protozoan pathogens that may lead to sight-threatening keratitis and fatal granulomatous encephalitis. The successful prognosis requires early diagnosis and differentiation of pathogenic *Acanthamoeba* followed by aggressive treatment regimen. The plasma membrane of *Acanthamoeba* consists of 25% phospholipids (PL). The presence of C20 and, recently reported, 28- and 30-carbon fatty acyl residues is characteristic of amoeba PL. A detailed knowledge about this unusual PL composition could help to differentiate *Acanthamoeba* from other parasites, e.g. bacteria and develop more efficient treatment strategies. Therefore, the detailed PL composition of *Acanthamoeba castellanii* was investigated by ^{31}P nuclear magnetic resonance spectroscopy, thin-layer chromatography, gas chromatography, high performance liquid chromatography and liquid chromatography-mass spectrometry. Normal and reversed phase liquid chromatography coupled with mass spectrometric detection was used for detailed characterization of the fatty acyl composition of each detected PL. The most abundant fatty acyl residues in each PL class were octadecanoyl (18:0), octadecenoyl (18:1 Δ9) and hexadecanoyl (16:0). However, some selected PLs contained also very long fatty acyl chains: the presence of 28- and 30-carbon fatty acyl residues was confirmed in phosphatidylethanolamine (PE), phosphatidylserine, phosphatidic acid and cardiolipin. The majority of these fatty acyl residues were also identified in PE that resulted in the following composition: 28:1/20:2, 30:2/18:1, 28:0/20:2, 30:2/20:4 and 30:3/20:3. The PL of amoebae are significantly different in comparison to other cells: we describe here for the first time unusual, very long chain fatty acids with Δ^5-unsaturation ($30:3^{5,21,24}$) and $30:2^{21,24}$ localized exclusively in specific phospholipid classes of *A. castellanii* protozoa that could serve as specific biomarkers for the presence of these microorganisms.

Editor: Tilmann Harder, University of New South Wales, Australia

Funding: This work was supported by Deutsche Forschungsgemeinschaft (German Research Council) DFG: FU 771/1-2 (http://www.dfg.de/). The funders had no role in study design, data collection and analysis, decision to publish, or preparation of the manuscript.

Competing Interests: The authors have declared that no competing interests exist.

* Email: Beate.Fuchs@medizin.uni-leipzig.de

Introduction

Acanthamoeba castellanii is a small amoeba that has been isolated from various natural environmental sources such as soil, fresh water, dust, air as well as from anthropogenic ecosystems. This indicates the ubiquitous occurrence of this organism. Its wide distribution in nature brings humans into contact with this protozoan, and massive evidence for this microorganism is provided by the presence of antibodies to *Acanthamoeba* in healthy individuals [1]. *A. castellanii* cause granulomatous amoebic encephalitis (GAE), a fatal infection of the central nervous system (CNS) that occurs in immuno-compromised patients. The amoeba causes also painful keratitis that can result in blindness also occurring in healthy individuals; the infection is often misdiagnosed and difficult to treat (amoebae are resistant to many therapeutic agents) [2,3,4]. The limited success in the treatment of GAE is also most likely due to the inability of drugs to cross the blood-brain barrier into the central nervous system (CNS) to target the pathogen. However, phospholipid (PL) analogues such as hexadecylphosphocholine (HPC) possess anti-*Acanthamoeba* properties and have the ability to cross the blood-brain barrier [5,6,7].

Many aspects of *A. castellanii* are still unknown. Therefore, we will use here a "non-targeted" approach based on the characteristic lipid composition of this microorganism. In our opinion, lipids serve this approach much better than proteins because lipids enable the comparison of different species (such as mice and humans): the occurrence of the individual (phospho)lipid classes is rather similar between different vertebrates, while that of microorganisms is normally significantly different from mammalians. The plasma membrane of *Acanthamoeba* consists of about 25% phospholipids [6]. A characteristic trait of amoeba phospholipids is the presence of C20 [8,9,10] and, recently reported, 28- and 30-carbon fatty acids (FA) [11]. These are quite unusual fatty acyl residues and may be, thus, useful to screen potential host organisms for the presence of this microorganism.

In addition to different mass spectrometric methods, we applied here for the first time high-resolution ^{31}P NMR for the quantitative analysis of phospholipids in *A. castellanii* membranes. The advantages of the ^{31}P NMR approach include the unequivocal identification of the individual PL species even in relatively complex mixtures (since only a limited number of P-containing resonances need to be assigned) and the lack of solvent signals, which avoids the requirement for their selective suppression. This is a significant advantage regarding quantitative analysis. Theoretically, ^{31}P NMR should permit the differentiation between PLs based not only on the headgroup, but also on the linkage type (acyl-acyl-, alkyl-acyl- or alkenyl-acyl) and the acyl chain composition [12]. In practice, however, even if the differentiation between saturated, moderately unsaturated (18:1 or 18:2) and highly unsaturated (20:4 or 22:6) acyl residues is basically possible, detailed acyl residue analysis fails [13]. NMR is therefore not the method of choice to analyze the fatty acyl distribution within the individual PL classes. However, mass spectrometry, particularly if combined with chromatographic methods such as HPLC or thin-layer chromatography is a powerful analytical tool to provide details of the fatty acyl composition [14,15,16]. To combine the advantages of the individual methods, the phospholipid composition of *A. castellanii* cells will be determined by a variety of different analytical techniques, in particular by one-dimensional thin-layer chromatography, HPLC and tandem mass spectrometry with electrospray ionization (ESI) in the positive and the negative ion mode to identify specific fatty acyl residues. Such a detailed characterization of the phospholipid composition of *A. castellanii* cells by combining different analytical tools of lipid characterization is in our opinion very important since using just one method could lead to erroneous or incomplete data.

We suppose that (phospho)lipids possess a significant potential in the rapid diagnosis of *Acanthamoeba* in clinical and environmental specimens because their unusual very long fatty acyl chain pattern is specific of a given microorganism and distinct from vertebrate (phospho)lipids and other microorganisms such as bacteria, viruses and fungi. Since the infection with *A. castellanii* is problematic and often misdiagnosed as bacterial, viral or fungal infection in keratitis, a specific lipid biomarker might help to diagnose and treat an *A. castellanii* infection more efficiently.

Therefore, the aim of this study is to investigate the overall and positional fatty acyl compositions of the phospholipids from *A. castellanii*. In particular, the distribution of the long chain fatty acyl residues was of interest, and it will be explicitly shown that certain fatty acyl residues are exclusively located in one dedicated phospholipid class. Finally, we describe here for the first time unusual, very long chain fatty acids with Δ^5-unsaturation ($30:3^{5,21,24}$) and $30:2^{21,24}$ localized exclusively in specific phospholipid classes of *A. castellanii* protozoa.

Materials and Methods

A. castellanii culture

A. castellanii (ATCC 3034) was kindly donated by Dr W. Balamuth (Department of Zoology, University of California). The trophozoites of *A. castellanii* were grown as axenic cultures in peptone-yeast extract-glucose (PYG) medium on a shaker incubator set at 120 rpm and 28°C as described previously [11]. The flasks were inoculated with a 3-day-old amoeba culture to obtain an initial population of approximately 5×10^3 organisms/ml. Amoebae from the early stationary growth phase were harvested by centrifugation at 300×g for 10 min, washed in amoeba saline prepared according to Band [17] and lyophilized.

Chemicals

All chemicals for mass spectrometry (chloroform and methanol in the highest commercially available purity as well as ammonium acetate), NMR spectroscopy (sodium cholate, EDTA, deuterated water with an isotopic purity of 96.6%), and buffer preparation (NaCl, TRIS) were obtained from Sigma-Aldrich Chemie GmbH (Taufkirchen, Germany). The components for *A. castellanii* cell

culture (proteose peptone and yeast extract) were obtained from Becton, Dickinson and Company (Sparks, USA). Chemicals for LC/MS system (hexane, isopropanol, acetonitrile LC grade) and for TLC were obtained from Merck (Darmstadt, Germany).

PLs used as standards were purchased as 10 mg/ml solutions in CHCl$_3$ from Avanti Polar Lipids (Alabaster, MA, USA) and used without purification.

Analysis of phospholipids

Total lipids were extracted using a modified Bligh and Dyer [18] method, i.e. chloroform/methanol/aqueous HCl (1:2:0.1, v/v/v). The addition of HCl is important to avoid losses of acidic phospholipids. After separation of the chloroform and the aqueous phases, the chloroform phase was isolated by a glass syringe and evaporated to dryness in a centrifugal evaporator (Jouan, Germany). The precipitate at the interphase between the organic and aqueous layer (primarily proteins) was discarded and not further investigated, so that lipids tightly bound to proteins were excluded from the study. The dried organic phase was purified from lipophilic proteins by an additional extraction with a mixture of hexane:isopropanol (3:2, v/v). The extracts were dried under nitrogen before weighing and then dissolved in chloroform (5.6 mg/ml) for further analysis. Lipid classes were separated into neutral and polar fractions on a silica gel column as described previously [19].

Neutral lipids were eluted with 7 ml of chloroform. The phospholipids were eluted with the following solvents: 5 ml of chloroform-acetone (1:1, v/v), 5 ml of acetone, 3 ml of chloroform-methanol (8:2, v/v), 4 ml of chloroform-methanol (1:1, v/v) and 5 ml of chloroform-methanol (1:50, v/v). The separation was performed twice. Neutral lipid fractions from both separations were discarded. After evaporation of solvents, polar fractions from the second separation were dissolved in chloroform, pooled, and stored at −20°C. The polar lipid fraction was used for ^{31}P NMR spectroscopy, thin-layer chromatography (TLC) and mass spectrometry.

To evaluate the reproducibility of the data, all samples were investigated in four independent replicates.

PL analysis by ^{31}P NMR spectroscopy

The dried organic layers were re-dissolved according to [12,20] in 50 mM TRIS (pH 7.65) containing 200 mM sodium cholate and 5 mM EDTA. After vortexing ^{31}P NMR spectra were recorded on 0.5 ml samples in 5 mm NMR tubes on a Bruker DRX-600 spectrometer operating at 242.88 MHz for ^{31}P. All measurements were performed using a direct, selective ^{31}P NMR probe at 37°C with composite pulse decoupling (Waltz-16) to eliminate ^{31}P - ^{1}H coupling. Pulse intervals of the order of T$_1$ permit the quantitative analysis of PL integral intensities [21].

Other NMR parameters were as follows: acquisition time: 1 s, data size: 8–16 k, 60° pulse (5 µs), pulse delay 2 s and a line-broadening (LB) of 1 Hz. Chemical shifts were referenced to the most intense resonance of diacyl-PC at δ = −0.60 ppm. All peak assignments were confirmed by comparison with the shift of commercially (Avanti Polar Lipids, Alabaster, USA) available PL11. Spectra were processed using the software "1D WIN-NMR" version 6.2 (Bruker Analytische Messtechnik GmbH, Rheinstetten) including the deconvolution (II) routine for peak area determination. Integral intensities of the selected phosphorus-containing metabolites were divided by the total areas of all detected peaks.

PL analysis by thin-layer chromatography

Phospholipids were separated by one-dimensional TLC on silica gel 60 F254 plates (Merck, 20×20 cm). The plates were washed twice with chloroform/methanol (1:1, v/v) and activated at 180°C before use. Phospholipids were applied onto the TLC plates and developed with chloroform:methanol:acetic acid:acetone:water (35:25:4:14:2) [22]. Lipids were visualized with 5% sulfuric acid in methanol after charring at 140°C for 15 min or with iodine vapor. Phosphorus was detected using the Bochner method [23], carbohydrates were detected with α-naphthol, amino groups with ninhydrin, and quaternary nitrogen (choline) with the Dragendorff reagent [24]. To determine the relative moieties of the individual lipids in the total mixture, the developed TLC chromatograms were visualized with 5% sulfuric acid in methanol, subsequently scanned and the optical densities of the spots integrated using Bio-Profil v. 99.01 (Image Analysis Software). The relative content of each phospholipid class was determined on digitalized densito-grams corrected according to calibration curves for authentic standards. Cardiolipin from bovine heart, 1,2-dipalmitoyl-sn-glycero-3-phosphoethanolamine, 1,2-dipalmitoyl-sn-glycero-3-phosphocholine (Sigma-Aldrich Chemical Co., St. Louis, MO, USA), 1-hexadecanoyl-2-(9,12-octadecadienoyl)-sn-glycero-3-phospho-L-serine and 1-hexadecanoyl-2-(9-octadecenoyl)-sn-glycero-3-phosphoinositol, 1-hexadecanoyl-2-lyso-sn-glycero-3-phosphocholine, and sphingomyelin 16:0 (Avanti Polar Lipids, Alabaster, MA, USA) were used as standards.

Phospholipids (about 1 mg) were applied about 1 cm from the bottom of the silica gel plate as a narrow band. The chromatogram was developed with chloroform:methanol: acetic acid:acetone:water (35:25:4:14:2). Bands were detected by iodine staining and identified by co-migration with the standards. Subsequently, individual lipid bands were scraped off and transferred to screw-capped tubes and extracted from silica gel with a mixture of chloroform/methanol 1:2 (v/v). The ratio between silica gel and solvent was 1:9 (m/m).

Preparation of fatty acid methyl esters and gas chromatographic analysis

Fatty acids were released from the corresponding lipid classes by heating a dried aliquot of each lipid class with 0.8 M NaOH in 50% methanol for 1 h at 80°C. This method provided in our hands better results than applying acidic conditions in the presence of BF_3. The free fatty acids extracted with $CHCl_3$ from the acidified samples were then converted into methyl esters by methanolysis (0.5 M HCl in methanol, 80°C, 1 h). The solution was then cooled to room temperature and the solvent evaporated. Fatty acid methyl esters were recovered by extraction with chloroform/water (1:2, v/v) and analyzed by gas-liquid chromatography.

GLC-MS was carried out on a Hewlett-Packard gas chromatograph (model HP 5890A) equipped with a capillary column (HP-5MS, 30 m×0.25 mm) and connected to a mass selective detector (MSD model HP5971). Helium (0.7 ml/min) was used as the carrier gas and the temperature program was initially 150°C for 5 min, subsequently increased to 310°C at a ramp rate of 3°C min^{-1}, final time of 20 min.

HPLC MS analysis of phospholipids

Normal phase (NP) and reversed phase (RP) liquid chromatographic methods were tested for identification of phospholipids isolated from A. castellanii cells. The NP LC/MS analysis was carried out using a High-Performance Liquid Chromatograph Prominence LC-20 (Shimadzu, Japan) coupled with tandem mass spectrometer 4000 Q TRAP (Applied Biosystems Inc, USA). The mass spectrometer was equipped with an electrospray (ESI) ion source (Turbo Ion Spray) and triple quadrupole/linear ion trap mass analyzer. LC separations were performed using a 4.6×150 mm Zorbax SIL RX (5 μm) column. Solvent A was hexane/isopropanol (3:2, v/v) and solvent B was isopropanol/hexane/ 5 mM aqueous solution of ammonium acetate (38:56:5, v/v/v). The following elution program was used: from 53% B to 80% B in 23 min, 80% B maintained by 4 min, 80% B to 100% B for 9 min and 100% B maintained by 14 min. The flow rate was 1 ml/min.

RP LC/MS analysis as a complementary method for the identification of PI and PC species was done using a Finnigan LCQ Advantage Max ion-trap mass spectrometer (ThermoElectron Corporation, San Jose, CA, USA) with ESI source. The phospholipid separation was carried out on a cyano column (5 μm, 4.6 mm×250 mm). Milli-Q water (component A), acetonitrile (component B), 0.1% w/v formic acid in acetonitrile (component C) and 1% w/w ammonia in acetonitrile (component D) were used as the mobile phase. The following gradient was employed for LC lipid separation: 12 min isocratic elution using a mixture composed of 40% A, 50.5% B, 9.2% C, 0.3% D and then a linear gradient of B up to 89.5% at the expense of A over 8 min. The final mobile phase composition (0.5% A, 89.5% B, 9.2% C, 0.3% D) was held for 40 min. The eluent flow rate and the column temperature were 0.5 ml min^{-1} and 55°C, respectively.

The low resolution spectra and collision induced decomposition (CID) spectra of identified compounds (the corresponding parent ions were isolated by using optimized settings of the ion trap) were performed in the positive and the negative ion mode. Nitrogen was used as the nebulizer, curtain and collision gas. The tip voltage was kept at 5500 V and -4500 V in the positive and the negative ion mode, respectively. The declustering potential was set to 30 V in both modes. The CID spectra were recorded at collision energies ranging between 35–60 eV.

Results

The chloroform-methanol extractable lipids of A. castellanii grown on the PYG medium accounted for 14.5% of the cell material (122 mg dry weight), out of which polar lipids (56 mg) constituted 46% of the total lipids. The polar lipids were used for ^{31}P NMR spectroscopy, thin-layer chromatography (TLC) and LC/MS analyses.

The organic extracts of A. castellanii were analyzed by ^{31}P NMR spectroscopy in order to be able to quantify all PL classes. A selected ^{31}P NMR spectrum of A. castellanii is shown in **Fig. 1** (left hand). Peak assignments and the corresponding chemical shifts are given on the top of the spectra. This spectrum exhibits signals of nine different phospholipid classes comprising phosphatidylcholine (PC, δ = −0.60 ppm), phosphatidylinositol (PI, δ = −0.41 ppm), phosphatidylserine (PS, δ = −0.2 ppm), lysophosphatidylcholine (LPC, δ = −0.15 ppm), phosphatidylethanolamine (PE, δ = −0.03 ppm), ether-linked glycerophosphoethanolamine (ether-GPE, δ = 0.06 ppm), cardiolipin (CL, δ = 0.38 ppm), and lysophosphatidylethanolamine (LPE, δ = 0.41 ppm). Although these resonances could be unequivocally assigned, another resonance was detected at δ = −0.3 ppm that could not be convincingly characterized. However, it is claimed that this resonance is caused by a PI species with a very remarkable fatty acyl composition. However, a more detailed structure elucidation would go beyond the scope of this manuscript.

Quantitative ^{31}P NMR data were obtained by integration of the individual ^{31}P NMR resonances: PC (47.5±2%), PE (29±2%), PS

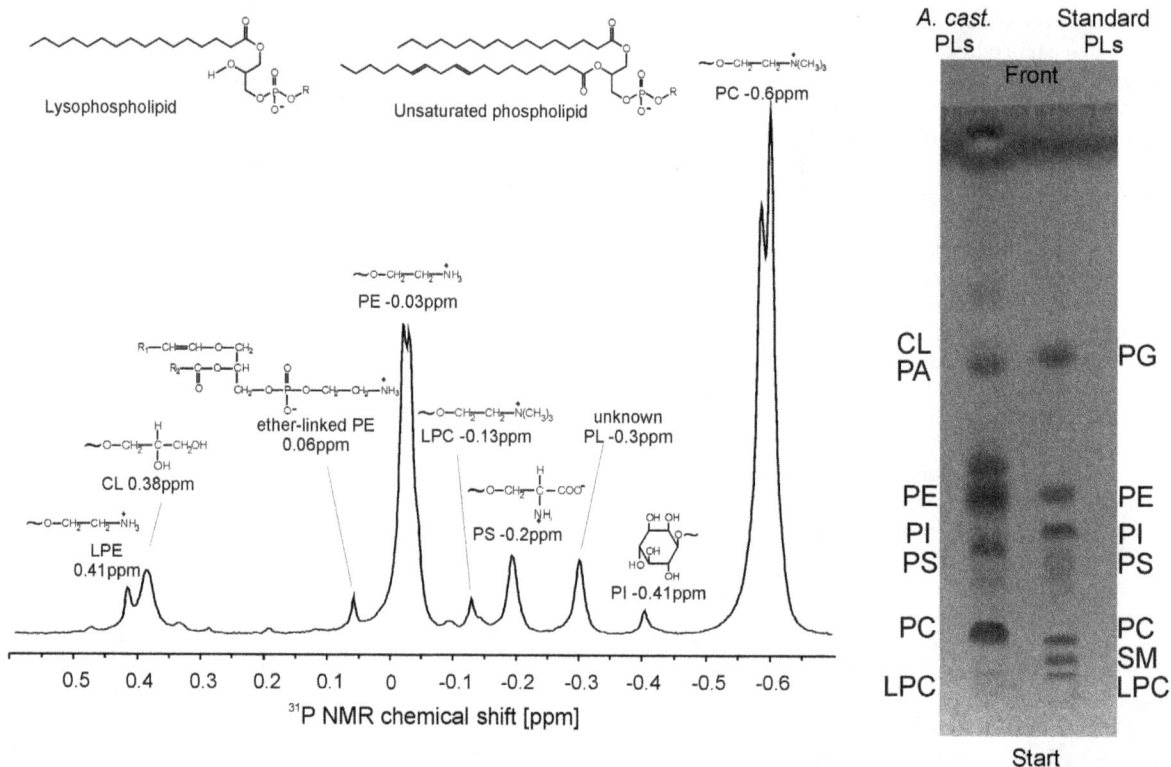

Figure 1. 31P NMR spectrum and one-dimensional TLC profile of *A. castellanii* phospholipids. Left: 242.88 MHz ³¹P NMR spectrum of the organic extract of *A. castellanii*. Assignments and chemical shifts are indicated in the spectrum. The chemical structures of a phospholipid and a lysophospholipid is shown. R stands for the different head groups, while R₁ and R₂ denote the different carbon chains in the *sn*-1 and *sn*-2 position of the glycerol back bone. Chemical structures of the head groups are shown on top of the corresponding peak. Right: One-dimensional TLC profile of *A. castellanii* phospholipids in comparison with a standard PL mixture that contains the same amounts of all indicated PL. TLC plates were developed in a solvent mixture consisting of chloroform:methanol:acetic acid:acetone:water (35:25:4:14:2, v/v/v/v/v). Abbreviations used in peak/spot assignments: CL, cardiolipin; LPC, lyso-phosphatidylcholine; LPE, lyso-phosphatidylethanolamine; PA, phosphatidic acid; PC, phosphatidylcholine; PE, phosphatidylethanolamine; PG, phosphatidylglycerol; PI, phosphatidylinositol; PS, phosphatidylserine; SM, sphingomyelin. For details see text.

(5.6±0.2%), and CL (6.9±0.3%) were the most abundant phospholipids. Other minor phospholipids comprised PI (1.2±0.05%), LPC (1.8±0.1%), LPE (1.5±0.05%), and ether-GPE (1.7±0.08%).

TLC analysis

Fig. 1 shows the one-dimensional TLC profile of the phospholipids of *A. castellanii* (right hand side), displaying seven bands. Individual polar lipid components were identified by comparison of their R_f values with those of the authentic standards and specific staining properties.

The following components were identified in the seven bands:

Band 1 contained lysophosphatidylcholine (LPC) and band 2 phosphatidylcholine (PC) as revealed by staining with the Dragendorff reagent and co-migration with standards. The predominant fatty acids obtained by hydrolysis of the PC fraction were octadecanoic (33%) and octadecenoic (25%) acid, as shown by GC/MS analysis of band 2 extracted from silica gel. However, long chain unsaturated FAs could also be detected in the PC fraction: eicosatetraenoic (20:4 $\Delta^{5,8,11,14}$), eicosatrienoic (20:3 $\Delta^{8,11,14}$) and eicosadienoic (20:2 $\Delta^{11,14}$), which constituted 19 mol% of the PC species (**Table 1**). The fatty acids of LPC were primarily hexadecanoic (41 mol%) and octadecanoic (29 mol%) acid.

Band 3 contained phosphatidylserine (PS). This phospholipid stained positively with ninhydrin and phosphomolybdate. This

phospholipid exhibited a higher content of unsaturated (60 mol%) than saturated fatty acids. Hexadecanoic and octadecanoic acid, which were less abundant in the other PL classes, were the dominant saturated acids, while octadecenoic acid (31 mol%) was the main unsaturated one in the PS fraction. Additionally, there were also C20 acids: eicosatetraenoic (20:4 $\Delta^{5,8,11,14}$), eicosatrienoic (20:3 $\Delta^{8,11,14}$) and eicosadienoic (20:2 $\Delta^{11,14}$), C28 acids - octacosenoic (28:1 Δ^{21}) and C30 acids - triacontadienoic (30:2 $\Delta^{21,24}$) and triacontatrienoic (30:3 $\Delta^{5,21,24}$) acids (**Table 1**).

Band 4 contained phosphatidylinositol (PI). This phospholipid was identified by staining with phosphomolybdate and co-migration with the standard. Saturated acids, mainly hexadecanoic and octadecanoic acid, dominated in this phospholipid class.

Band 5 corresponding to phosphatidylethanolamine (PE) was identified by its characteristic R_F value and staining with ninhydrin and phosphomolybdate. Octadecanoic acid was predominant in this class. Unsaturated acids constituted approximately half of the total FAs (50 mol%). As in the case of PS, PE contained 20-, 28-, and 30-carbon acids, with only 2 mol% of octacosenoic (28:1 Δ^{21}) and 2 mol% of triacontadienoic (30:2 $\Delta^{21,24}$) acid.

Band 6 contained phosphatidic acid (PA). PA was identified by its position relative to other lipids, the co-elution with the standard, and the absence of ninhydrin staining. PA belongs to the least abundant phospholipids and was presumably therefore not detectable by ³¹P NMR. PA was characterized by the presence

Table 1. Content (given in mol %) of the fatty acyl residues of each phospholipid class of A. castellanii.

Fatty acid	LPC	PC	PS	PI	PE	PA	Cardiolipin
14:0	13	4	7	2	6	26	6
16:1 Δ^{7*}	8	4	5	11	5	6	4
16:0	41	9	19	36	13	28	17
17:0	tr	1	1	1	1	1	1
18:2 $\Delta^{9,12}$	tr	5	9	2	5	5	5
18:1 Δ^{9}	9	25	31	19	21	17	28
18:0	29	33	12	28	29	15	28
20:4 $\Delta^{5,8,11,14}$		7	1	tr	4		tr
20:5		tr	tr	tr	tr		tr
20:3 $\Delta^{8,11,14}$		7	4	tr	4	tr	2
20:2 $\Delta^{11,14}$		5	6	1.0	6	2	3
20:1			tr		tr		
28:2 $\Delta^{5,21}$							tr
28:1 Δ^{21}			1		2	tr	2
28:0			1		1		2
30:3 $\Delta^{5,21,24}$			1		1		2
30:2 $\Delta^{21,24}$			2		2	tr	tr
sum of saturated	83	47	40	67	50	70	54
sum of unsaturated	17	53	60	33	50	30	46

Data were determined by GC/MS analysis of the FAs methyl esters subsequent to alkaline hydrolysis of the corresponding lipid class. Standard deviations of all measurements are estimated to be of the order of ±5%. The position of double bonds was already determined in a previous paper [11], tr – trace (<0.5%).

*The position of double bonds was already determined in a previous paper [11], tr – trace (<0.5%).

of saturated tetradecanic (26 mol%) and hexadecanoic (28 mol%) acid. The GC/MS analysis of the methyl esters demonstrated also minor amounts of octacosenoic (28:1 Δ^{21}) and triacontadienoic acid (30:2 $\Delta^{21,24}$), which were also identified by LC/MS.

Band 7 represented cardiolipin (CL), which was phosphomolybdate-positive. Cardiolipin included considerable amounts of octadecanoic (18:0) and octadecenoic (18:1 Δ^9) acids. Moreover, CL contained eicosatrienoic (20:3 $\Delta^{8,11,14}$) and eicosadienoic (20:2 $\Delta^{11,14}$), as well as octacosenoic (28:1 Δ^{21}), octacosanoic (28:0) and triacontatrienoic (30:3 $\Delta^{5,21,24}$) residues.

In order to confirm and extend the TLC analysis of the polar lipids of *A. castellanii*, all identified phospholipid classes were additionally analyzed by LC/MS.

HPLC/MS analysis

Normal (NP) and reversed phase (RP) liquid chromatography coupled with mass spectrometry detection were used to perform a more detailed analysis of the individual PL species of *A. castellanii* cells. The NP LC/MS method facilitates separation of the PLs into their respective classes and enables the characterization of the individual molecular species (data not shown). This is an indispensable method for PL identification, particularly for complex extracts from biological sources, where many different lipid classes with strongly varying fatty acyl compositions can be expected. In contrast to NP LC/MS, RP LC/MS analysis is capable of separating PLs according to their fatty acyl compositions. Although it is a useful tool for the identification of the individual molecular species in isolated phospholipid fractions, in this study the total mixture of PA, PS, PE, PC and PI was directly analyzed by RP LC/MS technique.

Identification of phospholipid classes

Phospholipid classes were identified by the NP LC/MS method using mainly precursor ion scan (PI) and neutral loss scan (NL) techniques as recently essentially reviewed [25]. These scanning methods can be utilized to analyze selectively dedicated phospholipid headgroups in a mixture. Defined phospholipid standards were also used to determine the retention time and fragmentation patterns of the PLs of interest.

In the precursor ion scan, the ion at *m/z* 184 (positive ion mode) corresponding to protonated phosphocholine was used to identify the individual PC species in the sample (data not shown). In contrast, phosphatidic acids were analyzed in the negative ion mode by the characteristic fragment at *m/z* 153 corresponding to deprotonated phosphoglycerol.

In the neutral loss scan technique, the tendency of PS and PE species to loose serine (87 Da) or phosphoethanolamine (141 Da), respectively, was used to identify these PL classes in the positive ion mode.

In contrast to other PL classes, PI species were identified by the RP HPLC/MS method only. Using NP HPLC conditions (described in Materials and Methods) exclusively the monophosphorylated but not diphosphorylated derivatives of inositol-containing lipids were identified. The detailed structural characterization of these compounds was beyond the scope of this study.

Identification of individual phospholipids

Many PL classes possess a net negative charge at neutral pH. PC and PE molecules are zwitterionic ions; therefore, positive and negative-ion mass spectra of these PLs are accessible. PI and PS species also can be analyzed in the positive ion mode giving the sodiated and the protonated adducts, whereas in the negative ion mode deprotonated PI and PS are directly observed [26].

In this work, low-resolution MS spectra of individual PL classes were recorded in the positive and the negative ion mode. In the negative ion mode, deprotonated PE, PS, PI and PA species were directly observed while for PC-containing compounds the ions [M+CH$_3$COO]$^-$ were also detectable. In the positive ion mode, the mass spectra of PC, PE and PS were dominated by the protonated species.

To determine the structures of the analyzed compounds, CID spectra of individual PLs were performed and compared to the CID spectra of the PL standards and literature data [26]. Since the fragment ion spectra recorded in the negative ion mode were easier to interpret (due to the reduced interference of the different adducts obtained in the positive ion mode), these CID spectra mainly served to establish the structures of the analyzed PLs (**Table 2**).

Loss of fatty acids as neutral molecules or as ketenes occurs readily during the fragmentation of PLs. The high intensities of ions corresponding to lysophospholipid-like products ([M-H-RCOOH]$^-$ or [M-H-R'CH=C=O]$^-$ and anions of free fatty acids ([RCOO$^-$]) have been suggested to be useful for the determination of the *sn*-1 and *sn*-2 positions of the fatty acyl residues in the individual PL [26]. In this work, the following rule was adopted for positional identification of the acyl residues: for PC-, PE- and PI-derived lipids, the ions corresponding to neutral losses of the *sn*-2 substituents (as a FA and a ketene) are more abundant than the ions reflecting neutral losses of the analogous substituents from the *sn*-1 position. In the case of PS- and PA-derived PL, the carboxylate anion arising from the *sn*-1 FA is more prominent than the carboxylate anion derived from the *sn*-2 FA.

To identify the positions of the fatty acyl residues in the PL structures, the CID spectra of the analyzed PLs with well-defined fragment ions were chosen (**Table 2**).

Most PC species were identified by HPLC/ESI-MS. This is consistent with the TLC results, which showed that PC is the main polar lipid in the PL extract of *A. castellanii* cells. PC species containing mainly octadecanoyl or octadecenoyl residues were detected. **Fig. 2** shows the negative ion fragmentation spectra of PS (trace A) and PE (trace B) species containing 30:2 (*m/z* 447.4), 28:1 (*m/z* 421.4) and 20:2 (*m/z* 307.3) fatty acids as representative examples. The loss of eicosadienoic and octadecenoic acids as neutral molecules or ketene is obvious from these CID spectra. These data help to unequivocally identify a given lipid.

Regarding the PA and PI classes, molecular species with pentadecanoyl, hexadecanoyl and heptadecanoyl residues were identified. The 30- and 28-carbon fatty acids were found particularly in combination with octadecenoic and eicosadienoic acids, respectively in PA-containing compounds. The PE class was dominated by octadecenoic and octadecadienoic acids. Thus, there are significant differences between the different PL classes in the content of long and short chain fatty acyl residues.

Discussion

Upon separation into neutral and polar fractions using the Bligh and Dyer extraction method and analysis of the polar fractions by ^{31}P NMR spectroscopy, 1D TLC and LC/MS, the following PL classes could be identified in *A. castellanii*: PC, PE, PS, PI, LPC, cardiolipin (not discussed in this manuscript) and PA. The PL composition was thus similar to that of the Neff *A. castellanii* strain obtained by Ulsamer *et al.* [8]. However, slight quantitative differences were observed. The two predominant lipid classes were PC and PE accounting for approximately 75% of the total PL. A significant content of these PL is characteristic of most parasitic protozoans [27].

Table 2. Survey of the most abundant molecular species of *A. castellanii* phospholipids determined by LC/MS operating in the negative (−) and positive (+) ion mode.

PL class	m/z	Fatty acyl composition
LPC	480.2 (+)	16:0
	506.2 (+)	18:1
	508.3 (+)	18:0
PC	750.8 (−)[a]	16:0/14:0
	758.6 (+)	16:1/18:1[c]
	760.6 (+)	16:0/18:1[c]
	764.8 (−)	17:0/14:0[c]
	778.8 (−)	14:0/18:0
	784.6 (+)	18:1/18:2
	786.6 (+)	18:1/18:1
	788.6 (+)	18:0/18:1
	806.6 (+)	18:1/20:5[c]
	812.6 (+)	18:1/20:2
	816.4 (−)	14:0/20:2; 16:1/18:1
	846.4 (−)	18:0/18:1
	850.8 (−)	18:2/20:4
	852.8 (−)	18:1/20:4
	854.8 (−)	18:0/20:4
	854.7 (−)	18:1/20:3
	856.8 (−)	18:0/20:3
	858.9 (−)	18:0/20:2
	866.5 (−)	18:2/20:3; 18:1/20:4
	868.7 (−)	18:0/20:4[c]
PE	688.4 (−)	14:0/18:1
	714.5 (−)	16:1/18:1; 16:0/18:2; 14:0/20:2
	716.4 (−)	16:0/18:1
	742.5 (−)	16:0/20:2; 18:0/18:2; 18:1/18:1
	744.5 (−)	18:1/18:0
	764.8 (−)	18:1/20:4; 18:2/20:3
	766.8 (−)	18:2/20:2; 18:1/20:3; 18:0/20:4
	768.8 (−)	18:0/20:3; 18:1/20:2
	768.7 (+)	18:0/20:4
	882.6 (−)	**30:2/16:0; 28:1/18:1[b]**
	908.6 (−)	**28:1/20:2; 30:2/18:1**
	910.6 (−)	**30:1/18:1; 28:0/20:2**
	930.6 (−)	**30:2/20:4**
	932.9 (+)	**30:3/20:3; 20:4/30:2**
	934.6 (−)	**30:2/20:2**
	938.5 (+)	**20:1/30:2**
PS	758.7 (+)	16:1/18:1[c]
	804.6 (−)	18:2/20:5
	806.4 (−)	18:2/20:4; 18:1/20:5
	808.5 (−)	18:1/20:4
	810.7 (−)	18:1/20:3[c]
	812.9 (−)	18:0/20:3
	812.6 (+)	18:1/20:2[c]
	972.6 (−)	**30:3/20:4**
	978.6 (−)	**30:2/20:2**
PI	831.5 (−)	16:1/18:2

Table 2. Cont.

PL class	m/z	Fatty acyl composition
	833.5 (−)	16:1/18:1
	835.5 (−)	16:0/18:1
	847.4 (−)	17:0/18:2
	859.5 (−)	18:1/18:2
	863.6 (−)	18:1/18:0
	889.6 (−)	18:0/20:2
PA	645.2 (−)	16:0/16:1
	661.5 (−)	15:0/18:1
	673.5 (−)	16:0/18:1
	675.5 (−)	16:0/18:0
	689.9 (−)	18:0/17:0[c]
	695.5 (−)	18:1/18:2
	865.6 (−)	**18:1/30:2[c]**
	865.6 (−)	**28:1/20:2[c]**

[a]Phosphatidylcholines were detected as acetate [M+CH$_3$COO]$^-$ or formate [M+HCOO]$^-$ adducts.
[b]bold-print - PL with long chain FAs.
[c]PL species without assignments to sn-1/sn-2 position.

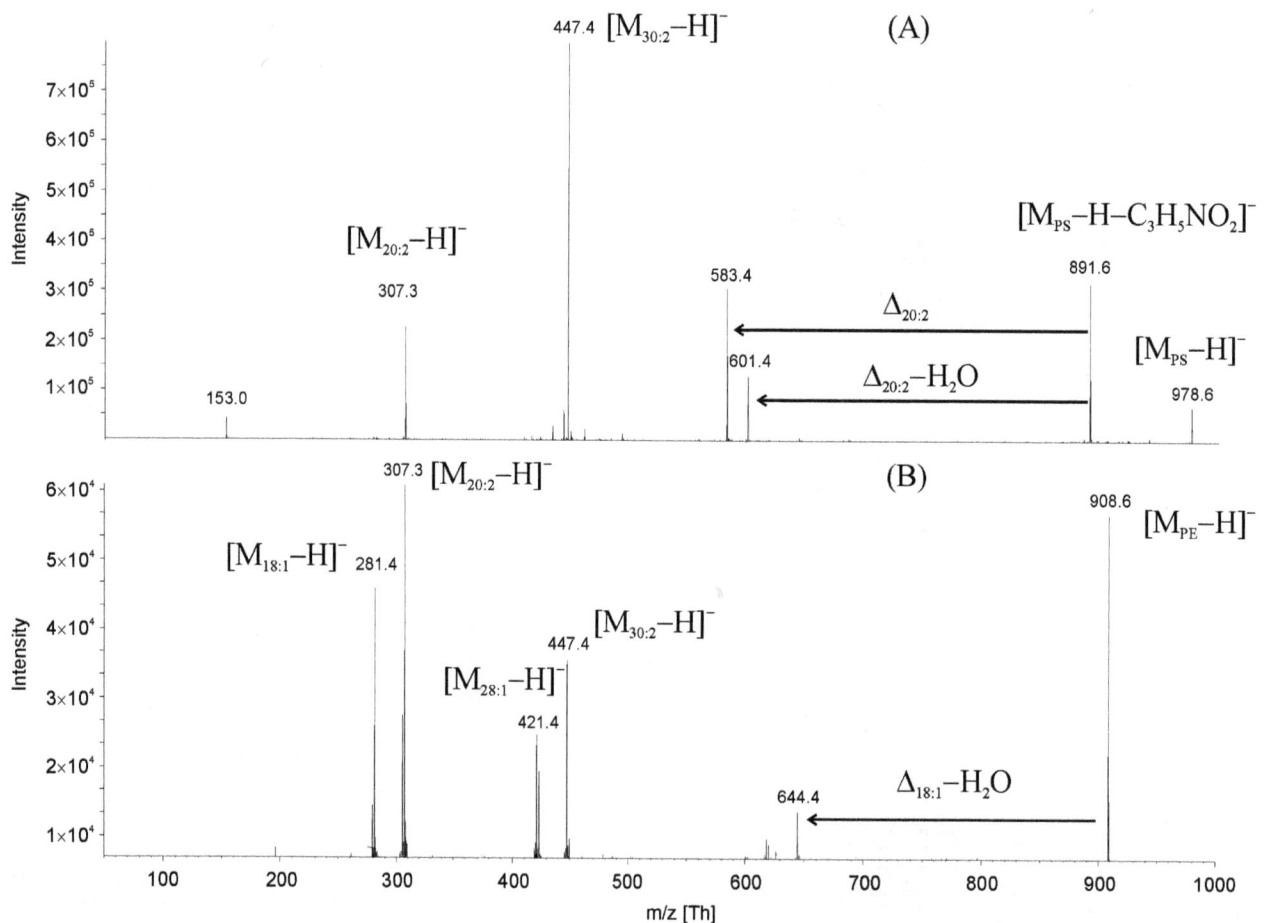

Figure 2. CID spectrum (recorded in the negative ion mode) of (A) a dedicated PS containing 30:2 (*sn*-1) and 20:2 (*sn*-2) fatty acyl residues. Δ 20:2 elimination of eicosadienoic acid, Δ 20:2-H$_2$O–elimination of eicosadienoic ketene and (B) a given PE species with the following fatty acyl combinations: 28:1/20:2; 30:2/18:1.

Seventeen different C14-C30 FAs of *A. castellanii* were identified by GC-MS analyses of their methyl esters. They included both saturated FAs such as C14:0, C16:0, and C18:0 as well as unsaturated FAs: C18:1 Δ^9, 18:2 $\Delta^{9,12}$, C20:4 $\Delta^{5,8,11,14}$, C20:3 $\Delta^{8,11,14}$, and C20:2 $\Delta^{11,14}$. The investigated PLs also contained long-chain C28 and C30 acyl residues. These long chain fatty acids are a characteristic feature of *A. castellanii* since PL of most multicellular organisms contain nearly exclusively C16-22 FAs. However, very long-chain C23-32 FAs were also found in marine sponges [28]. The 30:2 $\Delta^{5,9}$ acid is present in *Cinachyrella schulzei*, and 30:3 $^{\Delta 5,9,23}$ in *Chondrilla nucula*. Although the chain length and the number of double bonds are comparable, the corresponding *A. castellanii* fatty acids differ regarding their positions of the double bonds.

Fatty acid analysis of the individual phospholipid fractions provided evidence that certain long chain FAs of *A. castellanii* are associated with specific phospholipids. The presence of the 28- and 30-carbon acids was confirmed in PE, PS, PA and cardiolipin. PLs with the compositions 28:1/20:2, 30:2/18:1, 28:0/20:2, 30:2/20:4 and 30:3/20:3 were mainly identified in the PE fraction. They were less abundant in PS species such as 30:2/20:2, 30:3/20:4. The specific location of long chain fatty acids in certain PLs has been observed also in sponges [28], where the PE fraction contains large amounts of 26:2 and 26:3, for instance, in *M. prolifera* and *Halicondria panacea*. Most of the 30:3 fatty acids in *C. nucula* are also present in PE. PS contains high levels of 26:2 and 26:3 in *M. prolifera* and large amounts of 30:4 and 30:5 in *C. celata* while only small amounts of 30:3 $\Delta^{5,9,23}$ were detectable in *C. nucula* [29].

The LC/MS/MS analysis of the fatty acyl distributions in *A. castellanii* PLs confirmed the general principle that the *sn*-1 position is usually occupied by a saturated or monoenoic residue, whereas polyunsaturated residues are normally located in the *sn*-2 position of mammalian PLs. However, in the case of selected *A. castellanii* PLs, the *sn*-1 position contains a higher unsaturated fatty acyl residue, for example, PE 18:2/20:2, 30:2/20:4, 30:3/20:3 and PS 30:2/20:2, 18:2/20:4, 30:3/20:4. It seems likely this high double bond content affects the replication of *A. castellanii* in the same way as it affects the development of other organisms. For instance, spermatozoa are also characterized by a considerable content of unsaturated fatty acyl residues within their phospholipids because this is a prerequisite for successful fusion with the female oocyte [30]. Furthermore, the high level of unsaturation of *A. castellanii* membranes affects their fluidity significantly and, thus, is a prerequisite of effective phagocytosis. Acanthamoeba phagocytosis may be both, an efficient way of obtaining nutrients for amoeba, and a significant aspect regarding the pathogenesis of acanthamoeba infections [31].

In a nutshell our findings have proven that the PL class composition of *A. castellanii* does not differ significantly from the PL class composition of mammalian cells [32] or other parasitic protozoan. However, we have also shown that *A. castellanii* has not only an unusual fatty acyl composition but also an unusual distribution of these fatty acyl residues in the glycerol backbone of selected phospholipids. We describe here for the first time unusual, very long chain fatty acids with Δ^5-unsaturation (30:3^{5,21,24}) and 30:2^{21,24} localized exclusively in specific phospholipid classes of *A. castellanii* protozoa. This unusual fatty acyl composition and distribution could serve as a biomarker for the identification of *A. castellanii*. Since the majority of potential hosts are not expected to possess such long fatty acyl residues, the presence of *A. castellanii* can be easily screened by the presence of these characteristic fatty acids. Little is so far known about the distribution of *A. castellanii* within the body of the affected host and further studies are needed to clarify this aspect. As far as we can say, the following biopsy materials from infected patients might useful to check for the presence of typical *A. castellanii* fatty acids: corneal scrapings, the content of the nasal cavity, the throat, the internal ear, skin, as well as cerebrospinal fluid. It can be expected that *A. castellanii* lipids will be present only to a very small amount in physiologically relevant samples. Therefore, it is presumably advantageous to use the high sensitivity and the significant resolving power of GC/MS to screen selected body fluids and/or tissues (vide infra) for their contents of unusual very long chain, polyunsaturated free fatty acids (such as 30:3 and 30:2) which are characteristic of *A. castellanii* cells. A detailed lipidomics analysis of the intact (phospho)lipids would, however, provide much more information because the intact lipids can be analyzed. In contrast, all lipids have to be hydrolyzed and derivatized if GC/MS analysis is performed. Therefore, all information regarding the fatty acyl compositions of the different (phospho)lipid classes is lost.

As the unusual very long fatty acyl chain composition of A. castellanii are presumably absent in vertebrate, bacterial, viral and fungal phospholipids, this may have significant potential regarding the rapid diagnosis of *Acanthamoeba* in clinical and environmental specimens.

Finally, since synthetic PC analogues were recently suggested as promising drugs against *Acanthamoeba* species [7,33], the results obtained may help to find more effective antimicrobials to treat *Acanthamoeba* infections. The goal of our future studies is to assess the biological significance of these fatty acids and their sensitivity/resistance to antiprotozoal agents.

Acknowledgments

Beate Fuchs is indebted to Dr. Jürgen Schiller for many fruitful discussions and his kind and helpful advices.

Author Contributions

Conceived and designed the experiments: MPS BF MK. Performed the experiments: MPS BF MK. Analyzed the data: MPS BF MK RR ATS WD. Contributed reagents/materials/analysis tools: BF MK MPS ATS RR WD. Wrote the paper: BF MPS MK.

References

1. Brindley N, Matin A, Khan NA (2009) *Acanthamoeba castellanii*: High antibody prevalence in racially and ethnically diverse populations. Exp Parasitol 121: 254–256.

2. Marciano-Cabral F, Cabral G (2003) *Acanthamoeba spp.* as agents of disease in humans. Clin Microbiol Rev 16: 273–307.

3. Khan NA (2006) *Acanthamoeba*: biology and increasing importance in human health. FEMS Microbiol Rev 30: 564–595.

4. Visvesvara GS, Moura H, Schuster FI (2007) Pathogenic and opportunistic free-living amoebae: *Acanthamoeba spp.*, *Balamuthia mandrillaris*, *Naegleria fowleri*, and *Sappinia diploidea*. FEMS Immunol Med Microbiol 50: 1–26.

5. Siddiqui R, Iqbal J, Maugueret MJ, Khan NA (2012) The role of Src kinase in the biology and pathogenesis of *Acanthamoeba castellanii*. Parasit Vectors 5: 112.

6. Siddiqui R, Khan NA (2012) Biology and pathogenesis of *Acanthamoeba*. Parasit Vectors 5: 6.

7. Croft SL, Seifert K, Duchêne M (2003) Antiprotozoal activities of phospholipid analogues. Mol Biochem Parasitol 126: 165–172.

8. Ulsamer AG, Smith FR, Korn ED (1969) Lipids of *Acanthamoeba castellanii*. Composition and effects of phagocytosis on incorporation of radioactive precursors. J Cell Biol 43: 105–114.

9. Sayanova O, Haslam R, Guschina I, Lloyd D, Christie W, et al. (2006) A bifunctional Δ12, Δ15-desaturase from *Acanthamoeba castellanii* directs the synthesis of highly unusual *n*-1 series unsaturated fatty acids. J Biol Chem 281: 36533–36541.

10. Korn ED (1964) Biosynthesis of unsaturated fatty acids in *Acanthamoeba sp.* J Biol Chem 239: 396–400.

11. Palusinska-Szysz M, Turska-Szewczuk A, Karaś M, Russa R, Drozanski W (2009) Occurrence of new polyenoic very long chain acyl residues in lipids from *A. castellanii*. Acta Protozool 48: 63–72.

12. Pearce JM, Komoroski RA (2000) Analysis of phospholipid molecular species in brain by ^{31}P NMR spectroscopy. Magn Reson Med 44: 215–223.

13. Schiller J, Müller M, Fuchs B, Arnold K, Huster D (2007) ^{31}P NMR spectroscopy of phospholipids: From micelles to membranes. Curr Anal Chem 3: 283–301.

14. Schiller J, Arnold K (2000) Mass spectrometry in structural biology. In: Meyers RA, editor. Encyclopedia of Analytical Chemistry. Chichester: Wiley. pp. 559–585.

15. Schiller J, Süß R, Arnhold J, Fuchs B, Leßig J, et al. (2004) Matrix-assisted laser desorption and ionization time-of-flight (MALDI-TOF) mass spectrometry in lipid and phospholipid research. Prog Lipid Res 43: 443–478.

16. Schiller J, Süß R, Fuchs B, Müller M, Zschörnig O, et al. (2007) MALDI-TOF MS in lipidomics. Front Biosci 12: 2568–2579.

17. Band RN (1959) Nutritional and related biological studies on the free-living soil amoeba, *Hartmannella rhysodes*. J Gen Microbiol 21: 80–95.

18. Bligh EG, Dyer WJ (1959) A rapid method of total lipid extraction and purification. Can J Biochem Physiol 37: 911–917.

19. Kunsman JE (1970) Characterization of the lipids of *Butyrivibrio fibrisolvens*. J Bacteriol 103: 104–110.

20. London E, Feigenson GW (1979) Phosphorus NMR analysis of phospholipids in detergents. J Lipid Res 20: 408–412.

21. Schiller J, Arnold K (2002) Application of high resolution ^{31}P NMR spectroscopy to the characterization of the phospholipid composition of tissues and body fluids - a methodological review. Med Sci Monit 8: 205–222.

22. Xu G, Waki H, Kon K, Ando S (1996) Thin-layer chromatography of phospholipids and their lyso forms: application to determination of extracts from rat hippocampal CA1 region. Microchem J 53: 29–33.

23. Bochner BR, Mano DM, Ames BN (1981) Detection of phosphate esters on chromatograms: an improved reagent. Anal Biochem 117: 81–83.

24. Kates M (1986) Isolation, analysis and identification of lipids. In: Burdon RH, Van Knippenberg PH, editors. Techniques in lipidology. Amsterdam: Elsevier. pp. 232–254.

25. Lam SM, Shui G (2013) Lipidomics as a principal tool for advancing biomedical research. J Genet Genomics 40: 375–390.

26. Larsen A, Hvattum E (2005) Analysis of phospholipids by liquid chromatography coupled with on-line electrospray ionization mass spectrometry and tandem mass spectrometry. In: Byrdwell WC, editor. Modern methods for lipid analysis by liquid chromatography/mass spectrometry and related techniques. Urbana: AOCS Press.

27. Smith JD (1993) Phospholipid biosynthesis in protozoa. Prog Lipid Res 32: 47–60.

28. Rod'kina SA (2005) Fatty acids and other lipids of marine sponges. Russ J Mar Biol 31: 49–60.

29. Litchfield C, Tyszkiewicz J, Dato V (1979) 5, 9, 23- triacontatrienoic acid, principal fatty acid of the marine sponge *Chondrilla nucula*. Lipids 15: 200–203.

30. Fuchs B, Müller K, Paasch U, Schiller J (2012) Lysophospholipids: potential markers of diseases and infertility? Mini-Rev Med Chem 12: 74–86.

31. Khan NA (2001) Pathogenicity, morphology, and differentiation of Acanthamoeba. Curr Microbiol 43: 391–395.

32. Fuchs B, Schiller J, Cross M (2007) Apoptosis-associated changes in the glycerophospholipid composition of hematopoietic progenitor cells monitored by ^{31}P NMR spectroscopy and MALDI-TOF mass spectrometry. Chem Phys Lipids 150: 229–238.

33. McBride J, Mullen AB, Carter KC, Roberts CW (2007) Differential cytotoxicity of phospholipid analogues to pathogenic *Acanthamoeba* species and mammalian cells. J Antimicrob Chemother 60: 521–525.

17

Lipidomic Assessment of Plasma and Placenta of Women with Early-Onset Preeclampsia

Henri Augusto Korkes[1,2]*, **Nelson Sass**[1,2], **Antonio F. Moron**[1], **Niels Olsen S. Câmara**[3], **Tatiana Bonetti**[4], **Ana Sofia Cerdeira**[5], **Ismael Dale Cotrim Guerreiro Da Silva**[4], **Leandro De Oliveira**[1,2,3]

1 Department of Obstetrics – Federal University of Sao Paulo, Sao Paulo, Sao Paulo, Brazil, 2 Laboratory of Clinical and Experimental Investigation – School Maternity Vila Nova Cachoeirinha, Sao Paulo, Sao Paulo, Brazil, 3 Department of Immunology – University of Sao Paulo, Sao Paulo, Sao Paulo, Brazil, 4 Department of Gynecology - Federal University of Sao Paulo, Sao Paulo, Sao Paulo, Brazil, 5 Department of Medicine, Beth Israel Deaconess Medical Center, Harvard Medical School, Boston, Massachusetts, United States of America

Abstract

Introduction: Adipose tissue is responsible for triggering chronic systemic inflammatory response and these changes may be involved in the pathophysiology of preeclampsia.

Objective: To characterize the lipid profile in the placenta and plasma of patients with preeclampsia.

Methodology: Samples were collected from placenta and plasma of 10 pregnant women with preeclampsia and 10 controls. Lipids were extracted using the Bligh–Dyer protocol and were analysed by MALDI TOF-TOF mass spectrometry.

Results: Approximately 200 lipid signals were quantified. The most prevalent lipid present in plasma of patients with preeclampsia was the main class Glycerophosphoserines-GP03 (PS) representing 52.30% of the total lipid composition, followed by the main classes Glycerophosphoethanolamines-GP02 (PEt), Glycerophosphocholines-GP01 (PC) and Flavanoids-PK12 (FLV), with 24.03%, 9.47% and 8.39% respectively. When compared to the control group, plasma samples of patients with preeclampsia showed an increase of PS ($p<0.0001$), PC ($p<0.0001$) and FLV ($p<0.0001$). Placental analysis of patients with preeclampsia, revealed the PS as the most prevalent lipid representing 56.28%, followed by the main class Macrolides/polyketides-PK04 with 32.77%, both with increased levels when compared with patients control group, PS ($p<0.0001$) and PK04 ($p<0.0001$).

Conclusion: Lipids found in placenta and plasma from patients with preeclampsia differ from those of pregnant women in the control group. Further studies are needed to clarify if these changes are specific and a cause or consequence of preeclampsia.

Editor: Pascale Chavatte-Palmer, INRA, France

Funding: Funding provided by National Council for Scientific and Technological Development; CNPq Number: 476486/2011-4, http://www.cnpq.br/en and Foundation for Research Support of the State of São Paulo; FAPESP Number: 12/02270-2, http://www.fapesp.br/en/. The funders had no role in study design, data collection and analysis, decision to publish, or preparation of the manuscript.

Competing Interests: The authors have declared that no competing interests exist.

* Email: korkes@me.com

Introduction

Preeclampsia is a systemic disease characterized by intense inflammatory response, endothelial injury, platelet aggregation, coagulation system activation and increase vascular resistance. It affects about 5–8% of all pregnant women [1–3]. The diagnosis of preeclampsia is based on the development of hypertension ($\geq 140/90$ mmHg) and significant proteinuria (≥ 300 mg/24 hours) after 20 weeks of gestation [4].

The systemic complications of preeclampsia are not limited to the gestational period and recent studies have shown long-term adverse outcomes, such as increased risk for developing chronic hypertension, ischemic heart disease, acute myocardial infarction and venous thromboembolism, requiring longer follow-up and surveillance of these patients throughout their lives [4,5]. Despite

its relevance, preeclampsia pathogenesis is not completely understood. It has been established that the trophoblast has a key role in this process and many other conditions related to chronic inflammation can be relevant in different stages of the disease [3].

Obesity and Preeclampsia

Obesity, defined by the World Health Organization (WHO) through the body mass index above 30 kg/m^2, is a growing epidemic problem and it affects 500 million adults across the world [6,7]. It represents an important health problem and it has an enormous impact on modern obstetrics.

Adipose tissue is responsible for triggering chronic systemic inflammatory response, with increased levels of inflammatory cytokines such as TNF-α, IL-6 and MCP-1. The inflammatory response related to obesity has been considered as the link between this condition and preeclampsia [6,8–13].

Although the link between obesity and inflammatory response is well recognized, the roles of lipids in the cell function are even more extended. These molecules are responsible for the control of important cellular processes, including proliferation, apoptosis, metabolism and migration. They also assist in the transmission of biological information across cell membranes, directly contributing to proper cell functioning [14–16]. An impairment in lipid signaling pathways may contribute to the progression of chronic inflammatory diseases, such as autoimmune, allergic, neoplastic, atherosclerosis, hypertension, myocardial hypertrophy and metabolic degenerative diseases [17,18] and may be also related to preeclampsia pathophysiology.

Lipid molecules are defined by the International Committee for the Classification and Nomenclature of Lipids (ICCNL) in eight categories, based on their chemical functions: Fatty Acyls (FA), Glycerolipids (GL), Glycerophospholipids (GP), Sphingolipids (SP), Sterol Lipids (ST), Prenol Lipids (PR), Saccharolipids (SL) and Polyketides (PK). Each category is further subdivided into lipid main classes and subclasses [19,20].

Historically, the study of the function and properties of lipids was always extremely complicated due to their structural diversity and large number of isomorphic species. Technically, the distinction between pathogenic and nonpathogenic lipid molecules represents a challenge that has become possible through lipidomics [18,21,22].

Lipidomic analysis

Lipidomic analysis is a global characterization of all kinds of lipid molecules in biological system. The methodology used is mass spectrometry (MS) [22–24]. A technique known as Matrix-Assisted Laser Desorption/Ionization - Mass Spectrometry (MALDI-MS), has been the preferred method to evaluate lipidomics because it is relatively easy to handle [25,26]. MALDI is an ionization technique enabled by a laser beam (light amplification by stimulated emission of radiation) that acts upon a sample mixed with a matrix. This process generates ionized molecules. For complete separation, the most widely used technology is the time of flight (TOF), which consists of a long pipe (tube flight) capable of separating the ionized molecules according to a ratio of mass to charge (m/z) [25].

Lipidomic analysis in preeclampsia is a new research line. Recently, we demonstrated that women with early-onset preeclampsia have particular lipids in their plasma when compared to those with healthy pregnancy [26]. Additionally, Baig et al. published their findings evaluating samples of syncytiotrophoblast microvesicles from human placenta [27]. These authors also demonstrated that there was a significant increase of some classes of lipids as well as a reduction of others in samples from preeclamptic women.

Given the strong association between obesity and early dyslipidemia with preeclampsia and the first reports associating distinct lipid species with the disease, this study aimed to find a specific lipid profile that may be characteristic for these patients. Here we evaluated plasma samples and placental tissues of women with early-onset preeclampsia and established an interesting panel of lipids in these different settings.

Material and Methods

Ethics Statement

All participants in this study have provided their written informed consent. This study was approved by the Research Ethics Committee of the Federal University of São Paulo with the number 297/027 and by the Research Ethics Committee of the School Maternity Vila Nova Cachoeirinha with the number 34/2011, linked with the National database (CEP/CONEP) under the protocol number CAAE - 18100813.1.0000.5505.

Study Population

This is a case-control study involving 20 pregnant women (10 women with early-onset preeclampsia and 10 women with healthy pregnancy). All samples were collected at the School Maternity Vila Nova Cachoeirinha from October 2011 to April 2013. Early-onset preeclampsia was defined as blood pressure $\geq 140/90$ mmHg and significant proteinuria (≥ 300 mg/24 hours) after 20 weeks of gestation and before 34 weeks.

Sample collection

Blood. Five milliliters of peripheral blood were drawn in EDTA-tube at the time of delivery. Immediately after collection, blood samples were centrifuged at 2000 rpm for 5 minutes and supernatants aliquoted and stored at $-80°C$ for subsequent lipid extraction.

Placenta. Immediately after cesarean delivery, one fragment was removed from the central region of the basal plate of the placenta, with a wedge shape of approximately 3.0 cm in diameter at its greatest diameter, obtained with a sterile scalpel blade N°15. This fragment was divided in 3 smaller pieces of 1×1 cm, washed in saline solution and immediately frozen and stored at $-80°C$ for further processing and lipid extraction. For lipid extraction the frozen placental sample of 1×1 cm was plunged into liquid nitrogen for 1 minute and crushed using marble stone until obtaining small fragments (powder). Powder was then placed in a dry tube containing 300 μl of Milli-Q water. This material was subjected to further mixing and homogenization in a mechanical processor for five minutes and the resulting material subjected to lipid extraction.

Lipid extraction

Lipids were extracted from each sample using the Bligh–Dyer protocol [28]. Immediately after thawing, each 50 μL of plasma and placental homogenated (described above) were dissolved in a mixture of chloroform–methanol (125:250 μL) and vortexed well. After vortexing, 125 μL of chloroform and 100 μL of deionized water were added to supernatant and centrifuged at 1000 rpm in a table-top centrifuge for 5 min at room temperature. Following this protocol a two-phase system (aqueous top, organic bottom) was achieved. The bottom phase containing lipids was gently recovered using a Pasteur pipette; they were dried and sealed to be stored at $-80°C$.

Reagents

All chemicals were of analytical reagent grade and they were used as received. Chloroform ($CHCl_3$) and methanol (MeOH) were purchased from Burdick & Jackson (Muskegon, MI, USA). 2,5-Dihydroxybenzoic acid (DHB) was purchased from ICN Biomedicals (Aurora, OH, USA). Distilled water was deionized on a Millipore Milli-Q water reagent system (Millipore, Bedford, MA, USA). EDTA-tubes were purchased from Sigma-Aldrich (St. Louis, MO).

Mass spectrometry analysis

MALDI-MS spectra were acquired in the positive ion and reflector modes using the equipment MALDI TOF-TOF - Ultraflex model - Bruker and matrix 2,5 - DHB - White (2,5-Dihydroxy benzoic acid - 40 mg/ml in acetonitrile). The main operating condition used was 10 V (sample plate). The laser irradiation consisted of diverse shots during 60–90 seconds in the region where the sample had been placed.

Data processing

Raw data were analyzed by MarkerLynx (Waters, UK) for peak detection and alignment. The parameters were set as follows: mass tolerance was set at 100 ppm (suggested more than twice the instrument mass accuracy considering extreme value in signal saturation condition); peak width at 5% height and peak-to-peak baseline noise were calculated automatically by the software; mass window was set at 0.1 amu (atomic mass units); retention time window was set at 1 min (considering the maximum variation obtainable by a CapLC system); noise elimination level was 5; minimum intensity was set at 5%; peak intensity and retention time were normalized with the signal of the internal standard. This procedure allowed deconvolution alignment, and data reduction to give a table of mass and relative retention time pairs with associated relative intensities for all the detected peaks. Then data matrix was exported for partial least squares discriminant analysis (PLS-DA). In order to find differential circulating lipids, a VIP parameter (Variable Importance in the Projection) was employed to reflect the variable importance in the discriminant analysis. The major discriminant variables returned by the PLS-DA model were selected and underwent the Mann–Whitney U test to confirm the differential expression between groups. Those peaks showing p<0.05 were considered as having statistically significant differences.

Statistical Analysis

The lipid composition of the samples was established by the area of the peaks obtained for the main lipids identified. The data were normally distributed and statistical analysis was performed using the Student's t-test, using the GraphPad Prism version 6. Differences between the groups were considered statistically significant when p<0.05.

Results

Table 1 shows the demographic and obstetric characteristics of study participants.

Mass Spectrometry

Approximately 200 signals were identified between the lipid tracks acquisition from 600 to 1200 m/z (**Figure 1**). The identification of the different lipids found was carried out through the Lipid Database Search (http://www.lipidmaps.org), using the results of m/z analyzes.

Plasma samples

Table 2 and **Figure 2** show the lipid composition found in plasma samples from both groups. Plasma analysis of patients with preeclampsia revealed the main class of Glycerophosphoserines-GP03 (PS) as the most prevalent lipid, representing 52.30% of the total lipid composition, followed by Glycerophosphoethanolamines-GP02 (PEt), Glycerophosphocholines-GP01 (PC) and Flavanoids-PK12 (FLV). When compared to the control group, plasma samples of patients with preeclampsia showed an increased proportion of PS (p<0.0001), PC (p<0.0001) and FLV (p<0.0001).

Although in smaller proportion, other increased lipids in patients with preeclampsia were Glycerophosphates-GP10 (PAc) (p<0.0001), Phosphosphingolipids-SP03 (SM) (p<0.0001), Neutral glycosphingolipids-SP05 (p<0.0001) and Steroid conjugates-ST05 (p<0.0002). The main class PEt, was reduced in patients with preeclampsia when compared to the control group (p<0.0001). Other main lipid classes reduced in preeclampsia group were: Glycerophosphoglycerols-GP04 (PG) (p<0.003), Sterols-ST01 (p<0.0001) and Other acyl sugars-SL05 (p<0.0001).

Placental samples

Table 3 and **Figure 3** show the lipid composition found in placental samples from both groups. The placental analysis of patients with preeclampsia revealed the main class PS as the most prevalent lipid, representing 56.28% of the total composition. Other main classes found were: Macrolides and lactone polyketides-PK04 with 32.77%, both were increased in preeclamptic placentas when compared to the control group, PS (p<0.0001) and PK04 (p<0.0001). Some lipids found in placentas from patients with preeclampsia were reduced when compared to control group; PEt (p<0.0001), PG (p<0.0001), Glycerophosphoinositols-GP06 (PI) (p<0.0001), Glycerophosphoinositol monophosphates-GP07 (p<0.0001), Diradylglycerols-GL02 (p<0.0001), Triradylglycerols-GL03 (p<0.0001), Acidic glycosphingolipids-SP06 (GM3) (p<0.0001), Steroid conjugates-ST05 (p<0.0001), Other acyl sugars-SL05 (p<0.0001) and Flavonoids-PK12 (p<0.0001).

Discussion

The pathogenesis of preeclampsia has its roots on deficient trophoblastic invasion and failure in spiral artery remodeling. This incomplete transformation of the spiral arteries leads to inadequate placental perfusion and consequently to placental oxidative stress [29]. The altered placenta then releases great amount of microparticles, debris and antiangiogenic factors into the maternal circulation [30–32]. All these factors are supposed to act in synergy to initiate and to maintain an intense inflammatory response and an antiangiogenic state.

Maternal obesity has been considered to have important impact on the genesis of preeclampsia as obese women have higher risk for developing the disease. In addition, women with body-mass-index lower than 20 have lower risk to develop preeclampsia [33]. The complete link between preeclampsia and obesity has not been defined. However, it is known that the inflammatory aspect that characterize the lipotoxicity of adipose tissues leads to maternal endothelial dysfunction, decreases trophoblastic invasion and influences placental metabolism and function. In addition, the chronic inflammatory response of the "metabolic syndrome" can contribute to the systemic inflammation seen in preeclampsia [13].

Actually, lipids play important roles in cellular function as they are the main components of biological membranes [14]. Therefore, lipids can participate in the constitution of membrane receptors, ion channels and in cell signaling mechanisms. In addition, many lipids can act as endogenous ligands, binding to specific receptors and then initiating several immunological responses [34].

In this study we investigated and established the main composition of the lipid profile identified in plasma and placental tissue of normal pregnant and preeclamptic women. Either placental tissues or plasma from patients with preeclampsia expressed different lipid profile when compared to normal pregnant women. These findings suggest that specific lipid species may be more associated with risk of developing preeclampsia than

Table 1. Demographic and obstetric characteristics of study subjects.

	Control (n = 10)	Preeclampsia (n = 10)	P
Maternal age (years)	25.3±6.237	22.4±7.152	0.372
Gestational age (weeks)	38.6±1.174	35.5±3.504	0.026
Weight gain (Kg)	10.77±5.094	20.22±5.037	0.001
BMI (Kg/m2)	22.59±6.912	26.73±5.379	0.154
Blood Pressure			
Systolic (mmHg)	116±10.666	140.4±4.971	<0.0001
Diastolic (mmHg)	70.6±7.306	91±5.436	<0.0001
Birth Weights (g)	3190.6±581.115	2607±881.328	0.0975
Proteinuria (mg)	NE	1624.4±1072.915	NE

Results are expressed as mean ± standard deviation. Significant at P<0.05.
BMI – Body Mass Index.
NE - not evaluated.

others. Additional functional studies will be necessary to clarify the involvement of these lipids in the pathophysiology of preeclampsia.

PS were the most prevalent species of lipids in preeclamptic women group. These lipids belong to Glicophospholipids category, representing the major lipid constituent of cell membranes and lipoproteins. They play different biological roles, acting as signaling molecules involved in the processes of oxidative stress,

apoptosis and coagulation, all exacerbated in preeclampsia [35–39].

PC species also belong to Glicophospholipids category and were increased in plasma of patients with preeclampsia. *PC* species are precursors of several molecules that act as lipid second messengers, including phosphatidic acid. Increased levels of *PC* have been associated with increased cell proliferation. These lipids have been

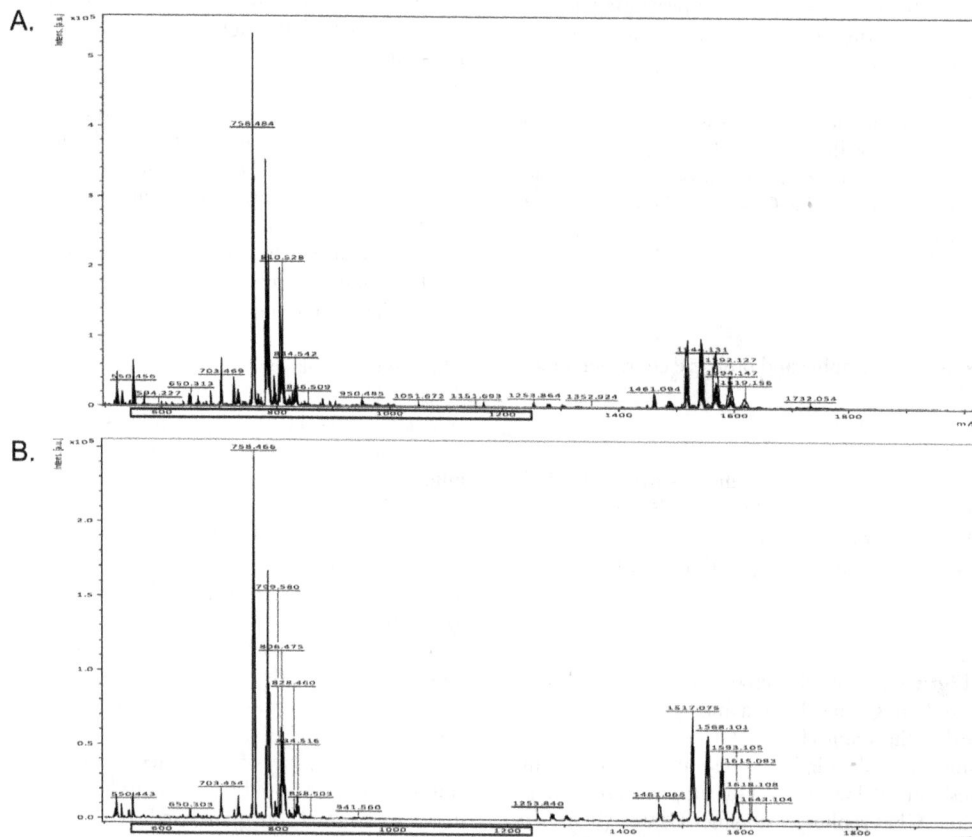

Figure 1. Representative lipid spectrum showing signals in plasma samples of control and preeclamptic patients. The signals were identified between the lipid tracks acquisition from 600 to 1200 m/z.

Table 2. Main classes of lipids in plasma samples.

Plasma	Group		P
[Main class]-(Common name)	Control (n = 10)	Preeclampsia (n = 10)	
Glycerophosphocholines [GP01]-(PC)	0.41%	9.47%	<0.0001
Glycerophosphoethanolamines [GP02]-(PEt)	77.38%	24.03%	<0.0001
Glycerophosphoserines [GP03]-(PS)	15.53%	52.30%	<0.0001
Glycerophosphoglycerols [GP04]-(PG)	3.14%	0.90%	0.003
Glycerophosphoinositols [GP06]-(PI)	0.62%	0.13%	NS
Glycerophosphates [GP10]-(PAc)	0.09%	1.40%	<0.0001
CDP-Glycerols [GP13]	0.02%	0.00%	NE
Triradylglycerols [GL03]-(TG)	0.22%	0.16%	NS
Phosphosphingolipids [SP03]-(SM)	0.00%	0.37%	<0.0001
Neutral glycosphingolipids [SP05]	0.00%	1.72%	<0.0001
Sterols [ST01]	0.31%	0.03%	<0.0001
Steroid conjugates [ST05]	0.00%	0.65%	0.0002
Polyprenols [PR03]	0.02%	0.00%	NE
Acylaminosugars [SL01]	0.00%	0.21%	NS
Other acyl sugars [SL05]	0.38%	0.00%	<0.0001
Flavonoids [PK12]	1.33%	8.39%	<0.0001
Fatty esters [FA07]	0.55%	0.24%	NS

Results are expressed as percentage. Significant at P<0.05.
NE - not evaluated.
NS – not significant.
Common names used: (PC) Phosphatidylcholines, (PEt) Phosphatidylethonolamines, (PS) Phosphatidylserines, (PG) Phosphatidylglycerols, (PI) Phosphatidylenositos, (PAc) Phosphatidic acid, (TG) Triradylglycerols, (SM) Sphingomyelins.

recently correlated with different tumor behaviors and cancer progressions. They are probably important for treatment considerations [40–42].

Oxidative stress generally affects lipid function due to changes in their native behavior. It can generate numerous different lipids that have diverse biological activities [43,44]. Glicerophospholipids, when oxidized, induce platelet aggregation, monocyte adhesion to endothelial cells, present in atherosclerotic lesions and they play an important role in signaling inflammatory response [45,46]. Thus, this increase in PS and PC in preeclampsia suggests a role in the inflammatory and oxidative phenomena observed in these patients.

There was a curious reduction of PEt in plasma samples of the preeclampsia group, compared to the control group. Apparently

the reduction of PEt in the endoplasmic reticulum is associated with araquidonic acid release [47,48]. This is the precursor of prostaglandins, thromboxanes and prostacyclins by the cyclooxygenase pathway and leukotrienes by the lipoxygenase pathway. Prostaglandins cause vasodilation, inhibition of platelet aggregation and pain. Thromboxane A2 promotes vasoconstriction and platelet aggregation. It is possible that these lipids act in opposing mechanisms in different patients.

Although also associated with apoptotic processes, cell proliferation and differentiation, the PI lipid class was found in small amounts in the plasma lipid composition in both groups. The SM lipid species, found only in plasma samples from patients with preeclampsia, act in vascular reactivity and mediate cell growth due to intrinsic properties of these vasoactive species [49].

A. Control plasma

B. Preeclamptic plasma

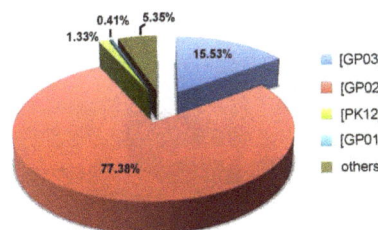 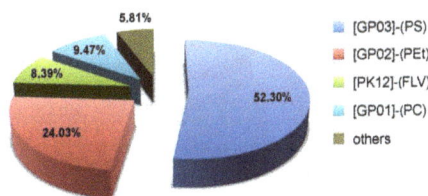

Figure 2. Lipid composition detected in plasma from control and preeclamptic patients. Comparison of relative distribution of main class of lipids in plasma of normal (A) and preeclamptic (B) patients, established by the area of the peaks obtained for the main lipids identified. N = 10 in each group. [GP03]-(PS): Glycerophosphoserines or Phosphatidylserines; [GP02]-(PEt): Glycerophosphoethanolamines or Phosphatidylethonolamines; [PK12]-(FLV): Flavonoids; [GP01]-(PC): Glycerophosphocholines or Phosphatidylcholines.

Table 3. Main classes of lipids in placental samples.

Placenta	Group		P
[Main class]-(Common name)	Control (n = 10)	Preeclampsia (n = 10)	
Glycerophosphocholines [GP01]-(PC)	0.00%	0.02%	NE
Glycerophosphoethanolamines [GP02]-(PEt)	9.69%	0.00%	<0.0001
Glycerophosphoserines [GP03]-(PS)	34.85%	56.28%	<0.0001
Glycerophosphoglycerols [GP04]-(PG)	13.57%	1.92%	<0.0001
Glycerophosphoinositols [GP06]-(PI)	13.28%	4.39%	<0.0001
Glycerophosphoinositol monophosphates [GP07]	0.57%	0.00%	<0.0001
Glycerophosphates [GP10]-(PAc)	1.30%	0.24%	NS
Diradylglycerols [GL02]-(DG)	7.73%	1.10%	<0.0001
Triradylglycerols [GL03]	1.04%	0.08%	<0.0001
Neutral glycosphingolipids [SP05]	0.07%	0.00%	NE
Acidic glycosphingolipids [SP06]- (GM3)	0.26%	0.06%	<0.0001
Sterols [ST01]	0.00%	0.12%	NS
Steroid conjugates [ST05]	1.68%	0.00%	<0.0001
Other acyl sugars [SL05]	0.42%	0.04%	<0.0001
Flavonoids [PK12]	11.68%	2.98%	<0.0001
Macrolide lactone polyketide [PK04]	3.86%	32.77%	<0.0001

Results are expressed as percentage. Significant at P<0.05.
NE - not evaluated; NS – not significant.
Common names used: (PC) Phosphatidylcholines, (PEt) Phosphatidylethonolamines, (PS) Phosphatidylserines, (PG) Phosphatidylglycerols, (DG) Diradylglycerols; (PI) Phosphatidylenositos, (PAc) Phosphatidic acid, (TG) Triradylglycerols, (SM) Sphingomyelins, (GM3) Gangliosides.

Experiments on hypertensive mice have showed significant increase in SM [50]. Apparently it is involved in processes of endothelial dysfunction, increased production of angiotensin II, elevated levels of thromboxane A2 and hence hypertensive disorders, which could explain its exclusively occurrence in the population of patients with preeclampsia [50–52]. The predominant lipid in preeclampsia placental samples was PS, and this aspect can be correlated with the oxidation process.

The Flavonoids are known for their antioxidant properties [53,54]. These lipids are not present in mammalian lipid composition, and their presence in our study is probably derived from the diet. These lipids interact in the signaling pathways of apoptotic processes, operating in both promotion and inhibition [55,56]. Evidences support the Flavonoids as protective factors against cardiac ischemic [57,58]. In our study, we found a

reduction in the amount of Flavonoids in samples of placentas from patients with preeclampsia when compared to controls. This did not happen when we performed the analysis of plasma samples that identified an increase of these lipids in patients with preeclampsia.

The Macrolides polyketides-lactone-PK04 have shown greater incidence in placental samples of preeclampsia group. They are not a natural part of mammalian lipid composition and they can be found in bacteria, fungi and plants. Rapamycin, also called Sirolimus, is an important and known polyketide with many biological and pharmacological activities, including antifungal, immunosuppressive, antitumor, neuroprotective and antiaging activities [59–63]. Recently, rapamycin has attracted interest for the clinical treatment of organ transplant rejection and autoimmune diseases [64].

A. Control placenta   B. Preeclamptic placenta

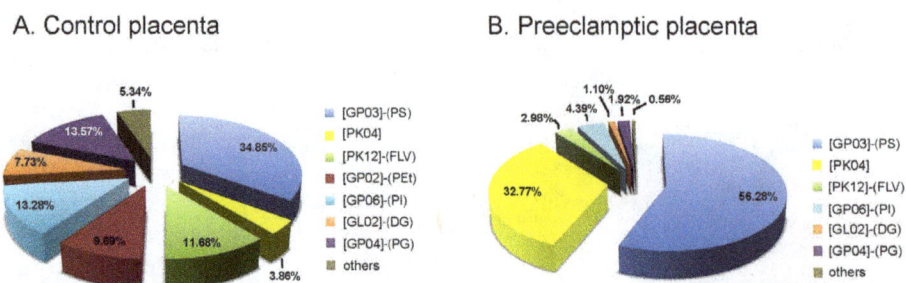

Figure 3. Lipid composition detected in placenta from control and preeclamptic women. Comparison of relative distribution of main class of lipids in placenta of normal (A) and preeclamptic (B) patients, established by the area of the peaks obtained for the main lipids identified. N = 10 in each group [GP03]-(PS): Glycerophosphoserines or Phosphatidylserines; [PK04] = Macrolide and lactone polyketides; [PK12]-(FLV): Flavonoids; [GP02]-(PEt): Glycerophosphoethanolamines or Phosphatidylethonolamines; [GP06]-(PI): Glycerophosphoinositols or Phosphatidylinositols; [GL02]-(DG): Diradylglycerols; [GP04]-(PG): Glycerophosphoglycerols or Phosphatidylglycerols.

Rapamicyn use has been associated with the development or exacerbation of proteinuria [65–67]. The pathogenesis of proteinuria is likely multifactorial and may involve tubular and glomerular contributions. Recent data derived from biopsy sub studies of clinical trials, which compared cyclosporine with rapamycin, demonstrated that rapamycin use is associated with tubular damage and tubular proteinuria [68]. In the glomerular compartment, other ones have demonstrated reduced nephrin expression [69] and reduced VEGF, particularly in patients with significant proteinuria [70,71].

The immunosuppressive effects of rapamycin result from its ability to inhibit proliferation by interfering with the function of the mammalian target of rapamycin (mTOR) [72]. mTOR is a serine/threonine protein kinase which controls the cellular processes of growth, proliferation, transcription, protein biosynthesis and ribosomal biogenesis [59,73]. mTOR exists in two distinct protein complexes referred to as mTOR complex 1 and mTOR complex 2. The inactivation of mTOR complex 1 kinase activity by rapamycin results in the inhibition of the activities of ribosomal S6 kinase and the eukaryotic translation initiation factor 4E-binding proteins, which have roles in ribosome biogenesis and protein translation, respectively. In contrast, apoptosis and autophagy are also stimulated by rapamycin [72,74].

Studies in immortalized cell lines originating from human trophoblast suggest a key role for mTOR in the regulation of trophoblast proliferation and it is suggested that the mTOR pathway is a regulator of invasive trophoblast differentiation [75,76]. In the mature placenta mTOR is expressed at the mRNA level, however, the cellular localization of mTOR and the functional role of this signaling pathway in the placenta after implantation and early placental development remains unknown [75,77,78].

Our study did not include other groups, such as gestational hypertension or intrauterine growth restriction, which would be important to evaluate the specificity of this association. In addition it would be interesting to compare the lipid profile between obese pregnant patients with normal outcomes and patients with preeclampsia with and without obesity to better understand if the lipid changes are a reflection of obesity, increased BMI or true PE.

Conclusion

We identified a different pattern of lipids and distinct concentrations of some lipid species in plasma and placenta samples of preeclamptic patients. Further studies are needed to clarify if these changes are specific to preeclampsia and whether and how they could be related to its pathogenesis.

Acknowledgments

We thank the Department of Obstetrics of Federal University of Sao Paulo, the Laboratory of Clinical and Experimental Investigation from School Maternity Vila Nova Cachoeirinha, the Department of Immunology of University of São Paulo and the Department of Gynecology of Federal University of Sao Paulo.

Author Contributions

Conceived and designed the experiments: HAK NS TB IDCGS LO AFM NOSC. Performed the experiments: HAK LO. Analyzed the data: HAK NS LO. Contributed reagents/materials/analysis tools: IDCGS TB. Contributed to the writing of the manuscript: HAK LO ASC NS.

References

1. Rana S, Cerdeira AS, Wenger J, Salahuddin S, Lim KH, et al. (2012) Plasma concentrations of soluble endoglin versus standard evaluation in patients with suspected preeclampsia. PLoS One 7: e48259.
2. Borzychowski AM, Sargent IL, Redman CW (2006) Inflammation and pre-eclampsia. Semin Fetal Neonatal Med 11: 309–316.
3. de Oliveira LG, Karumanchi A, Sass N (2010) [Preeclampsia: oxidative stress, inflammation and endothelial dysfunction]. Rev Bras Ginecol Obstet 32: 609–616.
4. Ghulmiyyah L, Sibai B (2012) Maternal mortality from preeclampsia/eclampsia. Semin Perinatol 36: 56–59.
5. Bellamy L, Casas JP, Hingorani AD, Williams DJ (2007) Pre-eclampsia and risk of cardiovascular disease and cancer in later life: systematic review and meta-analysis. BMJ 335: 974.
6. Walsh SW (2007) Obesity: a risk factor for preeclampsia. Trends Endocrinol Metab 18: 365–370.
7. Kral JG, Kava RA, Catalano PM, Moore BJ (2012) Severe Obesity: The Neglected Epidemic. Obes Facts 5: 254–269.
8. (2000) Obesity: preventing and managing the global epidemic. Report of a WHO consultation. World Health Organ Tech Rep Ser 894: i–xii, 1–253.
9. Hogan JL, Maguire P, Farah N, Kennelly MM, Stuart B, et al. (2011) Body mass index and blood pressure measurement during pregnancy. Hypertens Pregnancy 30: 396–400.
10. Bodnar LM, Catov JM, Klebanoff MA, Ness RB, Roberts JM (2007) Prepregnancy body mass index and the occurrence of severe hypertensive disorders of pregnancy. Epidemiology 18: 234–239.
11. Sohlberg S, Stephansson O, Cnattingius S, Wikstrom AK (2012) Maternal body mass index, height, and risks of preeclampsia. Am J Hypertens 25: 120–125.
12. Pou KM, Massaro JM, Hoffmann U, Vasan RS, Maurovich-Horvat P, et al. (2007) Visceral and subcutaneous adipose tissue volumes are cross-sectionally related to markers of inflammation and oxidative stress: the Framingham Heart Study. Circulation 116: 1234–1241.
13. Jarvie E, Hauguel-de-Mouzon S, Nelson SM, Sattar N, Catalano PM, et al. (2010) Lipotoxicity in obese pregnancy and its potential role in adverse pregnancy outcome and obesity in the offspring. Clin Sci (Lond) 119: 123–129.
14. Gross RW, Han X (2011) Lipidomics at the interface of structure and function in systems biology. Chem Biol 18: 284–291.
15. Shevchenko A, Simons K (2010) Lipidomics: coming to grips with lipid diversity. Nat Rev Mol Cell Biol 11: 593–598.
16. Yetukuri L, Ekroos K, Vidal-Puig A, Oresic M (2008) Informatics and computational strategies for the study of lipids. Mol Biosyst 4: 121–127.
17. Wymann MP, Schneiter R (2008) Lipid signalling in disease. Nat Rev Mol Cell Biol 9: 162–176.
18. Bou Khalil M, Hou W, Zhou H, Elisma F, Swayne LA, et al. (2010) Lipidomics era: accomplishments and challenges. Mass Spectrom Rev 29: 877–929.
19. Fahy E, Subramaniam S, Brown HA, Glass CK, Merrill AH, Jr., et al. (2005) A comprehensive classification system for lipids. J Lipid Res 46: 839–861.
20. Fahy E, Subramaniam S, Murphy RC, Nishijima M, Raetz CR, et al. (2009) Update of the LIPID MAPS comprehensive classification system for lipids. J Lipid Res 50 Suppl: S9–14.
21. Han X, Gross RW (2005) Shotgun lipidomics: electrospray ionization mass spectrometric analysis and quantitation of cellular lipidomes directly from crude extracts of biological samples. Mass Spectrom Rev 24: 367–412.
22. Schmelzer K, Fahy E, Subramaniam S, Dennis EA (2007) The lipid maps initiative in lipidomics. Methods Enzymol 432: 171–183.
23. Harkewicz R, Dennis EA (2011) Applications of mass spectrometry to lipids and membranes. Annu Rev Biochem 80: 301–325.
24. Want EJ, Cravatt BF, Siuzdak G (2005) The expanding role of mass spectrometry in metabolite profiling and characterization. Chembiochem 6: 1941–1951.
25. Postle AD (2012) Lipidomics. Curr Opin Clin Nutr Metab Care 15: 127–133.
26. De Oliveira L, Camara NO, Bonetti T, Lo Turco EG, Bertolla RP, et al. (2012) Lipid fingerprinting in women with early-onset preeclampsia: a first look. Clin Biochem 45: 852–855.
27. Baig S, Lim JY, Fernandis AZ, Wenk MR, Kale A, et al. (2013) Lipidomic analysis of human placental syncytiotrophoblast microvesicles in adverse pregnancy outcomes. Placenta 34: 436–442.
28. Bligh EG, Dyer WJ (1959) A rapid method of total lipid extraction and purification. Can J Biochem Physiol 37: 911–917.
29. Burton GJ, Woods AW, Jauniaux E, Kingdom JC (2009) Rheological and physiological consequences of conversion of the maternal spiral arteries for uteroplacental blood flow during human pregnancy. Placenta 30: 473–482.
30. Redman CW, Sargent IL (2008) Circulating microparticles in normal pregnancy and pre-eclampsia. Placenta 29 Suppl A: S73–77.
31. Levine RJ, Maynard SE, Qian C, Lim KH, England LJ, et al. (2004) Circulating angiogenic factors and the risk of preeclampsia. N Engl J Med 350: 672–683.
32. Rajakumar A, Cerdeira AS, Rana S, Zsengeller Z, Edmunds L, et al. (2012) Transcriptionally active syncytial aggregates in the maternal circulation may contribute to circulating soluble fms-like tyrosine kinase 1 in preeclampsia. Hypertension 59: 256–264.
33. Duckitt K, Harrington D (2005) Risk factors for pre-eclampsia at antenatal booking: systematic review of controlled studies. BMJ 330: 565.

34. Leavy O (2012) Inflammation: Trauma kicks up a storm. Nat Rev Immunol 12: 3.

35. Hopkins PN, Heiss G, Ellison RC, Province MA, Pankow JS, et al. (2003) Coronary artery disease risk in familial combined hyperlipidemia and familial hypertriglyceridemia: a case-control comparison from the National Heart, Lung, and Blood Institute Family Heart Study. Circulation 108: 519–523.

36. Leventis PA, Grinstein S (2010) The distribution and function of phosphatidyl-serine in cellular membranes. Annu Rev Biophys 39: 407–427.

37. Domingues MR, Reis A, Domingues P (2008) Mass spectrometry analysis of oxidized phospholipids. Chem Phys Lipids 156: 1–12.

38. Kagan VE, Borisenko GG, Tyurina YY, Tyurin VA, Jiang J, et al. (2004) Oxidative lipidomics of apoptosis: redox catalytic interactions of cytochrome c with cardiolipin and phosphatidylserine. Free Radic Biol Med 37: 1963–1985.

39. Li M, Huang SJ (2009) Innate immunity, coagulation and placenta-related adverse pregnancy outcomes. Thromb Res 124: 656–662.

40. Nakagami K, Uchida T, Ohwada S, Koibuchi Y, Suda Y, et al. (1999) Increased choline kinase activity and elevated phosphocholine levels in human colon cancer. Jpn J Cancer Res 90: 419–424.

41. Hernando E, Sarmentero-Estrada J, Koppie T, Belda-Iniesta C, Ramirez de Molina V, et al. (2009) A critical role for choline kinase-alpha in the aggressiveness of bladder carcinomas. Oncogene 28: 2425–2435.

42. Jagannathan NR, Kumar M, Seenu V, Coshic O, Dwivedi SN, et al. (2001) Evaluation of total choline from in-vivo volume localized proton MR spectroscopy and its response to neoadjuvant chemotherapy in locally advanced breast cancer. Br J Cancer 84: 1016–1022.

43. Fruhwirth GO, Loidl A, Hermetter A (2007) Oxidized phospholipids: from molecular properties to disease. Biochim Biophys Acta 1772: 718–736.

44. Spickett CM, Dever G (2005) Studies of phospholipid oxidation by electrospray mass spectrometry: from analysis in cells to biological effects. Biofactors 24: 17–31.

45. Leitinger N (2003) Oxidized phospholipids as modulators of inflammation in atherosclerosis. Curr Opin Lipidol 14: 421–430.

46. Zhang W, Salomon RG (2005) Oxidized phospholipids, isolevuglandins, and atherosclerosis. Mol Nutr Food Res 49: 1050–1062.

47. Meikle PJ, Christopher MJ (2011) Lipidomics is providing new insight into the metabolic syndrome and its sequelae. Curr Opin Lipidol 22: 210–215.

48. Andreyev AY, Fahy E, Guan Z, Kelly S, Li X, et al. (2010) Subcellular organelle lipidomics in TLR-4-activated macrophages. J Lipid Res 51: 2785–2797.

49. Hannun YA, Obeid LM (2008) Principles of bioactive lipid signalling: lessons from sphingolipids. Nat Rev Mol Cell Biol 9: 139–150.

50. Spijkers LJ, van den Akker RF, Janssen BJ, Debets JJ, De Mey JG, et al. (2011) Hypertension is associated with marked alterations in sphingolipid biology: a potential role for ceramide. PLoS One 6: e21817.

51. Berry C, Touyz R, Dominiczak AF, Webb RC, Johns DG (2001) Angiotensin receptors: signaling, vascular pathophysiology, and interactions with ceramide. Am J Physiol Heart Circ Physiol 281: H2337–2365.

52. Fenger M, Linneberg A, Jorgensen T, Madsbad S, Sobye K, et al. (2011) Genetics of the ceramide/sphingosine-1-phosphate rheostat in blood pressure regulation and hypertension. BMC Genet 12: 44.

53. Terao J, Kawai Y, Murota K (2008) Vegetable flavonoids and cardiovascular disease. Asia Pac J Clin Nutr 17 Suppl 1: 291–293.

54. Boots AW, Haenen GR, Bast A (2008) Health effects of quercetin: from antioxidant to nutraceutical. Eur J Pharmacol 585: 325–337.

55. Ramos S (2007) Effects of dietary flavonoids on apoptotic pathways related to cancer chemoprevention. J Nutr Biochem 18: 427–442.

56. Mandel S, Weinreb O, Amit T, Youdim MB (2004) Cell signaling pathways in the neuroprotective actions of the green tea polyphenol (−)-epigallocatechin-3-gallate: implications for neurodegenerative diseases. J Neurochem 88: 1555–1569.

57. Akhlaghi M, Bandy B (2009) Mechanisms of flavonoid protection against myocardial ischemia-reperfusion injury. J Mol Cell Cardiol 46: 309–317.

58. Akhlaghi M, Bandy B (2012) Preconditioning and acute effects of flavonoids in protecting cardiomyocytes from oxidative cell death. Oxid Med Cell Longev 2012: 782321.

59. Park SR, Yoo YJ, Ban YH, Yoon YJ (2010) Biosynthesis of rapamycin and its regulation: past achievements and recent progress. J Antibiot (Tokyo) 63: 434–441.

60. Sehgal SN, Baker H, Vezina C (1975) Rapamycin (AY-22,989), a new antifungal antibiotic. II. Fermentation, isolation and characterization. J Antibiot (Tokyo) 28: 727–732.

61. Calne RY, Collier DS, Lim S, Pollard SG, Samaan A, et al. (1989) Rapamycin for immunosuppression in organ allografting. Lancet 2: 227.

62. Steiner JP, Connolly MA, Valentine HL, Hamilton GS, Dawson TM, et al. (1997) Neurotrophic actions of nonimmunosuppressive analogues of immunosuppressive drugs FK506, rapamycin and cyclosporin A. Nat Med 3: 421–428.

63. Harrison DE, Strong R, Sharp ZD, Nelson JF, Astle CM, et al. (2009) Rapamycin fed late in life extends lifespan in genetically heterogeneous mice. Nature 460: 392–395.

64. Kahan BD (2000) Efficacy of sirolimus compared with azathioprine for reduction of acute renal allograft rejection: a randomised multicentre study. The Rapamune US Study Group. Lancet 356: 194–202.

65. Ko HT, Yin JL, Wyburn K, Wu H, Eris JM, et al. (2013) Sirolimus reduces vasculopathy but exacerbates proteinuria in association with inhibition of VEGF and VEGFR in a rat kidney model of chronic allograft dysfunction. Nephrol Dial Transplant 28: 327–336.

66. Letavernier E, Pe'raldi MN, Pariente A, Morelon E, Legendre C (2005) Proteinuria following a switch from calcineurin inhibitors to sirolimus. Transplantation 80: 1198–1203.

67. Chapman JR, Rangan GK (2010) Why do patients develop proteinuria with sirolimus? Do we have the answer? Am J Kidney Dis 55: 213–216.

68. Franz S, Regeniter A, Hopfer H, Mihatsch M, Dickenmann M (2010) Tubular toxicity in sirolimus- and cyclosporine-based transplant immunosuppression strategies: an ancillary study from a randomized controlled trial. Am J Kidney Dis 55: 335–343.

69. Biancone L, Bussolati B, Mazzucco G, Barreca A, Gallo E, et al. (2010) Loss of nephrin expression in glomeruli of kidney-transplanted patients under m-TOR inhibitor therapy. Am J Transplant 10: 2270–2278.

70. Vogelbacher R, Wittmann S, Braun A, Daniel C, Hugo C (2007) The mTOR inhibitor everolimus induces proteinuria and renal deterioration in the remnant kidney model in the rat. Transplantation 84: 1492–1499.

71. Vuiblet V, Birembaut P, Francois A, Cordonnier C, Noel LH, et al. (2012) Sirolimus-based regimen is associated with decreased expression of glomerular vascular endothelial growth factor. Nephrol Dial Transplant 27: 411–416.

72. Sehgal SN (2003) Sirolimus: its discovery, biological properties, and mechanism of action. Transplant Proc 35: 7S–14S.

73. Dowling RJ, Topisirovic I, Fonseca BD, Sonenberg N (2010) Dissecting the role of mTOR: lessons from mTOR inhibitors. Biochim Biophys Acta 1804: 433–439.

74. Abraham RT, Eng CH (2008) Mammalian target of rapamycin as a therapeutic target in oncology. Expert Opin Ther Targets 12: 209–222.

75. Wen HY, Abbasi S, Kellems RE, Xia Y (2005) mTOR: a placental growth signaling sensor. Placenta 26 Suppl A: S63–69.

76. Pollheimer J, Knofler M (2005) Signalling pathways regulating the invasive differentiation of human trophoblasts: a review. Placenta 26 Suppl A: S21–30.

77. Kim DH, Sarbassov DD, Ali SM, King JE, Latek RR, et al. (2002) mTOR interacts with raptor to form a nutrient-sensitive complex that signals to the cell growth machinery. Cell 110: 163–175.

78. Roos S, Jansson N, Palmberg I, Saljo K, Powell TL, et al. (2007) Mammalian target of rapamycin in the human placenta regulates leucine transport and is down-regulated in restricted fetal growth. J Physiol 582: 449–459.

18

Mass Spectrometry-Based Metabolite Profiling in the Mouse Liver following Exposure to Ultraviolet B Radiation

Hye Min Park[1], Jong Cheol Shon[2], Mee Youn Lee[1], Kwang-Hyeon Liu[2], Jeong Kee Kim[3], Sang Jun Lee[3], Choong Hwan Lee[1]*

1 Department of Bioscience and Biotechnology, Konkuk University, Gwangjin-gu, Seoul, Republic of Korea, 2 College of Pharmacy and Research Institute of Pharmaceutical Sciences, Kyungpook National University, Buk-gu, Daegu, Republic of Korea, 3 Food Research Institute, AmorePacific Corporation R&D Center, Giheung-gu, Yongin, Gyeonggi-do, Republic of Korea

Abstract

Although many studies have been performed on the effects of ultraviolet (UV) radiation on the skin, only a limited number of reports have investigated these effects on non-skin tissue. This study aimed to describe the metabolite changes in the liver of hairless mice following chronic exposure to UVB radiation. We did not observe significant macroscopic changes or alterations in hepatic cholesterol and triglyceride levels in the liver of UVB-irradiated mice, compared with those for normal mice. In this study, we detected hepatic metabolite changes by UVB exposure and identified several amino acids, fatty acids, nucleosides, carbohydrates, phospholipids, lysophospholipids, and taurine-conjugated cholic acids as candidate biomarkers in response to UVB radiation in the mouse liver by using various mass spectrometry (MS)-based metabolite profiling including ultra-performance liquid chromatography-quadrupole time-of-flight (TOF)-MS, gas chromatography-TOF-MS and nanomate LTQ-MS. Glutamine exhibited the most dramatic change with a 5-fold increase in quantity. The results from altering several types of metabolites suggest that chronic UVB irradiation may impact significantly on major hepatic metabolism processes, despite the fact that the liver is not directly exposed to UVB radiation. MS-based metabolomic approach for determining regulatory hepatic metabolites following UV irradiation will provide a better understanding of the relationship between internal organs and UV light.

Editor: Kyoung-Heon Kim, Korea University, Republic of Korea

Funding: This study was supported by a grant of the Korea Health Technology R&D Project, Ministry of Health & Welfare, Republic of Korea (Grant No. HN10C0017 and HN13C0076). The funder had no role in study design, data collection and analysis, decision to publish, or preparation of the manuscript.

Competing Interests: Two co-authors (Jeong Kee Kim and Sang Jun Lee) are employed by AmorePacific Corporation R&D center, but it has no connection with competing interests and financial disclosure.

* Email: chlee123@konkuk.ac.kr

Introduction

Ultraviolet (UV) radiation exerts a few beneficial effects (e.g., vitamin D production and use in treatment for jaundice, psoriasis, eczema, and vitiligo) but it also induces harmful responses such as sunburn, tanning, premature skin aging, suppression of the immune system, damage to the eyes, and cancer [1–4]. UV light is composed of UVA (320–400 nm), UVB (280–320 nm), and UVC (200–280 nm) light. Depending on its wavelength, UV light penetrates the skin and interacts with different cells located at different depths. UVB light is mostly absorbed in the epidermis of the skin and acts on DNA through direct excitation of its aromatic heterocyclic nucleobases. It causes direct damage to DNA, initiating the formation of the photoproducts cyclobutane pyrimidine dimers (CPDs) and pyrimidine pyrimidone (6-4PP). In addition to the direct effects of UVB radiation on DNA, UVB rays indirectly cause the production of free radicals, including reactive oxygen and nitrogen species [5–8]. The majority of previous studies on UV light have focused on the effects of acute/chronic exposure in skin tissue by using biochemical and molecular biological techniques. We also recently reported mass spectrometry (MS)-based metabolite profiling for time-dependent skin biomarkers in UVB-irradiated mice [9]. However, only a few reports have actually described the effects of UV radiation on non-skin tissues. Recently, Svobodová et al. [10] demonstrated that acute exposure to UVA/UVB light results in significant changes in oxidative stress-related biomarkers in the skin, liver, and blood of SKH-1 hairless mice. The liver has several functions; it regulates the levels of most chemicals in the blood, excretes bile, and detoxifies harmful compounds. Although the liver is not directly exposed to UV light, superoxide dismutase (SOD) activity and glutathione (GSH) levels are significantly altered following acute UVB exposure [10]. However, no study has yet been performed on metabolite alterations caused by chronic UVB exposure in the liver.

Identifying liver metabolites that change after exposure to UV is necessary to understand the indirect effect of UV irradiation on the liver in detail. Metabolites are critical in biology due to their involvement in cellular and physiological energetics, structure, and signaling [11]. Recently, metabolite profiling of the liver from mouse and human models was performed using nuclear magnetic resonance (NMR) and MS-based high-throughput techniques. The use of analytical instrument alone may not be sufficient because each instrument is limited in the chemical species it can analyze. Thus, the combined application of various analytical instruments is better suited to fully understand the correlation between liver and altered metabolites by analyzing the several types of compounds. Together with the use of MS, multivariate statistical analysis, such as principal component analysis (PCA), partial least-squares (PLS)-discriminant analysis (DA), and orthogonal PLS (OPLS)-DA, examines the relationship between experimental groups and involves data with several variables [12–14].

In the current study, the metabolites in the mouse liver after UVB irradiation for 6 weeks were profiled using various MS-based techniques including ultra-performance liquid chromatography (UPLC)-quadrupole time-of-flight (Q-TOF)-MS, gas chromatography (GC)-TOF-MS and nanomate LTQ-MS analyses with multivariate statistical analysis. We tentatively identified metabolites involved with various liver metabolism processes suggesting the hypothesis that could be used to distinguish liver tissue of non-exposed and UVB-exposed mice.

Materials and Methods

Reagents

Methanol, water, and acetonitrile were purchased from Fisher Scientific (Pittsburgh, PA, USA). Formic acid, dichloromethane, methoxyamine hydrochloride, and N-methyl-N-(trimethylsilyl)tri-fluoroacetamide (MSTFA) were obtained from Sigma Chemical Co. (St. Louis, MO, USA). All chemicals and solvents were of analytical grade and are commercially available.

Animals

Six-week-old female albino hairless mice (Skh:hr-1) weighing 18–22 g were obtained from Charles River Laboratories (Seoul, Korea). The animals were acclimatized for 1 week in an animal facility prior to the experiments and housed under controlled conditions of temperature ($23 \pm 2^\circ C$), relative humidity (55% $\pm 10\%$), and 12-hr light/dark light. The animals had free access to the laboratory diet (Purina, Seoul, Korea) and ion-sterilized tap water. All animal experiments were approved by the Amorepacific Institutional Animal Care and Use Committee (AP11-009-PE007) and adhere to the OECD guidelines.

Experimental design

The hairless mice were divided into the following two groups, with 10 mice in each group: normal group and UVB group. Each group of ten mice was housed in a cage. The average amount of feed consumed daily in each group was calculated statistically as mice were individually weighed once per week through the entire experimental period. The hairless mice of the normal group were sham irradiated, while those of the UVB group were exposed to UVB radiation 3 times per week (12 AM) starting with 1 minimal erythema dose (MED, 1 MED = 55 mJ/cm^2) for the first week. Then, the intensity was increased by 1 MED per week for up to 4 weeks, after which the mice were exposed to 4 MED for the duration of the experiment. The mice could move around freely in the cage during the period of exposure in a steel irradiation chamber. To mimic UV rays from sun, we used 10 fluorescent lamps (TL 20W/12RS; peak emission, 320 nm; wavelength, 275–390 nm; Philips, Amsterdam, Netherlands), and the UVB emission was monitored with a UV radiometer (VLX-3W; Vilber Lourmat, France). The irradiation intensity was measured at the bottom of the cage. After exposing mice to UVB radiation during week 6, mice of each group were sacrificed by cervical dislocation and liver tissue samples were collected.

Biochemical analyses

The hepatic cholesterol and triglyceride (TG) concentrations were determined with commercial kits (Abcam plc, Cambridge, UK). The statistical analysis was performed by an independent t-test.

Sample preparation

The liver extracts were prepared according to the modified method by Masson et al. [15] The mixture solvent (1.2 mL) with MeOH and water (1:1, v/v) was added to the frozen liver tissues (about 100 mg), which were then homogenized (30 frequency) for 2 min using a mixer mill (MM400; Retsch, Haan, Germany). The suspension was centrifuged at 4°C and 13,500 g for 10 min, and the resulting supernatant (MeOH/water (MW) extracts) was transferred to a 2 mL microcentrifuge tube. The remaining pellets were extracted again with 1 mL of the mixture solvent (dichloromethane:MeOH, 3:1, v/v). The supernatants (dichloromethane/MeOH (DM) extracts) were collected in new microcentrifuge tubes following centrifugation. Each extract solution was evaporated with a speed-vacuum machine. Dried samples were stored at − 80°C until UPLC-Q-TOF-MS and GC-TOF-MS analyses. Dried samples were resuspended with methanol/water (1:1, v/v) and were filtered through a 0.2-μm PTFE filter for the UPLC-Q-TOF-MS analysis. For GC-TOF-MS analysis, liver tissue extracts were oximated with 50 μL of methoxyamine hydrochloride (20 mg/mL) in pyridine at 30°C for 90 min. As a second derivatizing agent, 50 μL of MSTFA was added to the mixture, which was then incubated at 37°C for 30 min. Each liver extract was prepared with the same concentration for normalization of the different amount of tissue. The final concentration of each analyzed sample was 5 mg/mL.

For nanomate LTQ-MS analysis, total lipids in liver sample were extracted using a standard Bligh and Dyer's method. Liver (150 mg) was homogenized in the Tissue Lyser (frequency 1/s : 30, 3 min, 1 time, Qiagen) with 900 μL chloroform/methanol (1:2, v/v) and kept at room temperature for 1 h. Phase separation was achieved by adding 300 μL chloroform and 450 μL water, and the mixture was centrifuged at 10 min at 4°C and 100 g. The lower organic phase was transferred into a clean tube. The upper aqueous phase was reextracted with 600 μL chloroform and the mixture was centrifuged at 10 min at 4°C and 100 g. The resultant lower phase was combined with the previous organic phase extracts, and dried under a gentle stream of nitrogen. Dried samples were resuspended in 100 μL of chloroform/methanol (1:9, v/v) and diluted 10-fold with chloroform/methanol (1:9, v/v) containing 7.5 mM ammonium acetate. Aliquots were subjected to the direct infusion nanoelectrospray tandem mass spectrometry system to profile lipids in the samples.

GC-TOF-MS analysis

GC-TOF-MS analysis was performed on an Agilent 7890 GC system (Agilent, Atlanta, GA) coupled with a Pegasus HT TOF-MS (Leco Corp., St. Joseph, MI, USA) using an Agilent 7693 autosampler (Agilent, Atlanta, GA). The system was equipped with an Rtx-5MS column (29.8 m × 0.25 mm i.d.; particle size of 0.25 μm; Restek Corp., Bellefonte, PA, USA). The front inlet and transfer line temperatures were set at 250°C and 240°C, respectively. The helium gas flow rate through the column was 1.5 mL/min, and ions were generated by a −70 eV electron impact (EI). The ion source temperature was set at 230°C, and the mass range was 50–800 m/z. Column temperature was maintained isothermally at 75°C for 2 min, increased to 300°C at a rate of 15°C/min, and then held constant at 300°C for 3 min. One microliter of reactant was injected into the GC-TOF-MS with a split ratio of 10:1.

UPLC-Q-TOF-MS analysis

UPLC-Q-TOF-MS was performed on a Waters Q-TOF Premier system (Micromass MS Technologies, Manchester, UK) with a Waters Acquity UPLC System (Waters Corp., Milford, MA, USA) that was equipped with a Waters Acquity HPLC BEH C_{18} column (100 × 2.1 mm., i.d.; particle size of 1.7 μm). The samples were separated using a linear gradient consisting of water (A) and acetonitrile (B) with 0.1% v/v formic acid under the following conditions: 5% B for 1 min, gradually increased to 100% B for 10 min, held at 100% B for 1 min, decreased to 5% B over 1 min, and finally held at 5% B for 1 min. The injection volume of samples was 5 μL, and the flow rate was maintained at 0.3 mL/min. The TOF-MS data was collected in the range of 100–1,000 m/z with a scan time of 0.2 s and interscan time of 0.02 s in the negative ion mode. The capillary and cone voltages were set at 3.0 kV and 60 V, respectively. The desolvation gas flow was set to 600 L/h at a temperature of 200°C, and the cone gas flow was set to 50 L/h. The ion source temperature was 200°C.

Direct infusion-MS analysis

Liver lipid profiling was performed on a LTQ XL mass spectrometer (Thermo Fischer Scientific, West Palm Beach, FL) equipped with an automated nanospray source (TriVersa Nanomate, Advion Biosciences, Ithaca, NY) using nanoelectrospray chips with 5.5-μm diameter spraying nozzles. The ion source was controlled using the Chipsoft 8.3.1 software (Advion Biosciences). Ionization voltage was −1.4 kV in negative mode and backpressure was set at 0.4 psi. Ion transfer capillary temperature and tube voltage were 200°C and 100 V, respectively. For the lipid analysis, five microliters of each sample were loaded into a 96-well plate (Eppendorf, Hamburg, Germany), and placed on the Nanomate cooling plate, which was set to 5°C to prevent solvent evaporation. Full scan spectra were collected at the m/z 400–1,000 in positive ion mode. The mass spectra of each sample were acquired in profile mode over 2 min. A collision-induced dissociation (CID) was performed with over an isolated width of 3 m/z units, with 35% collision energy. The tandem mass spectrometry (MS/MS) triggering threshold was set to 1,000, with a default charge state of 1. All spectra were recorded with the Thermo Xcalibur software (version 2.1., Thermo Fisher Scientific). MS/MS spectra were analyzed for the identification of lipid species using LipidBlast [16] and in-house library.

Data processing and multivariate statistical analysis

Data processing for GC-TOF-MS and UPLC-Q-TOF-MS was performed using ChromaTOF software (Leco Corp., St. Joseph, MI, USA) and MassLynx software, respectively, and raw data files were converted to the network common data form (netCDF, *.cdf). After conversion, the MS data were processed using the Metalign software package (http://www.metalign.nl) to obtain a data matrix containing retention times, accurate masses, and normalized peak intensities. Metalign parameters were set according to the specific scaling requirements as follows: a peak slope factor of 1.0, peak threshold factor of 2, peak threshold of 500 or 1,000, and average peak width at half height of 10, which corresponds to a retention time of 3–20 min and mass range of 50–800 for GC-TOF-MS; a peak slope factor of 1.0, peak threshold factor of 2, peak threshold of 3, and average peak width

at half height of 10, which corresponds to a retention time of 1–10 min and mass range of 100–1,000 for UPLC-Q-TOF-MS. The resulting data were exported to Microsoft Excel (Microsoft, Redmond, WA, USA).

For nanomate LTQ-MS data, the nominal ion mass spectra, which arranged the scans between 0.5–1.0 min, was extracted using Xcalibur software (ThermoFisher Scientific, San Jose, CA, USA). We excluded the m/z values that showed peak intensities below 300. To normalize the spectrum, the average of the sum of intensities from the QC samples was divided by the sum of the intensities of each sample spectra, and then each value (fold) was multiplied by the intensity of each lipid species in that sample. The resulting data were exported to Microsoft Excel (Microsoft, Redmond, WA, USA).

Multivariate statistical analysis was performed using SIMCA-P+ software (version 12.0, Umetrics, Umea, Sweden). PCA, PLS-DA and OPLS-DA were performed to obtain information on differences in the metabolite profiles between two groups. The potential variables were selected based on variable importance in the projection (VIP) value (>0.7) that estimates the importance of each variable in the projection used in a PLS or OPLS model and p value (<0.05) using SIMCA-P+ software and Statistica 7 (StatSoft Inc., Tulsa, OK, USA). P-value was determined by single sample t-test for normality of two groups and Student's t-test for significance between two groups. Following multivariate statistical analysis, the corresponding peaks as selected variables were confirmed in the original chromatogram and were positively/tentatively identified using either commercial standard compounds in comparison with the mass spectra and retention time or on the basis of the NIST mass spectral database (National Institute of Standards and Technology, FairCom, Gaithersburg, MD, USA), in-house library, and references for GC-TOF-MS. For UPLC-Q-TOF-MS, assignment of metabolites contributing to the observed variance was performed by elemental composition analysis software with the calculated mass, mass tolerance (mDa and ppm), double bond equivalents (DBEs), and iFit algorithm implemented in the MassLynx and by commercial standard compounds, the Human Metabolome Database (HMDB, http://www.hmdb.com) and Lipid Maps Database (http://www.lipidmaps.org).

Results

Animal characteristics

Each female hairless mouse weighed between 27.59±1.41 and 28.16±2.05 g before the study began and was then fed ad libitum. Throughout the 6 weeks of the study, all animals appeared in good shape and gained weight without significant change. Exposure to UVB irradiation led to slight increases in the hepatic cholesterol and triglycerides compared with those for the normal mice, but there were no significant differences between the groups (**Table 1**).

Metabolite profiling of the livers from mice exposed to UVB irradiation

In this study, we profiled the changes seen in mice hepatic metabolites on UVB exposure by using UPLC-Q-TOF-MS, GC-TOF-MS and nanomate LTQ-MS with multivariate statistical analysis. According to analytical GC and UPLC, 7,664 (MW extract) and 3,040 (DM extract) variables were detected in both the normal and UVB group in GC while 836 (MW extract) and 571 (DM extract) variables were detected in UPLC, respectively. For lipid profiling, 168 variables were detected in two experimental groups in nanomate LTQ. These variables were applied to

Table 1. Metabolic parameters of hairless mice exposed to non-UVB and UVB irradiation for 6 weeks.

	Group	
	normal	UVB
Triglyceride (mM/g)	7.02±2.29	8.61±4.84
Total cholesterol (mg/g)[a]	3.04±0.21	3.17±0.32

[a]The assay detects total cholesterol, including cholesterol and cholesteryl esters.
Data were presented as mean ± SD. The statistical analysis was performed by an independent t-test.

PCA, PLS-DA score plots (**Fig. 1** and **Table S1 in File S1**) and OPLS-DA score plots to identify discriminable variables between the two experimental groups (**Fig. S1 in File S1**). Both groups were separated from each other by t [1], the predictive component, and t_0 [1], the first orthogonal component, based on the model of R^2X_{cum} and R^2Y_{cum} values of 0.38–0.61 and 0.81–1.00, respectively, and with Q^2Y_{cum} values of 0.59–0.84 in OPLS-DA model (**Table 2**). S-plots were generated using pareto scaling to visualize the metabolites responsible for the separation

between groups, selected on the basis of the VIP value (>0.7) and p value (<0.05), respectively (**Fig. 2**).

Metabolites contributing to the discrimination between groups from the livers of mice exposed to UVB irradiation

All metabolites, including 42 metabolites in GC-TOF-MS analysis, 33 metabolites in UPLC-Q-TOF-MS analysis and 28 metabolites in nanomate LTQ-MS analysis determined by the VIP and p values were significantly affected by the exposure to UVB radiation for 6 weeks. In addition, the number and changes

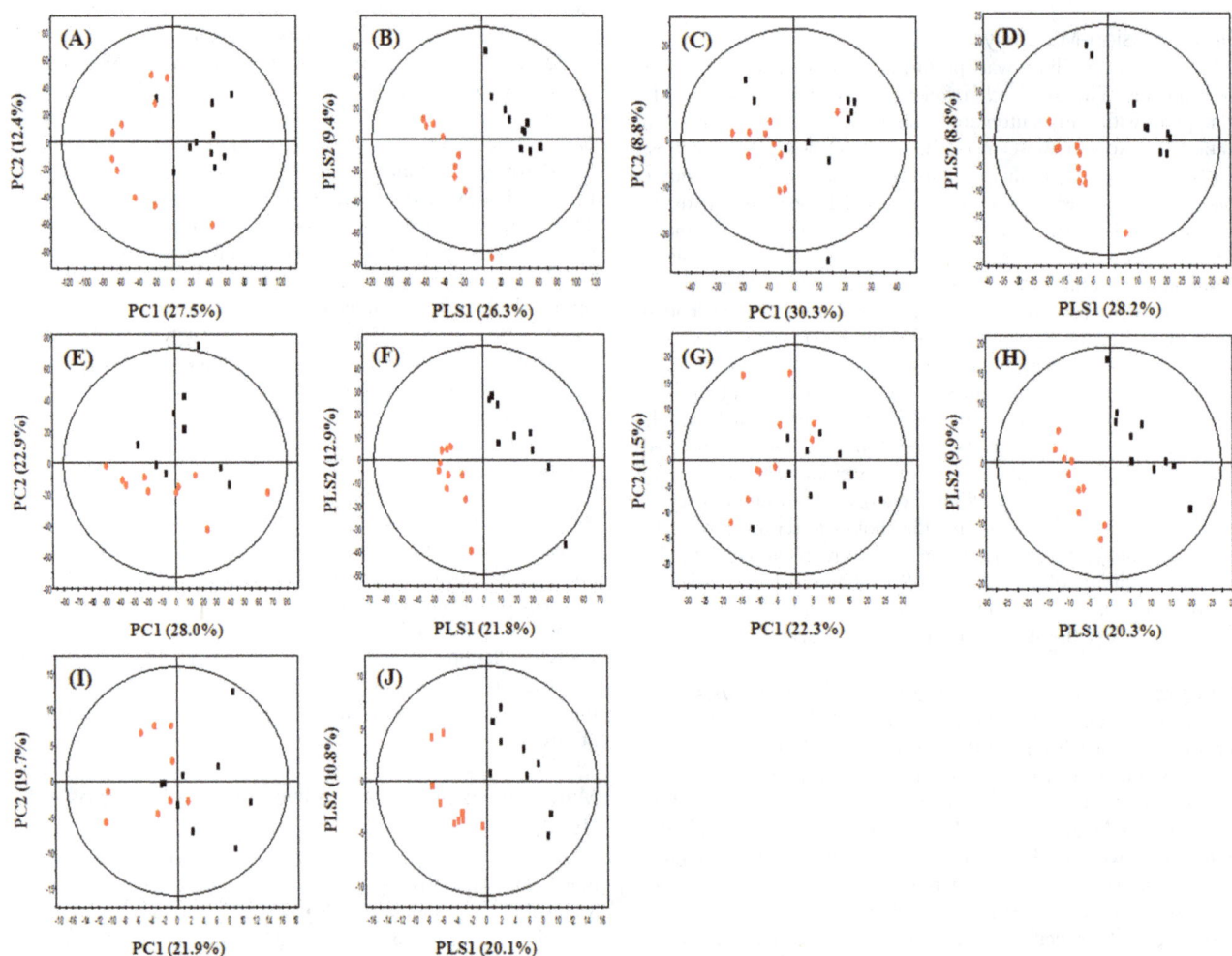

Figure 1. PCA score plot (A, C, E, G, I) and PLS-DA score plots (B, D, F, H, J) derived from GC-TOF-MS (A, B, E, F), UPLC-Q-TOF-MS (C, D, G, H) and Nanomate LTQ-MS (I, J) data sets for MW (A–D), DM (E–H) and lipid (I, J) extracts of mouse liver tissue after the exposure to UVB radiation for 6 weeks. ■ - normal, ● - UVB.

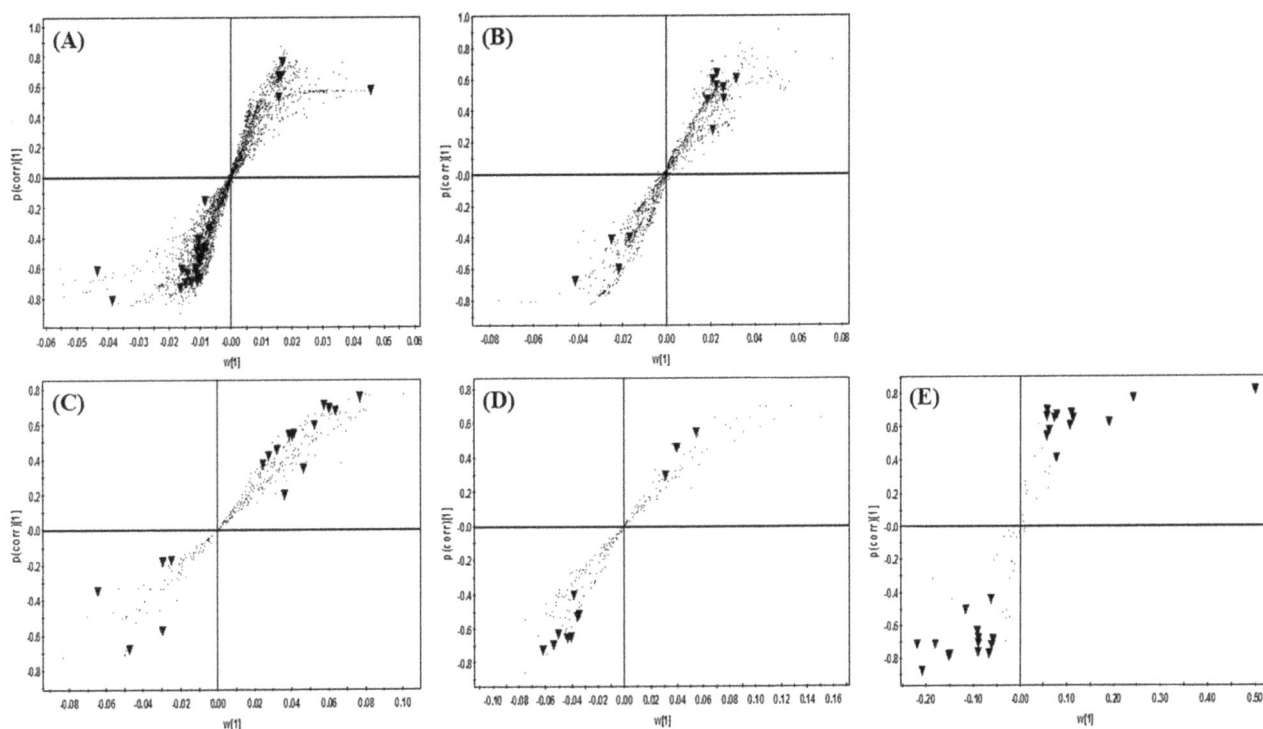

Figure 2. S-plots associated with OPLS-DA score plots derived from GC-TOF-MS (A, B), UPLC-Q-TOF-MS (C, D) and nanomate LTQ-MS (E) data sets for MW (A, C), DM (B, D) and lipid (E) extracts of mouse liver tissue after the exposure to UVB radiation for 6 weeks. The selected variables (▼, VIP>0.7 and $p<0.05$) are highlighted in S-plots. Each metabolites presented by a inverted triangle (▼) were the same metabolites presented in **Tables 3–5**.

of liver metabolites by UVB exposure were generally greater in MW extracts than in DM extracts from GC and UPLC. Among them, 686 metabolites were tentatively identified as important hepatic metabolites for determining the difference between normal and UVB-irradiated mice and are summarized in **Tables 3–5**.

In GC-TOF-MS analysis, the levels of 19 metabolites (aspartic acid, pyroglutamic acid, glutamic acid, L-glutamine, urea, nicotinamide, palmitic acid, linoleic acid, oleic acid, arachinodic acid, docosahexaenoic acid, *cis*- and *trans* oleamide, monopalmitine, uridine, inosine, glycerol and 2 saccharides) significantly increased after exposure to UVB rays, whereas the levels of 7 metabolites, that is, L-alanine, fumaric acid, taurine, elaidic acid, glucose, and 2 saccharides decreased (**Table 3**). Of these, L-

glutamine exhibited a 5.05-fold increase and glucose exhibited a 0.49-fold decrease. A VIP value greater than 3.0 was used to designate the major liver metabolites that contributed to discrimination between normal and UVB-irradiated mice.

UPLC-Q-TOF-MS analysis showed that the levels of 10 lysophosphatidylcholines (lysoPCs) (with two forms of C16:0, 18:1, 20:3, 20:4 and 22:6) and of 3 lysophosphatidylethanolamines (lysoPEs) with C 16:0, C18:1, and 20:4 related with glycerophospholipid metabolism significantly decreased, while 3 metabolites associated with taurine-conjugated bile acid metabolism were positively affected by UVB irradiation in the mouse liver (**Table 4**).

Table 2. Summary of parameters for assessment of the quality of OPLS-DA models.

	Extracts	R^2X_{cum}[a]	R^2Y_{cum}[a]	Q^2Y_{cum}[b]	p[c]
GC-TOF-MS	MW[d]	0.417	0.993	0.843	0.001
	DM[e]	0.379	0.984	0.834	<0.001
UPLC-Q-TOF-MS	MW	0.606	1.000	0.737	0.090
	DM	0.466	1.000	0.734	0.022
Nanomate LTQ-MS	Lipid[f]	0.521	0.812	0.587	0.015

[a]R^2X_{cum} and R^2Y_{cum} are the cumulative modeled variation in X and Y matrix, respectively.
[b]Q^2Y_{cum} is the cumulative predicted variation in Y matrix.
[c]P is p value obtained from cross validation ANOVA of OPLS-DA.
[d]MW, methanol/water (1:1, v/v).
[e]DM, dichloromethane/methanol (3:1, v/v).
[f]The lipid extract for nanomate LTQ-MS analysis was prepared as mentioned in M&M section.

Table 3. Metabolites in MW and DM extracts from the mouse liver that were significantly different between the normal and UVB groups after 6 weeks and were tentatively identified using GC-TOF-MS analysis.

t_R (min)[a]	Identified ion (m/z)	Metabolites	Derivatized	Fold Changes[b]	VIP[c]	P-value	Related metabolism	ID[d]
5.25	116	L-Alanine	$(TMS)_2$	0.68	1.48	<0.001	Amino acid metabolism	STD/MS[e]
6.71	171	Urea	$(TMS)_2$	1.93	3.80	0.004	Urea cycle	STD/MS
7.06	205	Glycerol	$(TMS)_3$	1.31	1.14	0.001	Glycerolipid, Carbohydrate metabolism	STD/MS
7.68	245	Fumaric acid	$(TMS)_2$	0.52^{DM}	2.29^{DM}	0.006	TCA cycle	STD/MS
8.91	179	Nicotinamide	TMS	1.16	0.88	0.001	Nicotinate and nicotinamide metabolism	STD/MS
9.25	232	Aspartic acid	$(TMS)_3$	1.36	1.30	<0.001	Amino acid metabolism	STD/MS
9.30	156	Pyroglutamic acid	$(TMS)_2$	1.22	0.84	0.024	Amino acid metabolism	STD/MS
10.04	246	Glutamic acid	$(TMS)_3$	1.44	1.43	<0.001	Amino acid metabolism	STD/MS
10.47	326	Taurine	$(TMS)_3$	0.67	1.43	<0.001	Taurine and hypotaurine metabolism	STD/MS
11.19	156	L-Glutamine	$(TMS)_3$	5.05	3.37	<0.001	Amino acid metabolism	STD/MS
12.02	204	Saccharide*		1.52^{DM}	1.75^{DM}	0.009	Carbohydrate metabolism	MS
12.08	205	Saccharide*		0.70	1.37	<0.001	Carbohydrate metabolism	MS
12.18	205	Glucose	MeOX, $(TMS)_5$	0.49	4.01	0.008	Carbohydrate metabolism	STD/MS
12.67	204	Saccharide*		0.59	1.37	0.019	Carbohydrate metabolism	MS
12.93	313	Palmitic acid	TMS	1.21	0.97	0.001	Lipid metabolsim	STD/MS
13.00	204	Saccharide*		1.23	0.89	0.011	Carbohydrate metabolism	MS
13.95	337	Linoleic acid	TMS	1.63	1.39	0.008	Lipid metabolsim	STD/MS
13.98	339	Oleic acid	TMS	1.37	0.97	0.041	Lipid metabolsim	STD/MS
14.00	339	Elaidic acid	TMS	0.84^{or}	1.19^{or}	0.007	Lipid metabolsim	STD/MS
14.86	91	Arachidonic acid	TMS	1.16	0.72	0.037	Lipid metabolsim	STD/MS
15.10	338	cis-Oleamide	TMS	$1.20^{MW}/1.21^{DM}$	$0.86^{MW}/1.24^{DM}$	$0.015^{MW}/0.007^{DM}$	Lipid metabolsim	STD/MS
15.14	338	trans-Oleamide	TMS	$1.25^{MW}/1.28^{DM}$	$0.88^{MW}/1.41^{DM}$	$0.030^{MW}/0.009^{DM}$	Lipid metabolism	MS
15.40	224	Uridine	$(TMS)_4$	1.22	0.94	0.010	Nucleic acid metabolism	STD/MS
15.88	91	Docosahexaenoic acid	TMS	1.26	1.02	0.005	Lipid metabolsim	STD/MS
16.00	371	Monopalmitin	$(TMS)_2$	$1.39^{MW}/1.21^{DM}$	$1.24^{MW}/1.25^{DM}$	$0.004^{MW}/0.008^{DM}$	Lipid metabolsim	STD/MS
16.05	217	Inosine	$(TMS)_4$	$1.23^{MW}/1.42^{DM}$	$0.97^{MW}/1.57^{DM}$	$0.003^{MW}/0.046^{DM}$	Nucleic acid metabolism	STD/MS

Variables were determined by using the VIP value (>0.7) and p-value (<0.05) from the OPLS-DA model.

MW; MeOH/water extracts, DM; dichloromethane/MeOH extracts.

[a]t_R was the retention time.

[b]Fold change was calculated by dividing the mean of the peak intensity of each metabolite from UVB group relative to the normal group.

[c]VIP, variable important in the projection.

[d]ID, identification.

[e]Metabolites were identified using commercial standard compounds (STD) in comparison with the mass spectra (MS) and retention time.

*Saccharide was not successfully identified, but its mass fragments were similar to general mass fragments of saccharides.

Table 4. Metabolites in MW and DM extracts from the mouse liver that were significantly different between the normal and UVB groups after 6 weeks and were tentatively identified using UPLC-Q-TOF-MS analysis.

t_R (min)[a]	Tentative metabolites[b]	Measured MS (m/z) Negative	Positive	M.W.[c]	HMDB Formula	Error (mDa)	Fold change[d]	VIP[e]	P-value
Glycerophospholipid metabolism									
8.09	LysoPC 22:6*	612.3304	568.3361	567	$C_{30}H_{50}NO_7P$	0.3	0.49	1.74	<0.001
8.12	LysoPE 20:4	500.2770	502.2991	501	$C_{25}H_{44}NO_7P$	−0.7	0.81^{DM}	0.83^{DM}	0.019
8.15	LysoPC 20:4*	588.3293	544.3316	543	$C_{28}H_{50}NO_7P$	−0.8	0.52^{MW}/0.81^{DM}	1.67^{MW}/0.85^{DM}	<0.001^{MW}/0.016^{DM}
8.21	LysoPC 22:6*	612.3292	568.3396	567	$C_{30}H_{50}NO_7P$	−0.9	0.6	1.2	0.012
8.28	LysoPC 20:4*	588.3292	544.3357	543	$C_{28}H_{50}NO_7P$	−0.9	0.65	1.12	0.013
8.49	LysoPC 20:3*	590.3464	546.3596	545	$C_{28}H_{52}NO_7P$	0.6	0.33^{MW}/0.69^{DM}	2.22^{MW}/1.20^{DM}	<0.001^{MW}/0.003^{DM}
8.54	LysoPC 16:0*	540.3307	496.3435	495	$C_{24}H_{50}NO_7P$	0.6	0.68^{MW}/0.79^{DM}	0.93^{MW}/0.97^{DM}	0.043^{MW}/0.002^{DM}
8.65	LysoPC 20:3*	590.3467	546.3624	545	$C_{28}H_{52}NO_7P$	0.9	0.46	1.52	0.006
8.74	LysoPE 16:0	452.2762	454.2962	453	$C_{21}H_{44}NO_7P$	−1.5	0.67^{DM}	1.29^{DM}	<0.001
8.79	LysoPC 16:0*	540.3285	496.3407	495	$C_{24}H_{50}NO_7P$	−1.6	0.77^{DM}	1.04^{DM}	0.002
8.80	LysoPC 18:1*	566.3464	522.3554	521	$C_{26}H_{52}NO_7P$	0.6	0.42	1.84	0.001
8.98	LysoPC 18:1*	566.3470	522.3516	521	$C_{26}H_{52}NO_7P$	1.2	0.61	1.17	0.014
8.99	LysoPE 18:1	478.2933	480.3109	479	$C_{23}H_{46}NO_7P$	−0.1	0.61^{DM}	1.48^{DM}	<0.001
Bile acid metabolsim									
4.78	Taurine conjugated cholic acid#	514.2838	538.2732	515	$C_{26}H_{45}NO_5S$	−1.6	1.39^{DM}	0.94^{DM}	0.047^{or}
5.29	Taurine conjugated cholic acid#	514.2843	538.2755	515	$C_{26}H_{45}NO_7S$	0.4	1.25	0.86	0.008
6.00	Taurine conjugated deoxycholic acid§	498.2893	522.2805	499	$C_{26}H_{45}NO_6S$	0.7	1.63^{MW}/1.63^{DM}	1.37^{MW}/1.30^{DM}	<0.001^{MW}/0.012^{DM}

Variables were selected by VIP value (>0.7) and p-value (<0.05) from OPLS-DA model.

MW; MeOH/water extracts, DM; dichloromethane/MeOH extracts.

FA; formic acid, LysoPC; lysophosphatidylcholine, LysoPE; lysophosphatidylethanolamine.

[a] t_R was retention time.

[b] Assignment of metabolites contributing to the observed variance was performed by elemental composition analysis software with calculated mass, mass tolerance (mDa and ppm), double bond equivalent (DBE), and the iFit algorithm was implemented in the MassLynx, and by either commercial standard compounds compared with the retention time and mass spectra or HMDB (The Human Metabolome Data Base (http://www.hmdb.ca/)).

[c] M.W.; molecular weight.

[d] Fold change was calculated by dividing the mean of the peak intensity of each metabolite from UVB group relative to normal group.

[e] VIP, variable important in the projection.

* Asterisk means the two forms of lysoPC, with the fatty acyl groups at positions 1 (sn-1) or 2 (sn-2) on the glycerol backbone.

Cholic acid derivatives with taurine were not successfully identified, but it was predicted to be one of the following compounds: taurocholic acid, taurallocholic acid, tauro-b-muricholic acid, taurohyocholate, or tauroursocholic acid.

§ Deoxycholic acid derivatives with taurine were not successfully identified, but it was predicted to be one of compounds such as tauroursodeoxycholic acid, taurodeoxycholic acid ortaurochenodesoxycholic acid.

Table 5. Metabolites in lipid extracts from the mouse liver that were significantly different between the normal and UVB groups after 6 weeks and were tentatively identified using nanomate LTQ-MS analysis.

ID	m/z (+)	Adduct	Fold change[a]	VIP[b]	p value
LPC 16:0	496.5	[M+H]+	0.75	2.83	0.004
LPC 18:1	522.5	[M+H]+	0.63	1.95	0.001
LPC 18:0	524.5	[M+H]+	0.63	2.68	<0.001
LPC 20:4	544.4	[M+H]+	0.74	1.17	0.019
LPC 20:3	546.5	[M+H]+	0.66	0.81	0.014
LPC22:6	568.5	[M+H]+	0.69	1.16	0.006
PE 34:2	716.5	[M+H]+	1.16	0.76	0.030
PC 32:0	734.6	[M+H]+	1.10	1.02	0.028
PE 36:4	740.5	[M+H]+	1.06	0.82	0.022
PE 36:2	744.5	[M+H]+	1.11	0.75	0.043
PC 32:0	756.6	[M+Na]+	1.21	1.44	0.005
PC 34:2	758.6	[M+H]+	1.23	6.49	<0.001
PE 38:6	764.5	[M+H]+	1.18	3.14	<0.001
PE 38:5	766.5	[M+H]+	1.09	1.49	0.006
PC 34:6	772.6	[M+Na]+	1.13	1.04	0.002
PC 34:3	778.6	[M+Na]+	1.15	0.75	0.005
PE 39:6		[M+H]+			
PE 38:5	788.7	[M+Na]+	0.93	1.19	0.030
PE 39:0	790.5	[M+H]+	1.11	1.40	0.005
PC 35:1	796.6	[M+Na]+	1.20	2.48	0.018
PE 39:5	802.6	[M+Na]+	1.15	0.97	0.002
PE 39:0	812.7	[M+Na]+	0.90	1.95	0.004
PE 40:6	814.6	[M+Na]+	0.92	0.87	0.008
PC 38:4	832.8	[M+Na]+	0.92	1.50	0.045
PC 38:3	834.8	[M+Na]+	0.89	2.32	0.010
PC 38:2	836.8	[M+Na]+	0.90	1.18	0.004
TG 56:4	928.6	[M+NH4]+	0.74	0.82	0.048

Variables were selected by VIP value (>0.7) and p value (<0.05) from OPLS-DA models.
[a]Fold change was calculated by dividing the mean of the peak intensity of each metabolite from the UVB-radiated group by that of the normal group.
[b]VIP, variable important in the projection.

In direct infusion-MS analysis, the levels of 6 lysoPCs with C16:0, C18:0, C18:1, C20:3, C20:4 and C 22:6 significantly decreased and this result was in agreement with the result of UPLC analysis. And, the levels of 5 phosphatidylcholines (PCs) and 7 phosphatidylethanolamines (PEs) significantly increased, whereas the levels of 3 PCs and 3 PEs significantly decreased after exposure to UVB light (**Table 5**).

Discussion

According to comprehensive MS-based metabolite profiling, we determined that several hepatic metabolites contributed to the differences seen on chronic exposure to UVB light and were associated with various metabolisms, including amino acid, lipid, glycerolipid, and nucleic acid metabolism. It is widely accepted that some of these metabolites regulate liver function and homeostasis.

The liver is a major organ for amino acid metabolism and is largely responsible for maintaining circulating amino acid homeostasis. Amino acids, including aspartate, glutamate, glutamine, glycine, and alanine, are detected at high concentrations in the liver [17]. Our data showed significant increases of most amino acid levels, especially L-glutamine by UVB exposure. Glutamine and glutamate metabolism are closely related with maintenance and promotion of cell function in diverse tissues and cells, especially in the liver. L-Glutamine is both the most abundant extracellular amino acid and the most significant nitrogen transporter between tissues, whereas glutamate is the most abundant intracellular amino acid and plays a specific role in the transamination of most amino acids, glucose homeostasis, lipid metabolism, and the regulation of the TCA cycle and urea cycle [18–21]. Glutamine synthetase uses glutamate and NH$_3$ to synthesize glutamine in the perivenous cells of the liver. In periportal cells, L-glutamine, as a precursor of glutamate, is associated with urea and glucose syntheses. Furthermore, glutamine acts as a key precursor for nucleic acid and nucleotide synthesis [19]. And high glutamine use in liver induces the production of glucose and urea [22]. Considering the multiple roles of glutamine, the increase of glutamine levels in the liver from exposure to UVB may have positive or negative influences on glutamine-related liver metabolisms.

This study showed that the levels of aspartate and urea increased, whereas the fumarate levels decreased following UVB irradiation for 6 weeks. These metabolites are associated with TCA and urea cycles, which are closely linked to each other. Urea is the major end product of nitrogen metabolism in humans and mammals. For urea formation, one of the two nitrogen atoms in urea comes from oxidative deamination of glutamate while the other nitrogen atom originates from aspartate. Aspartate is regenerated from fumarate produced by the urea cycle. Fumarate is oxidized to oxaloacetate by TCA cycle enzymes and is then converted by transamination into aspartate [23]. We predict that the increase in the urea level was probably induced, in part, by the alterations in aspartate, glutamate, and fumarate. Together with these metabolites, the levels of glycerol, glucose and some of the unidentified saccharides were altered in the liver of UVB-irradiated hairless mice. The liver also plays a unique role in controlling carbohydrate metabolism by regulating glucose production as well as glucose consumption. Glucose is produced either by breaking down glycogen (glycogenolysis) or by *de novo* synthesis of glucose (gluconeogenesis) from non-carbohydrate precursors, including lactate, amino acids, and glycerol, and is underutilized as the fuel of muscles and other organs such as the brain and kidney. The concentration of glucose in the liver is related to blood glucose levels. Excess glucose in blood converts into glycogen and is then stored in the liver. When blood sugar levels drop, the liver reconverts the glycogen back to glucose [24,25]. Although glucose levels were not measured in blood, our finding that glucose levels showed the greatest decline in the liver may affect glucose homeostasis.

In addition, the levels of glucose affect fatty acid (FA) and glycerolipid homeostasis [25,26]. In the liver of UVB-irradiated hairless mice, several FAs, including palmitic, linoleic, oleic, arachidonic and docosahexaenoic acid, increased, except for elaidic acid, the *trans* isomer of oleic acid. Lysophospholipids (e.g., lysoPC and lysoPE) decreased in liver (**Tables 4**). And, some PCs and PEs increased whereas others decreased in liver (**Table 5**). However, the levels of total TGs, which are formed by combining glycerol with 3 FAs, and the concentration of total cholesterol, including cholesterol and cholesteryl esters, were not altered by UVB exposure. FAs are the main component of phospholipids, TGs, and cholesterol esters. FAs and TGs are primarily an energy source for most organisms. FA metabolism in the liver involves 3 main pathways, that is, catabolism by β-oxidation, synthesis from acetyl CoA, and esterification into TGs [26,27]. Some FAs are used in the synthesis of phospholipids and eicosanoids, including prostaglandins, thromboxanes, and leukotrienes [28]. Eicosanoids, which are key mediators and regulators of inflammation, are usually generated from arachidonic acid, one of the *n*-6 polyunsaturated FAs (PUFAs). Eicosapentaenoic acid and docosahexaenoic acid as the *n*-3 PUFAs possess anti-inflammatory activity [29,30]. Inflammation also correlates with reactive oxygen species (ROS) production. The exposure of hepatocytes to fatty acids containing palmitic and oleic acid induced increased oxidative stress through ROS generation [31]. Furthermore, SFAs, such as palmitic and stearic acid, increase the saturation of membrane phospholipids and deregulate TCA cycle metabolism, leading to ROS accumulation [32]. Recent reports demonstrated acute exposure to UVB radiation caused significant increases in oxidative stress-related parameters, SOD activity and the level of GSH in the liver of hairless mice [10]. Together with FAs, the levels of either lysophospholipids (lysoPCs and lysoPEs) or partial phospholipids (PCs and PEs) diminished following UVB exposure in the liver. PC has been shown in numerous studies to protect liver cells from damage from a variety of toxins, such as ethanol,

carbon tetrachloride. PC can also be generated *via* choline pathway or methylation of PE by the enzyme PE *N*-methyltransferase. Li *et al.* [33] demonstrated that the alteration of PC/PE ratio influences membrane integrity and can lead to liver failure since the ratio of PC to PE is a key regulator of cell membrane integrity. In this study, the mouse liver of UVB group had an increased ratio of PC to PE compared to control livers. And, in particular, the changes of unsaturated forms of lysoPC and lysoPE were remarkable. Lysophospholipids are found in small amounts in most tissues. Notably, lysoPC, as a potent chemotactant, controls initiation of the adaptive cellular immune response [34] and induces pro-inflammatory cytokine production [35]. Furthermore, saturated acyl lysoPCs such as C16:0 and C18:0 induce inflammation while polyunsaturated acyl lysoPCs including C20:4 and C 22:6 inhibit lysoPC-induced inflammation *in vivo* [36]. Based on the literature for FAs, our results with the increased level of FAs and the decreased level of lysophospholipids indicate chronic exposure to UVB radiation may also be closely related to ROS production and inflammation in the liver of UVB-irradiated mice.

Uridine and inosine were altered by chronic 6-week exposure to UVB radiation. The pyrimidine precursor uridine undergoes degradation by being essentially cleared in a single pass through the liver and the formation by *de novo* synthesis in the liver. Le *et al.* [37] demonstrated that uridine modulated liver protein acetylation profiles in reference to the regulation of cellular energy metabolism in liver tissue. Inosine, a naturally occurring purine formed from the breakdown of adenosine, exerts potent anti-inflammatory effects by inhibiting the production of pro-inflammatory cytokines, both *in vitro* and *in vivo* [38,39]. The alterations in purine and pyrimidine nucleosides in the liver of UVB-irradiated mice can directly or indirectly influence the metabolism and immune-mediated processes of the liver.

This study demonstrated that taurine levels decreased following UVB irradiation, while those of taurine-conjugated bile acids increased, suggesting the changes in taurine and its conjugated metabolites are closely associated with major hepatic metabolism pathways. Taurine, a sulfur-containing amino acid, is found in high concentrations in all mammalian tissues and is involved in various physiological processes such as osmoregulation, antioxidation, detoxification, and bile acid conjugation in the liver [40–43]. Primary bile acids (e.g., cholic and chenodeoxycholic acid) and secondary bile acids (e.g., deoxycholic and lithocholic acid), which are synthesized from cholesterol and primary bile acids in the liver, primarily exist as *N*-acyl conjugated forms with glycine and taurine. The proportion of bile acid conjugated with taurine correlates with hepatic taurine concentrations [44–46]. These bile acids normally regulate cholesterol homeostasis, glucose metabolism, lipid solubilization, and metabolic signaling [47–48].

By measuring metabolite changes with a combination of MS analytic techniques in the mouse liver after UVB irradiation, we revealed that chronic exposure to UVB light may affect hepatic functions in liver metabolism processes. Although the liver is not directly exposed to UVB radiation, glutamine and glutamate metabolism, glucose and lipid homeostasis, and bile acid metabolism, were indirectly impacted through altering various kinds of metabolites. L-Glutamine exhibited the largest changes, indicating its potential as an indirect photodamage-related biomarker in the liver. In addition, because there are only a few in-depth reports on effects of UV light in the liver, further biochemical and molecular studies on the relationship between UVB radiation and liver tissue and between UVB radiation and altered metabolites are warranted. Nevertheless, this study suggests that the MS-based metabolomic approach for determining regulatory hepatic metabolites following the exposure to UV radiation will lead to a better understanding of the correlation between liver and UV exposure.

Supporting Information

File S1 Supporting information. Figure S1. OPLS-DA score plot derived from GC-TOF-MS (A, B), UPLC-Q-TOF-MS (C, D) and nanomate LTQ-MS (E) data sets for MW (A, C), DM (B, D) and lipid (E) extracts of mouse liver tissue after the exposure to UVB radiation for 6 weeks. ■ - normal, • – UVB. **Table S1.**

Summary of parameters for assessment of the quality of PLS-DA models

Author Contributions

Conceived and designed the experiments: HMP KHL CHL. Performed the experiments: HMP JCS MYL. Analyzed the data: HMP JCS MYL. Contributed reagents/materials/analysis tools: JKK SJL KHL CHL. Contributed to the writing of the manuscript: HMP JCS.

References

1. Piltingsrud HV, Odland LT, Fong CW (1976) An evaluation of fluorescent light sources for use in phototherapy of neonatal jaundice. Am Ind Hyg Assoc J 37: 437–444.
2. Bishop SC (1979) DNA repair synthesis in human skin exposed to ultraviolet radiation used in PUVA (psoralen and UV-A) therapy for psoriasis. Br J Dermatol 101: 399–405.
3. Midelfart K, Stenvold SE, Volden G (1985) Combined UVB and UVA phototherapy of atopic eczema. Dermatologica 171: 95–98.
4. El-Zawahry BM, Bassiouny DA, Sobhi RM, Abdel-Aziz E, Zaki NS, et al. (2012) A comparative study on efficacy of UVA1 vs. narrow-band UVB phototherapy in the treatment of vitiligo. Photodermatol Photoimmunol Photomed 28: 84–90.
5. Ravanat JL, Douki T, Cadet J (2001) Direct and indirect effects of UV radiation on DNA and its components. J Photochem Photobiol B 63: 88–102.
6. Trautinger F (2001) Mechanisms of photodamage of the skin and its functional consequences for skin ageing. Clin Exp Dermatol 26: 573–577.
7. Svobodova A, Walterova D, Vostalova J (2006) Ultraviolet light induced alteration to the skin. Biomed Pap Med Fac Univ Palacky Olomouc Czech Repub 150: 25–38.
8. Verschooten L, Claerhout S, Van Laethem A, Agostinis P, Garmyn M (2006) New strategies of photoprotection. Photochem Photobiol 82: 1016–1023.
9. Park HM, Shin JH, Kim JK, Lee SJ, Hwang GS, et al. (2013) MS-based metabolite profiling reveals time-dependent skin biomarkers in UVB-irradiated mice. Metabolomics. In press Doi: 10.1007/s11306-013-0594-x.
10. Svobodová AR, Galandáková A, Sianská J, Doležal D, Ulrichová J, et al. (2011) Acute exposure to solar simulated ultraviolet radiation affects oxidative stress-related biomarkers in skin, liver and blood of hairless mice. Biol Pharm Bull 34: 471–479.
11. Vinayavekhin N, Homan EA, Saghatelian A (2010) Exploring disease through metabolomics. ACS Chem Biol 5: 91–103.
12. Garrod S, Humpher E, Connor SC, Connelly JC, Spraul M, et al. (2001) High-resolution ¹H NMR and magic angle spinning NMR spectroscopic investigation of the biochemical effects of 2-bromoethanamine in intact renal and hepatic tissue. Magn Reson Med 45: 781–790.
13. Kim HJ, Kim JH, Noh S, Hur HJ, Sung MJ, et al. (2011) Metabolomic analysis of livers and serum from high-fat diet induced obese mice. J Proteome Res 10: 722–731.
14. Kim J, Choi JN, Choi JH, Cha YS, Muthaiya MJ, et al. (2013) Effect of fermented soybean product (Cheonggukjang) intake on metabolic parameters in mice fed a high-fat diet. Mol Nutr Food Res 57: 1886–1891.
15. Masson P, Alves AC, Ebbels TM, Nicholson JK, Want EJ (2010) Optimization and evaluation of metabolite extraction protocols for untargeted metabolic profiling of liver samples by UPLC-MS. Anal Chem 82: 7779–7786.
16. Kind T, Liu KH, Lee do Y, DeFelice B, Meissen JK, et al. (2013) LipidBlast in silico tandem mass spectrometry database for lipid identification. Nat Methods 10: 755–758.
17. Brosnan ME, Brosnan JT (2007) The Textbook of Hepatology: From Basic Science to Clinical Practice, 3rd Edition. Wiley-Blackwell, UK. 142–149.
18. Watford M (2000) Glutamine and glutamate metabolism across the liver sinusoid. J Nutr 130: 983S–987S.
19. Yang D, Brunengraber H (2000) Glutamate, a window on liver intermediary metabolism. J Nutr 130: 991S–994S.
20. Newsholme P, Procopio J, Lima MM, Pithon-Curi TC, Curi R (2003) Glutamine and glutamate–their central role in cell metabolism and function. Cell Biochem Funct 21: 1–9.
21. Brosnan ME, Brosnan JT (2009) Hepatic glutamate metabolism: a tale of 2 hepatocytes. Am J Clin Nutr 90: 857S–861S.
22. Watford M, Chellaraj V, Ismat A, Brown P, Raman P (2002) Hepatic glutamine metabolism. Nutrition 18: 301–303.
23. Shambaugh GE 3rd (1977) Urea biosynthesis I. The urea cycle and relationships to the citric acid cycle. Am J Clin Nutr 30: 2083–2087.
24. Sherwin RS (1980) Role of the liver in glucose homeostasis. Diabetes Care 3: 261–265.
25. Postic C, Dentin R, Girard J (2004) Role of the liver in the control of carbohydrate and lipid homeostasis. Diabetes Metab 30: 398–408.
26. Nguyen P, Leray V, Diez M, Serisier S, Le Bloc'h J, et al. (2008) Liver lipid metabolism. J Anim Physiol Anim Nutr (Berl) 92: 272–283.
27. Bechmann LP, Hannivoort RA, Gerken G, Hotamisligil GS, Trauner M, et al. (2012) The interaction of hepatic lipid and glucose metabolism in liver diseases. J Hepatol 56: 952–964.
28. Bradbury MW (2006) Lipid metabolism and liver inflammation. I. Hepatic fatty acid uptake: possible role in steatosis. Am J Physiol Gastrointest Liver Physiol 290: G194–198.
29. Calder PC (2006) n-3 polyunsaturated fatty acids, inflammation, and inflammatory diseases. Am J Clin Nutr 83: 1505S–1519S.
30. Calder PC (2009) Polyunsaturated fatty acids and inflammatory processes: New twists in an old tale. Biochimie 91: 791–795.
31. Chavez-Tapia NC, Rosso N, Tiribelli C (2012) Effect of intracellular lipid accumulation in a new model of non-alcoholic fatty liver disease. BMC Gastroenterol 12: 20.
32. Leamy AK, Egnatchik RA, Young JD (2013) Molecular mechanisms and the role of saturated fatty acids in the progression of non-alcoholic fatty liver disease. Prog Lipid Res 52: 165–174.
33. Li Z, Agellon LB, Allen TM, Umeda M, Jewell L, et al. (2006) The ratio of phosphatidylcholine to phosphatidylethanolamine influences membrane integrity and steatohepatitis. Cell Metab 3: 321–331.
34. Perrin-Cocon L, Agaugué S, Coutant F, Saint-Mézard P, Guironnet-Paquet A, et al. (2006) Lysophosphatidylcholine is a natural adjuvant that initiates cellular immune responses. Vaccine 24: 1254–1263.
35. Olofsson KE, Andersson L, Nilsson J, Björkbacka H (2008) Nanomolar concentrations of lysophosphatidylcholine recruit monocytes and induce pro-inflammatory cytokine production in macrophages. Biochem Biophys Res Commun 370: 348–352.
36. Hung ND, Sok DE, Kim MR (2012) Prevention of 1-palmitoyl lysophosphatidylcholine-induced inflammation by polyunsaturated acyl lysophosphatidylcholine. Inflamm Res 61: 473–483.
37. Le TT, Ziemba A, Urasaki Y, Hayes E, Brotman S, et al. (2013) Disruption of uridine homeostasis links liver pyrimidine metabolism to lipid accumulation. J Lipid Res 54: 1044–1057.
38. Haskó G, Kuhel DG, Németh ZH, Mabley JG, Stachlewitz RF, et al. (2000) Inosine inhibits inflammatory cytokine production by a posttranscriptional mechanism and protects against endotoxin-induced shock. J Immunol 164: 1013–1019.
39. Liaudet L, Mabley JG, Soriano FG, Pacher P, Marton A, et al. (2001) Inosine reduces systemic inflammation and improves survival in septic shock induced by cecal ligation and puncture. Am J Respir Crit Care Med 164: 1213–1220.
40. Vessey DA (1978) The biochemical basis for the conjugation of bile acids with either glycine or taurine. Biochem J 174: 621–626.
41. Häussinger D (2004) Neural control of hepatic osmolytes and parenchymal cell hydration. Anat. Rec. A Discov. Mol Cell Evol Biol 280: 893–900.
42. Refik Mas M, Comert B, Oncu K, Vural SA, Akay C, et al. (2004) The effect of taurine treatment on oxidative stress in experimental liver fibrosis. Hepatol Res 28: 207–215.
43. Das J, Roy A, Sil PC (2012) Mechanism of the protective action of taurine in toxin and drug induced organ pathophysiology and diabetic complications: a review. Food Funct 3: 1251–1264.
44. Murphy GM, Signer E (1974) Bile acid metabolism in infants and children. Gut 15: 151–163.
45. Hardison WG, Proffitt JH (1977) Influence of hepatic taurine concentration on bile acid conjugation with taurine. Am J Physiol 232: E75–79.
46. Hardison WG (1978) Hepatic taurine concentration and dietary taurine as regulators of bile acid conjugation with taurine. Gastroenterology 75: 71–75.
47. Hofmann AF (1999) The continuing importance of bile acids in liver and intestinal disease. Arch Intern Med 159: 2647–2658.
48. Staels B, Fonseca VA (2009) Bile acids and metabolic regulation: mechanisms and clinical responses to bile acid sequestration. Diabetes Care 32: S237–245.

19

The Relation of Rapid Changes in Obesity Measures to Lipid Profile - Insights from a Nationwide Metabolic Health Survey in 444 Polish Cities

Bernhard M. Kaess[1,2,3], Jacek Jóźwiak[4,5], Christopher P. Nelson[1,6], Witold Lukas[7], Mirosław Mastej[5], Adam Windak[8], Tomasz Tomasik[8], Władysław Grzeszczak[9], Andrzej Tykarski[10], Jerzy Gąsowski[11], Izabella Ślęzak-Prochazka[4], Andrzej Ślęzak[4], Fadi J. Charchar[1,12], Naveed Sattar[13], John R. Thompson[6,14], Nilesh J. Samani[1,6], Maciej Tomaszewski[1,6]*

1 Department of Cardiovascular Sciences, University of Leicester, Leicester, United Kingdom, 2 Deutsches Herzzentrum, Technische Universität, Munich, Germany, 3 Department for Internal Medicine II, University Hospital Regensburg, Regensburg, Germany, 4 Department of Public Health, Częstochowa University of Technology, Częstochowa, Poland, 5 Silesian Analytical Laboratories, Katowice, Poland, 6 The NIHR Leicester Biomedical Research Unit in Cardiovascular Disease, Leicester, United Kingdom, 7 Department of Family Medicine, Medical University of Silesia, Zabrze, Poland, 8 Department of Family Medicine, Jagiellonian University Medical College, Kraków, Poland, 9 Department of Internal Diseases, Diabetology and Nephrology, Medical University of Silesia, Zabrze, Poland, 10 Department of Hypertension, Vascular Diseases, and Internal Diseases, University of Medical Sciences, Poznań, Poland, 11 Department of Internal Medicine and Gerontology, Jagiellonian University Medical College, Kraków, Poland, 12 School of Health Sciences, University of Ballarat, Ballarat, Australia, 13 Institute of Cardiovascular and Medical Sciences, University of Glasgow, Glasgow, United Kingdom, 14 Department of Health Sciences, University of Leicester, Leicester, United Kingdom

Abstract

Objective: The impact of fast changes in obesity indices on other measures of metabolic health is poorly defined in the general population. Using the Polish accession to the European Union as a model of political and social transformation we examined how an expected rapid increase in body mass index (BMI) and waist circumference relates to changes in lipid profile, both at the population and personal level.

Methods: Through primary care centres in 444 Polish cities, two cross-sectional nationwide population-based surveys (LIPIDOGRAM 2004 and LIPIDOGRAM 2006) examined 15,404 and 15,453 adult individuals in 2004 and 2006, respectively. A separate prospective sample of 1,840 individuals recruited in 2004 had a follow-up in 2006 (LIPIDOGRAM PLUS).

Results: Two years after Polish accession to European Union, mean population BMI and waist circumference increased by 0.6% and 0.9%, respectively. This tracked with a 7.6% drop in HDL-cholesterol and a 2.1% increase in triglycerides (all $p<0.001$) nationwide. The direction and magnitude of the population changes were replicated at the personal level in LIPIDOGRAM PLUS (0.7%, 0.3%, 8.6% and 1.8%, respectively). However, increases in BMI and waist circumference were both only weakly associated with HDL-cholesterol and triglycerides changes prospectively. The relation of BMI to the magnitude of change in both lipid fractions was comparable to that of waist circumference.

Conclusions: Moderate changes in obesity measures tracked with a significant deterioration in measures of pro-atherogenic dyslipidaemia at both personal and population level. These associations were predominantly driven by factors not measureable directly through either BMI or waist circumference.

Editor: Vladimir N. Uversky, University of South Florida College of Medicine, United States of America

Funding: LIPIDOGRAM project was supported by a research grant from Schwarz Pharma, Silesian Analytical Laboratories (the executive medical faculty of the study) and the The College of Family Physicians in Poland. BK's research fellowship in the Department of Cardiovascular Sciences, University of Leicester, was supported by the Cardiogenics project of the European Union (LSHM-CT-2006-037593). The funding bodies had no role in study design, collection, analysis and interpretation of data, writing the report, or submitting the paper for publication.

Competing Interests: Silesian Analytical Laboratories and Schwarz Pharma are commercial funders. JJ and MM are employees of Silesian Analytical Laboratories, a commercial clinical chemistry laboratory.

* E-mail: mt142@le.ac.uk

Introduction

Clinical measures of obesity such as body mass index (BMI), waist circumference and waist/height ratio are strong risk factors of cardiovascular and metabolic diseases [1–4]. The central role of obesity in cardiovascular disorders is undisputed [5]. It is also widely acknowledged that the effect of obesity on cardiovascular disease is at least partly mediated by its traditional risk factors including lipids [6,7]. However, the exact mechanisms underlying relationships between BMI/waist indices and circulating levels of lipids remain elusive. In fact, there is a surprising paucity of epidemiological studies that have investigated the impact of changes in abdominal and general obesity measures on lipids. In particular, it is not clear to what extent increasing obesity at the population level is associated with a deteriorating lipid profile. It is also unknown whether increasing visceral obesity (i.e. best approximated simply by waist circumference) at the population levels may have a stronger relationship with deteriorating lipid profile than changes in general obesity (assessed by BMI).

To examine the extent of association between changes in measures of obesity and lipids we took advantage of data collected in an exceptional "experiment of history" – the accession of Poland to the European Union in 2004. We hypothesised that the economic and social consequences of this major political development in Europe would lead to rapid changes in measures of obesity and that these changes may have a significant impact on circulating levels of lipids in the Polish population. We chose to quantify and relate the short-term (2-year) changes in both measures of obesity to lipids initially at the population (cross-sectional analysis) and then seek independent replication of the findings at the personal level (prospective level).

Materials and Methods

Study Populations

LIPIDOGRAM was a nationwide survey of cardiovascular risk factors carried out through primary care outpatient centres in Poland right after Polish accession to the EU (2004) and two years later (2006). In 2004, a total of 700 doctors in 16 major administrative regions in Poland were invited to recruit up to 30 consecutive adults attending their primary care practices into this study. Between October and December 2004 a total of 675 primary care specialists actively enrolled 17,522 individuals in 444 cities. Of these 15,404 were recruited into the cross-sectional arm (no follow-up; LIPIDOGRAM2004) and 2,118 into a separate prospective observation with a planned follow-up in 2006 (LIPIDOGRAM PLUS).

Between October and December 2006, 556 primary care practitioners (82.4% of those recruiting in 2004) from 402 Polish cities recruited a total of 15,465 individuals into another cross-sectional analysis (LIPIODGRAM2006). In addition, individuals recruited into LIPIDOGRAM PLUS in 2004 had their follow-up in 2006. None of the participants recruited into the cross-sectional analysis in 2004 were included in the population survey in 2006. There was also no overlap in individuals recruited into the cross-sectional and the prospective studies.

The number of recruitment centres across 16 major administrative regions of Poland as well as the number of individuals recruited in each region and included in analysis was a direct function of demographic size of each region (**Table S1**). These proportions were retained in 2004 and 2006 in the cross-sectional surveys as well as in the prospective analysis (**Tables S1–2 and Figures S1–3**). All recruiting medical practitioners underwent a formal training in the study protocol, procedures and collection of the standardised questionnaires and blood samples.

Eligible for recruitment were individuals aged at least 30 years who attended an appointment in the primary care during the recruitment period. The exclusion criteria were dementia/mental disease resulting in inability to give an informed consent, and incomplete clinical or biochemical information. Each individual was phenotyped according to the protocol introduced in LIPIDOGRAM2004 and described in detail before [2]. In brief, phenotyping included collection of basic demographic and clinical data by standardised coded questionnaires, anthropometric measurements (weight, height, waist circumference) and fasting blood sample for further biochemical analyses. Weight and height measurements were carried out without heavy clothing and shoes. Waist circumference was measured over the unclothed abdomen at a level of the midpoint between the lower margin of the ribs and the anterior superior iliac crest spine [2]. Smoking was defined as current smoking of at least 1cigarette/day. All patients were of white Polish ethnicity.

All participants provided written informed consent. The study was approved by the Bioethical Committee of the Polish Chamber of Physicians (No 51/2004/U) and conforms to the principles outlined in the Declaration of Helsinki.

Biochemical Analyses

Blood samples were taken under fasting (for at least 12 hours) conditions. Blood samples were centrifuged locally and separated serum samples were transferred by one contracted courier company to the central facility where all biochemical analyses were conducted within 12 hours after blood collection. All biochemical measurements (serum) were conducted in the same International Organization for Standardization (ISO) certified laboratory and under the same experimental conditions on an automatic bio-analyzer (ARCHITECT-c8000, ABBOTT Laboratories, USA). Serum concentrations of total cholesterol were measured using a photometric method with cholesterol oxidase (DiaSys – Diagnostic Systems, Germany). HDL-cholesterol (HDL-C) and triglycerides were analyzed by immuno-separation-based homogenous assay and colorimetric enzymatic test with glycerol-3-phosphateoxidase, respectively (both DiaSys – Diagnostic Systems, Germany). The intra-assay and inter-assay coefficients of variation for total cholesterol, HDL-C and triglycerides were low (0.61, 1.22 and 0.41, 0.83 and 1.23, 1.60, respectively). LDL-cholesterol (LDL-C) was calculated using the Friedewald formula in those whose triglycerides were measured <4.5 mmol/L.

Internal tests of accuracy and precision were conducted by using CDC-certified lyophilised human serum control samples (including those with high, normal and low values for each directly measured lipid fraction) together with participants' samples in each run. Total error (TE – a measure that combines analytical imprecision and systematic bias) for each of these lipid fractions was calculated using the formula: $TE = Bias(\%)+2CV$ and never exceeded the total allowable error calculated at 9% for TC, 14% for TG and 16% for HDL-C. The national external audit of quality control (National Research Centre for Quality in Laboratory Diagnostics, Lodz, Poland) and the international accreditation of the external control (Labquality, Helsinki, Finland and INSTAND e.V., Düsseldorf, Germany) were conducted both in 2004 and 2006 and confirmed good quality standards for lipids measurements.

Statistical Analysis

Changes in quantitative metabolic variables were analysed using linear models fitted by standard regression to the cross-sectional data and by Generalized Estimating Equations to the prospective data. Triglycerides were log-transformed prior to regression analysis and back-transformed into the original measurement scale (data are presented as geometric means and their standard errors). All models included a term for the survey year (coded as 2004 = 0, 2006 = 1). Fully adjusted regression models included terms for age, age^2, sex, height, region of recruitment, current smoking and education (coded as primary, secondary or higher).

Sex differences in changes in measures obesity and lipids were examined by comparison of corresponding regression coefficients using Wald (z-statistic) tests.

In the prospective LIPIDOGRAM PLUS study, adjusted and unadjusted tests for linear trend were used to examine association between changes in lipids (as a continuous measure) across categorised change in BMI/waist circumference.

Results

The general and sex-stratified clinical characteristics of the study groups are summarised in **Table 1** and **Tables S3–4**. A total of 14,849 and 15,453 individuals recruited into the cross-sectional survey in 2004 and 2006 were included in this analysis after excluding those with missing or inconsistent information. A total of 1,840 individuals with full demographic and biochemical data recruited in 2004 into the prospective LIPIDOGRAM PLUS were surveyed again in 2006 (follow-up rate –86.9%).

In the cross-sectional study adjusted BMI and waist circumference increased between 2004 and 2006 by approximately 0.6% [0.18 (0.05) kg/m^2, P<0.001] and 0.9% [0.86 (0.14) cm, P<0.001], respectively (**Table 2**). These changes in the population were mirrored at the individual level; BMI increased by 0.7% [0.2 (0.05) kg/m^2, P<0.001] and waist circumference by 0.3% [0.28 (0.16) cm, P = 0.09] in LIPIDOGRAM PLUS (**Table 2**). The direction of change in both indices of obesity between 2004 and 2006 was similar in both sexes, although the increasing trend was generally more prominent in men (**Table S5**).

Of four lipid fractions the most significant change over the two-year observation period was observed in circulating concentrations of HDL-C. After adjustment, HDL-C dropped by 7.6% [0.125 (0.004) mmol/L, P<0.001] nationwide. The increase in circulating concentrations of triglycerides was calculated at 2.1% (P<0.001) (**Table 2**). Serum levels of LDL-C levels remained essentially constant whilst total cholesterol dropped by 1.7% (P<0.001) in the surveyed population (**Table 2**). The direction and magnitude of changes in lipids in the prospective LIPIDO-GRAM PLUS were very similar; a 8.6% drop in HDL-C (P<0.001) was accompanied by a 1.8% increase in triglycerides (P<0.001), a 2.8% decrease in total cholesterol (P<0.001) and no changes in LDL-C (**Table 2**). These trends were also apparent in

sex-specific analysis with very little difference in the magnitude of the changes in lipids between men and women over the two-year observation period (**Table S5**).

Table 2. Changes in obesity measures and lipids between 2004 and 2006 in the cross-sectional and prospective LIPIDOGRAM Studies.

Measure	Model	Cross-sectional β (SE)	P-value	Prospective β (SE)	P-value
BMI (kg/m^2)	Basic	0.20 (0.05)	<0.001	0.33 (0.04)	<0.001
	Full	0.18 (0.05)	<0.001	0.20 (0.05)	<0.001
Waist (cm)	Basic	0.72 (0.15)	<0.001	0.89 (0.15)	<0.001
	Full	0.86 (0.14)	<0.001	0.28 (0.16)	0.09
HDL-C (mmol/L)	Basic	−0.119 (0.004)	<0.001	−0.147 (0.006)	<0.001
	Full	−0.125 (0.004)	<0.001	−0.145 (0.006)	<0.001
TG (mmol/L)	Basic	0.027 (0.005)	<0.001	0.034 (0.009)	<0.001
	Full	0.031 (0.005)	<0.001	0.026 (0.009)	<0.001
TC (mmol/L)	Basic	−0.104 (0.013)	<0.001	−0.151 (0.025)	<0.001
	Full	−0.100 (0.013)	<0.001	−0.162 (0.025)	<0.001
LDL-C (mmol/L)	Basic	0.008 (0.011)	0.46	−0.016 (0.022)	0.05
	Full	0.015 (0.011)	0.19	−0.026 (0.022)	0.24

Data are changes between 2004 and 2006 are expressed as β-coefficients with standard errors (SE); BMI – body mass index; HDL-C – high-density lipoprotein cholesterol; TG – triglycerides; TC – total cholesterol; LDL-C – low-density lipoprotein cholesterol; Basic – unadjusted model; Full – model adjusted for age, age^2, sex, region of recruitment, height, education and smoking.

Table 1. General clinical characteristics of LIPIDOGRAM2004, LIPIDOGRAM2006 and LIPIDOGRAM PLUS Studies.

Characteristic	Cross-sectional 2004	2006	P-value	Prospective 2004	2006	P-value
N	14849	15453	–	1840	1840	–
Men (%)	6004 (40.4)	5806 (37.6)	<0.001	712 (38.7)	712 (38.7)	–
Age (years)	55.4 (10.7)	55.5 (11.1)	0.46	53.3 (9.7)	55.3 (9.7)	–
Height (cm)	166.4 (8.6)	166.1 (8.4)	<0.001	166.7 (8.5)	166.7 (8.5)	–
Weight (kg)	78.3 (15.1)	78.5 (14.9)	0.14	78.0 (14.9)	78.9 (15.1)	<0.001
BMI (kg/m^2)	28.2 (4.8)	28.4 (4.6)	<0.001	28.0 (4.6)	28.3 (4.6)	<0.001
Waist (cm)	92.4 (13.1)	93.1 (13.0)	<0.001	91.3 (12.5)	92.1 (12.5)	<0.001
HDL-C (mmol/L)	1.65 (0.39)	1.53 (0.38)	<0.001	1.68 (0.41)	1.54 (0.38)	<0.001
TG (mmol/L)	1.49 (0.69)	1.53 (0.64)	<0.001	1.47 (0.69)	1.52 (0.65)	<0.001
TC (mmol/L)	5.76 (1.12)	5.66 (1.13)	<0.001	5.82 (1.15)	5.67 (1.13)	<0.001
LDL-C (mmol/L)	3.36 (0.95)	3.37 (0.99)	0.47	3.40 (0.98)	3.38 (0.99)	0.42
Smokers (%)	3126 (21.1)	2879 (18.6)	<0.001	361 (19.6)	317 (17.2)	<0.001
Treatment (%)	4555 (30.7)	4898 (31.7)	0.06	512 (27.8)	706 (38.4)	<0.001

Data are means and standard deviations, geometric means and standard deviations (triglycerides) or counts and percentages; BMI – body mass index; HDL-C – high-density lipoprotein cholesterol; TG – triglycerides; TC – total cholesterol; LDL-C – low-density lipoprotein cholesterol; treatment – lipid-lowering medication.

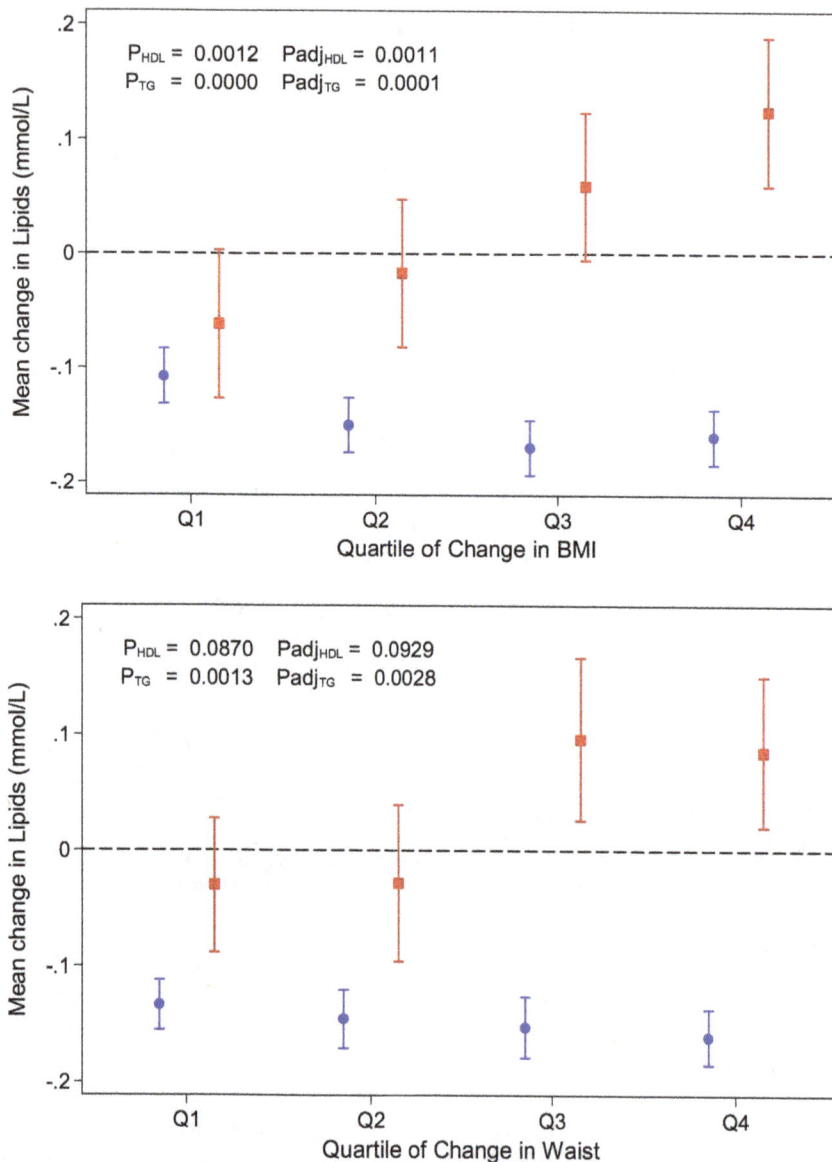

Figure 1. Relation of changes in obesity measures to changes in lipids in the LIPIDOGRAM PLUS Study. Mean changes in HDL-C (blue squares) and triglycerides (red circles) across quartiles of changes in BMI (top panel) and waist circumference (bottom panel) between 2004 and 2006. Data are means and standard errors, the bottom quartile represents subjects from the lowest 25% of distribution in BMI or waist circumference increase between 2004 and 2006. P-values for trend across quartiles; Padj - adjusted (for age, sex, region of recruitment, height, education and smoking) level of statistical significance from test for trend.

We then conducted a sensitivity analysis by selecting 21,983 subjects who were not on pharmacological lipid-lowering treatment (n = 10,294, n = 10,555 and n = 1,134 individuals in LIPIDOGRAM2004, LIPIDOGRAM2006 and LIPIDOGRAM PLUS, respectively). As expected, after exclusion of patients on therapy, LDL-C showed a small increase between 2004 and 2006 in both the cross-sectional and prospective analysis (**Table S6**). The trends in other lipids in this sensitivity analysis were consistent with those in the entire cross-sectional and prospective samples.

Using the data from the LIPIDOGRAM PLUS study we next assessed whether changes in obesity measures were directly related to the observed change in HDL-C, triglycerides, and total cholesterol over the two-year observation period. There was no direct association between changes in either BMI or waist circumference and total cholesterol levels (**Figure S4** and **Table**

S7). The linear trends in changes of HDL-C and triglycerides across the quartiles of BMI/waist circumference change distribution (**Figure 1** and **Table S7**) were modest. Changes in waist did not show a greater magnitude of association with HDL-C and triglycerides than BMI. In fact, the statistical significance of the association with changes in lipids was weaker for waist than BMI (**Figure 1** and **Table S7**).

Discussion

We investigated changes in two indices of obesity and main lipid fractions in a large sample of Polish individuals recruited after a major political and economic transformation in Europe – the accession of Poland to the EU in 2004. Economic growth is one the most important environmental drivers of obesity [8,9]. Indeed, socioeconomic transformations of the 1990's exposed the Eastern European populations to food market globalization that promotes processed, high in calories food. Changes in eating cultures and increasing rates of physical inactivity in the absence of appropriate health campaigns may have made this region particularly vulnerable to obesity. We hypothesized that the fast economic transition related to the European integration would provide an excellent epidemiological model to examine the relations of an increasing obesity burden with lipids in the Polish population.

We report several important findings. Firstly, our data demonstrate the expected increase in major anthropometric measures of obesity, tracking with a marked deterioration in HDL-C and triglycerides 2 years after the Polish accession to the EU. Secondly, the relative changes in these lipid fractions were actually greater than those of BMI or waist circumference, and only weakly associated with the increase in indices of obesity. Thirdly, our survey demonstrates that waist circumference and BMI are largely comparable as weak metabolic proxies of changes in lipids in both men and women. Finally, we show for the first time that the changes in obesity measures and lipids observed in the nationwide population are largely reproducible at the personal level.

Very few studies reported how changes in BMI and other measures of obesity translate into changes in levels of lipids in individual patients and populations. An increase in measures of obesity over a period of six–ten years has been associated with a deterioration of all major lipid fractions [10], components of metabolic syndrome [11] and incident hypercholesterolemia [12] in different studies. In contrast, a ten-year observation in the Framingham Heart Study revealed that an overall increase in BMI in the population was accompanied by an unexpected increase in HDL-C and a drop in triglycerides [13]. In our study, a short-term increase in BMI/waist circumference was paralleled by a significant deterioration in triglycerides and most apparently – HDL-C. However, neither the increase in BMI nor the increase in waist circumference could statistically explain the deterioration in HDL-C. This means that the observed drop in HDL-C is to a large extent independent of the parallel increase in BMI or waist circumference. Prevalence of smoking, another well-known correlate of low HDL-C [14] actually decreased between 2004 and 2006; thus it is unlikely to explain our findings. Diet, alcohol consumption and physical activity on their own may also impact on HDL-C and other lipid fractions, independent on their contribution to weight balance [15–17]. Insulin resistance and low-grade inflammation that cluster with excessive weight have also been shown to affect HDL-C and/or triglycerides, independent of adiposity [18,19]. Unfortunately, information on either of these phenotypes was not available in the LIPIDOGRAM project. A significant drop in HDL-C largely unexplained by measures of obesity was reported before in longer observations of general population [20,21]. These studies relied on BMI as the only measure of obesity. Our data show that these associations are apparent even short-term and that using waist circumference does not offer a better explanation for the change in HDL-C than BMI and that other factors, not captured directly through either of these anthropometric indices, may be responsible here.

Parallel to the significant changes in HDL-C, LDL-C was essentially stable in both the cross-sectional and prospective study. In the absence of a major change in LDL-C levels the (somewhat paradoxical) decrease in total cholesterol tracking with increasing measures of obesity is most likely related to decreasing HDL-C levels. Our data indicate that the observed drop in total cholesterol is not related to the use of statins. Indeed, after exclusion of participants on statin-based treatment, there was an apparent

decreasing trend in total cholesterol despite slightly increasing LDL-C.

To the best our knowledge this is the first investigation that has examined changes in metabolic health at both individual and population level simultaneously using identical protocols. The availability of prospective personal-level observation in addition to the two cross-sectional surveys provided also an independent replication. A consistency in the direction and magnitude of metabolic changes in both surveys also suggests that there is a good correlation between personal and nationwide trends in metabolic health. This consistency is also reassuring given that changes in cardio-metabolic risk factors from repeated observational studies are often criticised for providing overestimates due to unmeasured confounding. To this end our data point to common environmental roots of pro-atherogenic dyslipidaemia in individual patients and populations.

It would be fair to acknowledge that our data provide only a snapshot of the changes driven by the long-term process of political, economic and social transformation in Poland initiated by the Polish Solidarity movement in 1980s and transition to market economy in the early 1990s. We also recognise that LIPIDOGRAM surveys were recruited amongst individuals older than 30 years and thus may not be fully representative of the general population structure. The recruitment through primary care may also lead to over-representation of individuals with ill health, although a positive or negative bias towards any particular disease condition is unlikely given the wide inclusion criteria. Furthermore, the LIPIDOGRAM protocols did not include questionnaires on eating habits or physical activity. Finally, our study is observational, and definite causal inferences cannot be made.

However, our analysis has several strengths. Firstly, the recruitment strategy, phenotyping, data collection and processing of all LIPIDOGRAM data were carried out using essentially identical protocols. To minimise heterogeneity between 2004 and 2006 surveys no new recruitment sites were included in LIPIDOGRAM2006 and a majority of recruiting investigators from LIPIDOGRAM2004 participated again in 2006. Secondly, all surveys were recruited at the same time of the year, minimising differences related to seasonal variation i.e. in lipids. Thirdly, all recruiting doctors were trained prior to recruitment to maximise the standardisation of data collection. Furthermore, to ensure appropriate representation of individuals from all parts of the country the number and regional distribution of recruitment sites in LIPIDOGRAM studies was carefully matched to the demographic structure of individual geographic regions of Poland. A unique advantage of our study is that it used a nationwide rather than community samples. Regional variation in obesity measures across one country is a well-recognised source of potential confounding and the recruitment restricted to one community may not necessarily reflect the nationwide trends in obesity changes [7]. Finally, to eliminate inter-laboratory variation in biochemical measurements all blood analyses were conducted in one central laboratory that conforms to all national and international quality control standards. Therefore, we have no reason to suspect that a systematic bias in methodology may have influenced the findings of our study.

Conclusions

In summary, using data collected in Poland at the time and shortly after its accession to EU we demonstrate a deterioration of metabolic health at both personal and population level. The rapid increase in clinical measures of obesity tracked with a parallel deterioration in measures of pro-atherogenic dyslipidaemia, namely a drop in HDL-C and an increase in triglycerides. However, the magnitude of change in lipids was far greater than that of BMI or waist circumference and either measure of obesity were only weakly associated with changes in lipids. Thus, the observed deterioration in lipid profile was predominantly driven by other factors not measureable through BMI or waist circumference.

Supporting Information

Figure S1 Correlation between numbers of individuals recruited in 2004 and 2006 into the cross-sectional LIPIDOGRAM Studies.

Figure S2 Correlation between numbers of individuals recruited in 2004 into the cross-sectional and prospective LIPIDOGRAM Studies.

Figure S3 Correlation between numbers of individuals recruited in 2006 into cross-sectional and prospective LIPIDOGRAM Studies.

Figure S4 Mean changes in total cholesterol across 4 quartiles of changes in BMI (top panel) and waist circumference (bottom panel) between 2004 and 2006 in LIPIDOGRAM PLUS Study. Data are means and standard errors, the lowest quartile (Q1) - bottom 25th percentile of distribution in BMI or waist circumference increase between 2004 and 2006; P-values – level of statistical significance from test for linear trend; Padj - level of statistical significance (test for linear trend) after adjustment for age, sex, region of recruitment, height, education and smoking, the dotted horizontal line – no change.

Table S1 Number of recruiting cities, physicians, recruited and included individuals in cross-sectional LIPIDOGRAM2004 and LIPIDOGRAM2006. Population data obtained from the Polish National Main Statistical Office (Glowny Urząd Statystyczny; GUS) –2003.

Table S2 Individuals recruited into the prospective LIPIDOGRAM PLUS Study.

Table S3 Clinical characteristics of cross-sectional LIPIDOGRAM2004 and LIPIDOGRAM2006 Studies – sex-stratified analysis. Data are means and standard deviations, geometric means and standard deviations (triglycerides) or counts and percentages; BMI – body mass index; HDL-C – high-density lipoprotein cholesterol; TG – triglycerides; TC – total cholesterol; LDL-C – low-density lipoprotein cholesterol; treatment – lipid-lowering medication; P-value – level of statistical significance for comparison of men versus women.

Table S4 Clinical characteristics of LIPIDOGRAM PLUS Study – sex-stratified analysis. Data are means and standard deviations, geometric means and standard deviations (triglycerides) or counts and percentages; BMI – body mass index; HDL-C – high-density lipoprotein cholesterol; TG – triglycerides; TC – total cholesterol; LDL-C – low-density lipoprotein cholesterol; treatment – lipid-lowering medication; P-value – level of statistical significance for comparison men versus women.

Table S5 Changes in metabolic parameters between 2004 and 2006 in the LIPIDOGRAM studies – sex-stratified analysis. Changes between 2004 and 2006 are expressed as β-coefficients (β) with respective standard errors (SE) from linear regression or generalized estimation equations-based models; BMI – body mass index; HDL-C – high-density lipoprotein cholesterol; TG – triglycerides; TC – total cholesterol; LDL-C – low-density lipoprotein cholesterol; Basic – unadjusted model; Full – model adjusted for age, age^2, sex, region of recruitment, height, education and smoking; P-value – level of statistical significance from crude and adjusted analysis; P-value* – level of statistical significance for a difference in each lipid fraction change between men and women recruited in the same year; M vs W – men versus women.

Table S6 Changes in lipids between 2004 and 2006 in the cross-sectional and prospective LIPIDOGRAM Studies – sensitivity analysis after exclusion of subjects on lipid lowering medication. Changes between 2004 and 2006 are expressed as β-coefficients with respective standard errors (SE) from regression or generalized estimating equations-based models; HDL-C – high-density lipoprotein cholesterol; TG – triglycerides; TC – total cholesterol; LDL-C – low-density lipoprotein cholesterol; Basic – unadjusted model; Full – model adjusted for age, age^2, sex, region of recruitment, height, education and smoking; P-value – level of statistical significance from basic or fully adjusted analysis.

Table S7 Association between mean changes in lipids and body mass index/waist circumference between 2004 and 2006 in LIPIODGRAM PLUS Study. Data are expressed as means and standard errors of mean changes in each lipid fraction across quartile distribution of change in body mass index/waist circumference between 2004 and 2006, ranges of changes in obesity measures are shown per each quartile; TC – total cholesterol; HDL-C – high-density lipoprotein cholesterol; TG – triglycerides; P-value* – adjusted for age, sex, region of recruitment, height, education and smoking.

Author Contributions

Conceived and designed the experiments: BMK MT JJ JRT NJS NS FJC. Analyzed the data: BMK MT CPN JRT NS NJS JJ. Wrote the paper: BMK MT. Data acquisition: JJ WL MM AW TT WG AT JG IS-P AS. Manuscript revision: JJ WL MM AW TT WG AT JG IS-P AS NJS CPN NS FJC.

References

1. Biggs ML, Mukamal KJ, Luchsinger JA, Ix JH, Carnethon MR, et al. (2010) Association between adiposity in midlife and older age and risk of diabetes in older adults. JAMA 303: 2504–2512.
2. Kaess BM, Jozwiak J, Mastej M, Lukas W, Grzeszczak W, et al. (2010) Association between anthropometric obesity measures and coronary artery disease: a cross-sectional survey of 16,657 subjects from 444 Polish cities. Heart 96: 131–135.
3. Lee CM, Huxley RR, Wildman RP, Woodward M (2008) Indices of abdominal obesity are better discriminators of cardiovascular risk factors than BMI: a meta-analysis. J Clin Epidemiol 61: 646–653.
4. Yusuf S, Hawken S, Ounpuu S, Bautista L, Franzosi MG, et al. (2005) Obesity and the risk of myocardial infarction in 27,000 participants from 52 countries: a case-control study. Lancet 366: 1640–1649.

5. Poirier P, Giles TD, Bray GA, Hong Y, Stern JS, et al. (2006) Obesity and cardiovascular disease: pathophysiology, evaluation, and effect of weight loss: an update of the 1997 American Heart Association Scientific Statement on Obesity and Heart Disease from the Obesity Committee of the Council on Nutrition, Physical Activity, and Metabolism. Circulation 113: 898–918.

6. Brown JR, O'Connor GT (2010) Coronary heart disease and prevention in the United States. N Engl J Med 362: 2150–2153.

7. Emerging Risk Factors Collaboration, Wormser D, Kaptoge S, Di Angelantonio E, Wood AM, et al. (2011) Separate and combined associations of body-mass index and abdominal adiposity with cardiovascular disease: collaborative analysis of 58 prospective studies. Lancet 377: 1085–1095.

8. Kumanyika S, Jeffery RW, Morabia A, Ritenbaugh C, Antipatis VJ (2002) Obesity prevention: the case for action. Int J Obes Relat Metab Disord 26: 425–436.

9. Swinburn BA, Sacks G, Hall KD, McPherson K, Finegood DT, et al. (2011) The global obesity pandemic: shaped by global drivers and local environments. Lancet 378: 804–814.

10. Norman JE, Bild D, Lewis CE, Liu K, West DS (2003) The impact of weight change on cardiovascular disease risk factors in young black and white adults: the CARDIA study. Int J Obes Relat Metab Disord 27: 369–376.

11. Hillier TA, Fagot-Campagna A, Eschwege E, Vol S, Cailleau M, et al. (2006) Weight change and changes in the metabolic syndrome as the French population moves towards overweight: the D.E.S.I.R. cohort. Int J Epidemiol 35: 190–196.

12. Williams PT (2008) Changes in body weight and waist circumference affect incident hypercholesterolemia during 7 years of follow-up. Obesity (Silver Spring) 16: 2163–2168.

13. Ingelsson E, Massaro JM, Sutherland P, Jacques PF, Levy D, et al. (2009) Contemporary trends in dyslipidemia in the Framingham Heart Study. Arch Intern Med 169: 279–286.

14. Craig WY, Palomaki GE, Haddow JE (1989) Cigarette smoking and serum lipid and lipoprotein concentrations: an analysis of published data. BMJ 298: 784–788.

15. Fletcher B, Berra K, Ades P, Braun LT, Burke LE, et al. (2005) Managing abnormal blood lipids: a collaborative approach. Circulation 112: 3184–3209.

16. Lee DC, Sui X, Church TS, Lavie CJ, Jackson AS, et al. (2012) Changes in fitness and fatness on the development of cardiovascular disease risk factors hypertension, metabolic syndrome, and hypercholesterolemia. J Am Coll Cardiol 59: 665–672.

17. Sopko G, Leon AS, Jacobs DR Jr., Foster N, Moy J, et al. (1985) The effects of exercise and weight loss on plasma lipids in young obese men. Metabolism 34: 227–236.

18. Desroches S, Archer WR, Paradis ME, Deriaz O, Couture P, et al. (2006) Baseline plasma C-reactive protein concentrations influence lipid and lipoprotein responses to low-fat and high monounsaturated fatty acid diets in healthy men. J Nutr 136: 1005–1011.

19. Sinaiko AR, Steinberger J, Moran A, Hong CP, Prineas RJ, et al. (2006) Influence of insulin resistance and body mass index at age 13 on systolic blood pressure, triglycerides, and high-density lipoprotein cholesterol at age 19. Hypertension 48: 730–736.

20. Derby CA, Feldman HA, Bausserman LL, Parker DR, Gans KM, et al. (1998) HDL cholesterol: trends in two southeastern New England communities, 1981–1993. Ann Epidemiol 8: 84–91.

21. Sprafka JM, Burke GL, Folsom AR, Luepker RV, Blackburn H (1990) Continued decline in cardiovascular disease risk factors: results of the Minnesota Heart Survey, 1980–1982 and 1985–1987. Am J Epidemiol 132: 489–500.

Permissions

List of Contributors

Heidi R. Pethybridge, Barry D. Bruce, Jock W. Young, Peter D. Nichols
CSIRO Wealth from Ocean Flagship, Division of Marine and Atmospheric Research, Hobart, Australia

Christopher C. Parrish
Department of Ocean Sciences, Memorial University of Newfoundland, St. John's, Newfoundland, Canada

Joanna Bacon, Jon A. Allnutt, Robert Watson, Kim A. Hatch, Simon O. Clark, Rose E. Jeeves, Alice Marriott, Emma Rayner, Howard Tolley, Geoff Pearson, Graham Hall, Ann Williams, Philip D. Marsh
Public Health England, Microbiology Services, Porton Down, Salisbury, Wiltshire, United Kingdom

Luke J. Alderwick, Gurdyal S. Besra
Institute of Microbiology and Infection, School of Biosciences, University of Birmingham, Birmingham, United Kingdom

Evelina Gabasova, Lorenz Wernisch
MRC Biostatistics Unit, Institute of Public Health, Cambridge, United Kingdom

Masaki Ishikawa, Keiko Maekawa, Kosuke Saito, Yuya Senoo, Masayo Urata, Mayumi Murayama, Yoko Tajima, Yoshiro Saito
Division of Medicinal Safety Science and Disease Metabolome Project, National Institute of Health Sciences, Setagaya, Tokyo, Japan

Yuji Kumagai
Clinical Trial Center, Kitasato University East Hospital, Sagamihara, Kanagawa, Japan

Dennis Eggert
Heinrich-Pette-Institute, Leibniz Institute for Experimental Virology, University of Hamburg, Hamburg, Germany
Institute of Physical Chemistry, University of Hamburg, Hamburg, Germany

Kathrin Rösch, Rudolph Reimer and Eva Herker
Heinrich-Pette-Institute, Leibniz Institute for Experimental Virology, University of Hamburg, Hamburg, Germany

Giuseppe Astarita
Health Sciences, Waters Corporation, Milford, Massachusetts, United States of America

Department of Biochemistry and Molecular & Cellular Biology, Georgetown University, Washington, DC, United States of America

Jennifer H. McKenzie, Bin Wang and Jing X. Kang
Laboratory for Lipid Medicine and Technology, Department of Medicine, Massachusetts General Hospital and Harvard Medical School, Boston, Massachusetts, United States of America

Katrin Strassburg, Thomas Hankemeier and Rob J. Vreeken
Analytical Biosciences, Leiden Academic Centre for Drug Research, Leiden University, Leiden, The Netherlands
Netherlands Metabolomics Centre, Leiden University, Leiden, The Netherlands

Angela Doneanu, Jay Johnson, Andrew Baker, James Murphy and James Langridge
Health Sciences, Waters Corporation, Milford, Massachusetts, United States of America

Julian N. Rosenberg
Department of Chemical & Biomolecular Engineering, Johns Hopkins University, Baltimore, Maryland, United States of America
Synaptic Research LLC, Baltimore, Maryland, United States of America

Naoko Kobayashi, Austin Barnes, Eric A. Noel
Department of Biochemistry, University of Nebraska–Lincoln, Lincoln, Nebraska, United States of America

Michael J. Betenbaugh
Department of Chemical & Biomolecular Engineering, Johns Hopkins University, Baltimore, Maryland, United States of America

George A. Oyler
Synaptic Research LLC, Baltimore, Maryland, United States of America,
Department of Chemical & Biomolecular Engineering, Johns Hopkins University, Baltimore, Maryland, United States of America
Department of Biochemistry, University of Nebraska–Lincoln, Lincoln, Nebraska, United States of America

Kaan Yilancioglu, Inanc Pastirmaci, Batu Erman, Selim Cetiner
Faculty of Engineering and Natural Sciences, Sabanci University, Orhanlı, Istanbul, Turkey

Murat Cokol
Sabanci University Nanotechnology Research and Application Center, Orhanlı, Istanbul, Turkey
Faculty of Engineering and Natural Sciences, Sabanci University, Orhanlı, Istanbul, Turkey

Leandro Fernández-Pérez, Borja Guerra, Carlos Mateo-Díaz, Juan Carlos Díaz-Chico
Department of Clinical Sciences, University of Las Palmas de Gran Canaria - Biomedical and Health Research Institute (IUIBS), Molecular and Translational Endocrinology Group, Las Palmas de Gran Canaria, Spain
Cancer Research Institute of The Canary Islands (ICIC), Las Palmas de Gran Canaria, Canary Islands, Spain

Ruymán Santana-Farré, Mercedes de Mirecki-Garrido
Department of Clinical Sciences, University of Las Palmas de Gran Canaria - Biomedical and Health Research Institute (IUIBS), Molecular and Translational Endocrinology Group, Las Palmas de Gran Canaria, Spain

Irma García
Department of Animal Biology, University of La Laguna, Laboratory of Membrane Physiology and Biophysics, La Laguna, Spain

Diego Iglesias-Gato
Molecular Endocrinology group, University of Copenhagen - Novo Nordisk Center for Protein Research, Copenhagen, Denmark

Stefan Wallner, Margot Grandl, Tatiana Konovalova, Alexander Sigrüner, Thomas Kopf, Markus Peer, Evelyn Orsó, Gerhard Liebisch, Gerd Schmitz
Institute of Clinical Chemistry and Laboratory Medicine, University Hospital Regensburg, University of Regensburg, Regensburg, Germany

Pei-Luen Jiang
Graduate Institute of Marine Biotechnology, National Dong-Hwa University, Pingtung, Taiwan
Taiwan Coral Research Center, National Museum of Marine Biology and Aquarium, Pingtung, Taiwan

Buntora Pasaribu
Graduate Institute of Biotechnology, National Chung-Hsing University, Taichung, Taiwan

Chii-Shiarng Chen
Graduate Institute of Marine Biotechnology, National Dong-Hwa University, Pingtung, Taiwan
Taiwan Coral Research Center, National Museum of Marine Biology and Aquarium, Pingtung, Taiwan
Department of Marine Biotechnology and Resources, National Sun Yat-Sen University, Kaohsiung, Taiwan

Thomas Kopf, Tatiana Konovalova, Gerd Schmitz
Institute for Clinical Chemistry and Laboratory Medicine, University Hospital Regensburg, Regensburg, Germany

Hans-Ludwig Schaefer, Mark Broenstrup
Sanofi-Aventis Germany, R&D DIAB Div./Biomarker & Diagnostics, Frankfurt, Germany,

Martin Troetzmueller, Harald Koefeler
Core Facility Mass Spectrometry, ZMF, Medical University Graz, Graz, Austria

Claudia Lamina, Margot Haun, Stefan Coassin, Anita Kloss-Brandstätter, Florian Kronenberg
Division of Genetic Epidemiology, Department of Medical Genetics, Molecular and Clinical Pharmacology, Innsbruck Medical University, Innsbruck, Austria

Christian Gieger
Institute of Genetic Epidemiology, Helmholtz Zentrum Mu''nchen - German Research Center for Environmental Health (GmbH), Neuherberg, Germany

Annette Peters
Institute of Epidemiology II, Helmholtz Zentrum München – German Research Center for Environmental Health, Neuherberg, Germany

Harald Grallert
Department of Molecular Epidemiology, Helmholtz Zentrum München, German Research Center for Environmental Health, Neuherberg, Germany

Konstantin Strauch
Institute of Medical Informatics, Biometry and Epidemiology, Chair of Genetic Epidemiology, Ludwig-Maximilians-Universität, Munich, Germany,

Thomas Meitinger
Institute of Human Genetics, Technische Universität München, Munich, Germany
Institute of Human Genetics, Helmholtz Zentrum München – German Research Center for Environmental Health, Neuherberg, Germany

Lyudmyla Kedenko, Bernhard Paulweber
First Department of Internal Medicine, Paracelsus Private Medical University Salzburg, Salzburg, Austria

Matti A. Kjellberg, Anders P. E. Backman, Henna Ohvo-Rekil, Peter Mattjus
Biochemistry, Department of Biosciences, A°bo Akademi University, Turku, Finland

Aihua Li
Plant Germplasm and Genomics Center, Germplasm Bank of Wild Species, Kunming Institute of Botany, Chinese Academy of Sciences, Kunming, China
University of Chinese Academy of Sciences, Beijing, China

Dandan Wang, Buzhu Yu, Xiaomei Yu
Plant Germplasm and Genomics Center, Germplasm Bank of Wild Species, Kunming Institute of Botany, Chinese Academy of Sciences, Kunming, China

Weiqi Li
Key Laboratory of Biodiversity and Biogeography, Kunming Institute of Botany, Chinese Academy of Science, Kunming, China
Plant Germplasm and Genomics Center, Germplasm Bank of Wild Species, Kunming Institute of Botany, Chinese Academy of Sciences, Kunming, China

Marta Palusinska-Szysz, Anna Turska-Szewczuk, Ryszard Russa
Department of Genetics and Microbiology, Maria Curie-Sklodowska University, Lublin, Poland

Magdalena Kania, Witold Danikiewicz
Mass Spectrometry Group, Institute of Organic Chemistry, Polish Academy of Sciences, Warsaw, Poland

Beate Fuchs
Institute of Medical Physics and Biophysics, Medical Faculty, University of Leipzig, Leipzig, Germany

Henri Augusto Korkes, Nelson Sass
Department of Obstetrics – Federal University of Sao Paulo, Sao Paulo, Sao Paulo, Brazil
Laboratory of Clinical and Experimental Investigation – School Maternity Vila Nova Cachoeirinha, Sao Paulo, Sao Paulo, Brazil

Antonio F. Moron
Department of Obstetrics – Federal University of Sao Paulo, Sao Paulo, Sao Paulo, Brazil

Niels Olsen S. Câmara
Department of Immunology – University of Sao Paulo, Sao Paulo, Sao Paulo, Brazil

Tatiana Bonetti, Ismael Dale Cotrim Guerreiro Da Silva
Department of Gynecology - Federal University of Sao Paulo, Sao Paulo, Sao Paulo, Brazil

Ana Sofia Cerdeira
Department of Medicine, Beth Israel Deaconess Medical Center, Harvard Medical School, Boston, Massachusetts, United States of America

Leandro De Oliveira
Department of Obstetrics – Federal University of Sao Paulo, Sao Paulo, Sao Paulo, Brazil
Laboratory of Clinical and Experimental Investigation – School Maternity Vila Nova Cachoeirinha, Sao Paulo, Sao Paulo, Brazil
Department of Immunology – University of Sao Paulo, Sao Paulo, Sao Paulo, Brazil

Hye Min Park, Mee Youn Lee, Choong Hwan Lee
Department of Bioscience and Biotechnology, Konkuk University, Gwangjin-gu, Seoul, Republic of Korea

Jong Cheol Shon, Kwang-Hyeon Liu
College of Pharmacy and Research Institute of Pharmaceutical Sciences, Kyungpook National University, Buk-gu, Daegu, Republic of Korea

Jeong Kee Kim, Sang Jun Lee
Food Research Institute, Amore Pacific Corporation R&D Center, Giheung-gu, Yongin, Gyeonggi-do, Republic of Korea

Index

www.ingramcontent.com/pod-product-compliance
Lightning Source LLC
Chambersburg PA
CBHW080251230326

41458CB00097B/4273